On Physical Adsorption

Electronmicrograph of thermally treated P-33 carbon black (courtesy *Journal of Colloid Science*). Each particle, as viewed from above, is represented by a central polygon (usually of six or eight sides) surrounded by trapezoids. Shadowed electronmicrographs show that the depth of the particle is comparable with the width. It is therefore concluded that the individual particles are doubly truncated polygonal (principally hexagonal and octagonal) bipyramids. Each particle consists of a group of minute radiating crystals. The single exposed surface of each crystal is the lowest-energy plane of graphite. The surface of the whole particle, or spherulite, is therefore entirely composed of the carbon-layer plane, or basal plane, of crystalline graphite.

ON PHYSICAL ADSORPTION

Sydney Ross
Professor of Colloid Science
Rensselaer Polytechnic Institute

James P. Olivier
Freeport Kaolin Company

with a foreword by J. H. De Boer

INTERSCIENCE PUBLISHERS

a Division of John Wiley & Sons, Inc.

New York • London • Sydney

Library of Congress Catalog Card Number: 63-19667
Printed in the United States of America

Foreword

This book is an unusual one. It is the result of theoretical and experimental studies on physical adsorption by Professor Sydney Ross and his collaborators in his laboratory at Rensselaer Polytechnic Institute, Troy, New York. For many years this group of workers has been trying to tackle the difficult task of obtaining a realistic picture and a workable mathematical set of equations for the physical adsorption of gases on real surfaces and to express the picture and the equations in terms of the molecular properties of the adsorbed molecules and of the surfaces of the adsorbents.

Sydney Ross and J. P. Olivier, who has been one of the collaborators almost from the beginning of this tremendous task, combined forces to present the results in this book. It is therefore on the one hand an extensive scientific publication, on the other, a book indicating a possible and promising way of penetrating deeper into the great "unknowns" of the surface of solids.

Several years ago (1950, in *Advances of Colloid Science*, III, p. 3) I gave three reasons to indicate that our calculations could, at the utmost, give only the right order of magnitude when we are dealing with adsorption forces. In short, these reasons are

1. we do not know the real structure of the surface;
2. we do not know, really, the repulsion forces, checking the attracting forces at short distance;
3. we do not know the real distance of the adsorbed atom from the surface.

Focusing our attention now on the experimental data, we may well state that our complete lack of knowledge with respect to the deviations from homogeneity of the real, actually occurring surfaces prevents us from drawing the correct numerical results from many experiments, especially if we want to translate the results into terms of molecular properties of the adsorbed species.

Sydney Ross and J. P. Olivier deserve our gratitude for having had the courage to try to attack this difficult problem.

Indeed, since one of the properties of real surfaces is their heterogeneous character, some method of tackling the mathematical treatment of surface heterogeneity has to be introduced before well-known equations describing adsorption isotherms in molecular terms can be applied. Various isotherm equations are known in the literature, all derived for surfaces of homogeneous character; we find them in Table I-1 of this book, where they are related to the corresponding two-dimensional equations of state.

Because the adsorbed molecules are not mathematical points and because certainly, they show mutual interaction forces, only two of those equations are used: the isotherm equation for mobile films, based on the two-dimensional van der Waals' equation, and a corresponding equation for immobile films.

The real action of the book comes to its climax in Chapter VII, where the whole machinery, explained in the preceding chapters, is applied to the Troy experiments. Before this, in Chapters I to IV, we find a useful introduction. In Chapter I the significance of the phase rule is stressed and extended to surface phenomena, including those on heterogeneous (real) surfaces. Any real sorbent "invariably has many different surfaces, which may be caused by surface roughness as well as by different crystal faces." The scope of the book is then laid down, and we learn that it is primarily concerned with gas-solid adsorption systems and is limited to monolayer adsorption. The title of the book does not indicate this restriction.

After this introductory chapter and another on experimental techniques (II), indicating again that the book is basically a large research paper, there follows a chapter (III) on the various heats of adsorption. The introduction to the main part of the work starts with Chapter IV, where attention is focused on the heterogeneity of surfaces. The surface is considered to be composed of infinitesimal patches of different energy that adsorb independently of one another. Each of these patches is a "homotattic" surface, a word coined by C. Sanford and Sydney Ross in 1954 for a surface of a submicroscopic patch or region, part of a larger surface, which acts as if its structure were uniform and homogeneous. The authors describe some points of the evolution of the concepts of treating heterogeneity of surfaces during the last 20 years (Section IV-3) and conclude that "no analytic solution has yet been, or probably can be, based on the models of mobile or localized monolayers, taking into account both interaction and surface heterogeneity." Considering that the surface of a solid is a nonequilibrium system with accidental energy variations, they postulate a random distribution of adsorptive potentials and they select a Gaussian probability function, which, however, they restrict to an energy range of 5 kcal/mole. For computational purposes integrations are then replaced by summations, subdividing the range of energy in very small homotattic sections (mostly 50, sometimes 100).

The applications start in Chapter V with the remark that, despite a heterogeneous character of the surface, isotherms can often be described by a Langmuir equation. Very rightly a warning is then given not to derive quantitative data from such curves by means of this equation because numbers obtained in this way are always seriously in error.

Sigmoidally shaped isotherms, found experimentally in studies of adsorption on graphitized carbon or other near-homotattic surfaces, may also lead to serious quantitative errors, for the heterogeneity of the surface which is still present with these surfaces leads to a decrease of the experimentally found heat of adsorption. In a polite way they criticize, rightly however, a conclusion drawn by Dr. Kruyer and me with respect to the repulsive character of interaction forces due to parallel-oriented dipoles. It is true that we should not have described the effect as due entirely to this polarization and we, indeed, ought to have taken into account the still existing heterogeneity of the graphite surface. Nevertheless, the polarization of adsorbed molecules by the substrate is a reality, and this polarization does give a deviation of the two-dimensional van der Waals constant from the "ideal" one, as also discussed in Chapter VI.

Chapter VII is the experimental chapter of the book; it is its climax. The experiments are performed with well-selected surfaces, mostly graphitized carbon blacks, homogenized by heating to various high temperatures. "In every case only the model of mobile adsorption has successfully produced descriptions of the experimental data that are consistent at two or more temperatures." From all the equations, therefore, the one based on the two-dimensional van der Waals' equation of state, which is referred to throughout the book as the Hill-de Boer equation, gives parameters that are the same at different temperatures. It is only below the temperature of liquid nitrogen that adsorbed films of argon, nitrogen, and krypton become localized.

Two of the constants are determined by shifting the experimental isotherms and well-chosen computed isotherms until they coincide. One of these constants, V_β, gives the "monolayer capacity" based on the molecular area as determined by the two-dimensional van der Waals' constant b (which is called β here). The authors compare V_β with the monolayer capacity V_m derived from multimolecular adsorption measurements. Since, according to their table, the point B method is mostly used in the cases that are compared, their remark that the B.E.T. theory is based on the Langmuir equation, is irrelevant. The molecular area used in B.E.T. determinations is based mostly on data derived from liquid nitrogen

In the first part of Chapter VII a possible polarization of the adsorbed molecules by the surface of the adsorbent is neglected. After having examined the adsorption of $CHCl_3$ and of $CFCl_3$ and having obtained a value for the electric field of the surface from the study of these adsorptions, the authors return to argon and nitrogen to correct for the influence of this field. Comparing Figure VII-36 with Figure VII-6, they say, "the theoretical description is now more precise for the low-pressure data." This is true, but it looks as if the data for somewhat higher pressures fit less well.

I make only a few remarks of this kind. There are many more such remarks that can be made. They do not endanger the main theme; they indicate only that we are, experimentally and theoretically, just moving along the border between what may be concluded with confidence and what may not. The authors undoubtedly have shifted that border further into the hitherto unknown.

The book ends with Chapter VIII, "The Nature of the Adsorptive Forces," in which excellent results are obtained with the methods developed in recent years.

Because of the extensive use that the authors make of the "Hill-de Boer" equation and from the observations laid down in my *Dynamical Character of Absorption*, Professor Ross in many of the discussions that we had on the subject of this book expressed his view that it could be considered a continuation, or a second part, of my earlier book. It is true that the authors have succeeded in making full use of some of the main features of my previous work. In their experimental results they have given beautiful examples of the two-dimensional gaseous state and of two-dimensional condensation phenomena. They have largely increased our scanty knowledge in this field. Nevertheless, the work is focused on points other than those discussed in my book. The present book may be based on my former work–in fact, it is–but it is not a direct continuation. One wishes, however, that the authors had used the same notations as I did. The different meaning of many letters in the present work, compared with my book, has given me some trouble. Another minor point is that the definition of mobile and localized layers is not exactly the same as in my book and in the literature, but this does not affect the conclusions at all.

As I said before, these are minor points. The whole of the work is conceived as one homogeneous structure, if I may use this adjective here.

J. H. DE BOER

The Hague
March 1964

Preface

This book presents, in more detail than is usually permitted in a technical journal, a new way to use the data of physical adsorption in order to measure the heterogeneity of a solid surface, as well as other properties of the adsorption system. The heterogeneity thus measured is not expressed in absolute terms but is relative to the gas used as adsorbate. What is original in the treatment is not so much the theory of adsorption on which it is based as the practical method that it teaches of how to analyze and interpret the experimental data of physical adsorption in order to characterize solid surfaces. To make the book more useful as a practical guide, we have included a chapter on laboratory techniques and apparatus, anticipating that many readers will be about to undertake research work in adsorption for the first time.

Our chief objective has been to obtain information about the solid surface, and physical adsorption has been of interest to us only so far as it throws light on that subject. We did not scruple, therefore, to leave entirely out of the present consideration the adsorption isotherm in the multilayer region, in which theories that describe the effect of the surface are still less completely developed. Thus it happens that only a few of the numerous published papers on adsorption report data suitable for our analysis; the majority of the data refer to relative pressures at which multilayers are formed. The building-up of a body of data in the monolayer region and referable to reproducible surfaces is a grand cooperative work that still largely waits to be done.

The present treatment has its antecedents, among which two were most significant: J. H. de Boer in his book *The Dynamical Character of Adsorption* (Clarendon Press, Oxford, 1953) predisposed us toward accepting the mobility of the adsorbed film as a more likely model of its behavior than the classical Langmuir concept of localized adsorption on fixed sites; second, the theoretical arguments were strengthened by the experimental findings of Sydney Ross and his co-workers who, by using as substrates the most uniform solid surfaces than available, had, in a series of researches commencing in 1947, explored and to a large extent confirmed the predictions of de Boer that for such ideal surfaces the two-dimensional van der Waals' equation of state would provide a description of the adsorbed film, both above and below its two-dimensional critical temperature.

The analyses of experimental data in terms of our (Ross-Olivier) theory, as reported in Chapter VII, are most successful for substrates that do not depart far from complete uniformity. With substrates of great heterogeneity, the theory can indeed provide a description, but it is not a unique description. The effects of what we have ascribed to the surface electric field and those arising from heterogeneity of the substrate tend to counterbalance to some degree, so that a range of possible values exists for the parameters derived. Much of this uncertainty can be reduced by extending the temperature range of the investigation; indeed, the meagerness of data rather than innate shortcomings of the theory is more frequently the reason for the lack of a definitive answer to an analysis. The number of parameters to be evaluated accounts for the existence of some uncertainty. Any additional information about the characteristics of the adsorption system is bound to be helpful in reducing this uncertainty. The monolayer capacity, for example, as determined by some independent method, such as the B.E.T. theory, although it gives an answer that is not likely to agree precisely with our V_β, could be enough additional assistance to orient the curve fitting and so pin down the rest of the parameters.

For extremely heterogeneous substrates our technique of using the adsorption isotherms as an instrunent by means of which we can probe the detailed character of the solid surface reaches its natural limit and begins finally to fail us. Too many variables are now acting and interacting; and various combinations of their effects can be recognized as expressing themselves in the same resulting behavior. Presented with variables whose quantitative effects cannot be definitively determined, theory may have recourse to a more simplified characterization: causes are to be less particularized and two or more are to be lumped together under a general name. A less powerful analysis does indeed permit the broad outline of the phenomena to be described succinctly, whereas efforts to obtain a more detailed view give rise to too many alternatives. The latter course suggests more possibilities, but the questions that it raises cannot be answered.

The Polanyi potential theory, in the form in which it has recently been developed by Dubinin and his co-workers, is a low-magnification view. The distribution of adsorptive potentials that it yields is an unanalyzable blend of various factors, which is further distorted by its simplifying approximation that molecules are adsorbed by the surface in the serial order of adsorptive potentials. This approximation is the condition that we mention on p. 128 as applying only at the absolute zero of temperature; it is also less in error, the less heterogeneous the substrate. If one accepts the approximation, it makes short work of the analysis of data. In return for this convenience, however, one is permitted only a misty, myopic vision and must abandon the hope, as too ambitious to be attained, of obtaining a more clear-sighted understanding. Perhaps it will turn out that for some heterogeneous substrates nothing better can be done. Meanwhile our duty has been to push our own method as far as we are able, believing that efforts toward the higher goal must be made, even at the risk of uncertainty or error. Not to do so is to be satisfied with ignorance.

Some readers will question our selection of the Gaussian distribution function to describe the heterogeneity of adsorptive potentials. The success of this function, they well may say, is not put to a severe test by investigating only substrates that have a narrow distribution of adsorptive potentials; furthermore, for very heterogeneous substrates the Gaussian function does *not* recommend itself as a probable description: one would be inclined to repudiate a function that was capable of handing out negative values of the adsorptive potential as part of its description. This anomalous result would occur in our treatment if the value of U_0' were less than 2.5 kcal/mole and the spread of heterogeneities were more than 5 kcal/mole. A distribution that is skewed toward higher adsorptive potentials would be more acceptable. In extenuation we plead that the symmetry of the Gaussian function allowed us one less empirical constant, which, had it been included would have enormously complicated the practical application of the mathematical description. Perhaps a more cogent reason lies in the insensitivity of the adsorption isotherm to the low-energy portion of the distribution, particularly to surface patches with energies less than about RT, which patches can contribute almost nothing to the amount adsorbed in the first monolayer before multilayers form on top of the more energetic patches—after which the experimental data are beyond the scope of our analysis. For this reason the actual distribution curve may have a form quite different from the symmetrical Gaussian function, perhaps lacking a narrow leading-edge at the high-energy end and falling off abruptly at the low-energy end, without these features interfering to any pronounced degree with the isotherm that would be obtained. The Gaussian distribution to which the data are matched therefore represents an equivalent distribution in Gaussian terms, the most important practical consequences of which are the evaluation of the midpoint patch of most frequent occurrence and the quantitative expression that is furnished for the degree of heterogeneity of the substrate. But the final answer to the objection must await experimental observations of many adsorption systems and the working out of computed isotherms based on alternative models. The theory presented here claims to be no more than a certain stage in the progress of our understanding.*

Another objection that can be raised is to our selection of the Hill-de Boer adsorption-isotherm equation, which is based on the description of the mobile adsorbed film by the two-dimensional van der Waals equation. The major emphasis of the present treatment has been placed on the mobile adsorbed film, although such a choice does not reflect the most popular model, if one were to poll the published opinions of previous investigators. Those opinions, we believe, have been unduly influenced by the mathematical difficulties posed by the model of a mobile film—difficulties that were indeed insuperable before electronic computers. The method that we demonstrate here is, however, not more difficult for one model than for another

* "No grand practical result of human industry, genius, or meditation, has sprung forth entire and complete from the master hand or mind of an individual designer working straight to its object, and foreseeing and providing for all details. As in the building of a great city, so in every such product, its historian has to record rude beginnings, circuitous and inadequate plans; frequent demolition, renewal and rectification; the perpetual removal of much cumbrous and unsightly material and scaffolding, the constant opening out of wider and grander conceptions; till at length a unity and nobility is attained, little dreamed of in the imagination of the first projector." J. F. W. Herschel, *Outlines of Astronomy* (1848).

and our emphasis on the mobile film is more rationally based on the discovery that the model answered better as a description of the observed data. We are aware, however, that we have investigated only a small number of substrates. Unequivocal examples of localized adsorbed films are certain to be discovered; they will require the computation of a different set of tables, but we anticipate that the methods we have developed in Chapter V will still prove useful for such an extension of the subject, when it is required.

Possibly more attention must be paid in the near future to the intermediate type of adsorbed film: that is, a mobile film that is partially localized. The mobility of adsorbate molecules on the surface is actually never completely unrestricted but is always affected to some extent by the crystalline structure of the substrate. The potential-energy barrier to translation, as represented in Fig. I-3*a*, may be sufficiently great to retard the surface mobility of an appreciable fraction. The progress of the molecule does not so much resemble the smooth gliding of a skater on a sheet of ice as the flitting of a bee from flower to flower (the simile of the *super*-bee was a happy fancy of de Boer).

For the substrates here investigated in some detail, particularly graphite and carbon black, the retardations of molecular velccities at temperatures even as low as 77°K. were not sufficiently pronounced to invalidate the description of the adsorbate by the model of a mobile film, but measurements at lower temperatures would be expected to reveal significant deviations from this model. Other substrates could well show such deviations even at 77°K. and above. The problem can be evaded by raising the temperature sufficiently, as long as we are still dealing with physical adsorption, not chemical adsorption. If we are to confront the problem, however, we should have to evaluate a new parameter—the one represented by χ in Fig. I-3—that is, the potential-energy barrier to surface diffusion. For the present, χ remains an undifferentiated part of the total adsorptive potential U_0. Something could be done with the model of a partially localized mobile film to describe adsorption on a homotattic substrate, possibly by measuring rates of surface mobility as a function of temperature to evaluate χ; but if, for a heterogeneous substrate, we are to postulate two independent distributions of χ and U_0, the analysis of adsorption isotherms would become hopelessly bogged in uncertainties.

We owe our readers an explanation of our choice of symbols, which break with tradition in this subject, are awkward to manipulate, and have wearied our printers and proof-readers. We believe, however, that if the reader will take the trouble to do so, he would find himself able to read the meaning of the symbols on their faces. The four positions around a letter-symbol, corresponding to the anterior and posterior superscripts and subscripts, are reserved, respectively, for designations of concentration, quality, phase, and temperature (°K.), as

$$^{\text{concentration}}_{\text{phase}}\times^{\text{quality}}_{\text{temperature}}$$

We are obliged to depart from this rule to accommodate the well-established mathematical convention of the running index in the lower right position: for example, x_i for the ith value of x, as also for particular values of a variable—T_c for critical temperature. We used corresponding letters of the Greek alphabet to indicate for the two-dimensional phase what are represented by letters of the English alphabet for the three-dimensional phase. Occasional exceptions to the above rules had to be conceded for what seemed to be good reasons, the cost of consistency coming too high.

SYDNEY ROSS
JAMES P. OLIVIER

Troy, New York
March 1964

Acknowledgments

The program of research on adsorption at Rensselaer Polytechnic Institute has been supported by grants from the National Science Foundation, the Petroleum Research Fund of the American Chemical Society, and from Esso Research and Engineering Company. To the last-mentioned donor we are particularly obliged for granting *continuous* support for a period of seven years, so that the work did not have to terminate during the lean intervals.

We are indebted to Mr. David Devoe and to Professor J. W. Hollingsworth and the staff of the Computer Laboratory of Rensselaer Polytechnic Institute for assistance with the computer program, and for the use of their computer for the tables up to $2\alpha/RT\beta = 6.70$, and to the Research Laboratories of the United Aircraft Corporation for assistance and use of their computer facilities for the remainder of the tables. We also thank Professor G. D. Halsey, Jr., and Dr. Victor R. Deitz for sending us experimental data before publication. Dr. Donald Graham kindly provided a glossy print of the electronmicrograph that we use as a frontispiece.

Acknowledgment of our indebtedness would not be complete without mentioning the young men who, as graduate students in the colloid laboratory at Renssaeler Polytechnic Institute at different times during the last eleven years, have contributed to the facts and theories reported in this work: Werner Winkler, Vernon Ballou, Hadden Clark, W. W. Pultz, E. W. Albers, E. S. Chen, W. D. Machin, and J. J. Hinchen.

Our thanks are due to the following holders of copyright for permission to reproduce diagrams and to quote extracts.

Academic Press, Inc., publishers of the *Journal of Colloid Science*, and Donald Graham for the frontispiece

The American Chemical Society, publishers of the *Journal of the American Chemical Society*, *The Journal of Physical Chemistry*, and *Advances in Chemistry Series;* and the following authors:

J. R. C. Brown for Fig. II-18
G. D. Halsey, Jr. for Fig. II-14, Fig. II-15, Fig. II-21, the extract on pp. 61–4, and the extract on p. 72
H. H. Podgurski for Table II-1
Sydney Ross for Figs. II-29, II-30, IV-2, V-1, VI-1, VII-1, VII-2, VII-3, VII-4, VII-13, VII-14, VII-21, VII-38, VII-44, VII-45, VII-46, VII-47, VII-1, Table VIII-1, Table VII-2, Table VII-11, Table VII-12, Table VIII-2 and Table VIII-3

The American Institute of Physics, publishers of the *Journal of Chemical Physics and the Review of Scientific Instruments;* and the following authors:

W. G. McMillan for Fig. II-12 and the extract on pp. 56–9
J. A. Morrison for Fig. II-13 and the extract on pp. 59–61
D. M. Young for Fig. II-10 and the extract on pp. 52–3

Chemical Rubber Publishing Co., publishers of the *Handbook of Chemistry and Physics*, for Table VI-1

The Faraday Society, publishers of the *Transactions of the Faraday Society* and the *Discussions of the Faraday Society;* and

J. A. Morrison for Fig. II-25 and Fig. VII-23

Interscience Publishers, Inc., publishers of *Advances in Colloid Science*, vol. 2 (1942) and *Surface Chemistry* (1949); and

A. E. Alexander for Fig. I-2
P. H. Emmett for Fig. II-17 and the extract on pp. 66–9

The National Research Council of Canada, publishers of the *Canadian Journal of Chemistry;* and

R. A. Beebe for Fig. II-24 and the extract on pp. 83–7
C. A. Winkler for Fig. II-23

Pergamon Press, publishers of *Surface Phenomena in Chemistry and Biology* (1958); and

M. M. Dubinin and B. N. Vasil'ev for Fig. II-22 and the extract on pp. 73–5

The Royal Society of London, publishers of the *Proceedings of the Royal Society* and

Sydney Ross for Figs. V-8, VII-24, VII-25, VII-26, VII-31, VII-32, VII-33, VII-41 and Table VII-13

Springer-Verlag, publishers of Landolt-Bornstein's *Zahlenwerte und Funktionen* (Berlin, 1951) for Table VI-2

Contents

CHAPTER I

Introduction

The adsorption of a gas or vapor by a solid surface may be thought of as an incipient condensation that can take place, to some extent, at any pressure, no matter how low. The adsorbed film, like the condensed phase into which it passes as the pressure is raised beyond the saturation vapor pressure, may be regarded as a thermodynamically separate phase—a two-dimensional phase; but a curious distinction appears: the adsorbed film has a close analogy to a solution, since the quantity adsorbed per unit area (i.e., surface concentration) varies with the superimposed pressure. A true condensed phase, which is a one-component system, cannot, of course, be described in these terms. The phase rule and some thermodynamic relations appropriate to a sorption system are developed in this introductory chapter.

1. The Two-Dimensional Phase

In a multicomponent system, molecules of one of the components moving into a phase boundary may be reflected immediately; or they may linger for a time, after which they will either penetrate into the bulk of the adjoining phase or they will return to the bulk of the original phase. The relative number of genuine reflections at a phase boundary is negligibly small, except for the lightest gas molecules, or at temperatures much higher than those in the range at which physical adsorption is normally measured; under normal conditions, therefore, one of the components will be found at the interface at a higher concentration than in either of the contiguous bulk phases. This phenomenon is called sorption (1), Latin, *sorbere*, to suck up. If the sorbed molecules are unable to penetrate into the second bulk phase but can only return to their original phase, we have the special case of sorption that is called adsorption.

We have mentioned so far only the two contiguous bulk phases, but the sorbed molecules at the phase boundary are also to be considered as constituting an additional phase; hence, there is a minimum of three phases in such a system: two bulk phases and a two-dimensional phase. The phase rule for such systems (i.e., when we wish to take into account the presence of phases residing at interfaces) is

$$P + F = C + 2 + i \tag{I-1}$$

where P, F, and C have their usual meanings of number of phases, degrees of freedom, and number of components, and i is the number of interfaces that we wish to consider. For the simplest systems, $C = 1$, $i = 1$, and $P = 3$. This could refer, for example, to the equilibrium between water and water vapor, in which the density of the interfacial region is to be considered as a variable (concentration of the two-dimensional phase): we have $F = C + 3 - P = 4 - 3 = 1$; the one degree of freedom could be temperature, which, being specified, fixes both the concentration of the vapor phase (vapor pressure) and that of the two-dimensional phase.

The concept of a sorption isotherm has no application, however, to this simple illustration: it enters only with the consideration of multicomponent systems. The sorption isotherm is a graphical representation of the concentration of a particular component, called the sorbate, in the two-dimensional phase, plotted against the concentration of the same component in one of the bulk phases, temperature being held constant. Any component capable of being a sorbate must be able to move about—sorption is a dynamic process: individual molecules of the sorbate continually transfer between the bulk and the two-dimensional phases, maintaining at equilibrium a constant concentration in each phase.

A gas in contact with a solid is a two-component system of two bulk phases; if sorption is not taken into account, the ordinary phase rule applies:

$$P + F = C + 2$$

hence, $F = 2$; i.e., only two of the three variables, p, T, and ${}_g n/V$, need to be specified to describe the system. Taking sorption into account, let us suppose an ideal solid having only one type of interface: the phase rule that applies is equation I-1 with $i = 1$. We now have three phases and two components; again, $F = 2$. The four variables are p, T, ${}_g n/V$, and the concentration of sorbate in the two-dimensional phase; of these, only two need be specified to describe the system. If we fix the temperature of the system and the pressure in the gas phase, the concentration, ${}_g n/V$, of sorbate in the gas phase as well as its concentration in the two-dimensional phase are fixed.

In the case of the partition of a solute between two immiscible liquids, we have three components; one interfacial and two bulk phases; for the condensed system

$$P + F = C + 1 + 1$$

hence $F = 2$, i.e., fixing the temperature, and the composition in one of the phases, the compositions in the remaining two are uniquely defined.

There may be more than one state for the two-dimensional phase. The molecules at the interface may translate in the same way as molecules of a gas, their motion being restricted, however, to two dimensions; this state is referred to as that of a two-dimensional gas. Two-dimensional condensed states would be analogous to liquid or solid states. The two-dimensional liquid would be produced by compression of the two-dimensional gas at a temperature less than critical; the two-dimensional solid state could arise by cooling the two-dimensional liquid or by the formation of a localized film, caused by points of anchorage, as it were, to the substrate. The coexistence of two sorbed phases at the same interface (e.g., a two-dimensional gas and a two-dimensional liquid) reduces the number of degrees of freedom of the system by one. For example, for the gas/solid system mentioned above, with two bulk and two sorbed phases, $F = 1$. Hence, at a given temperature, all the compositions are fixed; during the phase change the sorption isotherm will be discontinuous.

Another situation occurs when a gas is in equilibrium with two bulk solid phases, each of which is ideal in the sense that it has only one type of interface. Here there are two interfaces ($i = 2$), three bulk phases, and three components, with one sorbed phase at each interface, $P = 5$; hence, $F = 2$. At a given temperature and pressure, the surface concentration at each interface is fixed. The same conclusion is true for any number of solid components, each with one interface, in the system.

Now let us consider a solid phase of one component having two different interfaces, such as might be imagined, for example, for a crystal with two different crystal faces. Here $i = 2$, $C = 2$, and $P = 4$ (for one sorbed phase at each interface); hence, $F = 2$. Once again the surface concentration at each interface is fixed at a given temperature and pressure. The same conclusion is true for any number of interfaces on the same solid substrate. This situation is pertinent to real sorption systems where a single sorbent invariably has many different surfaces which may be caused by surface roughness as well as by different crystal faces; each of these surfaces of differing atomic arrangement creates a different environmental "force field" for a visiting molecule, thus producing interfacial phases of different concentrations. This situation is analogous to the partition of a single solute between a number of immiscible solvents—each surface acts as if it were a separate "solvent" for the sorbate.

The actual thickness of the two-dimensional phase is not always readily ascertained. Insoluble films on aqueous substrates are obtained by spontaneous spreading of certain surface-active materials that are placed on the surface; such films have the thickness of only a single molecule, i.e., they are monomolecular films. The same is true of chemically adsorbed films of a gaseous adsorbate on a solid surface. But an adsorbed vapor cannot be expected to remain only monomolecular in thickness as the pressure in the bulk phase is increased. Ultimately, when the pressure equals the saturation vapor pressure, the adsorbate must condense without limit on all available surfaces. A comparison of the monomolecular adsorption isotherm (A) of a gas, which reaches a saturation value when the surface is filled, and the multimolecular adsorption isotherm (B) of a vapor, which becomes infinite at a pressure equal to the vapor pressure, is shown in Figure I-1. When the bulk phase of adsorbate is liquid rather than gas, the thickness of the two-dimensional phase may be monomolecular or multimolecular; no good criteria have yet been developed that will give a definitive answer on this point.

In this book we shall be primarily concerned with gas/solid

adsorption systems, that is, those systems in which the adsorbate does not dissolve in the bulk solid; the only equilibria that we consider are those between bulk gas and interfacial phases. We limit this consideration still further by confining our attention to monolayer adsorption, because we wish to study directly the interaction between a solid surface and a visiting gas molecule. Even in the adsorption of vapors where multilayers are ultimately formed, there is a region of the isotherm at low relative pressures that can be treated as true monolayer adsorption. In Figure I-1 this region is depicted as extending to one- or two-tenths relative pressure, which is typical.

2. Behavior of the Two-Dimensional Phase

The most direct method for observing the behavior of a two-dimensional phase is that used for insoluble monolayers on a liquid surface. Quantitative measurements can be obtained by means of a film-balance, such as that introduced by Langmuir (2) in 1917, of which a modern version is shown in Figure I-2. The float *A*, of thin sheet metal, is subjected to a lateral thrust from the surface film; the thrust causes a turning of the bar *B*, on which a mirror

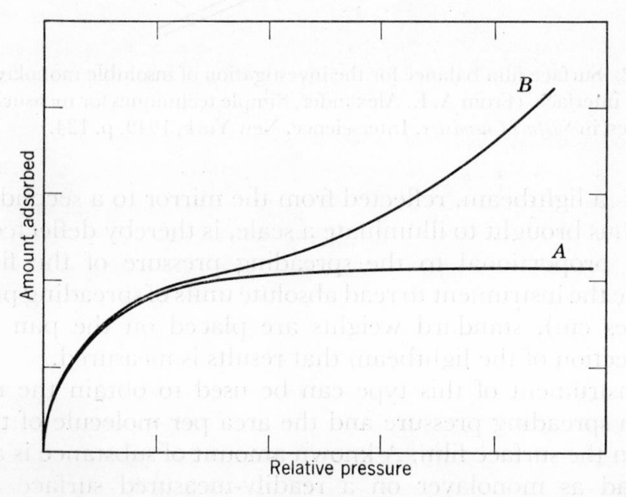

Fig. I-1. Typical adsorption isotherms: *A*, monomolecular adsorption; *B*, multimolecular adsorption.

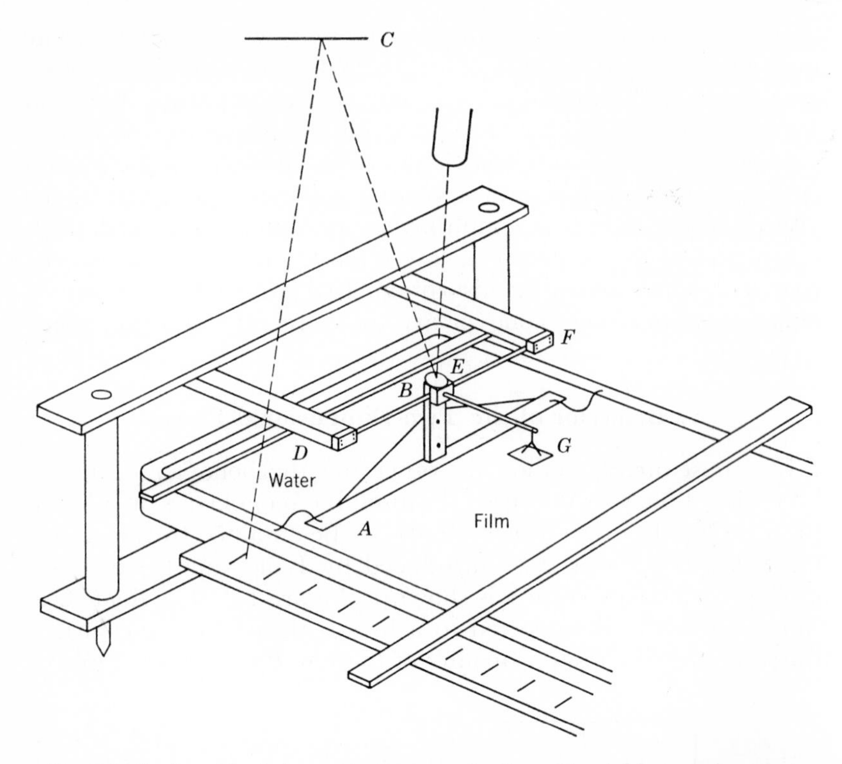

Fig. I-2. Surface film balance for the investigation of insoluble monolayers at an air/water interface. (From A. E. Alexander, Simple techniques for measuring force-area curves, in *Surface Chemistry*, Interscience, New York, 1949, p. 124.)

is fixed; a lightbeam, reflected from the mirror to a second mirror C and thus brought to illuminate a scale, is thereby deflected by an amount proportional to the spreading pressure of the film. To calibrate the instrument to read absolute units of spreading pressure, Π (dynes/cm), standard weights are placed on the pan G, and the deflection of the lightbeam that results is measured.

An instrument of this type can be used to obtain the relation between spreading pressure and the area per molecule of the substance in the surface film. A known amount of substance is allowed to spread as monolayer on a readily-measured surface area of water—the measurements of mass and area are the data for the calculation of σ, the area per molecule; the monolayer exerts a spreading pressure that can be measured by the instrument, as just described. Upon compressing the film, by reducing the area of aqueous surface available to it, the spreading pressure increases. An interesting analogue of the pressure-volume relation of a gas is thus enacted by a single layer of molecules adsorbed at an aqueous surface. An equation of state can be found, which under limiting conditions of area tending to infinity has the same form as the ideal gas law, i.e.,

$$\Pi\sigma = kT \tag{I-2}$$

where Π is the spreading pressure in dynes per centimeter, σ is the area per molecule, k is the Boltzmann constant (R/N), and T is the absolute temperature. Equation I-2 is called "the two-dimensional ideal-gas law." The behavior of the monomolecular film shows the same general kind of deviation from ideality that is found in real gases: the effect of molecular size is to *reduce* the measured area to an actual area available to each molecule, and the effect of intermolecular attraction is to make the measured spreading pressure fall short of the "ideal" value, so that a term has to be *added* to it to bring it up to the "ideal" spreading pressure. The proper corrections for nonideal behavior are, therefore, similar to those introduced by van der Waals for a real gas. An even more striking analogy between the two-dimensional state of matter and ordinary states is the phenomenon of two-dimensional condensation, in which the film shows the characteristic loss of one degree of freedom that results, by Gibbs' phase rule, equation I-1, from the presence of a new, additional phase. The Π-σ isotherms of a monolayer, in a nice analogy with the p-V isotherms of a gas below its critical temperature, have the abrupt discontinuities of a first-order phase transition; in addition, the range of the discontinuity becomes less with rising temperature, culminating in a point discontinuity at a critical temperature, exactly like the behavior described by Andrews for the condensation of carbon dioxide.

The late W. D. Harkins (3) always emphasized that the greatest part of our knowledge and understanding of adsorbed films has been obtained from insoluble monolayers on water or on aqueous solutions, and that the study of solid surfaces ought to start at that point. Upon moving from the aqueous substrate to that of a solid, we leave a homogeneous force field for an unknown heterogeneity of rigid surfaces that are unable to achieve an equilibrium among themselves. The theoretical approach to this more complex problem is to postulate an ideal solid surface that is supposed to be energetically uniform or homotattic* throughout for adsorption. This was also the point of view originally adopted by Langmuir in 1917, although at that time he did not have available the body of knowledge to which Harkins referred thirty years later.

The study of films on liquid substrates is based on Π-σ-T relations that describe the films by means of two-dimensional equations of state. For the adsorption of a gas or vapor on a solid, however, the spreading pressure Π cannot be measured directly; instead, the adsorption is followed by measuring the amount adsorbed as a function of the equilibrium pressure and temperature. These experimental quantities can be converted to Π and σ if the specific surface area Σ in cm^2/g of the adsorbent is known. Thus

$$\sigma = 22{,}400\ \Sigma/VN \tag{I-3}$$

where V is the amount adsorbed in cm^3 at STP per gram and N is Avogadro's number; also

$$\Pi = {}^{0}_{a}f - {}^{\theta}_{a}f = (RT/22{,}400\Sigma) \int_0^p V\, d \ln p \tag{I-4}$$

where ${}^{0}_{a}f$ is the free energy of one square centimeter of the clean surface in equilibrium with its own vapor and ${}^{\theta}_{a}f$ is the free energy of one square centimeter of the interface when V/Σ cm^3 of gas is adsorbed on it, so that a certain fraction θ of the surface is covered. Equation I-4 is Bangham's application (5) to the equilibrium between the gas phase and the adsorbed phase of Gibbs' adsorption theorem (6). By its use, any mathematical description of an adsorption isotherm can be converted into the corresponding two-dimensional equation of state, or conversely an equation of state previously selected can be converted into its corresponding adsorption-isotherm equation. The simplest of these interconversions is that of the two-dimensional ideal gas, $\Pi\sigma = kT$, which yields $p = K\theta$. This equation is Henry's law of gaseous solubility applied to adsorption. Other interconversions are shown later in Table I-1.

* This term was introduced by Sanford and Ross (4), who defined a homotattic surface as the surface of a submicroscopic patch or region, part of a larger surface, which acts as if its structure were uniform and homogeneous.

When the monolayer is compressed to its maximum extent, the area available per molecule is designated σ_0; surface concentration may be conveniently expressed by means of this reference point as $\theta = \sigma_0/\sigma$; thus, θ equals the fraction of the surface that is occupied by adsorbed molecules.

Using equation I-4 in the form

$$\Pi = (kT/\sigma_0) \int_0^p \theta \, d \ln p \tag{I-4a}$$

any two-dimensional equation of state can be converted to its corresponding $p - \theta$ isotherm. For $\Pi\sigma = kT$,

$$d\Pi = (kT/\sigma_0) \, d\theta = (kT/\sigma_0) \, \theta \, d \ln p$$

hence

$$d \ln \theta = d \ln p$$

or

$$p = K\theta \tag{I-5}$$

3. Some Mathematical Descriptions of the Dynamic Equilibrium between the Gas and the Adsorbed Phase

The earliest attempts to describe the adsorption isotherm analytically were entirely empirical. A parabolic equation of the following form was found to be suitable:

$$x/m = kc^{1/n} \tag{I-6}$$

where x is the grams of adsorbate per m grams of adsorbent, c is the concentration of the bulk phase (p may be substituted, representing pressure, when the bulk phase is a gas), and k and n ($n > 1$) are empirical constants. This equation, although not original with Freundlich, was so extensively used by that writer (7) that it is now known as the Freundlich adsorption isotherm. The Freundlich equation cannot, however, by its very nature, be a perfect representation of the physical reality, since it states that the value of x/m increases without limit as c increases; whereas in reality the saturation of the surface with a monomolecular layer would set an upper limit to the concentration in the two-dimensional phase.

Langmuir produced in 1916 the first theoretical treatment of an isothermal adsorption equilibrium (8). According to Langmuir, the rate of adsorption, u, is proportional (a) to the number of molecules that strike 1 cm² of surface per second, (b) to the fraction f_1 of that number that remain on the surface long enough for an exchange of kinetic energy to take place, and (c) to the fraction of the 1 cm^2 of surface not already occupied by adsorbed molecules. Expressed analytically:

$$u = (pN/\sqrt{2\pi MRT})\, f_1(1 - \theta) \qquad \text{(I-7)}$$

where M is the molecular weight of the adsorbate, and θ is the fraction of the adsorbent surface covered with adsorbate. The rate of desorption, v, is proportional to the fraction of the surface occupied by adsorbate and to the fraction of the total number of adsorbed molecules that have enough energy to leave the attractive forces of the surface, which are expressed as a uniform adsorptive energy of U cal/mole. Hence,

$$v = k_0\, \theta\, e^{-U/RT} \qquad \text{(I-8)}$$

At equilibrium,

$$u = v$$

from which we obtain

$$p = K(\theta/1 - \theta) \qquad \text{(I-9)}$$

where

$$K = (k_0/f_1)(\sqrt{2\pi MRT/N})\, e^{-U/RT} = Ae^{-U/RT} \qquad \text{(I-10)}$$

The concept underlying the use of θ, which is defined as the fraction of the surface covered by an adsorbed film, implies that when $\theta = 1$, the whole surface of the solid is covered with a monomolecular film. Let $\theta = V/V_m$, where V is the amount of gas adsorbed at an equilibrium pressure p, and V_m is the amount adsorbed at surface saturation. Equation I-9 then becomes

$$p = KV/(V_m - V) \qquad \text{(I-11)}$$

in which the experimentally measured quantities are p and V, and V_m and K are constants that can be evaluated if the experimental data are capable of being described by equation I-11.

Equation I-11 is called the Langmuir adsorption isotherm. Implicit in its development are the following assumptions about the model of adsorption: 1. the adsorbate in the bulk gaseous phase behaves as an ideal gas; 2. the amount adsorbed is confined to a monomolecular layer; 3. every part of the surface has the same energy of adsorption; 4. no adsorbate-adsorbate interaction is taken into account; presumably it is negligible; 5. the adsorbed molecules are localized; i.e., they have definite points of attachment to the surface. This assumption is made explicit only in the statistical derivation of the Langmuir equation, such as that given by Fowler and Guggenheim (9). These authors state that nonstatistical derivations of the Langmuir equation make assumptions about the mechanism of sorption and desorption, which assumptions are not strictly required. Equation I-11 "must hold whatever the kinetics of the processes, provided only that the molecules are adsorbed on to definite sites and do not interact with one another."

The first two of the preceding assumptions are acceptable as true of many gas/solid systems, but the second pair of assumptions are always false. No solid surface is ever ideally uniform. Even in the absence of different crystalline faces or adventitious impurities, some nonuniformity is introduced by edges and corners, cracks, dislocations, and other crystal imperfections. Also the assumption that at all surface concentrations the adsorbate-adsorbate interaction is negligible is never correct. Even a noble gas has enough energy of interaction to account for about 25% of the measured heat of adsorption at half saturation of the surface (see Section I-7). The effect of surface nonuniformity is to cause the energy of adsorption to decrease with θ, whereas the effect of adsorbate-adsorbate interaction is to cause it to increase. The net result of the two erroneous assumptions is, therefore, partially to cancel out each other's effects; the success, not always deserved, of the Langmuir equation as a description of the adsorption isotherm owes much to this coincidence. In spite of its grave faults, however, the development of the Langmuir model was of prime importance in the history of the subject, because it named for the first time the factors that are significant in the process of adsorption; all subsequent theories have used the concepts and the terminology introduced by Langmuir.

With the possibility of evaluating the saturation capacity, V_m, of a surface in terms of a monomolecular layer, a practical application of the adsorption isotherm became obvious, namely, the measurement of the degree of subdivision of a solid in terms of its specific surface. The solid is used as the adsorbent and V_m determined; all that is then required is a definitive value for the molecular cross sectional area, σ_0, of the adsorbate. The specific surface of the solid adsorbent, Σ, in m^2/g then follows:

$$\Sigma = (V_m/22{,}400) \times N \times \sigma_0 \times 10^{-20}$$

where V_m is in cm^3 of adsorbate, measured as if at STP, per gram of adsorbent, N is Avogadro's number, and σ_0 is the molecular cross section of the adsorbate in A^2/molecule.

A convenient laboratory measurement of an adsorption isotherm is possible with nitrogen as adsorbate, at the temperature of its boiling point, 77.5°K. Liquid nitrogen is readily available for use as a thermostat liquid, and the gas itself is easily obtained in a high state of purity. But this choice of conditions, so convenient for practice, introduces new difficulties in the theory. At 77.5°K, nitrogen is not a gas but a vapor; the adsorption, therefore, is multimolecular rather than monomolecular, and the Langmuir equation, which contains the parameter we want to measure, is applicable only to the low-pressure portion of the isotherm at which the adsorbed film is monomolecular. The theory was extended in 1938 by Brunauer, Emmett, and Teller (10), whose model of multimolecular adsorption, known as the B.E.T. theory, postulated a number of simultaneous Langmuir-type adsorptions between each two successive molecular layers. A few assumptions had to be added to the model in order to simplify the mathematical treatment, namely: the energy of adsorption of the bare solid surface is uniform and characteristic of the solid; the energies of adsorption in the second and succeeding layers of adsorbate are also uniform, and are all equal to the heat of liquefaction of the adsorbate. The final expression of the B.E.T. theory is

$$V = V_m C p / \{(p_0 - p)[1 + (C - 1)(p/p_0)]\} \tag{I-12}$$

where p_0 is the saturation vapor pressure of the adsorbate and V_m and C are constants; V_m has the same significance in the B.E.T. theory as it has in the Langmuir theory.

The B.E.T. theory is the basis of a method of measuring the specific surface of a solid, using the value of V_m derived from equation I-12. This method has been experimentally shown to be sufficiently accurate for most practical purposes and has successfully met an urgent requirement for a rapid estimate of particle size. By so doing, it has made possible great advances in both theoretical and practical studies of surface chemistry and physics. We must not, however, confuse the triumphs of the B.E.T. theory in this practical application with its claims as a model of the adsorption process—in this latter respect, it is obviously far from satisfactory. The Langmuir model with its two unjustifiable assumptions, the effects of which fortuitously cancel each other, is the basis of the B.E.T. model, and the additional assumptions introduced in the B.E.T. derivation merely add further improbabilities to the total picture. At the present time, in fact, an adequate model of multimolecular adsorption could hardly be produced, since no adequate model of the simpler phenomenon of monomolecular adsorption is yet available. Such a model, if it is to gain our intellectual assent, will have to take into account both the heterogeneity of the underlying solid substrate and also the variation of adsorbate-adsorbate interaction with surface concentration.

The fifth assumption implicit in the Langmuir isotherm introduced the concept of localization. As defined by the requirements of the statistical derivation, localization excludes any lateral movement of the adsorbed molecules; i.e., molecules are adsorbed at and desorbed from specific sites on the surface. The adsorbed molecules still have thermal energy, which must therefore manifest itself in three degrees of vibrational freedom about the site. The energy required for a molecule to move from site to site is the same as that required for desorption into the gas phase, so that there is no difference in the case of localized adsorption between translation along the surface and desorption.

4. Mobile and Localized Adsorbed Films

The surface of a solid is not a continuum but a two-dimensional array of atoms that causes maxima and minima in the interaction energy as a molecule moves along the surface. If the depth of these potential wells is equal to the energy for desorption we have the condition for a localized film, but we could imagine the situation where the energy required for a molecule to move from site to site is less than that required for its escape into the gas phase. Instead of using the concept of a site, which implies a fixed point of attachment to the surface, it would now be more exact to think of an energy barrier to translation along the surface. The important point is, however, that the molecule can move laterally within the two-dimensional phase and desorption can take place from any point on the surface. This sort of adsorbed film is said to be mobile.

Figure I-3 depicts the energy relations that we have described in the two preceding paragraphs. The adsorptive potential between a gas molecule and any point on a solid surface is expressed by the ordinate U; the depth of the potential well is represented by the amplitude χ of the wavy line; the hatched areas are the regions of the two-dimensional phase. Figure I-3a, where ${}_gP \gg \chi$, represents the condition for a mobile adsorbed film: when the average thermal energy, kT, of the adsorbed molecule is large enough, compared to χ, the molecule will translate freely inside the energetically continuous two-dimensional phase. Figure I-3b, where ${}_gP = \chi$, represents the condition for a localized film. It must not be thought, however, that a sharp natural boundary separates these two types of adsorbed

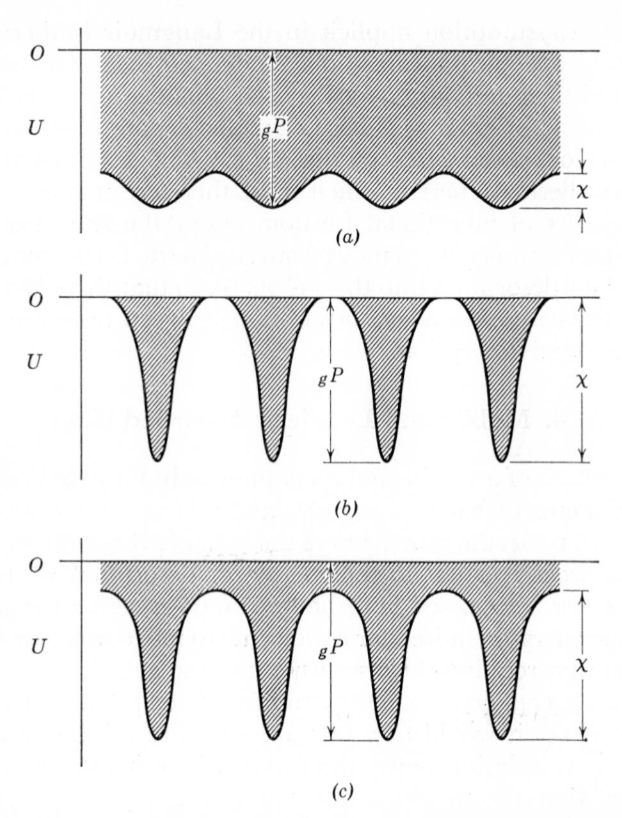

Fig. I-3. Schematic representation of energy barriers to the translation of adsorbed molecules along the surface.

film. Figure I-3c depicts a situation that is energetically similar to that of Figure I-3a, but the height of the energy barrier is so close to $_gP$ that the adsorbed film, though technically mobile, would behave effectively as a localized film. Molecules having an energy greater than χ but less than $_gP$ *can* translate without leaving the two-dimensional phase, but in Figure I-3c such molecules would be only a small fraction of those adsorbed; consequently the effective adsorption equilibrium would be that of localized adsorption.

In Figures I-3a, b, and c we have considered the potential energy relations between $_gP$ and χ and shown how they determine whether the adsorbed film is mobile or localized. These quantities are not

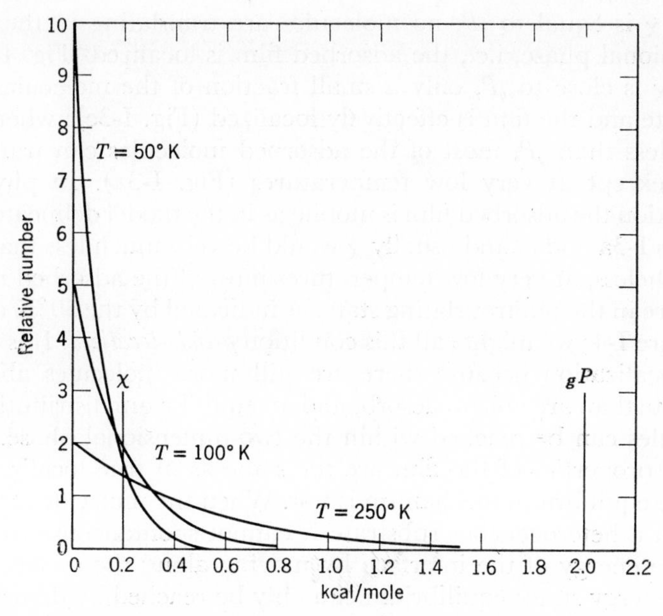

Fig. I-4. Boltzmann distribution of kinetic energies, normal to the surface, of adsorbed molecules at three different temperatures.

temperature dependent but are properties of the adsorbate-adsorbent system. What are the effects of thermal energy on the behavior of these systems?

Figure I-4 represents the distribution of kinetic energies normal to the surface among the adsorbed molecules at any instant at three successively higher temperatures. The number of molecules having energies greater than $_gP$ increases with temperature rise; these are the molecules in the process of desorbing into the gas phase. The number of molecules with energies less than χ decreases with temperature rise; these molecules, having insufficient energy to overcome the translational barrier, remain vibrating about fixed points on the surface of the adsorbent. Molecules that have energies in the range between χ and $_gP$ are confined to the two-dimensional phase but have translational freedom within it. The ratio of translating to non-translating molecules is given approximately by

$$\text{translating/nontranslating} = (e^{-\chi/RT} - e^{-{}_gP/RT})/(1 - e^{-\chi/RT}) \quad \text{(I-13)}$$

When χ is equal to $_gP$, no molecules are translating in the two-dimensional phase, i.e., the adsorbed film is localized (Fig. I-3b); when χ is close to $_gP$, only a small fraction of the molecules can translate and the film is effectively localized (Fig. I-3c); when χ is much less than $_gP$, most of the adsorbed molecules can translate freely except at very low temperatures (Fig. I-3a). In physical adsorption the adsorbed film is mobile as in the model delineated by Figures I-3a and c, and usually χ would be very much less than $_gP$. Nevertheless, at very low temperatures most of the adsorbed molecules are in the nontranslating state, as indicated by the 50°K curve in Figure I-4; we might call this condition *pseudo-localized*. It is not a true localization because there are still more molecules able to translate than are able to desorb, and an equilibrium distribution of molecules can be reached within the two-dimensional phase. The kinetic properties of the film are the same as in true localization, but the equilibrium mechanism is not. When the molecule is localized on a heterogeneous substrate it cannot spontaneously reduce the free energy of the interface by moving along the surface to a lower energy state; equilibrium can only be reached by desorption and chance readsorption in a lower energy state, at which it is held for a longer period of time. The attainment of dynamic equilibrium, therefore, is apt to be a slow process in localized adsorption. The mobile film, on the other hand, can reach the state of minimum free energy more readily, and hence, more rapidly.

5. Further Descriptions of the Dynamic Equilibrium between the Gas and the Adsorbed Phase

The method of deriving an adsorption-isotherm equation by means of a kinetic mechanism, such as was done for the Freundlich and Langmuir equations, does not readily lend itself to the introduction of all the necessary conditions that one would like to take into account in a complete description of the adsorption process. The methods of statistical mechanics have been used by several theorists to provide more sophisticated pictures of adsorption; an example of these results is the finding that a mobile adsorbed film on a homotattic substrate can be described by a two-dimensional virial equation of state:

$$\Pi = kT[(1/\sigma) + (B(T)/\sigma^2) + (C(T)/\sigma^3) + \ldots] \quad \text{(I-14)}$$

TABLE I-1
Mathematical Descriptions of Some Models of Adsorption on Homotattic Substrates

Equation number	Description of adsorbed film	Two-dimensional equation of state (a)	Isotherm equation (b)
I-16	Mobile film—no interaction Point molecules (two-dimensional ideal gas)	$\Pi\sigma = kT$ or $\Pi = \frac{kT}{\beta}\cdot\theta$	$p = K\theta$
I-17	Mobile film—no interaction (Volmer equation)	$\Pi(\sigma - \beta) = kT$ or $\Pi = \frac{kT}{\beta}\left(\frac{\theta}{1-\theta}\right)$	$p = K\frac{\theta}{1-\theta}\exp\left(\frac{\theta}{1-\theta}\right)$
I-18	Mobile film with interaction [two-dimensional van der Waals equation (11)]	$\left(\Pi + \frac{\alpha}{\sigma^2}\right)(\sigma - \beta) = kT$ or $\Pi = \frac{kT}{\beta}\left[\frac{\theta}{1-\theta} - \frac{\alpha\theta^2}{kT\beta}\right]$	$p = K\frac{\theta}{1-\theta}\exp\left(\frac{\theta}{1-\theta} - \frac{2\alpha\theta}{kT\beta}\right)$
I-19	Immobile film—no interaction (Langmuir equation)	$\Pi\sigma = kT\frac{\sigma}{\beta}\ln\left(\frac{\sigma}{\sigma-\beta}\right)$ or $\Pi = \frac{kT}{\beta}\ln\left(\frac{1}{1-\theta}\right)$	$p = K\frac{\theta}{1-\theta}$
I-20	Immobile film with interaction [Fowler and Guggenheim equation (9, 9a)]	$\Pi\sigma = kT\frac{\sigma}{\beta}\ln\left(\frac{\sigma}{\sigma-\beta}\right) - \frac{c\omega}{2}\frac{\beta}{\sigma}$ or $\Pi = \frac{kT}{\beta}\left[\ln\left(\frac{1}{1-\theta}\right) - \frac{c\omega\theta^2}{2kT}\right]$	$p = K\frac{\theta}{1-\theta}\exp - \left(\frac{c\omega\theta}{kT}\right)$

By means of the Gibbs transformation, equation I-4a, the adsorption isotherm corresponding to equation I-14 is:

$$p = K\theta \exp\left(\frac{2\theta}{\sigma_0} B(T) + \frac{3\theta^2}{2\sigma_0^2} C(T) + \frac{4\theta^3}{3\sigma_0^3} D(T)\ldots\right) \quad \text{(I-15)}$$

These equations emphasize once more that the mobile adsorbed film can be treated as a two-dimensional compressible fluid—a point of view that has been productively developed by de Boer (11).

In Table I-1 are collected the results of the mathematical treatments of some models of adsorption; these results are expressed both as adsorption-isotherm equations and as two-dimensional equations of state.

Equations I-16a and b refer to the two-dimensional ideal gas; the latter equation has already been derived (eq. I-5). When an attempt is made to correct the ideal-gas law for the effects of molecular size, we get Volmer's equation (I-17a) and its corresponding adsorption isotherm, equation I-17b.

The analogue of van der Waals' corrections for nonideality is shown in equation I-18a with its corresponding adsorption isotherm, equation I-18b. The constants α and β in these equations are not identical with the customary van der Waals constants of gases; for some simple adsorbates these two-dimensional constants are indeed directly related to the three-dimensional constants. For all adsorbates, however, they retain their theoretical significance as analogs of the van der Waals interaction constant a and size constant b. We do, in fact, henceforth substitute β for σ_0 (11).

The remainder of Table I-1 contains mathematical descriptions of two models of localized films: equations I-19a and b refer to the Langmuir isotherm previously discussed and equations I-20a and b describe Fowler and Guggenheim's modification of the Langmuir equation, in which modification adsorbate-adsorbate interactions are taken into account by means of the constants ω and c.

Each isotherm equation in Table I-1 is derived by a specific application of Gibbs' adsorption theorem. The general nature of the transformation is as follows. Let a two-dimensional equation of state be represented by

$$\Pi = kT f(\sigma, T) = (kT/\beta)\, \mathbf{f}(\theta, T) \quad \text{(I-21)}$$

Equation I-4 can be written in the form

$$\Pi = kT \int_0^p (1/\sigma) d \ln p = (kT/\beta) \int_0^p \theta d \ln p \qquad \text{(I-4a)}$$

Equating the derivatives of equations I-21 and I-4a with respect to σ at constant temperature gives

$$d \ln p = \sigma[\partial f(\sigma, T)/\partial\sigma]_T \, d\sigma = \mathbf{f}'(\theta, T)_T \, d \ln \theta \qquad \text{(I-22)}$$

Upon integrating,

$$\ln p = \int_0^\sigma \sigma f'(\sigma, T)_T \, d\sigma + \ln K \qquad \text{(I-23)}$$

where $\ln K$ is an integration constant; alternatively, in terms of θ,

$$\ln p = \int_0^\theta \mathbf{f}'(\theta, T)_T \, d \ln \theta + \ln K$$

or

$$p = K \, g(\theta, T) \qquad \text{(I-24)}$$

The constant K that appears in Table I-1 and in equation I-24 turns out to be an integration constant, hence independent of θ, but possibly dependent on temperature. Its nature is made clearer by the following argument. Consider the integral molar change of free energy of the isothermal adsorption process for one mole of gas in its standard state (760 mm) to the adsorbed state, defined by a relative surface concentration θ and equilibrium pressure p:

$$\Delta F^{\text{ads}} = RT \ln (p/760)$$

or

$$\Delta F^{\text{ads}} = RT \ln K + RT \ln g(\theta, T) - RT \ln 760 \qquad \text{(I-25)}$$

Also

$$\Delta F^{\text{ads}} = \Delta H^{\text{ads}} - T\Delta S^{\text{ads}} \qquad \text{(I-26)}$$

We show later that

$$\Delta H^{\text{ads}} = -q^{\text{diff}} - RT + (T\beta/\theta) \, (\partial\Pi/\partial T)_\theta \qquad \text{(III-62)}$$

The evaluation of ΔH^{ads} by means of equation III-62 for the five models of the adsorbed film described in Table I-1 has been worked out (see Table III-1). In its most general terms, however, we can write

$$\ln K = \cdot \cdot \frac{q^{\text{diff}}}{RT} - \frac{\Delta S^{\text{ads}}}{R} - \ln g(\theta, T) + \ln 760 - 1 + \frac{\beta}{R\theta}\left(\frac{\partial\Pi}{\partial T}\right)_\theta$$

or

$$K = A \exp(-q^{\text{diff}}/RT) \tag{I-27}$$

where

$$\ln A = -\frac{\Delta S^{\text{ads}}}{R} - \ln g(\theta, T) + \ln 760 - 1 + \frac{\beta}{R\theta}\left(\frac{\partial \Pi}{\partial T}\right)_\theta \tag{I-28}$$

The original introduction of K in equation I-23 requires that it should always be a term independent of θ, no matter what equation of state is used to describe the adsorbed phase. Even when q^{diff} is affected by adsorbate-adsorbate interaction, so as to increase with θ, the *net* θ-dependence of K will disappear by a cancelling out of all terms in θ. The way in which this happens will be illustrated later, when we come to apply equation I-27 to specific descriptions of the adsorbed phase in Section V-3. At present, we can recognize that the evaluation of K can be carried out at any value of θ. The most convenient value to take is that of the two-dimensional standard state, which can be defined in any arbitrary way. For the mobile adsorbed film the standard state of the surface phase, θ_s, is taken so that the molecules are the same distance apart as they would be in an ideal gas at 0 °C. For the localized adsorbed film the standard state is more conveniently taken as $\theta_s = 0.500$. Whatever condition is selected, one can calculate from basic theory the standard entropy change on adsorption, ΔS_s^{ads}, from the standard state of the gas to the standard state of the adsorbed film. Using standard state conditions, equations I-27 and I-28 become

$$K = A \exp(-q_s^{\text{diff}}/RT) \tag{I-27a}$$

where

$$\ln A = -\frac{\Delta S_s^{\text{ads}}}{R} - \ln g(\theta_s, T) + \ln 760 - 1 + \frac{\beta}{R\theta_s}\left(\frac{\partial \Pi}{\partial T}\right)_\theta \tag{I-28a}$$

The invariance of K with θ suggests that the constant K is related to interaction between the adsorbate and the surface, excluding all adsorbate-adsorbate interactions. These latter would be taken into account by the two-dimensional equation of state, equation I-21, and would, besides, be θ-dependent. In the adsorption-isotherm equation for a homotattic surface, $p = K\,g(\theta, T)$, the two types of interaction are clearly separated: the K-factor refers to adsorbate-adsorbent interaction only and adsorbate-adsorbate interaction, if included at all, is expressed within the $g(\theta, T)$ factor. The constant K, for this reason, has particular importance for the measurement of the adsorptive potential of a solid surface and the kinetic state of the adsorbed molecule.

6. Potential and Kinetic Energy Changes on Adsorption

The consideration of potential energy changes on adsorption is not required for a thermodynamic description of the process, but is essential for the construction of a model that takes into account the translational, rotational, and vibrational states of the adsorbed molecule, as well as the nature and origin of the adsorptive forces.

A molecule in the gas phase has potential energy with respect to its possible adsorbed state at an interface, just as a rubber ball on a table top has potential energy with respect to its possible position on the floor. This potential energy is fixed by the nature of the gas/solid interface, just as the potential energy of the ball is fixed by the height of the table top and the mass of the ball; in both processes the potential energy is independent of any kinetic energy possessed

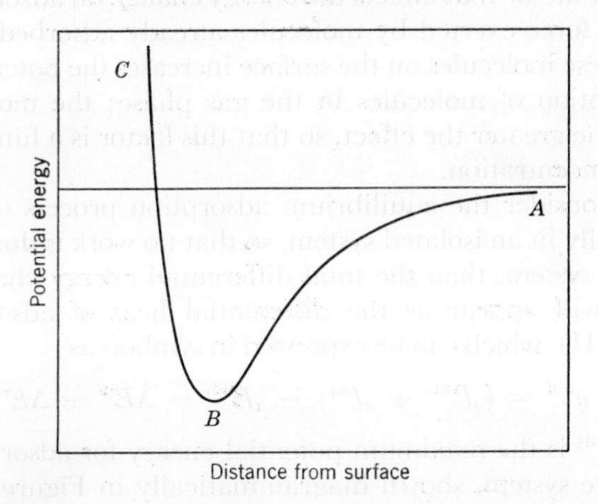

Fig. I-5. Change of potential energy of a gas molecule on approaching a solid surface.

by the particles either before or after the transition. The total energy evolved in the process is the sum of the changes of potential and kinetic energies.

As an illustration, a diatomic molecule in the gas phase would have three degrees of translational freedom plus two of rotation and one of internal vibration. Suppose the molecule approaches to and is adsorbed at a solid surface. The change of potential energy of the system follows the path *ABC* in Figure I-5 and would then return along the reverse path *CBA* to its original value unless, by collision, some of the molecule's momentum in the component normal to the surface is lost, the energy appearing as heat. The molecule is now detained in the two-dimensional phase until, by collision, it gains enough kinetic energy to escape from the potential well. In the two-dimensional phase the kinetic state of the molecule is governed by the conditions described in Section I-4. The molecule either has two degrees of translational freedom or none; if the former, then an additional (external) degree of vibrational freedom appears; if the latter, then three additional (external) degrees of vibrational freedom appear, these acquiring some of the energy of the system. The internal vibrations of the molecule are probably not affected by adsorption, but one or two of the rotational degrees of freedom may be restricted, and the energy they contained given up as heat.

Another factor that affects the energy change on adsorption is the attractive force exerted by molecules already adsorbed. The presence of these molecules on the surface increases the potential energy for adsorption of molecules in the gas phase; the more that are present, the greater the effect, so that this factor is a function of the surface concentration.

If we consider the equilibrium adsorption process taking place isothermally in an isolated system, so that no work is done either on or by the system, then the total differential energy change on adsorption will appear as the differential heat of adsorption (see Chapter III), which can be expressed in symbols as

$$q^{\text{diff}} = ({}_{g}P^{\text{ads}} + {}_{a}P^{\text{ia}}) - {}_{a}E^{\text{vib}} - \Delta E^{\text{tr}} - \Delta E^{\text{rot}} \qquad \text{(I-29)}$$

where ${}_{g}P^{\text{ads}}$ is the maximum potential energy for adsorption of the gas-surface system, shown diagrammatically in Figure I-5; ${}_{a}P^{\text{ia}}$ is the additional potential energy due to molecules already adsorbed; ${}_{a}E^{\text{vib}}$ is the average vibrational energy of a molecule in the adsorbed film; ΔE^{tr} is the change of translational energy for the adsorption process and ΔE^{rot} is the change of rotational energy for the adsorption process. The differential heat of adsorption, therefore, measures the energy necessary to *remove* an adsorbed molecule from its average vibrational state and from the attractive forces of its adsorbed neighbors to an infinite distance from the surface, plus energy equivalent to the degrees of freedom of the gaseous molecules in excess of those in the adsorbed state. A more rigorous proof of the relation described by equation I-29 is developed in Section III-6.

For purposes of theory, an energy function that is characteristic of the system and independent of temperature is desirable; such a quantity is conveniently defined by the potential energy difference between the lowest energy state of the molecule in the gas phase and its lowest energy state in the adsorbed phase, both at infinite dilutions. In this way we obtain a quantity that is independent of the kinetic states of the molecule in either phase and that measures most directly the adsorptive potential of the system. This adsorptive potential, U_0, is defined as

$$U_0 = {}_gP^{ads} - {}_aE_0^{vib} \tag{I-30}$$

where ${}_aE_0^{vib}$ is the zero-point vibrational energy of the adsorbed molecule with respect to the surface, or

$${}_aE_0^{vib} = (1/2)Nh\nu \tag{I-31}$$

Equation I-27 can be expressed in the terms of these newly defined functions:

$$K = A^0\, e^{-U_0/RT} \tag{I-32}$$

where

$$\ln A^0 = -\frac{\Delta S^{ads}}{R} + \frac{{}_aE^{vib} - {}_aE_0^{vib}}{RT} + \frac{\Delta E^{tr}}{RT} + \frac{\Delta E^{rot}}{RT} - \frac{{}_aP^{ia}}{RT} - \ln g(\theta, T) + \ln 760 - 1 + \frac{\beta}{R\theta}\left(\frac{\partial \Pi}{\partial T}\right)_\theta \tag{I-33}$$

In the next section we shall show how to calculate ${}_aP^{ia}$ from the adsorption-isotherm equation. When a model (presumably reasonable) of the adsorbed film has been constructed, every term in equation I-33 can be evaluated; also, when a value of K can be determined experimentally, the adsorptive potential U_0 may be calculated by equation I-32.

7. Cooperative Additions to the Adsorptive Potential

The influence of cooperative effects on the heat of adsorption was first observed experimentally when the isosteric heat of adsorption was found to increase with the surface concentration of adsorbed molecules. It was concluded that heterogeneity of the surface could not be the cause of this effect because, at a given surface concentration, the higher adsorptive energy portions are more densely occupied than the lower adsorptive energy portions of a nonuniform surface and this tendency increases as the temperature is lowered. Therefore, the ratio of surface of higher to that of lower adsorptive energy continuously decreases as the surface is progressively filled. Molecules already adsorbed, on the other hand, would be expected to increase the heat of adsorption by virtue of their attraction for one another and this effect would increase in magnitude as the surface population increases.

The isosteric heat of adsorption, by means of which this effect was observed, is defined by an application of the Clausius-Clapeyron equation to the adsorption equilibrium, giving (for derivation see antecedents of equation II-33):

$$q^{st} = RT^2 (\partial \ln p/\partial T)_\theta \tag{I-34}$$

Applied to equation I-24, this gives

$$q^{st}/RT^2 = (\partial \ln K/\partial T)_\theta + (\partial \ln g(\theta,T)/\partial T)_\theta \tag{I-35}$$

then

$$(1/RT^2)(dq^{st}/d\theta) = d^2 \ln g(\theta,T)/d\theta\, dT \tag{I-36}$$

since $\ln K$ is independent of θ. We show in Section III-5 that

$$q^{st} = q^{diff} + RT \tag{I-37}$$

The expression for q^{diff} given by equation I-29 tells us that ${}_aP^{ia}$ is the only term that can be influenced by a variation of θ; hence

$$dq^{st}/d\theta = d\, {}_aP^{ia}/d\theta \tag{I-38}$$

Equations I-36 and I-38 are true only for homotattic surfaces. Combining equations I-36 and I-38 and integrating gives:

$$ {}_aP^{\text{ia}} = RT^2 \int_0^\theta (d^2 \ln g(\theta,T)/d\theta\, dT)d\theta \tag{I-39}$$

Various expressions for the function $g(\theta, T)$ for uniform surfaces are given in Table I-1, corresponding to two-dimensional equations of state. Of the adsorbed films therein described only two include adsorbate-adsorbate interaction, namely equations I-18 and I-20, and consequently these are the only two in that table for which ${}_aP^{\text{ia}}$, calculated by equation I-39, is not equal to zero for all values of θ.

For a mobile adsorbed film with interaction as described by the two-dimensional virial equation of state, equation I-14, and its corresponding isotherm equation, equation I-15, the calculation of ${}_aP^{\text{ia}}$ by equation I-39 gives:

$$ {}_aP^{\text{ia}} = RT^2 \left[\frac{2\theta}{\sigma_0} B'(T) + \frac{3\theta^2}{2\sigma_0^2} C'(T) + \frac{4\theta^3}{3\sigma_0^3} D'(T) + \ldots\right] \tag{I-40}$$

When the mobile adsorbed film is described by a two-dimensional van der Waals equation, i.e., equation I-18a, and its corresponding adsorption-isotherm equation, equation I-18b,

$$ p = K(\theta/1 - \theta)e^{\theta/(1-\theta)}e^{-2\alpha\theta/RT\beta} \tag{I-18b}$$

the calculation of ${}_aP^{\text{ia}}$ by equation I-39 gives:

$$ {}_aP^{\text{ia}} = 2\alpha\theta/\beta \tag{I-41}$$

The constants α and β of the two-dimensional equation of state can under certain conditions be related to the customary and familiar three-dimensional van der Waals constants a and b. The conditions are that the molecules in the adsorbed state are neither oriented nor polarized by the substrate, so that both their effective diameter and their polarization remain the same as in the gas phase. Practically speaking, this restriction excludes all but spherical, isotropic molecules and the following consideration applies only to such molecules. For a mobile monolayer described by the two-dimensional van der Waals model, the constants α and β for spherical isotropic molecules are expressed (11) by the following equations: $\alpha = \pi C/4d^4$, where C is proportional to the square of the polarizability of the molecule; and $\beta = \pi d^2/2$. In these equations, d is the effective diameter of the molecule, and the corresponding expressions for the three-dimensional van der Waals constants a and b are $a = 2\pi C/3d^3$ and $b = 2\pi d^3/3$. It follows that

$$\alpha/\beta = a/2b \qquad \text{(I-42)}$$

For localized monolayers an equivalent approximation to the two-dimensional van der Waals equation used for a mobile monolayer, is described by equation I-20b:

$$p = K \frac{\theta}{1 - \theta} \exp - (c\omega\theta/kT) \qquad \text{(I-20b)}$$

The calculation of ${}_aP^{\text{ia}}$ by equation I-39 gives

$${}_aP^{\text{ia}} = c\omega\theta = 4\omega\,\theta/z \qquad \text{(I-43)}$$

where ω is defined (9a) as the interaction energy per pair of nearest neighbors, and z is the number of nearest neighbor sites to a given

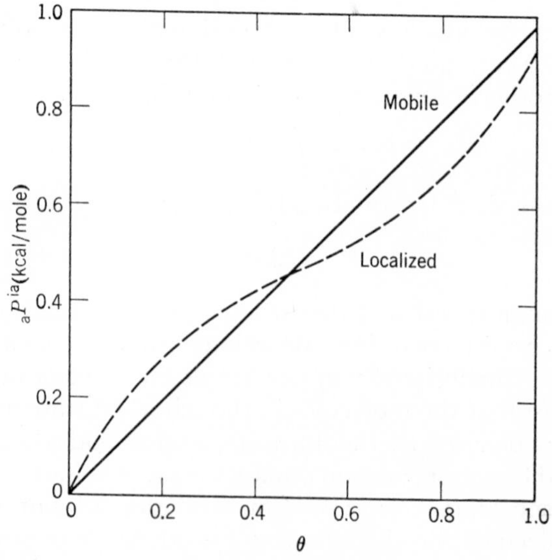

Fig. I-6. Lateral interaction potentials as function of surface concentration, calculated for mobile and localized adsorbed films of argon on {111} surface of potassium chloride.

site in an arbitrary lattice. These factors must take into account the nature of the adsorptive bond and the lattice character of the substrate. The value of $c\omega$ is not readily determined and is unique for each adsorbate-adsorbent system. A treatment of interaction in a localized monolayer was developed by Young (12), using the equation derived by Wang (13) for a statistical distribution of the adsorbed molecules. In Figure I-6 the variation of the differential interaction potential with surface concentration for a completely localized monolayer of argon on the {111} surface planes of potassium chloride, as calculated by Young, is compared with the result for a mobile monolayer, as calculated from the Hill-de Boer equation (I-18b). The significance of this comparison is that, for this surface at least, the differential interaction potential, ${}_aP^{\mathrm{ia}}$, is much the same whether the film is localized or mobile on a homotattic surface. For argon on other substrates the near equality of ${}_aP^{\mathrm{ia}}$ for the two types of film will not always be found, but presumably they are always close to one another. One cannot assume that ${}_aP^{\mathrm{ia}}$ is so small that it can be neglected. Experimental results for the adsorption of argon by a near-homotattic surface of graphite reveal that the differential heat of adsorption, q^{diff}, is 2–3 kcal/mole; at $\theta = 0.5$, ${}_aP^{\mathrm{ia}}$ equals 0.5 kcal/mole, which is a significant contribution to q^{diff}. The adsorbed film in this example is assumed to be mobile.

References

1. J. W. McBain, *Phil. Mag.*, (6), **18,** 916 (1909); *Idem, Z. Physik. Chem.*, **68,** 471 (1909).
2. I. Langmuir, *J. Am. Chem. Soc.*, **39,** 1848 (1917).
3. G. Jura and W. D. Harkins, *J. Am. Chem. Soc.*, **68,** 1941 (1946).
4. C. Sanford and S. Ross, *J. Phys. Chem.*, **58,** 288 (1954).
5. D. H. Bangham and N. Fakhoury, *J. Chem. Soc.*, 1324 (1931); D. H. Bangham, N. Fakhoury, and A. F. Mohamed, *Proc. Roy. Soc. (London)*, **138A,** 162 (1932); D. H. Bangham, *J. Chem. Phys.*, **14,** 352 (1946).
6. J. W. Gibbs, *Scientific Papers, Vol. 1*, Longmans, Green and Co., London, 1906, pp. 229 ff.
7. H. Freundlich, *Colloid and Capillary Chemistry* (transl.), Methuen, London, 1926, pp. 110–134.
8. I. Langmuir, *J. Am. Chem. Soc.*, **38,** 2267 (1916); *Idem. ibid.*, **40,** 1361 (1918); *Idem, Phys. Rev.*, **8,** 149 (1916).
9. R. H. Fowler and E. A. Guggenheim, *Statistical Thermodynamics*, Cambridge University Press, Cambridge, 1949, p. 431, equation 1007,10.

9a. T. L. Hill, *Introduction to Statistical Thermodynamics*, Addison-Wesley, Reading, Mass., 1960, pp. 246–9.
10. S. Brunauer, P. H. Emmett, and E. Teller, *J. Am. Chem. Soc.*, **60,** 309 (1938); P. H. Emmett in *Advances in Colloid Science, Vol. 1*, Interscience, New York 1942, pp. 1–36; S. Brunauer, *The Adsorption of Gases and Vapors, Vol. 1*, Princeton University Press, Princeton, 1945, pp. 149–179.
11. J. H. de Boer, *The Dynamical Character of Adsorption*, Clarendon Press, Oxford, 1953, p. 170ff.
12. D. M. Young, *Trans. Faraday Soc.*, **48,** 548 (1952).
13. J. S. Wang, *Proc. Roy. Soc. (London)*, **161A,** 127 (1937).

CHAPTER II

Experimental Techniques

1. Measurements of Adsorption Isotherms

In this chapter we propose to provide an outline of the techniques employed in the measurement of adsorption of gases by solids, for the benefit of the reader who is not acquainted with the variety of methods that have been developed. We do not propose to write a comprehensive review of experimental apparatuses but to present the principles common to all methods of measurement and a few examples; more detailed descriptions can be obtained from the references cited.

The two principal methods of measuring adsorption equilibria are classified as manometric and gravimetric, depending on whether the amount of adsorbed gas is determined by means of experimentally measured pressures and gas-law relationships of the gas phase, or by direct measurement of the weight gained by the adsorbent; both methods require a certain amount of vacuum technique.

A. SOME PRECAUTIONARY ADVICE

The measurement of the physical adsorption of gases does not usually require attaining pressures less than 10^{-7} mm of mercury, so that a good mechanical forepump and a two-stage mercury or oil diffusion pump suffices. Only the best high-vacuum stopcocks, which have precision fitted parts, should be used; the preferred styles, shown in Figure II-1, have a hollow, open plug with vacuum bulb. The best stopcock can be made useless, however, by improper lubrication. First, the right lubricant must be chosen, and then it must be applied correctly. The proper lubricant should have a negligible vapor pressure and a suitable viscosity at the temperature

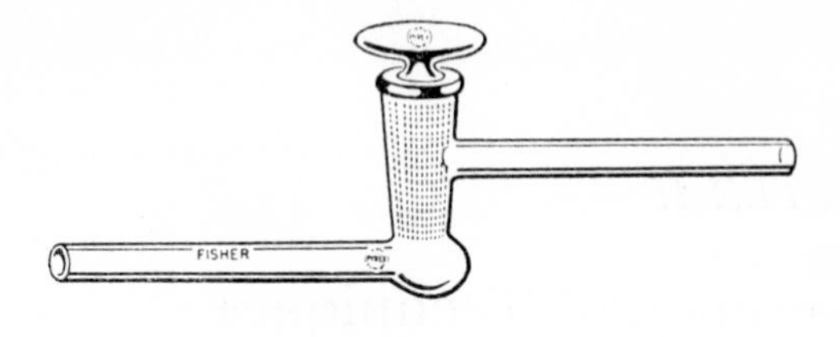

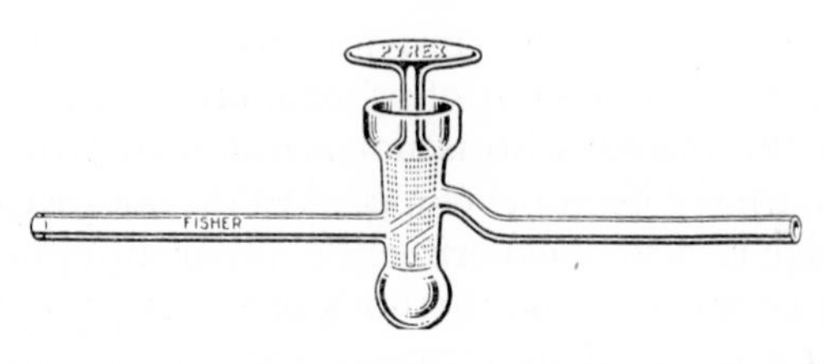

Fig. II-1. Preferred styles of high-vacuum stopcocks (courtesy of Fisher Scientific Company).

of its use. In the range of room temperatures, Apiezon N is suitable from 18 to 23 °C and Apiezon T from 23 to 35 °C. The use of the latter in a cool room introduces the danger of snapping off the handle of the stopcock; the use of the former in a warm room introduces the danger of air forcing its way through channels in the grease, and the certainty that the lubrication will fail prematurely.

Glass stopcocks should not be turned as casually and abruptly as one would turn the bathroom tap: the barrel should be supported with the free hand and the cock turned slowly and steadily with a slight inward pressure.

Mercury cutoffs are frequently used instead of glass stopcocks, where vapors that could dissolve in the stopcock lubricant are being handled. For this purpose, and all others in which mercury is present in the vacuum system, the mercury must be of high purity. This requirement is especially important when a measurement depends on the position of a mercury meniscus: oxide scum floating on the surface contaminates glassware and may actually allow gas to leak between the mercury and the glass, as was discovered long ago by Faraday. The method of purifying mercury is described by Wichers (1).

It should not be necessary to point out that the glass of which the apparatus is constructed must be clean and free from bits of cork, dust, etc., that might "outgas" in the vacuum. For further details of vacuum technique the following references may be consulted: W. E. Barr and V. J. Anhorn, *Scientific and Industrial Glass Blowing and Laboratory Techniques*, Instruments Pub. Co., Pittsburgh, 1949; R. E. Dodd and P. L. Robinson, *Experimental Inorganic Chemistry*, Elsevier, Amsterdam, 1954; S. Dushman, *Scientific Foundations of Vaccum Technique*, J. M. Lafferty, ed., second edition, Wiley, New York, 1962; Paul A. Faeth, *Adsorption and Vacuum Technique*, Inst. of Science and Technology, U. of Michigan, Ann Arbor, 1962; A. Farkas and H. W. Melville, *Experimental Methods in Gas Reactions*, Macmillan, London, 1939; S. Jnanananda, *High Vacua*, Longmans, Green and Co., London, 1947; R. T. Sanderson, *Vacuum Manipulation of Volatile Compounds*, Wiley, New York, 1948.

B. THE MANOMETRIC METHOD OF MEASURING ADSORPTION

1. *General Method and Calculation of Data*

The measurement of adsorption by manometric techniques is equivalent to the gasometric determination of the apparent volume of a sample bulb that contains a large surface. When the gas is ideal, the apparent volume V_s is defined as

$$V_s = n\,RT/p \tag{II-1}$$

where n is the total number of moles of adsorbate in the sample bulb and p, T, and R have their usual significance. The geometrical volume of the sample bulb, V_{geo}, is given by

$$V_{\text{geo}} = {}_g n\,RT/p \tag{II-2}$$

where ${}_g n$ is the number of moles of adsorbate in the gas phase side the sample bulb. The amount adsorbed is, therefore, $n - {}_g n$, which is often expressed as $V = V_s - V_{\text{geo}}$, corrected to standard conditions (STP) per gram of adsorbent.

The evaluation of the amount adsorbed is determined from pressure measurements in an apparatus of the type shown diagrammatically in Figure II-2, which consists, *inter alia*, of a thermostatted sample bulb of volume V_{geo} at temperature T, a container of accurately known volume V_c, and a manometer of such design (see Fig. II-5) that the mercury can be adjusted to a fixed reference point before a reading is taken, so that the volume of the adsorption system does not vary from one reading to another. The volume of the apparatus between stopcocks 1, 2, 3, and the manometer is V_1 at ambient temperature T_r; the volume between stopcock 3 and the thermostat bath is V_2, also at temperature T_r. These unknown

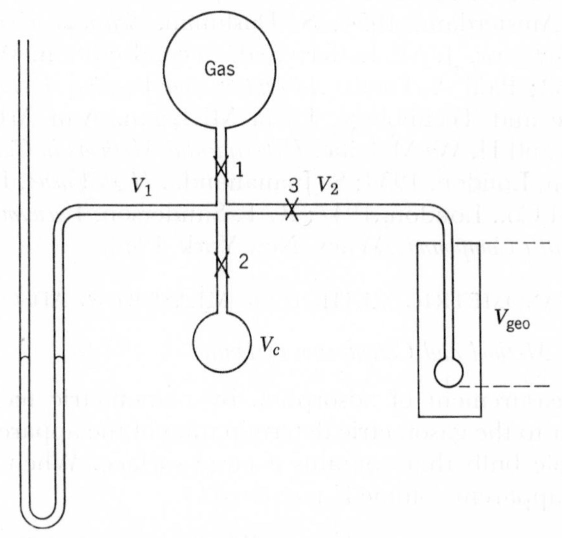

Fig. II-2. Schematic diagram of manometric adsorption system.

volumes are determined as follows. Gas from the reservoir is introduced into the volume V_1 of the previously evacuated system and the resulting pressure p_1 is measured; stopcock 2 is then opened, the gas expands into V_c to a pressure p_2; then at constant temperature,

$$V_1 = V_c[p_2/(p_1 - p_2)] \tag{II-3}$$

Finally, gas at a measured pressure may be expanded from V_1 into the sample-side of the system by opening stopcock 3. The volume of the sample-side, calculated from the results of this operation, is called the "dead-space"; it is not, however, the actual volume of the sample-side unless the thermostat bath is at ambient temperature. The volume of the dead-space, V_d, is given by

$$V_d = V_2 + (V_{geo} T_r)/T \qquad \text{(II-4)}$$

These volumes of the system, V_1 and V_d, must be measured with the adsorbent present inside the sample bulb; the gas used must not be measurably adsorbed at the temperature T: helium is a suitable gas at temperatures above 30 °K.

For the determination of an adsorption isotherm the volume V_c is not used; adsorbate gas is introduced into V_1 from the reservoir and the amount of the dose calculated as cubic centimeters at STP by the relation

$$\Delta V_{\text{dose}} = (273\ V_1\ \Delta p_{\text{dose}})/(T_r\ 760) = F_{\text{dose}}\ \Delta p_{\text{dose}} \qquad \text{(II-5)}$$

where Δp_{dose} is the increase of pressure in V_1. Next, stopcock 3 is opened and the gas allowed to equilibrate with the sample. The time required for equilibrium may vary from five minutes to a day; this time effect is discussed below. When equilibrium is reached, the amount of gas not adsorbed is calculated by the relation

$$V_{\text{unads}} = (273/T_r)[(V_1 + V_d)/760]\, p = F_{\text{eq}}\, p \qquad \text{(II-6)}$$

where p is the final equilibrium pressure. The specific amount adsorbed, expressed as cm^3 at STP per gram of adsorbent, is calculated by

$$V = [(\Sigma \Delta V_{\text{dose}}) - V_{\text{unads}}]/W \qquad \text{(II-7)}$$

where $\Sigma \Delta V_{\text{dose}}$ is the total amount of adsorbate in the system and W is the mass of the adsorbent. Successive points on the isotherm are determined by admitting another dose, ΔV_{dose}, and so reaching another equilibrium pressure. The adsorption isotherm thus determined is reported graphically by plotting V *vs.* p.

2. *Possible Errors in the Manometric Method*

Errors of principle. The calculations given in the previous section were based on the ideal-gas law. It may well happen, however, that, in the range of pressures and temperatures used during investigations, the gas phase deviates significantly from

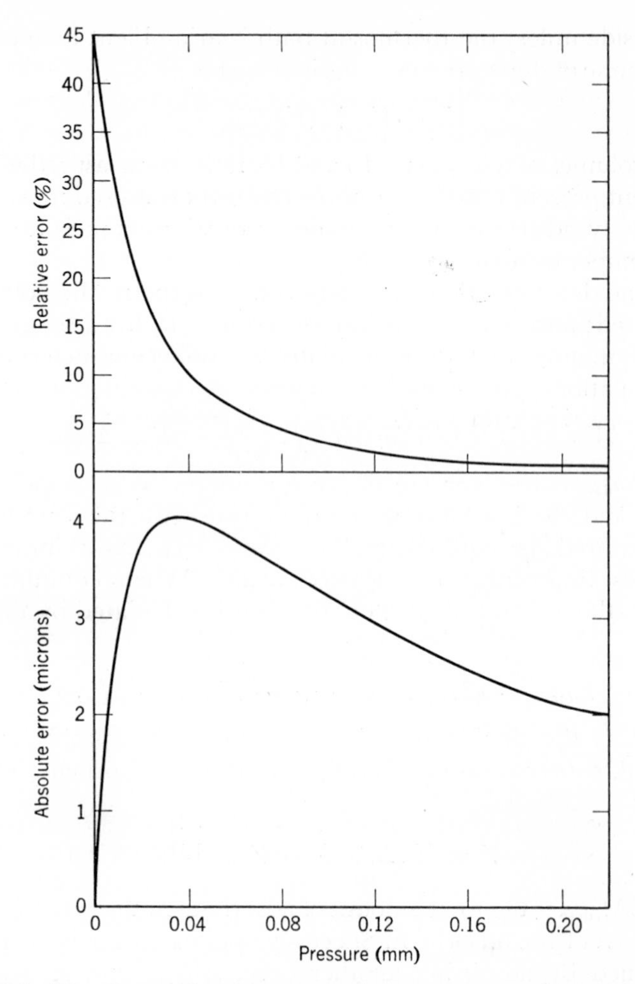

Fig. II-3. The effects of thermal transpiration on pressure readings at room temperature (T_2), when $T_1 = 78°K$, calculated by equations II-13 and II-13a, for krypton, in a tube of 6 mm internal diameter.

ideality. Every investigator has the responsibility to apply the requisite corrections (see p. 69) when needed.

Another type of correction is required when a pressure reading is obtained from a manometer that is at a different temperature from that of the sample. The gas tends to pass from the cooler to the warmer region, and a steady state is reached when the pressure difference between the two regions is sufficient to balance this thermal effusion or transpiration (2). In the determination of an adsorption isotherm, thermal transpiration results in the measured equilibrium pressure being greater or less than the true value, accordingly as the temperature of the sample is lower or higher than the temperature of the manometer. The magnitude of the absolute and relative errors as a function of pressure is illustrated in Figure II-3, from which it can be seen that the error is significant at low pressures.

When the mean free path of the gas molecules is very much greater than the diameter of the connecting tube that is subject to the temperature gradient, the following relation holds:

$$p_1/p_2 = (T_1/T_2)^{1/2} \tag{II-8}$$

where p_1 is the true equilibrium pressure over the sample at temperature T_1 and p_2 the pressure measured by the gauge at temperature T_2. Thus, for example, if a pressure gauge is at room temperature, e.g., 25 °C, and the system under investigation is at liquid nitrogen temperature (77.5 °K) then $p_1/p_2 = 0.51$, which would lead to 100% error unless correction is made for this effect. Increasing the pressure, thus shortening the mean free path, or increasing the diameter of the connecting tube, causes equation II-8 to fail; the failure occurs when the mean free path is less than about 20 times the diameter of the tube, which, practically speaking, restricts the use of the equation to pressures below a few microns.

The glass tubing used to construct adsorption apparatus has usually an inside diameter of several millimeters to allow rapid diffusion of gas and short pumpout times; consequently, the pressure range of many adsorption measurements lies beyond the scope of equation II-8, while still in the range where thermal transpiration causes significant errors.

A strictly theoretical correction for the effect of thermal transpiration can be obtained only for two extreme cases: (a) the mean free path is large compared to the diameter of the connecting tube, when equation II-8 is applicable; (b) the mean free path is small compared with the diameter of the tube, when the following equation due to Maxwell is applicable:

$$p_2^2 - p_1^2 = (c/\mathbf{d}^2)(T_2^2 - T_1^2) \tag{II-9}$$

where $\mathbf{d}$ is the diameter of the connecting tube, and c is a characteristic constant for each gas. Equations II-8 and II-9 can be expressed in terms of the pressure difference Δp across the connecting tube:

$$\Delta p = p_2[1 - (T_1/T_2)^{1/2}] \tag{II-10}$$

and

$$\Delta p = c(T_2^2 - T_1^2)/\mathbf{d}^2(p_2 + p_1) \tag{II-11}$$

These equations describe, respectively, the low pressure and the high pressure portions of the curve shown in Figure II-3 (lower half). All efforts to describe the intermediate region of pressure are empirical in their final expression. On the basis of a number of fundamental studies by Maxwell, Knudsen, and others, Weber (1937) developed an equation for the thermal transpiration of a gas along a closed cylindrical tube:

$$\frac{dp}{dT} = \frac{p}{2T}\frac{1}{\alpha y^2 + \beta y + \mu}, \qquad y = \mathbf{d}/\lambda \tag{II-12}$$

where p is the pressure, $\mathbf{d}$ is the inside diameter of the tube, λ is the mean free path of the gas molecules, and the coefficients are given by the relations $\alpha = \pi/128$; $\beta = \pi/12$; $\mu = (1 + gy)(1 + hy)$; $g = 2.5$ and $h = 2$.

A useful approximate solution of equation II-12 has been suggested by Miller (3):

$$\frac{dp}{dT}\frac{2T}{p} \simeq \frac{\Delta p}{\Delta T}\frac{2T}{p} \simeq \frac{1 - (p_1/p_2)}{1 - (T_1/T_2)^{1/2}} \equiv \mathbf{F}$$

$$\mathbf{F} = \frac{1}{\alpha y^2 + \beta y + \mu}, \qquad T_2 > T_1 \tag{II-13}$$

where y is some mean value between $y(p_1, T_1)$ and $y(p_2, T_2)$. The solution is exact in the limits where the theory is exact: i.e.,

$$\lim_{y\to 0} (p_1/p_2) = (T_1/T_2)^{1/2}$$

$$\lim_{y\to\infty} (p_2 - p_1) = 0 \tag{II-8}$$

The mean free path of the gas molecules, on which the parameter y is based, can be calculated from the kinetic theory of gases:

$$\lambda p = kT/\pi(2)^{1/2}\,d^2$$

where k is the Boltzmann constant and d is the molecular hard sphere diameter. The hard sphere diameter is based on the experimental value of the coefficient of viscosity at the average temperature, T; values of d as a function of T have been tabulated for a number of gases (3a). Usually we want to find p_1, knowing p_2, T_2, and T_1; the following expression for y avoids dependence on p_1:

$$y = \mathbf{d}/\lambda = p_2\,\mathbf{d}\pi(2)^{1/2}\,d/kT = (p_2\,\mathbf{d}d^2/2.33T) \times 10^3 \quad \text{(II-13a)}$$

where p_2 is in millimeters, $\mathbf{d}$ is in millimeters, d is in A, and $T = (T_1 + T_2)/2$. Equations II-13 and II-13a can be used to estimate the thermal transpiration ratio p_1/p_2 for any gas for which the prerequisite viscosity data are available.

Because of their immediate usefulness the following table of experimentally determined transpiration ratios for hydrogen, neon, argon, and xenon for T_1 at 77 and 90°K and for T_2 at 299

TABLE II-1
Pressure Shifting Factors
$p_2\,\mathbf{d}$

	$T_1 = 77°K$, $T_2 = 299 \pm 1°K$					$T_1 = 90°K$, $T_2 = 299 \pm 1°K$		
p_1/p_2	H_2	Ne	A	N_2	Xe	Ne	N_2	Xe
1.000	3.5	2.5	...	...	0.5	2.5	...	0.4
0.950	0.920	0.880	0.440	0.452	0.181	0.870	0.438	0.190
.900	.500	.520	.275	.258	.117	.505	.247	.118
.800	.216	.190	.124	.114	.0495	.177	.103	.0465
.700	.0930	.0690	.0555	.0478	.0210	.0600	.0375	.0179
.650	.0590	.0395	.0370	.0274	.0136	.0236	.0187	.0103
.625	.0420	.0250	.0295	.0198	.0103	.00830	.00119	.00690
.600	.0250	.0117	.0204	.0134	.00740	.00157	.00714	.00375
.575	.00950	.00335	.0117	.0084	.00450	.00021	.00290	.00130
.550	.00095	.00064	.00480	.00458	.00215	...	...	10^{-4}
.525	...	...	.00086	.00160	.00070	...	...	...
.508	...	...	...	...	10^{-4}	...	...	...

$\pm$ 1 °K are quoted from the results of Podgurski and Davis (3b). These ratios are based on measurements with specially designed McLeod gauges, with a maximum error of $\pm 2\%$ down to readings of 10^{-4} mm. In addition, the transpiration ratios for nitrogen have been calculated by means of equations II-13 and II-13a and included in Table II-1. To obtain the appropriate transpiration ratio for use as a correction factor, the observed value of $(p_2\mathbf{d})$ is interpolated in the vertical column for the gas in use, and the ratio is read off horizontally.

Errors of practice. Errors in the manometric determination of adsorption isotherms begin with possible errors in the calibration of the volumes of the apparatus. The reference volume V_c is usually determined by filling the bulb with a liquid of known density and measuring the increase in mass. This is done before the bulb is incorporated into the apparatus. The precision of this measurement is far better than any of the subsequent pressure measurements.

The error in the measurement of the dosing volume V_1 must be kept as low as possible, as any error here accumulates in the calculation of the amount adsorbed. From equation II-3 in the form

$$V_1/V_c = p_2/(p_1 - p_2) \qquad \text{(II-3)}$$

it follows that to ensure the relative error in V_1 being no greater than the relative error in p_2, the difference $p_1 - p_2$ must be equal to p_2; hence, V_c should be nearly the same as V_1. In addition, of course, the pressures p_1 and p_2 should both lie in the range of greatest precision for the pressure gauge used.

This above conclusion can be deduced as follows:

$$dV_1/V_1 = (dp/p_2) + [dp/(p_1 - p_2)] \qquad \text{(II-14)}$$

i.e., the relative error in V_1 is expressed as the sum of the relative errors in p_2 and $(p_1 - p_2)$. Eliminating p_2 gives

$$dV_1/V_1 = (dp/p_1)\ [(V_c + V_1)^2/V_cV_1] \qquad \text{(II-15)}$$

At any value of p_1, the relative error in V_1 will be a minimum when the expression

$$(V_c + V_1)^2/V_cV_1 \qquad \text{(II-16)}$$

is a minimum; which can readily be shown to occur when $V_1 = V_c$.

The same argument for minimizing the error in the determination of V_d would apply (i.e., that $V_d = V_1$), but that is not the sole consideration in determining the optimum value of V_d. The nature of equations II-6 and II-7 for the determination of the amount adsorbed required that V_{unads}, hence $(V_1 + V_d)$, be kept as small as possible to minimize the relative error in the calculation of V; at the same time the precision of the measurement of $(V_1 + V_d)$ must not be sacrificed. Fortunately the two requirements are

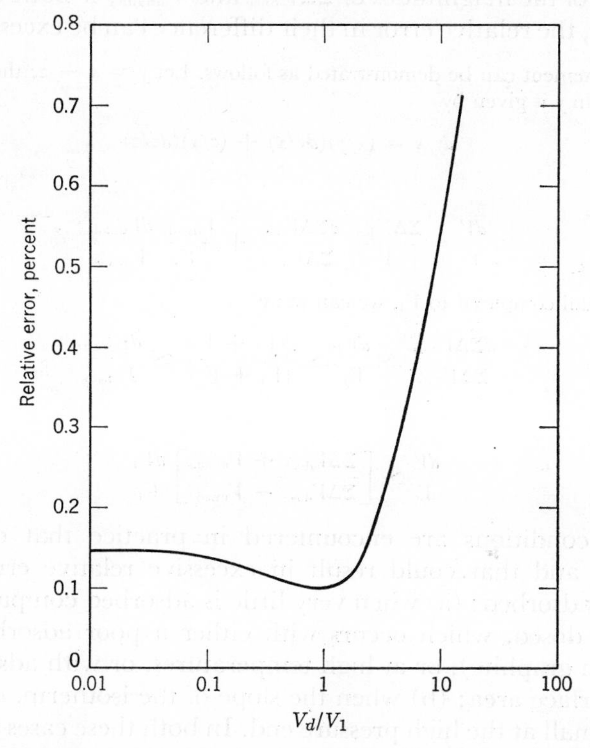

Fig. II-4. Relative error in the measurement of $(V_1 + V_d)$ as a function of the ratio V_d/V_1, assuming a relative error of 0.1% in measuring V_1.

compatible. Figure II-4 presents a typical curve for the relative error in measuring $(V_1 + V_d)$ as a function of the ratio V_d/V_1, assuming a relative error of 0.1% in V_1. Although the minimum relative error does occur for $V_d = V_1$, the reduction of V_d below this value does not exercise too harmful an effect on the precision of the determination. The slight increase in relative error that does occur by using, for example, a ratio $V_d/V_1 = 0.01$ is more than compensated by the increase thus obtained in the precision of V.

The amount adsorbed is calculated as a difference of two quantities (eq. II-7), each of which has a fixed relative error as long as the relative error of the pressure measurements does not change. The relative error in the amount adsorbed will vary, however, as a function of the magnitudes of $\Sigma\Delta V_{\text{dose}}$ and V_{unads}; if both quantities are large, the relative error in their difference can be excessive.

This statement can be demonstrated as follows. Let $y = x - z$; then the relative error in y is given by

$$dy/y = (x/y)(dx/x) + (z/y)(dz/z)$$

i.e.,

$$\frac{dV}{V} = \frac{\Sigma\Delta V_{\text{dose}}}{V}\frac{d\Sigma\Delta V_{\text{dose}}}{\Sigma\Delta V_{\text{dose}}} + \frac{V_{\text{unads}}}{V}\frac{dV_{\text{unads}}}{V_{\text{unads}}} \tag{II-17}$$

If V_d is small compared to V_1, we can write

$$\frac{d\Sigma\Delta V_{\text{dose}}}{\Sigma\Delta V_{\text{dose}}} \simeq \frac{dV_1}{V_1} \simeq \frac{d(V_1 + V_d)}{(V_1 + V_d)} \simeq \frac{dV_{\text{unads}}}{V_{\text{unads}}}$$

hence,

$$\frac{dV}{V} \simeq \left[\frac{\Sigma\Delta V_{\text{dose}} + V_{\text{unads}}}{\Sigma\Delta V_{\text{dose}} - V_{\text{unads}}}\right]\frac{dV_1}{V_1} \tag{II-18}$$

Two conditions are encountered in practice that cannot be avoided and that could result in excessive relative error in the volume adsorbed: (a) when very little is adsorbed compared to the amount dosed, which occurs with either a poor adsorbent (e.g., water on graphite), or at high temperatures, or with adsorbents of small surface area; (b) when the slope of the isotherm, dV/dp, becomes small at the high pressure end. In both these cases the bracketed term in equation II-18 becomes large.

Methods for evading this experimental error exist; the choice of the method to be used will depend on the ultimate purpose of the experiments. Thus, for example, if adsorption is being used as a tool to determine small surface areas of adsorbents, the adsorbate can be selected to give maximum precision: e.g., krypton, instead of nitrogen, because it saturates the surface at less than one-hundredth of the pressure required for nitrogen. This circumstance makes V_{unads} very small for a given amount dosed, thus reducing the relative error in V (see eq. II-18). It is not always permissible, however, to slide out of the difficulty by changing the adsorbate. We may actually want to find the adsorption isotherm of a system in which the adsorptive forces are very weak. For this purpose one must design the adsorption apparatus specially to minimize V_{unads}, which can be done by making V_1 very small and measuring the dose by an alternative method, as discussed later (see Section II-1-D).

C. PRESSURE MEASURING DEVICES

The measurement of adsorption may require pressure measurements that range from 10^5 to 10^{-4} mm of mercury: any one isotherm for the monolayer region of adsorption could well include a pressure range of three decades. No one gauge will be adequate for measurements throughout this range without a sacrifice of the desired precision. In this section we describe manometers suitable for adsorption measurements in different ranges of pressure; we take as a standard requirement that the maximum error of a properly made reading is 0.2%. Adsorption measurements at pressures much above 1 atmosphere are a special problem, which we do not consider here.

1. The Range 10^3 to 10 mm

The common *U*-tube manometer filled with mercury is the best instrument for pressure measurements in this range. A design suitable for adsorption work is illustrated in Figure II-5. A customary requirement of these measurements is that the volume of the adsorption system to which the manometer is attached be held constant from one reading to another. This requires that the mercury level be returned to a reference point during equilibration. The design shown in Figure II-5 provides a satisfactory method by which this can be done, by admitting or withdrawing air from the reservoir. The reservoir bulb should be lagged to slow down and reduce fluctuations in the position of the meniscus with changes in room temperature. A stopcock between the reservoir and the manometer would eliminate these fluctuations but has the more serious disadvantage of contaminating the mercury and the glass with the stopcock lubricant.

The legs of the manometer, if made of ordinary pyrex tubing, should be of the same bore and at least 10 mm in diameter, to make

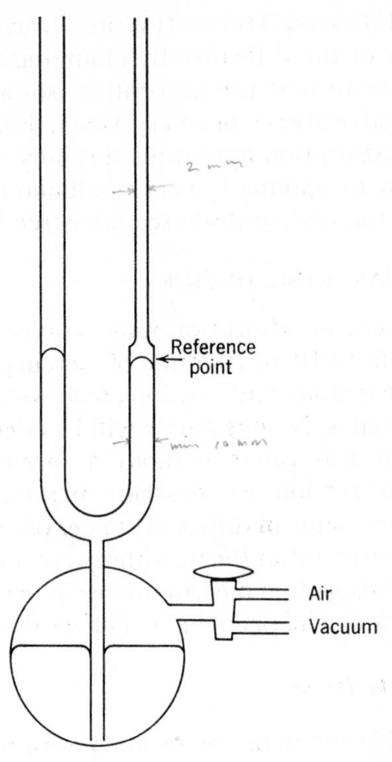

Fig. II-5. Design for mercury *U*-tube manometer.

negligible any variations in the capillary effects that could be introduced by lack of uniformity in the glass tubing. The diameter of the tubing above the reference point may be reduced, to decrease the volume V_1.

If the glass and mercury are clean, this instrument is capable of a precision of ± 0.02 mm in the pressure measurement; the residual error is inherent in the design and is not to be reduced by the use of a cathetometer, no matter how precise.

A refined type of *U*-tube manometer for adsorption measurements has been described (4), drawing on the techniques of gas thermometry, in which the error has been reduced to ± 0.005 mm in the pressure measurement. The instrument thus designed probably provides the highest precision that can be achieved with a *U*-tube manometer; its use is only required for special purposes, such as the measurement of small total amounts adsorbed. With measurements of this degree of precision a correction for the geographical variation of the acceleration due to gravity is warranted.

A consideration frequently neglected is the variation of the density of mercury with the temperature. Pressure expressed as millimeters of mercury is related to fundamental units only at 0 °C and standard gravity; the latter correction is negligible for most measurements but the temperature correction ought to be made for all measurements whose relative error is less than 0.5%.

Where the volume V_1 is very small, which as we have seen is a desirable feature for many systems, it may be necessary to take into account the co-volume of the mercury meniscus. The curvature of the mercury surface is affected by the diameter of the tube and the pressure of the gas: appropriate corrections based on the theory of capillarity are reported by Blaisdell (5).

The position of the mercury meniscus can be read on a meter stick, or preferably a mirrored scale, with the naked eye with a precision of ±0.3 mm. A measurement of a pressure greater than about 200 mm can therefore be made in this way with a relative error of less than ±0.2%. A cathetometer reading to ±0.05 mm would be required to obtain results of the same precision in the pressure range from 200 to 25 mm; and a cathetometer reading to ±0.01 mm would be required in the lowest range of pressure measurements down to 10 mm, below which the ordinary *U*-tube manometer is not a suitable pressure-reading device. For the high-precision design, if it were to be read as low as 2 mm with the same relative error of 0.2%, a cathetometer with a precision of ±0.001 mm would be required.

2. *The Range 50 to 0.2 mm*

The range of the *U*-tube manometer can be extended by substituting an oil of low vapor pressure for mercury. The advantage derived is due not only to the lower density of the fluid, which confers a magnification of about 15, but also to its lower surface tension and low contact angle against glass, which increase the inherent precision of readings of the meniscus.

A suitable design for an oil-filled, *U*-tube manometer for adsorption measurements is shown in Figure II-6. With this design the oil

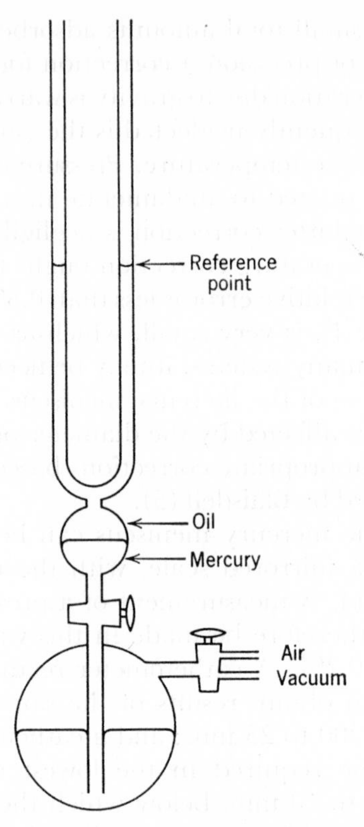

Fig. II-6. Design for an oil-filled *U*-tube manometer.

level can be returned reproducibly to a reference point in the constant-volume limb; a stopcock above the mercury reservoir is permissible and so makes unnecessary the thermal lagging of the bulb. The instrument may be calibrated either by measuring the density of the oil at the temperature of use, or preferably by comparison with a mercury manometer in the overlapping range of 10–50 mm. Errors due to capillary effects are less significant in the oil manometer, although tubing of the same bore for both legs is not a wasted precaution; capillary tubing is again not recommended, though now for a different reason: the drainage of oil from the wall, after the falling movement of the bulk oil, can entrap gas bubbles and so break the liquid thread. Tubing of any diameter great enough to prevent this effect (i.e., greater than about 2 mm di-

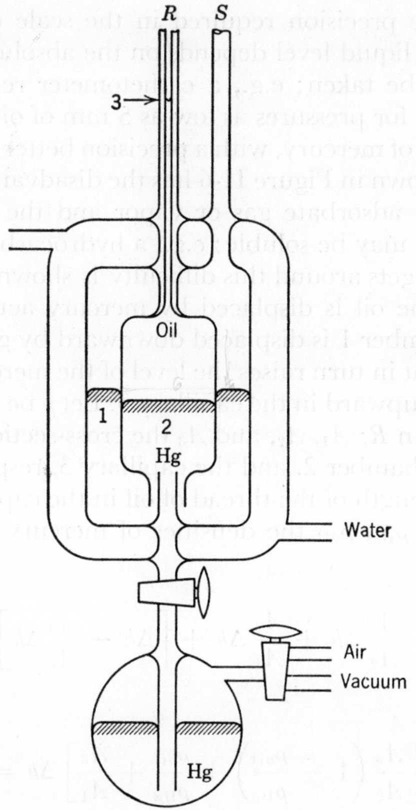

Fig. II-7. Design for an oil manometer in which the adsorbate gas is not in contact with the oil (by J. P. O.).

ameter) is suitable; the time required for the subsidence of the oil film through a given height is independent of the tube diameter.

Few oils are available that meet the requirements of this manometer for our purpose: the vapor pressure at room temperature should be of the order of 10^{-7} mm; the viscosity should be low enough to permit drainage in a convenient time, etc. The oils known as Octoil S (di-2-ethylhexyl sebacate) and Apiezon B are suitable. Dissolved air and other volatile matter in the oil must be pumped out of the instrument before use; this can be facilitated by warming gently with an infrared lamp during the latter stages of the out-gassing.

As before, the precision required in the scale or cathetometer used to read the liquid level depends on the absolute magnitude of the readings to be taken: e.g., a cathetometer reading to ± 0.01 mm can be used for pressures as low as 5 mm of oil, corresponding to about 0.3 mm of mercury, with a precision better than $\pm 0.2\%$.

The design shown in Figure II-6 has the disadvantage of bringing into contact the adsorbate gas or vapor and the manometer oil, in which the gas may be soluble: e.g., a hydrocarbon gas in Octoil S. A design that gets around this difficulty is shown in Figure II-7. In this gauge the oil is displaced by mercury acting as a piston. Mercury in chamber 1 is displaced downward by gas in the system S; this movement in turn raises the level of the mercury in chamber 2, displacing oil upward in the capillary 3. Let p be the gas pressure in S above that in R; A_1, A_2, and A_3 the cross-sectional areas of the annulus 1, the chamber 2, and the capillary 3, respectively; Δh the increase in the length of the thread of oil in the capillary due to the pressure p; and ρ_{Hg}, ρ_{oil} the densities of mercury and oil, respectively; then,

$$p = \frac{A_3}{A_2}\Delta h + \frac{A_3}{A_1}\Delta h + \left[\Delta h - \frac{A_3}{A_2}\Delta h\right]\frac{\rho_{\mathrm{oil}}}{\rho_{\mathrm{Hg}}}$$

or

$$p = \left[\frac{A_3}{A_2}\left(1 - \frac{\rho_{\mathrm{oil}}}{\rho_{\mathrm{Hg}}}\right) + \frac{\rho_{\mathrm{oil}}}{\rho_{\mathrm{Hg}}} + \frac{A_3}{A_1}\right]\Delta h = F\Delta h \qquad \text{(II-19)}$$

From equation II-19 one can see that this gauge is both absolute and linear in its response; the most convenient calibration, however, is to compare the readings with a mercury manometer throughout their common range, to determine the proportionality factor F. The sensitivity approaches that of the U-tube oil manometer as a limit, and in practice 90% of this sensitivity can be readily achieved.

The gauge is usually used with a vacuum as the reference pressure; that is, with R connected to the pumps and the zero point determined with S evacuated. The pressure is calculated from a determination of Δh, which is a change of level in the capillary; possible errors arising from meniscus and capillary effects are cancelled out in the process. As the theory of the gauge depends on a strict proportionality between the volume and height of oil in the capillary, that part of the gauge must be made of precision-bore tubing. The diameter of the capillary should not be less than about 2 mm for convenience in out-gassing; the small capacity of a capillary of this diameter makes negligible the volume change of that part of the gauge connected to the system volume V_1. The level of the mercury in chamber 1 need not, and in fact must not, be adjusted between measurements; after the initial adjustment of the oil level the stopcock to the mercury reservoir is not used again. If a correction for the volume change of V_1 be significant it can be readily taken into account, as it is equal to the volume of oil displaced up the capillary; this correction, however, will always be small compared to the internal volume of the gauge, which unfortunately is large of necessity. An intrinsic disadvantage of this design is the large quantity of oil and mercury enclosed, which, because of its thermal expansion, makes the instrument sensitive to fluctuations of temperature. This weakness is overcome by circulating water at constant temperature in the outer jacket; control of the water temperature to ± 0.05 °C suffices.

3. *The Range 20 to 10^{-4} mm*

Pressures in the range of 20 to 10^{-4} mm can be measured with a series of two or three McLeod gauges, each one of which is designed for optimum performance in measuring a portion of the range. A typical form of this gauge is illustrated in Figure II-8. In the operation of this gauge the mercury is first kept below the level of the cutoff, the gas in the bulb C then being at the same pressure as the gas in the system; to make a measurement of this pressure the mercury is brought up into the bulb C, so that all the gas that was formerly in the bulb is compressed into the measuring capillary A; the compression is continued until the mercury meniscus in capillary tube B is brought to a point that is, as closely as possible, level with the top of the bore of capillary A. Let V be the volume of the bulb C from the end of the cutoff to the top of the capillary A; a, the cross-sectional area of the precision-bore capillary tube used to construct the capillary tubes A and B; h, the length of capillary A containing the compressed gas, measured from the mercury meniscus to the top of the bore; and m, the vertical distance between

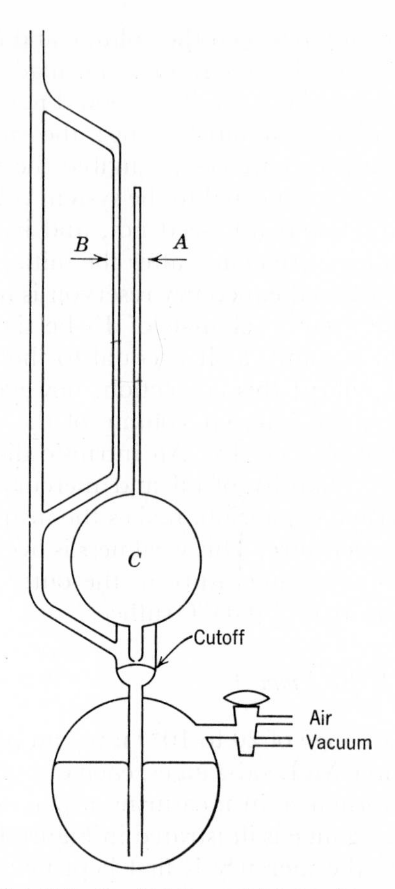

Fig. II-8. Typical form of McLeod gauge.

the mercury meniscus in tube B and the top of the bore of tube A; then,

$$pV = (h + m + p)ah$$

or

$$p = [(ah\,(h + m)]/(V - ah) \qquad \text{(II-20)}$$

In using equation II-20 the sign of the mis-set m is taken as positive if the meniscus in tube B overshoots the top of the bore; negative, if it undershoots. A skillful operator can usually keep the mis-set sufficiently small that equation II-20 reduces, if h is large, to

$$p = ah^2/(V - ah) \qquad \text{(II-21)}$$

Further, if the volume of the capillary A is small compared to V, the equation becomes

$$p = ah^2/V \qquad \text{(II-22)}$$

Also to be considered as factors that are possibly significant under certain conditions are: (a) the co-volume of the mercury meniscus in the measuring capillary A, which becomes increasingly significant as the height h approaches the order of magnitude of the tube diameter, and which contributes to the decrease of precision of measurements in this range; and (b) the nonideality of the gas on compression, which may be corrected by use of an appropriate equation of state. When dealing with a condensable gas, a McLeod gauge can only be used when h is less than the vapor pressure of the condensate at the temperature of the gauge.

Equation II-22 is useful in designing a McLeod gauge to measure a specified range of pressures with any required precision. Let us suppose that we want a gauge with which the lowest reading of 10^{-3} mm of mercury may be measured to within 1%; this requires that h for the lowest reading of the gauge must be read to within 0.5%. If the reading of h is made with a cathetometer of precision $\pm.05$ mm, the minimum reading of h within the assigned precision would be 10 mm. Substituting in equation II-22 gives $a/V = 10^{-5}$ mm^{-1}. With precision-bore tubing 1 mm in diameter, the bulb volume V would have to be 78.5 cm^3. The relative error decreases slowly as the pressure reading increases; thus, for the gauge just described, the relative error reaches 0.2% at a pressure reading of 25 microns of mercury. The maximum pressure for which this gauge may be used will now be determined by the length of the capillary tubing A: for a convenient length of, say, 300 mm, the maximum pressure reading would be 0.9 mm. The range of a given gauge may be extended by compressing the gas until it is all included in the capillary tube A, thus intentionally introducing a large positive mis-set; the pressure must then be calculated by means of equation II-20.

A McLeod gauge intended for accurate measurements must be calibrated before it is installed in an adsorption system. The volume V is measured by determining the mass of mercury required to fill it, and the cross sectional area a of the capillaries is measured by introducing a known mass of mercury and measuring the length of the thread. These measurements should be made with a much smaller relative error than that required in the final pressure measurement.

An error may be introduced in the reading of a mercury meniscus, especially when contained in a capillary, if the contact angle is not the same on the advance and recession of the meniscus; even clean mercury on a clean glass surface will show different advancing and receding contact angles. Tapping the glass tube prior to reading helps to promote an equilibrium position of the meniscus. A roughened glass surface, which can be obtained by etching lightly with hydrofluoric acid, also promotes this equality; with a properly etched tube and the practice of tapping, this source of error can be largely avoided. The real disadvantages of the McLeod gauge are that it does not give a continuous indication of the pressure and is, besides, slow and cumbersome; the gauge is fragile and offers an ever-present risk of breakage, particularly as a result of careless control of the influx of mercury. For some purposes the large internal volume of the gauge, which adds to V_1 of the adsorption system, may be a serious disadvantage, and is never helpful.

4. The Thermistor Gauge

A large number of mechanical, electrical, and electronic pressure gauges have been designed, many of them triumphs of ingenuity. For adsorption measurements, however, most of these gauges are not sufficiently precise; even when sufficiently precise their use is seldom warranted on account of their complexity and expense. A simple and inexpensive gauge that is precise enough for many useful purposes is the thermistor type of thermal conductivity gauge.

This gauge consists of a matched pair of thermistors, one of which is in the gas whose pressure is to be measured; the other, for reference, sealed in a highly evacuated glass bulb. These two units are made the adjacent arms of a Wheatstone bridge circuit; the bridge is balanced with the measuring thermistor *in vacuo;* when a gas is present around the measuring thermistor, its equilibrium temperature is lowered, its resistance increases and hence, the pressure of the gas becomes a function of the imbalance of the bridge. The circuit of the thermistor gauge is shown schematically in Figure II-9.

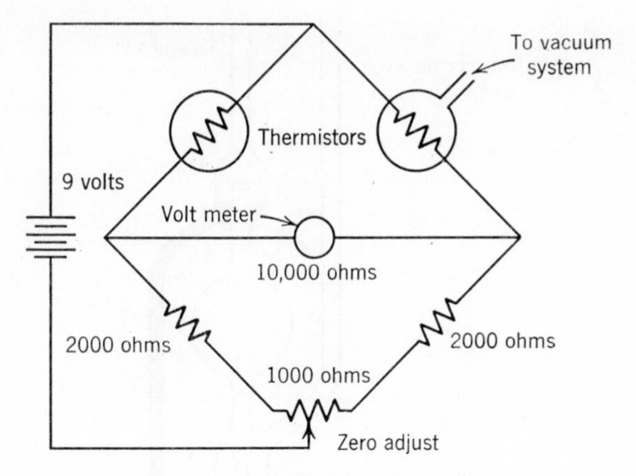

Fig. II-9. Schematic electrical circuit for thermistor gauge.

The thermistor gauge is not an absolute instrument and must be calibrated against an absolute gauge such as a McLeod, using the same gas whose pressure is to be measured. A plot of the logarithm of the imbalance voltage E *versus* the logarithm of the pressure p is found to be linear for pressures below about 1 mm; in practice a lower limit of 10^{-5} mm pressure is set by the limitations of commonly available voltmeters. The range of voltages to be measured extends from a few millivolts to 2 or 3 volts.

The relative error in the pressure measurement is approximately the same as the relative error in the voltage reading, which is not likely to be much less than 1%.

The advantages of the thermistor gauge are its small internal volume, about 5 cm^3, and the possibility that it offers of a continuous measure of a changing pressure, which can be presented graphically by a recording potentiometer. It has the advantage, compared to other thermal conductivity gauges, of a higher sensitivity and of a low temperature "filament," which is not likely to decompose organic molecules.

D. DOSING DEVICES

The method of determining the amount of gas "dosed" into the adsorption system that was mentioned previously in Section II-1-B depends on measuring the increase of pressure in the known volume

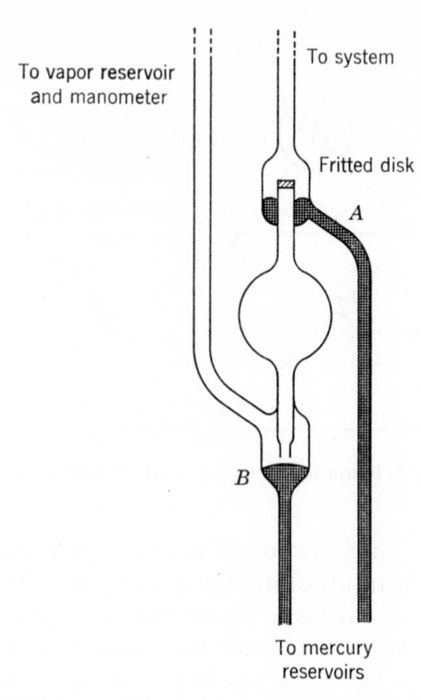

Fig. II-10. Young's dosing device (from D. M. Young, *Rev. Sci. Instr.*, **24**, 77 (1953)).

V_1 of the system. This internal-dosing method, although the simplest method, has the disadvantage of including V_1 in the calculation of the volume unadsorbed. An increase in precision could be made were the dosing volume external to the equilibrium system, which can be done by using a gas buret to measure the volume dosed. A flexible and convenient design that avoids stopcocks is Young's doser, shown in Figure II-10. The following description is taken from Young's paper.

The lower part of the doser terminates in a conventional mercury cutoff *B*; to the upper end is sealed a fritted-glass disk surrounded by a wider tube *A* into which mercury may be admitted. The method of operation is as follows: mercury is admitted into *A* so that the fritted disk is covered and the mercury in *B* is adjusted until the meniscus is just below the cutoff (as illustrated). Vapor is then admitted via the sidearm of *B*, its pressure measured on a manometer, and the temperature of the doser recorded. The head of mercury above the fritted disk must, of course, be greater than the pressure of vapor admitted. The doser is iso-

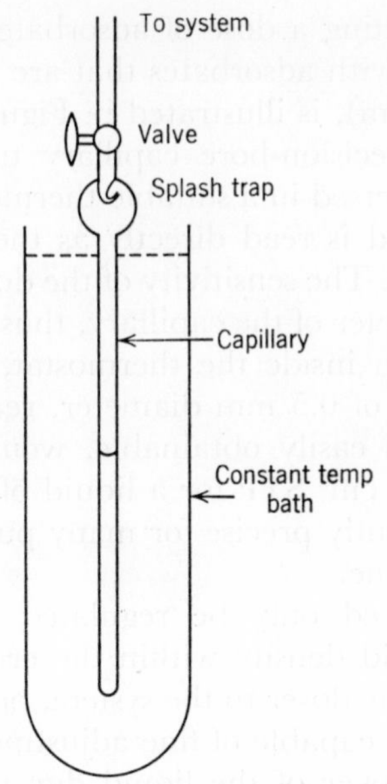

Fig. II-11. Design for a dosing buret.

lated by closing the cutoff *B*. The mercury in *A* is lowered and the charge of vapor expelled through the fritted disk by completely filling the doser with mercury. Subsequent doses are admitted by covering the fritted disk with mercury, withdrawing mercury from the doser to its original level in *B*, and repeating the above cycle.

The use of Young's doser has sometimes given rise to the complaint that it cannot be used to remove measured increments of gas from V_1, as is required when making measurements of desorption; for this purpose, an additional vessel has to be connected with V_1.

Although the capacity of the doser is fixed, the quantity of gas dosed is proportional to the pressure; the pressure can be anything less than the maximum of approximately 40 mm that is determined by the design of the buret, as long as an appropriate pressure gauge is available to measure it. A doser of about 20 cm^3 capacity will, therefore, admit accurately known quantities of gas ranging from 1 to 10^{-4} cm^3 at STP. The upper limit is not a serious limitation as the quantity of adsorbent placed in the sample bulb can be selected so as to make such large doses unnecessary.

A method of admitting a dose of adsorbate that is particularly well adapted for use with adsorbates that are liquids of low vapor pressure (below 20 mm), is illustrated in Figure II-11. The liquid is contained in a precision-bore capillary tube of known cross section, which is immersed in a suitable thermostat at temperature T; the quantity dosed is read directly as the decrease in height of the liquid meniscus. The sensitivity of the doser can be increased by reducing the diameter of the capillary, thus making it necessary to increase the length inside the thermostat, to retain the same capacity. A capillary of 0.5 mm diameter, read with precision of ± 0.05 mm, which is easily obtainable, would have a precision equivalent to ± 0.002 cm^3 STP for a liquid of 100 cm^3 molar volume; which is sufficiently precise for many purposes, e.g., adsorption of heptane or octane.

The thermostat need only be regulated sufficiently to keep variations of the liquid density within the error of measurement. The tap connecting the doser to the system, besides being resistant to the vapor, must be capable of fine adjustment, so as to prevent ebullition and carry-over of the liquid due to a sudden drop in pressure.

E. CONTROL OF THE TEMPERATURE OF THE SAMPLE

Variations in the temperature of the sample may, under certain conditions, become the largest single factor in reducing the accuracy of an adsorption measurement and the precision of the whole isotherm. The error that enters here is inherent in the temperature dependence of the amount adsorbed, and is not to be confused with the much smaller error that is due to the p, V, T relation of the gas in the sample bulb. A useful rule of thumb is to assume

$$dp/dT = 0.05\,p \tag{II-23}$$

Thus,

$$dp/p = 0.05\,dT = 5\,dT\,\% \tag{II-24}$$

Equation II-24 means that the relative error in p is five per cent per degree of temperature fluctuation; therefore, a control of $\pm 0.04\,^{\circ}$C is necessary to keep the relative error in p down to $\pm 0.2\%$. For measurements of the equilibrium pressure only, this would be adequate. The manometric method, however, uses the

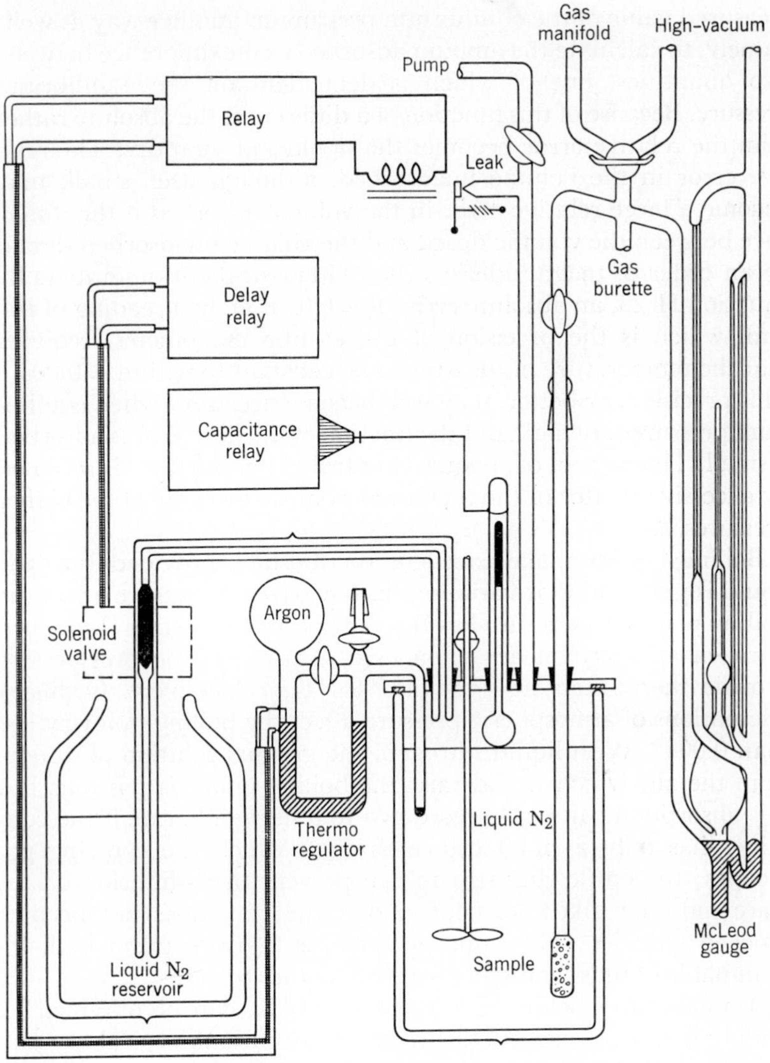

Fig. II-12. Adsorption apparatus showing three-range McLeod gauge and the constant-level liquid nitrogen regulator [from B. B. Fisher and W. G. McMillan, *J. Chem. Phys.*, **28,** 549 (1958)].

measured value of the equilibrium pressure in another way as well: namely, to calculate the amount adsorbed as the difference between two quantities, one of which is dependent on the equilibrium pressure. Because of this function of a difference, the absolute rather than the relative error becomes the significant quantity. The relative error in the volume unadsorbed, although itself small, may become a large relative error in the volume adsorbed if the difference between the volume dosed and the volume unadsorbed should be an order of magnitude less than either of these quantities. By equation II-23, an absolute error of ± 0.02 mm, in a reading of 100 mm, which is the precision of the *U*-tube manometer, requires that the temperature of the sample be constant to within $\pm 0.004°$. This precision, or better, may well become necessary when equilibrium pressures are high and the fraction of the dose that is adsorbed is small. The neglect of adequate control of temperature is revealed by excessive scatter of the measured points, especially at the higher pressures.

Boiling liquids make excellent thermostats, provided that the liquid is pure and that variations in the external pressure affect the boiling point temperature less than the permissible error. Liquefied nitrogen and oxygen are often used at atmospheric pressures to provide thermostats at 77.5 and 90.1 °K, respectively. Ordinary fluctuations of atmospheric pressure affect the boiling-point by less than 0.004°. With liquid nitrogen, the gradual solution of oxygen from the air, which would raise the boiling point, is possible, but this absorption can be slowed down to a negligible rate if the container has only a small orifice through which the evolving gas escapes; the outflowing stream can prevent back-diffusion of air. Care must be taken, of course, that the vent does not become blocked with ice. The same precautions that are taken with inflammable liquids should be used with liquid oxygen.

Temperatures below 77.5, to about 60 °K, can be obtained by boiling liquid nitrogen under reduced pressure. A typical cryostat based on this principle, designed for adsorption measurements from 64 to 85 °K by Fisher and McMillan (6), is shown in Figure II-12. The following description is taken from the paper by Fisher and McMillan (6).

"The novel feature of the liquid nitrogen thermostat is the automatic leveling device, which permits maintenance of a constant level under reduced pressure, despite a high rate of evaporation. As shown schematically in Figure II-12, the position of the metal-tipped float relative to the capacitance-sensing element determines the opening and closing of the solenoid valve by the relay. The difference between the vapor pressure of the condensed (solid) argon in the cold finger and the existing pressure in the external argon reserve bulb determines the position of the magnetically actuated leak, and thus the rate of evaporation of the liquid nitrogen. The argon vapor pressure, read on a separate vapor pressure thermometer (not shown), was converted to absolute temperature through the use of the NBS-NACA tables (7).

"Under normal operation the leak frequency was ~3–20 per minute depending on the adjustment of the leak valve; the solenoid frequency was ~0.5–2 per minute, depending mainly on the thermostat temperature. With moderate stirring, no variation in temperature (±0.05 °K) was observable during normal cycling, although a small long-period drift, correlated with changing room temperature, was ascribable to the concomitant change in pressure of the argon reservoir of the thermoregulator. Since the efficiency of the thermostat depended critically on the inleak of warm air, the 650-cm^3 capacity Dewar was fitted with a hermetically sealed cover through which leak-proof ports admitted the various tubes. The stirring shaft bore on a spherical ball joint atop the glass tube through which it passed.

"The Pyrex conduit between reservoir and thermostat was of double-wall construction, with the annular space silvered internally and evacuated. This jacketed tube was made by placing a 4-mm (o.d.) tube inside a 10-mm (i.d.) tube and centering with Dewar seals at both ends, a side tube for evacuation being provided. It was found that by bringing the outside to red heat, the double-walled tubing could be bent at right angles without collapsing the annular space. The solenoid armature was a 1-inch length of a 5d nail, whose conically-shaped ends were ground into constrictions in the 4-mm tube, allowing ~5-mm end play. A short plug of glass wool in the intake of the conduit prevented entry of ice, accumulated in the bottom of the reservoir, which might cause the armature to stick. The delay relay acts only as an overload protective device in case the sensing float would require the solenoid to be on for more than 20 seconds at a time. In the closed position, the armature

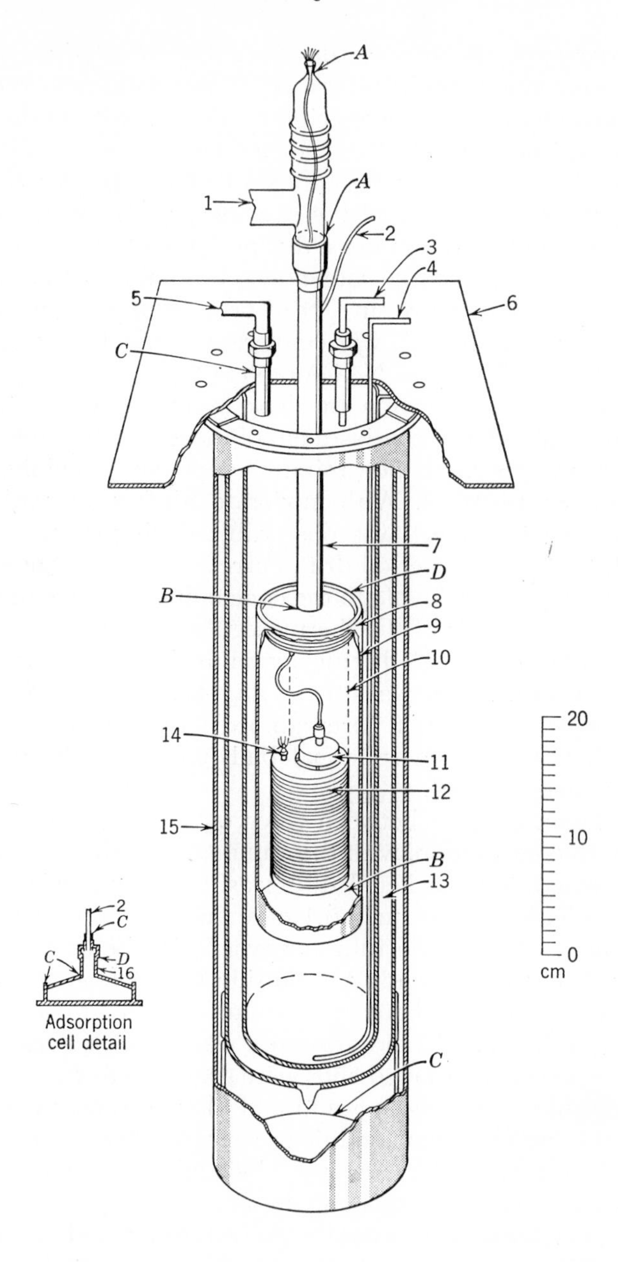
A
A
1
2
3
4
5
6
C
7
D
B
8
9
10
14
11
12
15
B
13
C
20
10
0
cm
2
C
C
D
16
Adsorption
cell detail

is held against the upper seat by the reduced internal pressure, the solenoid pulling it down to admit more liquid. Under usual conditions this thermostat required only ~250 cm^3 of liquid nitrogen per hour to maintain a temperature ~70 °K."

For thermostat temperatures in the range between 60 °K and room temperature more elaborate equipment is required. The simplest device for this range is that of Morrison and Young (8), who claim that "temperature control to 0.001 ° for a period of hours at any temperature between 60 and 300 °K can be readily attained with a relatively simple cryostat. The principle of the cryostat is to attach the adsorption cell to a block of metal of large heat capacity which is well insulated from its surroundings. In the model to be described, the block is hung within an evacuated can (pressure—1×10^{-5} mm Hg or less) immersed in liquid air in a Dewar vessel. Loss of energy from the block by radiation and conduction to its surroundings is compensated for by manual control of the current passed through a heater wound on the block. Since the thermal inertia of the block is large, automatic control of the heater current is hardly necessary. There are no critical dimensions in the design, hence the physical arrangement may be varied within wide limits.

"The cutaway drawing, Figure II-13, illustrates a form of the cryostat which we have found convenient to use. Constructional details are given in the diagram but a few comments on the operation may be made.

"1. The block is cooled by admitting a small pressure of exchange gas (helium or hydrogen) to the vacuum space. To cool from room temperature to 90 °K requires 4–5 hours. Temperatures below 90 °K are attained by pumping on the liquid air. (If temperatures below about 90 °K are not desired, the outer case assembly can be dispensed with.)

Fig. II-13. Cryostat of Morrison and Young (8). 1—High vacuum line. 2—Tube to adsorption cell (1 mm o.d., 0.1 mm wall thickness German silver). 3—Liquid air inlet. 4—Liquid air outlet. 5—Rough vacuum line for pumping on liquid air. 6—Brass mounting plate. 7—Stainless steel tube (13 mm o.d., 0.45 mm wall thickness). 8—Anchoring ring for lead wires. 9—Brass can (0.8 mm wall thickness). 10—Nylon suspension cords. 11—Adsorption cell (copper). 12—Copper block (3 kg). 13—Glass Dewar vessel. 14—Capsule type platinum resistance thermometer. 15—Galvanized iron can (No. 20 gauge). 16—German silver tube (6 mm o.d., 0.1 mm wall thickness). A—Vacuum wax seal. B—Silver solder joint. C—Soft solder joint. D—Wood's metal joint.

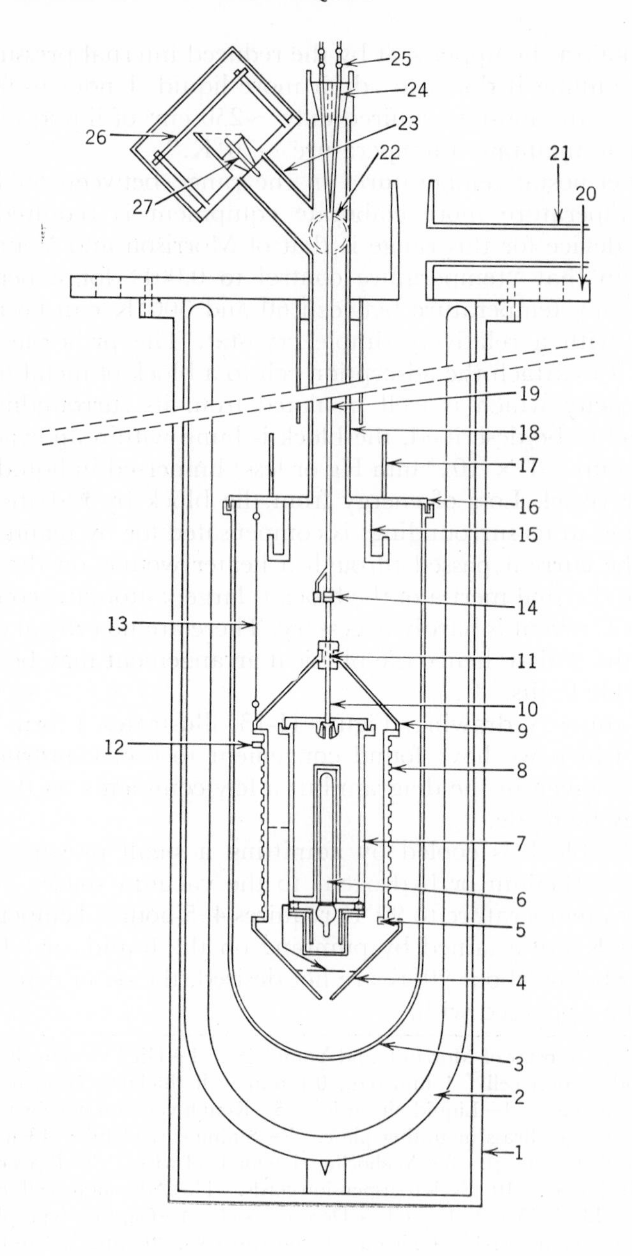
25
24
23
22
21
20
26
27
19
18
17
16
15
14
13
11
10
9
12
8
7
6
5
4
3
2
1

"2. Control of the current through the heater (100 ohms of No. 30 D.C.C. constantan) is achieved by means of a potential divider across a Variac. Using the ordinary laboratory power supply, it was found necessary to adjust the current slightly as often as every 10 or 15 minutes in order to maintain the temperature within 0.001 °. With a controlled voltage supply, virtually no adjustment was required. On several occasions the system was left overnight on controlled voltage and the total drift in temperature in this time never exceeded 0.02 °. This also shows the insensitivity of the temperature control to the level of the liquid air in the Dewar vessel.

"3. For temperature measurement a capsule-type platinum resistance thermometer was used because one was available. Other temperature measuring devices may equally well be used, e.g., a resistance element wound on the block, thermocouple, or vapor-pressure thermometer.

"4. The adsorption cell is fixed to the block with four screws and thermal contact is enhanced by a film of grease. In the experiments done so far with the cryostat there has been no requirement to heat the adsorbent in the cell; hence, construction of the cell with soft-solder and Wood's metal joints has been possible. Construction of metal cells which can be heated to high temperatures should not present difficult problems.

"5. The German silver tube goes directly to the adsorption cell without making contact with components at liquid air temperatures. Condensation of gas or vapor in the tube is prevented by a heater wound on that portion of the tube which passes through the central 13-mm tube."

A precision cryostat for the same range has been described by Constabaris, Singleton, and Halsey (9), as follows.

"The cryostat is shown in Figure II-14. The massive copper bulb,

Fig. II-14. Precision cryostat of Constabaris, Singleton, and Halsey (9). 1—Brass outer can (6 in × 28 in). 2—Glass Dewar flask. 3—Brass inner can. 4—Bottom radiation shield. 5—Inner binding post. 6—Platinum resistance thermometer. 7—Copper sample bulb. 8—Side radiation shield. 9—Top radiation shield. 10—German silver inlet tube. 11—Copper stud. 12—Outer binding post. 13—Nylon supporting cord. 14—Brass connector. 15—Copper heat interchanger. 16—Wood's metal gutter. 17—Water well. 18—German silver evacuation tube. 19—German silver gas inlet tube. 20—Brass cryostat top. 21—Refrigerant filling tube. 22—Evacuation tube. 23—Wire bundle exit tube. 24—Brass taper plug. 25—Brass connector. 26—External binding post. 27—Brass taper plug.

7, which contains the powder is evacuated at 150 °C outside the cryostat, filled to atmospheric pressure with C.P. argon, and soldered into position at the tubing connector 14. This connector leads to the manometric system through the gas inlet tube 19. (A spare inlet tube is shown in the figure.) The cryostat components are then assembled. The electrical leads which enter the radiation shields run through the spiral grooves on the outside of the side section 8 and are tied to the inner binding post 5. The shield

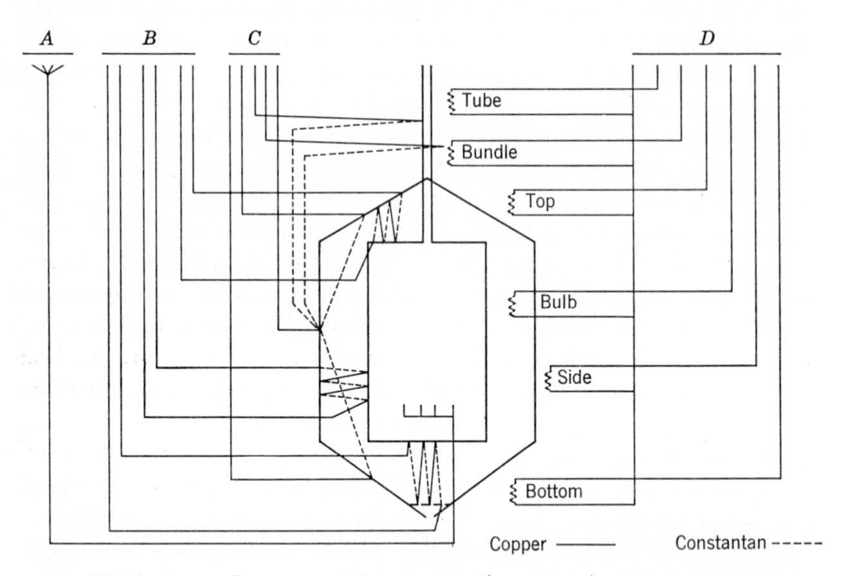

Fig. II-15. Internal wiring of the cryostat (Fig. II-14). Heater resistances in ohms: tube–85; bundle–25; top–100; side–140; bottom–120; bulb–80. A—Resistance thermometers. B—Controlling thermopiles. C—Difference thermocouples. D—Heaters.

heaters completely cover the outer surfaces. Above the shields the wires are tied in a bundle which is led through the grease-filled trough of the heat interchanger 15 and the evacuation tube 18 to the tapered plug 27, where it is split. The individual wires exit through grooves on 27, and are tied to the main binding post 26 where connections are made to the external circuits. The positions of the temperature controlling thermopiles are indicated by the arrows between the shields and the bulb. The junctions are held in place with small clips, from which they are insulated with one thickness of rice cigarette paper.

"The internal wiring is shown in Figure II-15 (which omits a series of thermocouples spaced at known distances of about 7 cm along the inlet tube). The wires that enter the cryostat are 32 swg double rayon wrapped copper. The heater wires are similarly insulated 32 swg constantan. (These wires were supplied by the Concordia Electric Wire and Cable Co., Ltd., Long Eaton, Nr. Nottingham, England.) The difference thermocouple junctions are soft-soldered together. Those of the controlling thermopiles are arc welded under a thin mica sandwich; the residual emf between the sets of junctions was found to be less than 0.1 microvolt when the sets were thermally equilibrated in a water-triple-point cell.

"The separate groups of wires of Figure II-15 are connected to various external circuits. The thermometer resistance is measured potentiometrically (10) in an all-copper circuit in which, on current reversal, thermal emf's larger than 0.2 microvolt are not observed. The circuit contains a Tinsley 4363A Vernier potentiometer and a Leeds and Northrup Type HS galvanometer with a 2 m scale distance. The lead wires from the difference thermocouples and controlling thermopiles are thrown onto the galvanometer with a set of Tinsley Series A copper selector switches. The lead wires of the thermopiles can be removed from this circuit and placed in an automatic controller circuit. The individual heaters draw current from a 500 ohm Helipot connected across an Automatic Electric 3 amp, 24 v dc, battery eliminator. In series with each heater are a two-range (0–100 and 0–250 ma) milliammeter, and a three-position switch which selects off, high, and low current. In the last position a variable resistance (0–100 ohm) is put in series with the heater. The heater circuit is fused at 250 ma.

"If the temperature of the shields is allowed to oscillate symmetrically about some temperature, that of the bulb will oscillate about the same temperature, but the extremely high thermal inertia of the system greatly chokes the amplitude of the oscillation at the bulb. Therefore, the problem of maintaining the temperature of the bulb to 0.001 ° is reduced to the simpler problem of keeping the temperature of the shields oscillating about the bulb temperature with an amplitude of, say, 0.01 °. These facts make the control of the cryostat fairly simple, although it is somewhat complicated by the poor thermal conductivity between the three shields, which necessitates individual control for each shield component. The inlet tube, wire bundle, shield assembly, and bulb are brought to the same temperature by use of the heaters and the two groups of thermocouples. The power to the inlet tube and wire bundle is then adjusted to maintain these components slightly above cryostat temperature. The controlling thermopiles are monitored, and the power to the shield heaters is adjusted alternately to high and low manually to give the desired temperature control. When temperature control has been obtained manually the thermopiles are switched into the automatic controller. In this, a motor driven (2 rpm) thermal-emf-free rotary switch is synchronized with a mechanical interlock relay switch. The former throws the thermopiles, in turn, onto a Type HS galvanometer, the light from which actuates a flip-flop phototube relay. This relay in turn actuates the selected interlock relay which puts the heater series resistance in and out of the heater circuit. The controller thus simulates manual control. The details of the device may be found elsewhere (11)."

Either of the above two cryostats can be used for temperatures up to room temperature. If the range from 240 °K and higher is the only one of interest, a simpler design, using a mechanical refrigerator to cool an ordinary thermostat bath, is adequate. Ordinary thermostats that operate up to 300 °C can be controlled to ±0.001 °C and are well suited for investigations of adsorption above room temperature.

A cryostat, at present in use in the authors' laboratories, which has been found to be versatile and easy to use, is illustrated in Figure II-16. The cryostat vessel consists of a modified Dewar flask partly filled with a bath fluid, in which the sample bulb is immersed; the bath liquid is heated by means of a coil of resistance wire, and stirred by bubbling dry nitrogen through it; cooling is provided by immersing the vessel in liquid nitrogen or a "dry ice"-acetone mixture. The current to the heater is controlled by a temperature-sensitive relay to maintain the temperature of the bath at any desired level. The relay operates as follows. The thermoelectric emf corresponding to the desired temperature is preset on a precision potentiometer, such as a Rubicon or a Leeds and Northrup Type K3; the galvanometer terminals of the potentiometer are connected to the input of a zero-center recording potentiometer,

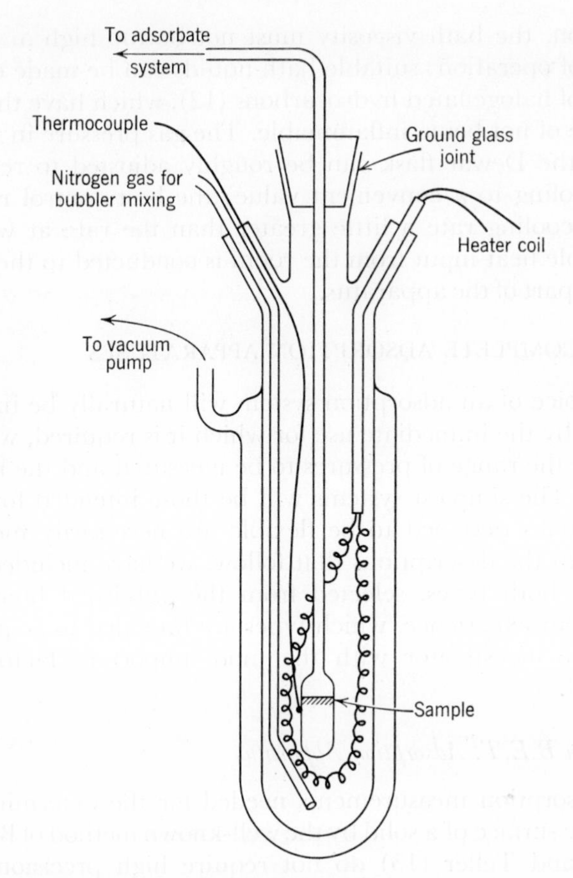

Fig. II-16. Cryostat.

reading −0.5 to +0.5 millivolts, and equipped with relay contacts; the contacts are arranged to close the heater circuit when the temperature of the bath is too low. The temperature will then rise until it reaches the point corresponding to the preset emf, when the recorder-controller will indicate zero and the contacts will then open the heater circuit. The apparatus as described will control the bath temperature to ±0.01° over the range 120–300°K. The lower temperature limit of operation is determined by the requirement that the bath remain liquid, coupled with the advisability of having a liquid whose boiling point is above room temperature.

In addition, the bath viscosity must not be too high at the temperature of operation; suitable bath liquids can be made of special mixtures of halogenated hydrocarbons (12), which have the further advantage of not being inflammable. The gas pressure in the outer jacket of the Dewar flask can be roughly adjusted to reduce the rate of cooling to a convenient value: the best control results by having a cooling rate a little greater than the rate at which the unavoidable heat input from the room is conducted to the bath by the upper part of the apparatus.

F. SOME COMPLETE ADSORPTION APPARATUSES

The choice of an adsorption system will naturally be influenced primarily by the immediate use for which it is required, which will determine the range of pressures to be measured and the necessary precision. The simplest systems will be those intended for routine work: systems designed to be flexible are necessarily more complicated. In the descriptions that follow, we have included typical designs of both types, selected from the published literature or based on our experience, which types are intended to acquaint the prospective investigator with the more important factors to be considered.

1. Nitrogen B.E.T. Adsorption Apparatus

The adsorption measurements needed for the determination of the specific surface of a solid by the well-known method of Brunauer, Emmett, and Teller (13) do not require high precision, except when the surface area is less than 10 m^2/g. The apparatus usually used consists of a mercury *U*-tube manometer, a gas buret, and a thermostat consisting of a Dewar flask filled with liquid nitrogen; such an apparatus is illustrated in Figure II-17, taken from one of Emmett's reports (14), whose operation is described as follows.

"The adsorption apparatus used in making the low-temperature adsorption measurements is of standard design. A typical setup is shown in Figure II-17. It includes an adsorption bulb, a calibrated gas buret, a manometer, a high-vacuum oil pump and Langmuir diffusion pump, and a McLeod gauge. The method of operation will be clear from the following brief description. The adsorption bulb containing a sufficient sample of adsorbent to

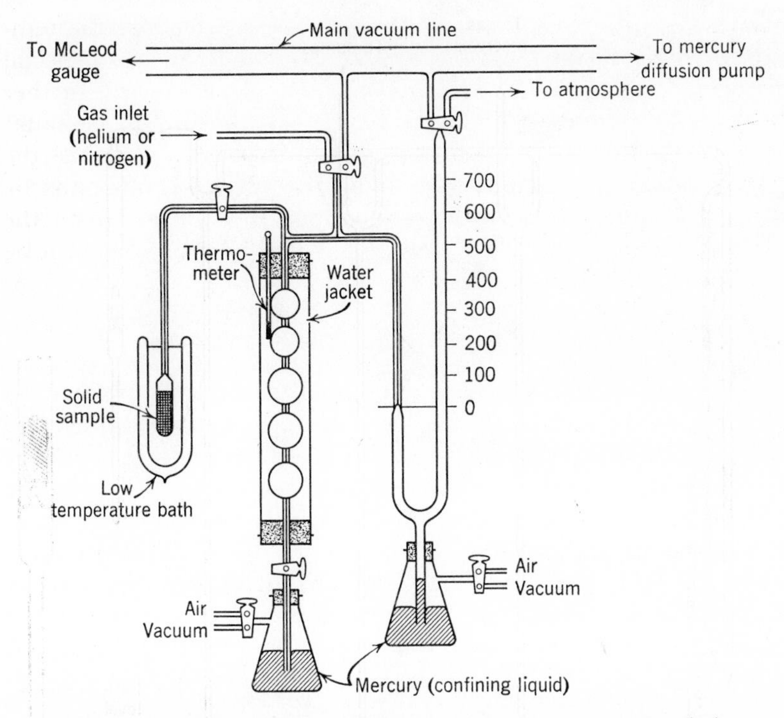

Fig. II-17. Nitrogen B.E.T. adsorption system, after Emmett (14).

furnish a total area of at least 2 m^2 is sealed to the rest of the system by 2-mm capillary tubing as indicated in Figure II-17. At the beginning of an experiment, the adsorbent is evacuated *in situ* at a temperature sufficient to remove water vapor and physically adsorbed gases (usually 110 °C). The dead space in the adsorption bulb up to the stopcock is then determined with pure helium. Such a calibration assumes, of course, that helium is not adsorbed by the adsorbent but merely fills in the space around it. This assumption is justified because at adsorption temperatures of 77 °K or higher the adsorption of helium will always be very small compared to that of the adsorbate being used. After the dead space is calibrated, one again evacuates the system to remove the helium and then, after closing the stopcock to the adsorbent, admits to the buret the adsorbate to be used (nitrogen, for example) and measures the quantity taken. The stopcock to the sample bulb is then

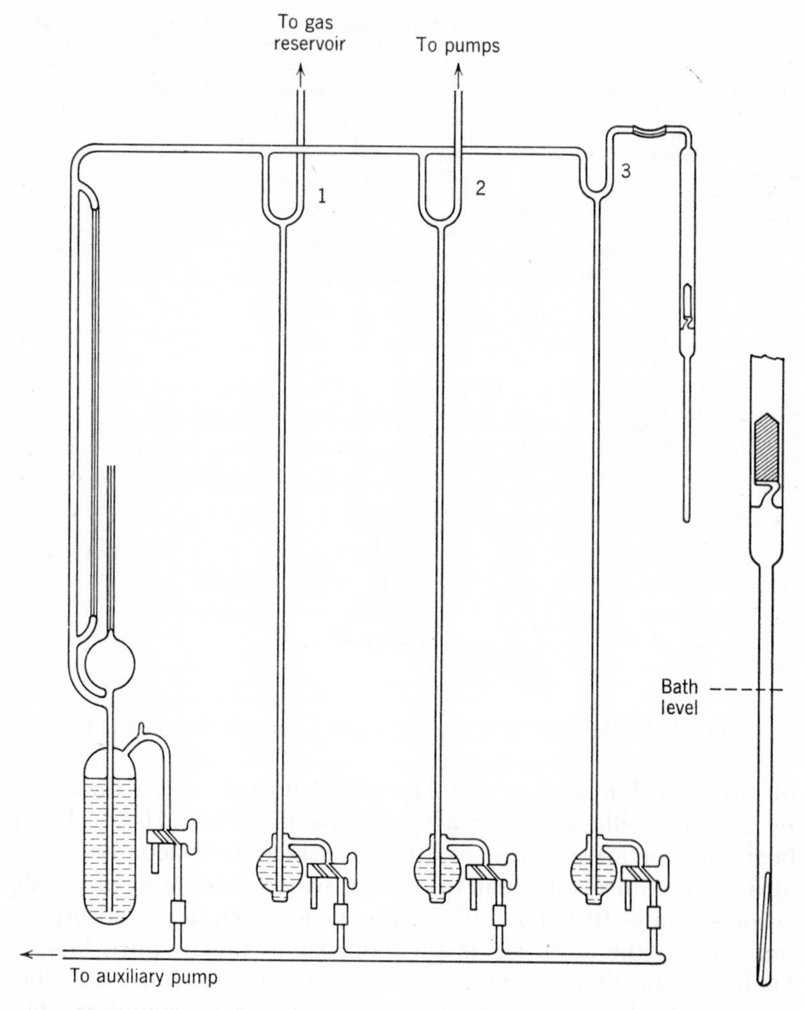

Fig. II-18. Low-pressure adsorption system with mercury cutoffs, after Wooten and Brown (16).

opened and the sample let stand long enough to effect equilibration. By repeatedly adding successive amounts of adsorbate through the buret system one can obtain a series of values for adsorption as a function of pressure at a given bath temperature of the adsorbent and thus obtain the data for an adsorption isotherm.

"The corrections (15) for the deviation of various simple gases from the perfect gas laws are as follows: nitrogen at −195.8 °C, 5% and at −183 °, 2.8%; oxygen at −183 °, 3.17%; argon at −195.8 °, 8.7%, and at −183 °, 3.0%; carbon monoxide at −183 °, 2.68%; methane at −140 °, 5.92%; NO at −140 °, 4.0%; N_2O at −78 °, 5.84%; CO_2 at −78 °, 2.09% and at 25 °, 0.58%; ammonia at −36 °, 2.64% and at 25 °, 1.185%; *n*-butane at 0 °, 10.8%, and at 25 °, 3.2%. All deviations refer to 760 mm and are assumed to vary linearly with pressure."

2. *Low-Pressure Adsorption Apparatus*

Wooten and Brown (16) designed an adsorption system, intended for B.E.T. surface area determinations of small-area solids for which they suggest the use of ethylene or ethane vapor as the adsorbate at liquid oxygen temperature: small amounts adsorbed are more readily measured at low equilibrium pressures. The adsorption system features mercury cutoffs and a McLeod gauge for the pressure measurement: the former are required because of the hydrocarbon vapor, which would dissolve in stopcock lubricant. The original design of Wooten and Brown is shown in Figure II-18:

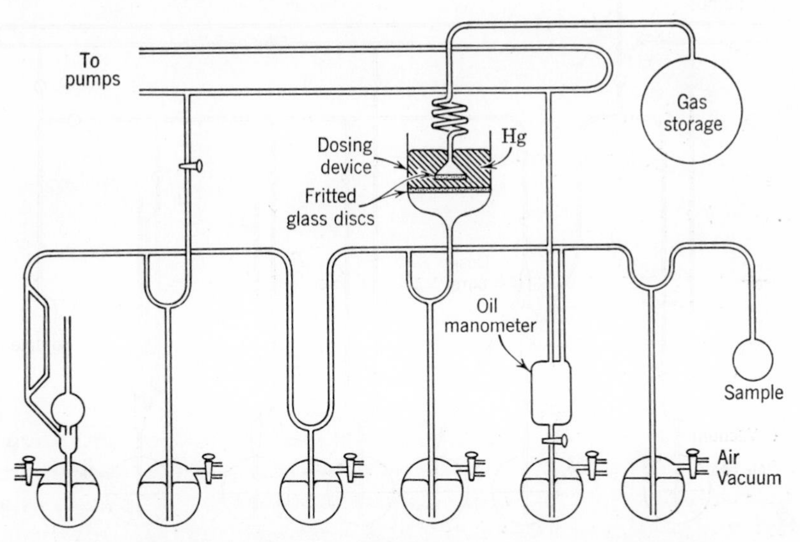

Fig. II-19. Low-pressure adsorption system with mercury cutoffs.

the adsorption of gas is measured by the usual difference method by use of the McLeod gauge capillary as a gas buret. The design shown in Figure II-18 has been made the basis of a more flexible adsorption system in the authors' laboratories by the incorporation of an oil-displacement manometer (Figure II-7) and a mercury *U*-tube manometer, as shown in Figure II-19. The large internal volume of the McLeod gauge, which is particularly undesirable as part of the system volume at higher pressures, can be cut off from the system when its range is exceeded and one of the other gauges is being used.

3. *Wide-Range Adsorption Apparatus*

A design of a wide-range adsorption system is shown in Figure II-20. This system has a McLeod gauge designed for the pressure range 1–600 μ; another (small volume) McLeod gauge designed for the pressure range 0.5–20 mm; and a mercury manometer for

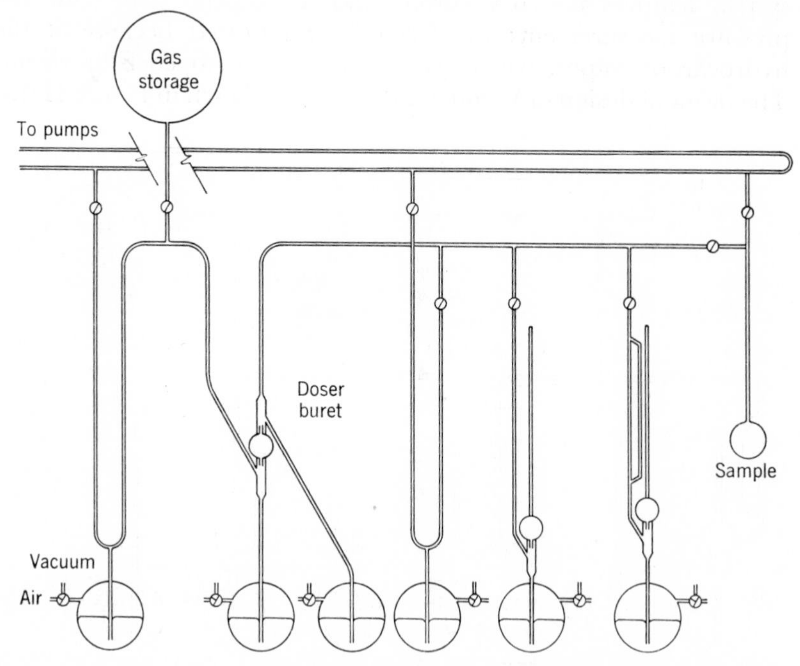

Fig. II-20. Wide-range adsorption system.

the range 20–760 mm. Only the gauge that is actually in use for the pressure measurements is part of the system volume, the other gauges being closed off; in this way the system volume is decreased at higher pressures, which is necessary if the same precision of measurement is to be maintained. By the use of Young's doser, the system volume with the *U*-tube manometer can be made quite small. The design has proved satisfactory for measurements of moderate precision. Since it includes stopcocks, it cannot be used with grease-soluble gases. The diagram also shows the provision made for the storage and transfer of a number of adsorbates.

4. *A High-Precision Adsorption Apparatus for 5–760 mm Pressure*

Constabaris, Singleton, and Halsey (9) have described a high-precision adsorption system, designed to obtain measurements that are more precise by an order of magnitude than those obtained

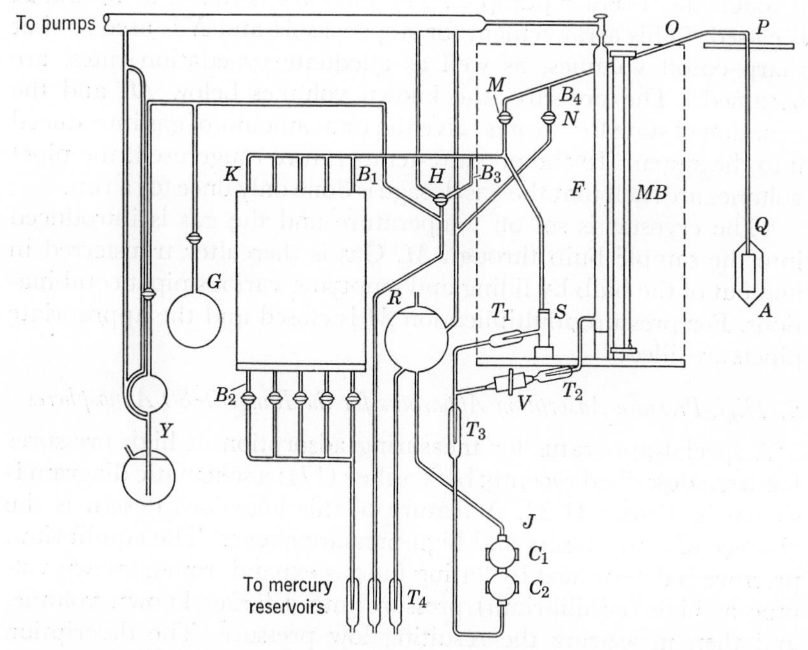

Fig. II-21. Adsorption system (on right side of diagram), and gas transfer system (on left side of diagram), after Constabaris, Singleton, and Halsey (9).

previously. The following description of the apparatus is taken from their paper, and illustrated in Figure II-21.

"The ice mantle *K* contains five mercury pipets with volumes between the calibration marks B_1 and B_2 of 10, 20, 40, 60, and 160 ml, respectively. Thus, combinations of these give volumes at 10-ml intervals between 0 and 310 ml. The direct connection between the pipets and the short legs makes the system unimanometric: both gas dose and run pressures are measured on the same precise manometer. At times, during a run, a large fraction of the gas is in the pipets. This does not introduce an appreciable error into the estimation of the amount of gas inside the sample bulb, for the temperature measurement and control of an ice mantle are virtually free of experimental error.

"After the system is evacuated, the mercury is run up to the mark B_4, and *M* and *N* are closed. The pipets that have the desired volume are emptied and the gas from *G* pumped into the system through the Toepler pump *Y*. The mercury is raised to B_3 and H is closed. (This arrangement for stopcocks *M* and *N* is used so that sharp cutoff volumes, as well as adequate evacuation rates, are obtained.) The pressure, the known volumes below *M*, and the equation of state for the gas, give the total amount of gas introduced into the system. In the pressure-temperature range used, the pipet volumes are such that the gas dosage is done only once for a run.

"The cryostat is set on temperature and the gas is introduced into the sample bulb through *M*. Gas is thereafter transferred in and out of the bulb by filling and emptying various pipet combinations. For pressure multiplication *M* is closed and the appropriate pipets are filled."

5. *High-Pressure Adsorption Apparatus for the Range 0–80 Atmospheres*

A special apparatus for measuring adsorption at high pressures has been described recently by Vasil'ev (17); a schematic diagram is shown in Figure II-22. A feature of this ingenious design is the absence of compressors and high-pressure gauges. The equilibrium pressure is determined by letting the gas expand from a known volume at high (equilibrium) pressure into a larger known volume, and then measuring the resulting low pressure. The description that follows is taken from Vasil'ev's paper in which he describes the adsorption of carbon dioxide at high pressures by silica gel.

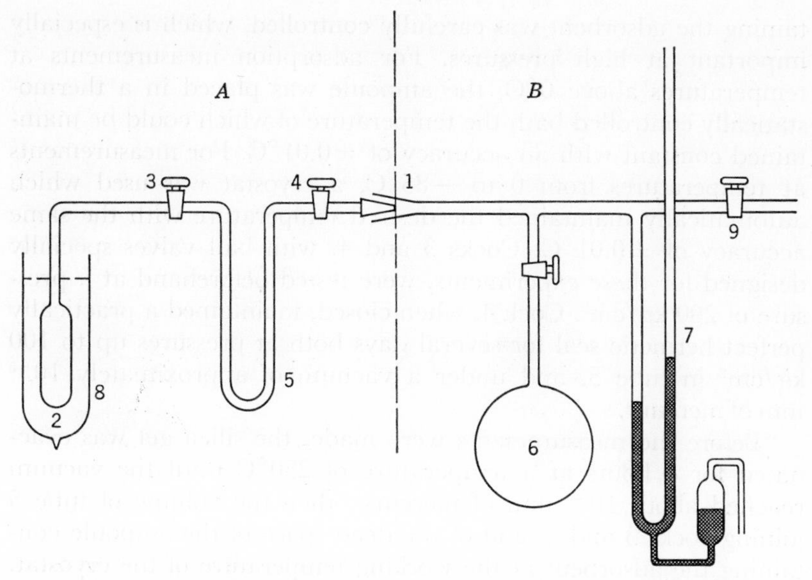

Fig. II-22. High-pressure adsorption system, After Vasil'ev (17).

"A characteristic feature of this apparatus is the absence of compressors and high-pressure gauges. A schematic diagram of the apparatus is shown in Figure II-22. It consisted of two principal parts: part *A* made entirely of metal and designed to withstand high pressures and part *B* made of glass. The parts were interconnected by means of a glass-to-metal ground joint, 1. Part *A* consisted of a brass ampoule, 2, containing the adsorbent, and two similar metal cocks (3 and 4) joined to each other by an intermediate *U*-shaped brass tube, 5. The glass part, *B*, consisted of a gas-buret, 6, connected to a *U*-tube mercury manometer, 7, and of ancillary devices (not shown), which served for evacuating the entire system to a high vacuum and for introducing helium or carbon dioxide into the buret.

"With the exception of ampoule 2, the entire apparatus shown in the diagram was housed in a thermostatically controlled air chamber whose temperature 32.0 ±0.1 °C, was selected to be above the critical temperature of carbon dioxide to eliminate the possibility of carbon dioxide condensing in tube 5 at high pressures. During the measurements, the temperature of the ampoule containing the adsorbent was carefully controlled, which is especially important at high pressures. For adsorption measurements at temperatures above 0 °C, the ampoule was placed in a thermostatically controlled bath the temperature of which could be maintained constant with an accuracy of ±0.01 °C. For measurements at temperatures from 0 to −85 °C, a cryostat was used which automatically maintained the desired temperature with the same accuracy of ±0.01 °C. Cocks 3 and 4, with ball valves specially designed for these experiments, were tested beforehand at a pressure of 200 kg/cm^2. Cock 4, when closed, maintained a practically perfect hermetic seal for several days both at pressures up to 100 kg/cm^2 in tube 5, and under a vacuum of approximately 10^{-5} mm of mereury.

"Before the measurements were made, the silica gel was evacuated for 4 hours at a temperature of 250 °C until the vacuum reached about 10^{-6} mm of mercury; then the volume of tube 5 joining cocks 3 and 4, and of the dead space of the ampoule containing the adsorbent at the working temperature of the cryostat, were determined by using helium. To carry out the experiment, a carbon dioxide sample was introduced into the gas-buret, whose volume was known, and the pressure of the gas was measured by means of a manometer and a cathetometer to an accuracy of ±0.01 mm. The measured gas sample was then quantitatively transferred to tube 5, which was cooled with liquid nitrogen, and cock 4 was closed. When tube 5 was heated, the carbon dioxide evaporated into the space between cocks 3 and 4, after which it was introduced into the vessel containing the adsorbent by opening cock 3. After the attainment of adsorption equilibrium, the part of the gas contained under high pressure in tube 5 was cut off by means of cock 3 and then gradually emitted through cock 4 into the gas buret, where that portion of the gas was measured by means of the mercury manometer. From the known ratio of the volumes of tube 5 and the buret, the volume concentration of the gas over the adsorbent could be calculated, and the value of the equilibrium pressure could be found from the well-known equation of state for the gas.

"To carry out the succeeding test, a fresh gas sample measured in the buret was transferred to tube 5 with cock 3 closed, and all the following operations were then carried out as described above.

In desorption tests, successive portions of the equilibrium gas were transferred from tube 5 to the buret and, after being measured, were removed from the apparatus."

G. SAMPLE PREPARATION

The specific surface of solids may lie anywhere in the range between a few cm^2/g (coarse crystals) to over 500 m^2/g (microporous particles). The amount of sample used for adsorption measurements should be enough to provide from 1 to 10 m^2 of surface; the sample weight may vary, therefore, from several grams to a few milligrams. A little experience with adsorbents of different types enables one to make a shrewd guess about the probable specific surface of an unknown sample. Materials that are gritty to the sense of touch or that have been prepared by grinding, if they are nonporous, would not exceed 10 m^2/g; precipitated salts do not exceed 20 m^2/g; unsintered hydrous oxides, such as alumina, silica, and magnesia, are likely to fall in the range 100–400 m^2/g.

Before making adsorption measurements it is always necessary to free the surface as much as possible from physically adsorbed air, water or organic contaminants, by heating the sample *in vacuo;* the temperature of this preliminary desorption, or out-gassing, should be as high as possible without causing chemical decomposition of the adsorbent or loss of its surface area by sintering. A useful guide to avoid sintering is to take one fourth of the melting point of the solid as an upper limit, until experiments show higher temperatures to be safe; the most thorough treatment on record was that given to tungsten wire, which was heated electrically to a white heat *in vacuo*.

The course of the out-gassing procedure should be followed by means of pressure measurements and continued until no further evolution of gas occurs. Out-gassing can be considered as complete when, the heating of the sample being continued with the adsorption system closed off from the pumps, a pressure of less than 10^{-5} mm is maintained for 12 hours; this result may be achieved within a few hours or it may require days of pumping and heating.

The effect of pretreatment of the sample is frequently one of the objects of study in an investigation. Among the physical effects of heating a solid we can distinguish mobility of the surface atoms, leading to a decrease of surface heterogeneity and sometimes to a loss of surface area; at higher temperatures the atoms in the crystalline lattice become sufficiently mobile to produce a more perfect crystal. The changes in the surface character of the solid can be measured by means of the adsorption isotherm, and the changes in the bulk crystal by X-ray diffraction. The effect of surface mobility in perfecting the crystallinity of the surface, at the same time tends to decrease the fraction of surface having high adsorptive potentials. An opposite effect, namely, an increase in the fraction of surface having high adsorptive potentials, is to be observed with surfaces such as glass and alumina, due to the removal of surface —OH groups.

H. THE GRAVIMETRIC METHOD OF MEASURING ADSORPTION

By the gravimetric method of determining the adsorption isotherm the amount of gas adsorbed is weighed using a vacuum microbalance; the equilibrium pressure is measured by one of the methods previously described. The two measurements are independent of each other, unlike the manometric method where the measurement of the equilibrium pressure is also used for the calculation of the amount adsorbed. The necessity of measuring doses of gas into the system is eliminated and, therefore, there is less chance of cumulative errors in the results.

Vacuum microbalances that are suitable for adsorption measurements have been classified by Rhodin (18) as follows: (a) cantilever type, (b) knife-edge type, (c) torsion type, (d) spring type. These basic designs are illustrated and discussed by Rhodin in terms of their total load, sensibility, ease of fabrication, installation, calibration, and operation. More recent modifications have introduced the principle of magnetic balancing (19) and the use of the linear differential transformer to measure extremely small deflections (20).

A simple form of a vacuum microbalance that combines the load-carrying capacity of the torsion type with the stability over a range of temperatures and pressures of the McBain-Bakr quartz spiral balance, was developed by Bushuk and Winkler (21) and is shown diagrammatically in Figure II-23. A weighed sample of the solid is suspended from one end of a glass beam, which is pivoted on a fine quartz thread; the other end of the beam is balanced by means of a quartz spring that can be extended from its lower end by an

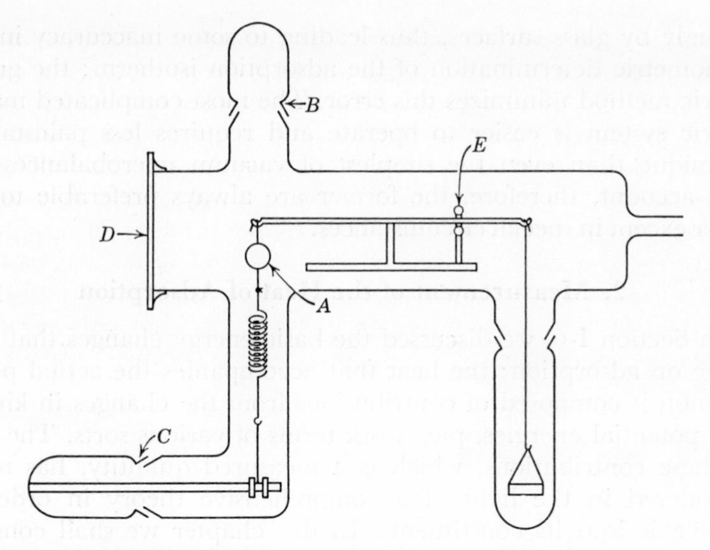

Fig. II-23. Quartz-Spiral Torsion Microbalance, after Bushuk and Winkler (21).

attached thread wound about a glass rod. The whole balance is enclosed in a glass case that is connected to a high vacuum manifold, which has the customary facilities for the introduction of gases and the measurement of pressure. The part of the glass case that surrounds the sample of adsorbent is immersed in a thermostat; the gaseous adsorbate is brought into the balance case, and after equilibrium has been reached the final pressure is measured. The quantity adsorbed at the equilibrium pressure is determined by the extension of the spring that is needed to bring the beam back to its null position. In this way, one point on the isotherm is obtained. By changing the amount of the adsorbate inside the balance case, other points at higher or lower equilibrium pressures can be obtained, until the whole adsorption isotherm is defined.

The gravimetric method is well suited to measure adsorption isotherms in the region where the manometric method is least accurate, namely, where the amount adsorbed is small compared to the amount unadsorbed; this condition may occur when the available surface area is small or the adsorptive forces are weak. Another, perhaps a special case, where the gravimetric method is advantageous is with the adsorption of water vapor, which is adsorbed strongly by glass surfaces, thus leading to some inaccuracy in the manometric determination of the adsorption isotherm: the gravimetric method minimizes this error. The most complicated manometric system is easier to operate and requires less painstaking technique than even the simplest of vacuum microbalances; on that account, therefore, the former are always preferable to the latter except in special circumstances.

2. Measurement of the Heat of Adsorption

In Section I-6, we discussed the basic energy changes that take place on adsorption; the heat that accompanies the actual phenomenon is composed of contributions from the changes in kinetic and potential energies, plus work terms of various sorts. The sum of these contributions, which is a measured quantity, has to be considered in the light of a comprehensive theory in order to resolve it into its constituents. In this chapter we shall consider the experimental methods used to measure heats of adsorption; in the next chapter we shall show the relation between the heats of adsorption as variously defined.

A. THE HEAT OF ADSORPTION DETERMINED FROM ADSORPTION ISOTHERMS

The isosteric heat of adsorption is a differential molar quantity derived from the temperature dependence of the isotherm, and can be related to fundamental thermodynamic quantities, as shown in Chapter III. A slightly simplified method of deriving the isosteric heat is similar to the derivation of the Clausius-Clapeyron equation for a phase change in a one-component system, and assumes that the adsorbent is inert; hence, we can write the equation $d_gF = d_aF$ at equilibrium. For a gas

$$d_gF = {}_gV\,dp - {}_gS\,dT + {}_g\mu\,d_gn \qquad \text{(II-25)}$$

For the adsorbed phase the corresponding equation is

$$d_aF = {}_aV\,dp - {}_aS\,dT + {}_a\mu\,d_an + A\,d\Pi \qquad \text{(II-26)}$$

where A is the total area of the adsorbed phase. At equilbrium, therefore,

$${}_gV\,dp - {}_gS\,dT + {}_g\mu\,d_gn = {}_aV\,dp - {}_aS\,dT + {}_a\mu\,d_an + A\,d\Pi \qquad \text{(II-27)}$$

For a constant amount adsorbed, equation II-27 becomes

$$_gV\,dp - {_gS}\,dT = {_aV}\,dp - {_aS}\,dT + A\,d\Pi$$

hence

$$\left(\frac{\partial p}{\partial T}\right)_{_an} = \left(\frac{_gS - {_aS}}{_gV - {_aV}}\right) + \frac{\beta}{\theta}\left(\frac{1}{_gV - {_aV}}\right)\left(\frac{\partial \Pi}{\partial T}\right)_{_an} \tag{II-28}$$

Neglecting the molar volume of the adsorbed phase compared to that of the gas, and assuming the gas to be ideal,

$$\left(\frac{\partial p}{\partial T}\right)_{_an} = \left(\frac{_gS - {_aS}}{RT}\right)p + \frac{Ap}{RT}\left(\frac{\partial \Pi}{\partial T}\right)_{_an}$$

or

$$\left(\frac{\partial \ln p}{\partial T}\right)_{_an} = \frac{_gH - {_aH}}{RT^2} + \frac{A}{RT}\left(\frac{\partial \Pi}{\partial T}\right)_{_an} = \frac{q^{st}}{RT^2} \tag{II-29}$$

where q^{st} is defined as

$$q^{st} = {_gH} - {_aH} + AT\left(\frac{\partial \Pi}{\partial T}\right)_{_an} \tag{II-30}$$

The quantity q^{st} defined by equation II-30 is called the *isosteric heat* of adsorption. As so defined, it is a positive quantity, in spite of the fact that heat is evolved during the process of adsorption. In equation II-30 the limitation of constant $_an$ on the partial derivative is equivalent to a constant volume adsorbed (hence the name *isosteric*) and is also equivalent to a constant relative surface concentration, θ, which is obtained by

$$_an/{_an_m} = V/V_m = \theta \tag{II-31}$$

provided that the amount required for completion of a monolayer of adsorbed phase, $_an_m$, is independent of temperature. For the purpose of this work, θ is *defined* by

$$\theta = \beta/\sigma \tag{II-32}$$

where β is the limiting area of an adsorbed molecule and σ is the area per adsorbed molecule at any point on the isotherm. The condition that V_m be temperature independent is, therefore, equivalent to the condition that β be temperature independent; now β

is related directly to the collision diameter of the molecule adsorbed, which may be assumed to be only slightly temperature dependent. Over moderate temperature intervals, therefore, we shall assume

$$(\partial \ln p/\partial T)_{an} = (\partial \ln p/\partial T)_{\theta} = q^{st}/RT^2 \qquad \text{(II-33)}$$

The isosteric heat of adsorption at any value of θ can be calculated graphically from equation II-33, by plotting the adsorption isostere as $\ln p$ *vs* $1/T$ and determining the slope, which equals $-q^{st}/R$. By this method, we do not need to assume that q^{st} is invariant with T, but the application of the method requires the measurement of a number of adsorption isotherms at different temperatures. The common practice, however, is to use the integrated form of equation II-33, applied to adsorption isotherms measured at two temperatures. Integrating equation II-33 gives

$$\ln (p_2/p_1) = (q^{st}/R)(1/T_1 - 1/T_2)$$

or

$$q^{st} = R[T_1T_2/(T_2 - T_1)][\ln (p_2/p_1)] \qquad \text{(II-34)}$$

As a first approximation, if T_1 and T_2 are close together, the value of q^{st} calculated by equation II-34 may be taken as applying to both isotherms; a better approximation is to take the value of q^{st} thus derived as pertaining to an average isotherm whose temperature is given by

$$1/T = [(1/T_1) + (1/T_2)]/2 \qquad \text{(II-35)}$$

and whose pressures are given by

$$p = (p_1p_2)^{1/2} \qquad \text{(II-36)}$$

The graphical determination of the isosteric heat is carried out by plotting the two isotherms as θ *vs* p with a scale appropriate for preserving the precision of the data; at values of θ taken at convenient intervals, the isosteric pressures p_1 and p_2 of the two isotherms are read off, and the values substituted in equation II-34. Isosteric heats are usually reported in the form of a plot q^{st} *vs* θ, which refers to the temperature defined by equation II-35.

The question now arises: what of the precision of the isosteric heats obtained in this way? We estimate the error inherent in the

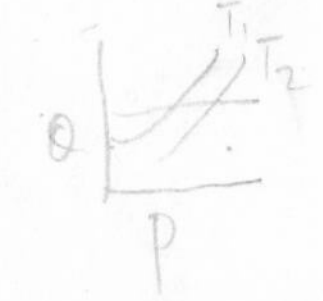

processes of plotting, curve-drawing, and interpolation, and in the graph paper itself, to be at least ±1% in the determination of p_1 and p_2, which makes the RMS error of their ratio 1.4%; therefore,

$$\ln [(p_2/p_1)(1 \pm 0.014)] \simeq \ln (p_2/p_1) \pm 0.014$$

hence

$$q^{st} = R[T_1T_2/(T_2 - T_1)] \ln (p_2/p_1) \pm R [T_1T_2/(T_2 - T_1)] \, 0.014 \tag{II-37}$$

The precision of q^{st} is seen, therefore, to depend entirely on the factor $T_1T_2/(T_2 - T_1)$: for isotherms with a 10° temperature interval near room temperature, the minimum error in q^{st} is ±250 cal/mole. The error is considerably less for isotherms determined at the boiling points of nitrogen and oxygen, i.e., 77.5 and 90.1 °K, respectively, which are two temperatures commonly used for the adsorption of nitrogen or argon. Here the minimum error in q^{st} is only ±15 cal/mole. It should be observed, however, that the 1% error estimated above may grow to 2 or 3% in those portions of the isotherm that intersect the constant-θ line at an acute angle.

B. THE CALORIMETRIC MEASUREMENT OF THE HEAT OF ADSORPTION

A review of adsorption calorimetry by R. A. Beebe was published in 1944 (22), covering developments up to the beginning of World War II. The review pays particular attention to experimental methods and sources of error.

As an illustration of the subject matter, take the case of the heat of adsorption of argon on a graphite whose specific surface is 10 m²/g. This would be a system of some interest, with a specific surface of a magnitude frequently found; but, in practice, adsorption calorimetry is rarely attempted with solids of specific surface less than 100 m²/g. Our calculation will show why: 10 g of graphite of specific surface 10 m²/g has a total surface of 100 m² which corresponds at $\theta = 1$ to a volume adsorbed of 30 cm³ STP; assume an average heat of adsorption of 3 kcal/mole, then, for a dose corresponding to 5% of the monolayer, the heat evolved is 0.2 calories;

from the specific heats of graphite and platinum, the heat capacity of the calorimeter vessel (10 g of platinum) and contents is about 0.45 cal/centigrade degree; therefore, the maximum temperature rise due to the adsorption of the dose is 0.50 centigrade degree. If the required precision is 1%, the temperature change must be measured to $\pm 5 \times 10^{-3}$ centigrade degree, which entails that the calorimeter be thermostatted to much less than $\pm 5 \times 10^{-3}$ centigrade degree: this approaches the limit of what can be done. With an adsorbent of specific surface of 100 m^2/g or greater, high precision is more readily achieved. The heart of the adsorption calorimeter then becomes the device used to measure a small temperature change with a precision of about 1%. The two devices most suitable for this purpose are the thermocouple and the platinum resistance thermometer.

For most effective thermal contact the thermocouple junction is welded to the metallic body of the calorimeter vessel. The use of more than one junction introduces the need for electrical insulation, which unfortunately is also thermal insulation, and has the further disadvantage of creating a larger thermal shunt between the calorimeter and its surroundings. For these reasons, a single junction is frequently deemed best. A copper-constantan couple at 77°K has a response of about 16 μv/centigrade degree; for the evolution of 0.2 cal, as in the above example, the output of the thermocouple would be about 8 μv; a precision of 1% would therefore require an instrument able to measure to ± 0.08 μv. This is again to approach the limit of what can be attained, even with a sensitive galvanometer or dc microvolt amplifier.

The platinum resistance thermometer when used with a precision Mueller bridge is capable of a precision of $\pm 0.01°$; about the utmost limit of precision now available, taking extraordinary precautions to avoid sources of error, approaches $\pm 0.001°$. To attain the postulated 1% precision the quantity of adsorbent surface in the calorimeter would have to be at least the 100 square meters postulated in our sample above.

Many factors conspire to reduce the precision from that calculated for our example, in which we postulated a heat input and estimated the maximum electrical output, assuming no losses. First, the heat is not evolved instantaneously but over a period of time, during which the loss to the surroundings may be appreciable.

Second, regardless of how fast the heat is actually evolved, the thermal equilibrium between the bulk of the powdered adsorbent and the temperature-sensing element is hampered by the poor thermal conductivity of the gas, particularly at low pressures, which again results in a temperature rise that is less than it should be, due to loss to the surroundings. Third, in those cases of chemisorption where the equilibrium pressure is "zero," i.e., the adsorbed molecules remain on the portion of the surface that they happen to strike first, the heat evolved is not an equilibrium heat of adsorption, and the temperature rise measured depends on the location of the sensing element with respect to that portion of the sample first in contact with the gas stream. Fourth, heat effects, other than that due to adsorption, such as the adiabatic compression of the gas in the calorimeter by an incoming dose, are registered—a source of error that becomes greater with increasing residual pressure, when the ratio of unadsorbed gas to adsorbed gas is large. Of these sources of error, the first two can be minimized by good design and proper calibration of the calorimeter; the third error is irremediable, although by having a partial pressure of helium in the system to distribute the evolved heat rapidly, a more reproducible result can be obtained; the fourth source of error can be minimized by means of a calculated correction (see Sections III-2 and III-3).

1. *The Amherst Calorimeter*

R. A. Beebe and his co-workers at Amherst College have developed a relatively simple vacuum adsorption calorimeter that is well adapted for rapid measurements of moderate precision. One of the later types is shown in Figure II-24, and is described by Beebe *et al.* (23) as follows.

"It consists of a platinum cylinder *B*, 7 cm long, 2 cm in diameter, and of 0.25 mm wall thickness. Into this was close-fitted a set of six platinum fins, *C*, symmetrically spaced about a central platinum tube of 3 mm diameter which serves as the housing for heater, *D*. A clearance of 2 mm is provided between the fins and the bottom of the cylinder to facilitate gas distribution. A platinum lid was silver-soldered onto the cylinder after having been joined, at *F*, to a Kovar-Pyrex seal, *E*. The Kovar inlet tube is 19 mm long and has an inside diameter of 5.5 mm. Surrounding the platinum cylinder is a glass jacket, *A*, the spacing between the two being

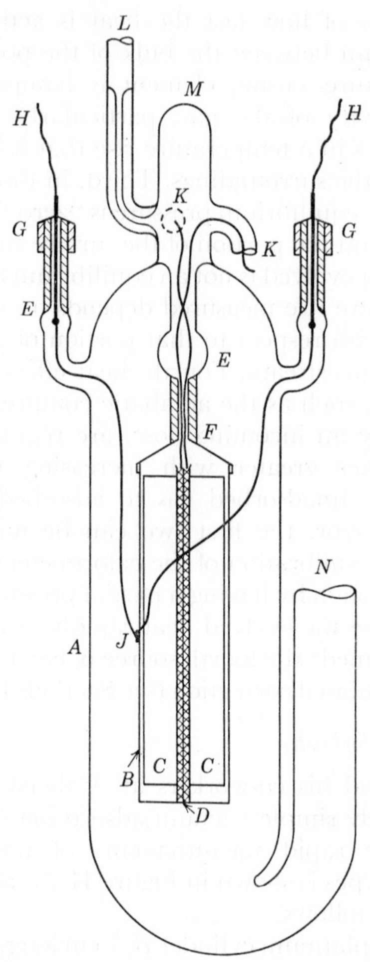

Fig. II-24. The Amherst Adsorption Calorimeter. A—Glass jacket. B—Platinum cylinder. C—Platinum fins. D—Heater. E—Kovar-pyrex seals. F—Kovar-platinum seal. G—Brass caps. H—Thermocouple leads. J—Copper-constantan thermal junction. K—Heater leads (side arms not fully shown). L—Inlet tube. M—Top seal. N—Connecting tube to pump and helium reservoir.

about 1 cm. One copper-constantan thermal junction, *J*, is silver-soldered onto the outside of the platinum cylinder. The thermocouple wires (B and S No. 30 and No. 31 for copper and constantan, respectively) are connected to short sections of heavier wire (B and S No. 18) and these in turn are sealed through side arms in the jacket as shown on the diagram at *G*. Here a brass cap is soldered to the Kovar tube of a Kovar-Pyrex seal and the heavier wire soldered through a narrow bore in the brass cap. This type of connection has been found preferable to the straight tungsten-through-glass seal previously employed, because it has greater mechanical stability and because the lead wires can be more easily replaced. The heater leads are similarly sealed through side arms at *K* (not fully shown in the diagram). The latter are turned downward to keep the total length of the instrument to a minimum. B and S No. 18 copper wire is used in this instance for the soldered connection through the brass cap at *G*.

"Gas inlet *L* consists of a short section of 3 mm capillary tubing, which was completely immersed in the constant temperature bath during runs; it is joined onto regular 8 mm tubing. Thermal transpiration was thus minimized because of the relatively large diameter of the inlet tube in the region of the temperature gradient. The top of the instrument, *M*, was open for the purpose of inserting the adsorbent and installing the heater and was subsequently sealed off. The 15 mm tube, *N*, is connected through stopcocks to the vacuum manifold as well as to a helium reservoir. Thus, the calorimeter jacket may be pumped out during measurements or else helium may be admitted to it, if rapid heat exchange is desired, as it is between separate admissions of adsorbate. Furthermore, the tube, *N*, serves as the main mechanical support of the instrument.

"The heater was was made from B and S No. 34 enamel covered "Advance" wire, by bending several lengths back and forth and using them as a core upon which to wind the rest of the wire. Roughly 20 feet of wire were used to give a resistance of 160 ohms. The coil was covered with a thin layer of aluminum foil which was held in place by four narrow strips of platinum, shaped so as to provide a sliding fit with the heater housing. This made it possible to drop the heater into position through *M* and also to achieve good thermal conduction.

"A close fitting brass plug with a long wire handle was employed to protect the heater housing from adsorbent during the filling operation. For the latter, performed before the installation of the heater, a long-stemmed funnel was introduced through *M* and the adsorbent tapped through.

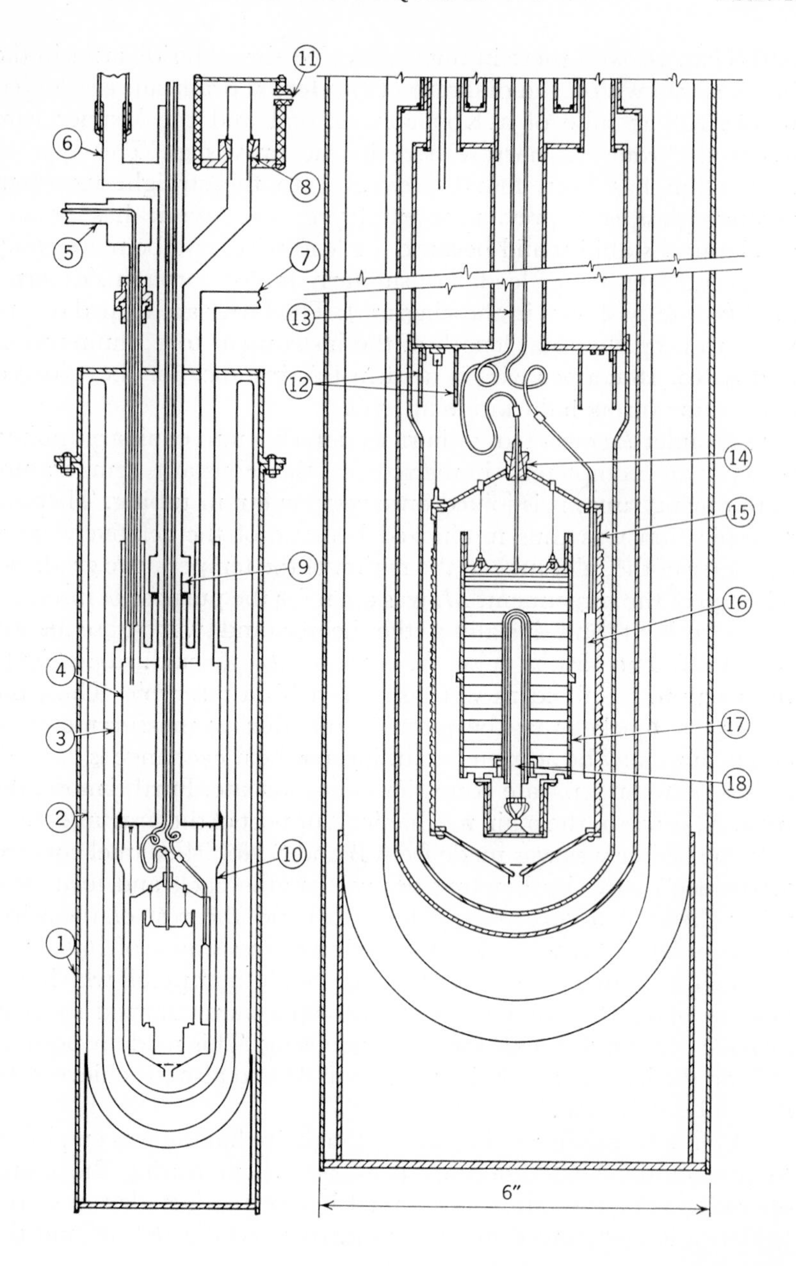

1
2
3
4
5
6
7
8
9
10
11
12
13
14
15
16
17
18
6"

"The reference thermal junction was housed in an 8 mm glass tube which, in turn, was held in a copper cylinder of 2 cm inside diameter by a rubber stopper. The whole unit, not shown in the diagram, was wired to the calorimeter jacket. For runs at −195 °C the inner glass tube was filled with pump oil. This, however, was discarded for runs at liquid oxygen temperature with no apparent change in the performance of the thermocouple.

"Before each run the whole calorimeter was out-gassed overnight at 200 °C. Next, the Dewar vessel containing liquid nitrogen or oxygen was raised around the apparatus and the system allowed to establish temperature equilibrium over a period of 2 to 3 hours. To accelerate this process, helium was admitted to the jacket to a pressure of about 1 cm or higher. Finally, the jacket was pumped out and the instrument was then ready for measurements."

This calorimeter has been used by Beebe and his associates both isothermally, and in a manner that is essentially adiabatic (24). When used as an isothermal calorimeter, sufficient helium is left in the outer jacket *A* to provide the thermal conductivity required for the temperature to return to its initial value in about 30 minutes after the liberation of heat inside the calorimeter. The variation of electrical output *versus* time is recorded and the area under the recorded curve, from the initial moment to its point of return, is proportional to the heat input; the proportionality is determined by calibrating the instrument with heat liberated electrically in the resistance wire *D*. When used adiabatically the outer vessel is highly evacuated so that cooling is very slow; from the cooling curve the Newton cooling coefficient is calculated and the adiabatic cooling curve is reconstructed. After a few minutes

Fig. II-25. The N.R.L. low-temperature adiabatic calorimeter. 1—Monel metal case. 2—Silvered pyrex Dewar flask, $4^1/_4$ inch int. diam. × 24 inch inside depth. 3—Brass container ($^3/_{64}$ inch wall thickness). 4—Brass liquid hydrogen container, 680 cm^3 volume. 5—Vacum-jacketed metal siphon. 6—Vacuum line connection. 7—Hydrogen vent line. 8—Vacuum seal for lead wires. 9—Brass anchoring ring at liquid nitrogen temperature. 10—Copper container ($^1/_{32}$ inch wall thickness). 11—Brass clamps for lead wires. 12—Anchoring rings at temperature of hydrogen container—inner ring $^1/_8$ inch thick brass, outer ring $^1/_{32}$ inch thick copper. 13—Copper-nickel filling tube (2 mm ext. diam. and 1 mm int. diam.). 14—Brass tapered joint. 15—Adiabatic shield, $^1/_{16}$ inch thick copper. 16—Vapor pressure thermometer bulb, 7 cm^3 volume. 17—Calorimeter vessel, $^1/_{16}$ inch thick aluminum, 90 cm^3 volume. 18—Platinum resistance thermometer.

the reconstructed curve reaches a constant value with time; this value is proportional to the heat input. Calibration is again done by electrically-derived heat. The Amherst calorimeter is rarely used in this way, however, because the calculations required take so much time, though the results obtained are more reproducible and reliable. Corrections for isothermal or adiabatic heats of compression, which should always be considered, will be discussed in the next chapter.

The experimental determination of calorimetric heats proceeds in the same way as the manometric determination of an adsorption isotherm; in fact, the adsorption isotherm is determined at the same time. The heat evolved per mole with each dose is plotted against the total amount adsorbed, thus producing a step-wise graph, which is smoothed out by means of the chord-area technique (25). The estimated experimental precision is $\pm 3\%$.

2. *The N.R.L. Adiabatic Calorimeter*

An adsorption calorimeter of high precision has been designed by J. A. Morrison and his associates at the National Research Laboratories, Ottawa (26). Because of the complexity of construction that is required to achieve high precision, it is inconvenient with this apparatus to change samples frequently; consequently, it is best suited for exhaustive studies of a single adsorbent with a variety of adsorbates. The design of the calorimeter incorporates a cryostat that is capable of maintaining constant temperatures anywhere between 15 °K and room temperature.

Adiabatic conditions inside the calorimeter are maintained by having a metal shield surrounding it and at a temperature at all times equal to that of the calorimeter vessel, which contains the adsorbent; thermocouple junctions are attached to the shield and to the calorimeter, to detect temperature differences between them; any such differences are equalized by adjusting the electrical energy admitted to the heater windings around the shield. Adiabatic conditions thus ensured, no thermal losses of the heat evolved in adsorption occur: consequently, the adiabatic heat of adsorption is directly measured by the temperature rise of the calorimeter vessel. In Figure II-25 the cryostat assembly and details of the calorimeter assembly are illustrated; for further information about the construction and operation of this calorimeter, the original paper (26) should be consulted.

Some measurements have been made with this calorimeter of the heat capacity of adsorbed argon at surface concentration $\theta = 0.4$: the absolute increment in the heat capacity from 14 to 75 °K was from 0.25 to 0.70 joules/degree, and this increment was measured to approximately 1% of its value; the measurements of heats of adsorption were internally consistent to better than 0.5%.

3. *Heats of Adsorption from Heats of Immersion*

Theory. Heat is evolved when a solid is immersed in a liquid: if the solid, prior to its immersion, has been in a vacuum the heat evolved per square centimeter of surface, h_s, is called the *total heat of immersion;* if the surface of the solid has a film adsorbed on it,

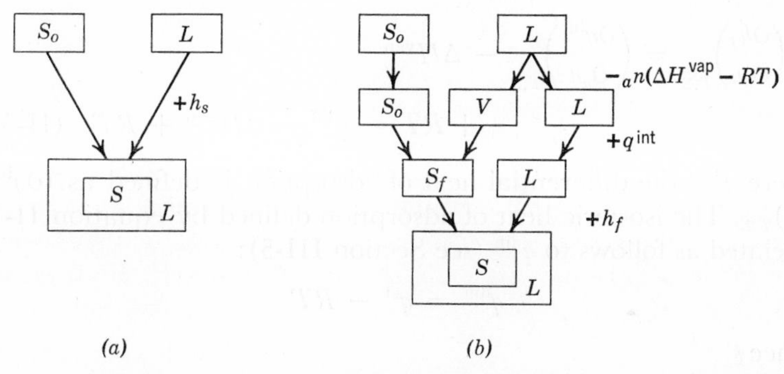

Fig. II-26. Relation between heats of immersion and integral heat of adsorption.

due to its having been in equilibrium with a partial pressure of the vapor of the liquid, the heat evolved per square centimeter of surface, designated h_f, we shall call the *residual heat of immersion*. The relation between these heats of immersion and the integral heat of adsorption is shown in Figure II-26, (*a*) and (*b*): in part (*a*) the total heat of immersion is represented by the mixing of the bare solid S_0 with the liquid *L*: in part (*b*) we start with S_0 and *L*, then isothermally vaporize $_an$ moles of liquid into a vacuum with the input of $_an(\Delta H^{vap} - RT)$ calories, where ΔH^{vap} is the molar heat of vaporization of the liquid; the $_an$ moles of vapor are then transferred isothermally to one square centimeter of the surface of the evacuated solid as an adsorbed film, with the evolution of the integral heat of adsorption (see Section III-1) q^{int} cal/cm^2; the solid containing the adsorbed film, and now designated S_f, is immersed in the liquid with the evolution of the residual heat of immersion, h_f cal/cm^2. The initial and final states of the processes represented in (*a*) and (*b*) are the same. Therefore,

$$h_s = h_f + q^{\text{int}} - {}_an(\Delta H^{\text{vap}} - RT)$$

or

$$h_f = h_s + {}_an(\Delta H^{\text{vap}} - RT) - q^{\text{int}} \qquad \text{(II-38)}$$

The residual heats of immersion measured calorimetrically for various amounts of preadsorbed gas are usually reported as a function of the amount preadsorbed per cm^2, ${}_an$: the slope of this curve is given mathematically by

$$-\left(\frac{\partial h_f}{\partial {}_an}\right)_{T,\Sigma} = \left(\frac{\partial q^{\text{int}}}{\partial {}_an}\right)_{T,\Sigma} - \Delta H^{\text{vap}} + RT = q^{\text{diff}} - \Delta H^{\text{vap}} + RT \qquad \text{(II-39)}$$

where q^{diff}, the differential heat of adsorption is defined as $(\partial q^{\text{int}}/\partial {}_an)_{T,\Sigma}$. The isosteric heat of adsorption defined by equation II-33 is related as follows to q^{diff} (see Section III-5):

$$q^{\text{diff}} = q^{\text{st}} - RT$$

hence

$$q^{\text{st}} = \Delta H^{\text{vap}} - (\partial h_f/\partial {}_an)_{T,\Sigma} \qquad \text{(II-40)}$$

The isosteric heat thus obtained should agree with that obtained from the adsorption isotherms by means of equation II-33. The isosteric heat is normally, though not necessarily, greater than the heat of liquefaction; therefore, the slope of h_f *vs* ${}_an$ is usually negative. The exact shape of the curve depends on the nature of the system.

Typical curves for the residual heats of immersion as a function of the amount preadsorbed are shown in Figure II-27, (*a*) and (*b*). (*a*) is calculated for a completely homotattic surface; such surfaces are so rare that no experimental example of this type of heat curve has yet been reported, although some near-homotattic surfaces have produced curves that crudely approximate it. Curve II-27(*b*) repre-

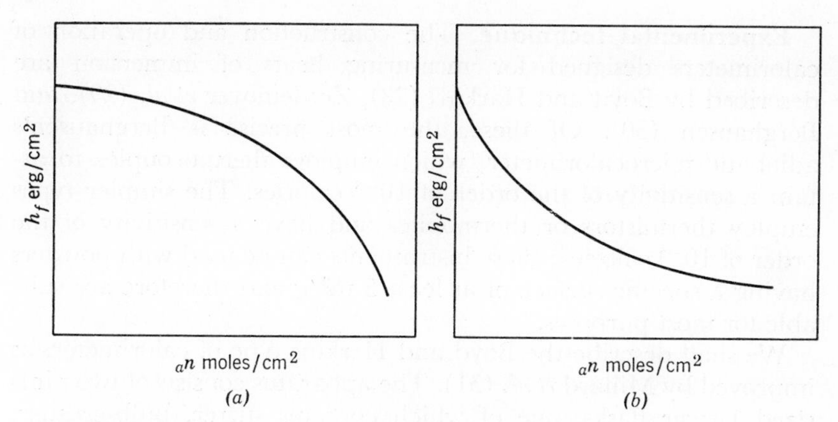

Fig. II-27. Typical curves for residual heat of immersion as a function of the amount preadsorbed.

sents h_f *vs* $_an$ for a heterogeneous surface, such as are commonly found experimentally, e.g., water on TiO_2. Chessick and Zettlemoyer (27) have erroneously described the heat of immersion of a homotattic solid surface as decreasing *linearly* with $_an$; actually, such a linear curve is still representative of a rather heterogeneous substrate. The true shape of the heat of immersion curve of a homotattic surface can be calculated from the shape of the isosteric heat curve for such a surface. As described in Section I-7, the cooperative additions to the adsorptive energy of a homotattic surface result in isosteric heats of adsorption constantly increasing with surface concentration. Let us suppose that such an isosteric heat curve has the form

$$q^{st} = A + B\ {}_an$$

hence

$$q^{diff} = A - RT + B\ {}_an$$

and

$$q^{int} = \int_0^{_an} (A - RT + B\ {}_an)\ d_an = (A - RT)\ {}_an + B\ {}_an^2/2$$

substituting in equation II-38 gives:

$$h_f = h_s + (\Delta H^{vap} - RT)\ {}_an - (A - RT)\ {}_an - B\ {}_an^2/2$$

or

$$h_f = h_s - A'\ {}_an - B'\ {}_an^2 \qquad \text{(II-41)}$$

Equation II-41 is illustrated in Figure II-27(*a*); the curve suggested by Chessick and Zettlemoyer would evidently result from an isosteric heat that was independent of surface concentration.

Experimental technique. The construction and operation of calorimeters designed for measuring heats of immersion are described by Boyd and Harkins (28), Zettlemoyer *et al.* (29), and Berghausen (30). Of these, the most precise is Berghausen's adiabatic microcalorimeter, which employs thermocouples to attain a sensitivity of the order of 10^{-3} calories. The simpler types employ thermistors or thermopiles and have a sensitivity of the order of 10^{-2} calories; these instruments can be used with powders having a specific surface of at least 5 m^2/g and therefore are suitable for most purposes.

We shall describe the Boyd and Harkins type of calorimeter, as improved by Millard *et al.* (31). The apparatus consists of two pint-sized Dewar flasks, one of which contains stirrer, bulb-crusher, calibration heater, and the "hot" junctions of a fifty-junction, copper-constantan thermopile. The "cold" junctions are contained in the other Dewar flask. Both flasks, sealed with rubber stoppers, are immersed in the same thermostat bath, which is controlled to $\pm 10^{-3}$ degree; the temperature control inside the Dewars is of the order of $\pm 10^{-5}$ degree by this arrangement. The amount of liquid in each Dewar is 100 ml and the amount of sample 2 or 3 grams. A constant-speed stirrer in a calorimeter of this type is important as variations in the speed of stirring cause erratic drifts in the time-temperature cooling curves; a synchronous motor is recommended. Sample containers are made by blowing thin-walled bulbs of about 10 ml volume from 10 mm Pyrex tubing. After thermal equilibrium is established in the calorimeter, the sample bulb is broken: this is done by a one-half turn of a screw, which forces a metal bar down on the bulb and shatters it completely. Millard *et al.* (31) found that blank runs on empty bulbs evolved 0.15 ± 0.02 calories, and this amount of heat was subtracted from each of the measurements of the heat of immersion. The output of the thermopile is plotted or recorded *versus* time to get the time-temperature cooling curve; the extrapolated zero time deflection measures the heat evolved, after calibration with the electrical heater. The sensitivity of this apparatus as used by Millard *et al.*, was such that the liberation of 0.01 calories gave a deflection of 1 mm with the galvanometer scale at 2 meters. The electrical sensitivity was 0.75 μv/mm at this scale distance. The heat of immersion data were reproducible to 2–3%.

C. ISOSTERIC HEAT OF ADSORPTION FROM GAS-SOLID CHROMATOGRAPHY

1. *Experimental Technique*

The technique of gas-solid chromatography uses the differences in adsorptive potentials to separate constituents of a mixture of gases. The gas mixture flows through a column that is packed with a solid adsorbent; each constituent of the mixture has a different time of retention on the adsorbent surface, and the resulting differences in the rates of diffusion through the column make each constituent appear at the other end after different intervals of time. The pulse flow technique is to transport a pulse of the gas mixture through the adsorbent column in a stream of inert (i.e., non-adsorbable) carrier gas. At room temperature and above, helium or neon are suitable carrier gases, as they are not appreciably adsorbed.

Apparatus suitable for the purpose is sold commercially. It consists of a thermostat into which the packed column is fitted and a detector, which may be a conductivity cell that responds to changes

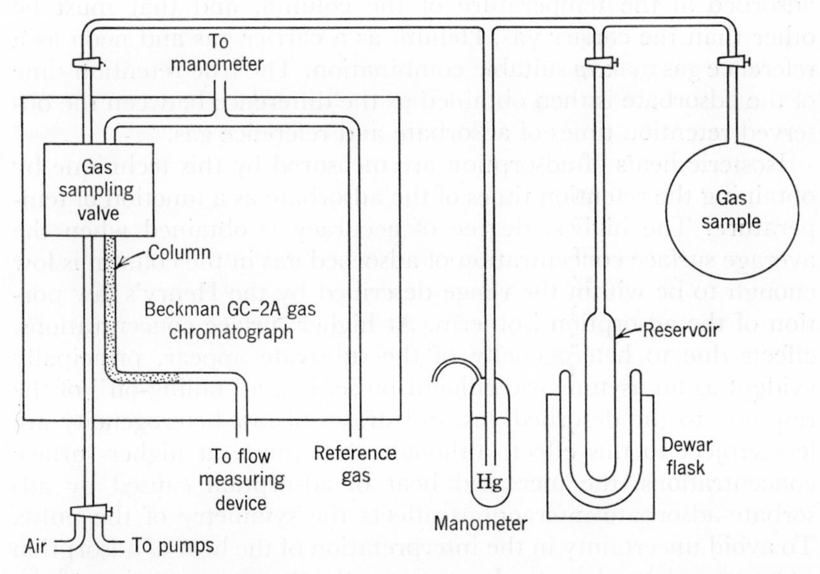

Fig. II-28. Arrangement for introduction of gas or vapor adsorbates and for flow and pressure measurements.

in the thermal conductivity of the adsorbate-carrying gas eluted from the column relative to a stream of the pure carrier gas. The output of the detector circuit is usually applied to a recording potentiometer. Gas or vapor is introduced into the instrument by evacuating the sample valve and then filling it with the gas at any predetermined pressure. A suitable device is shown in Figure II-28. Also shown in the diagram are the points at which the pressure head on the column and the flow rate of the gas going through it are measured: the pressure head is measured at a *T*-joint near the entrance to the column and the flow rate is measured at the column outlet. The temperature of the column is measured by any suitable means. The adsorbent material is packed in a column, usually *ca.* $^1/_4$ inch diameter and about 3 feet in length. The adsorbent should be in the form of aggregates between approximately 100 and 60 mesh, in order to avoid an excessive pressure drop across the column, which would result from the packing of fine particles.

The retention time of the adsorbate must be corrected for the time required by the carrier gas to sweep through the column. This correction can be obtained by the use of a reference gas that is not adsorbed at the temperature of the column, and that must be other than the carrier gas. Helium as a carrier gas and neon as a reference gas make a suitable combination. The true retention time of the adsorbate is then obtained as the difference between the observed retention times of adsorbate and reference gas.

Isosteric heats of adsorption are measured by this technique by obtaining the retention times of the adsorbate as a function of temperature. The highest degree of accuracy is obtained when the average surface concentration of adsorbed gas in the column is low enough to be within the range described by the Henry's law portion of the adsorption isotherm. At higher surface concentrations, effects due to heterogeneity of the substrate appear, principally evident as an asymmetric effluent pulse, i.e., a "tailing-off" of the response to the desorbed gas. Substrates of low heterogeneity are less subject to this effect, although even there, at higher surface concentrations, the increased heat of adsorption caused by adsorbate-adsorbate interactions affects the symmetry of the pulse. To avoid uncertainty in the interpretation of the heat of adsorption as measured by the gas chromatograph, the concentration of the sample gas is kept so low that the heat of adsorption can be referred

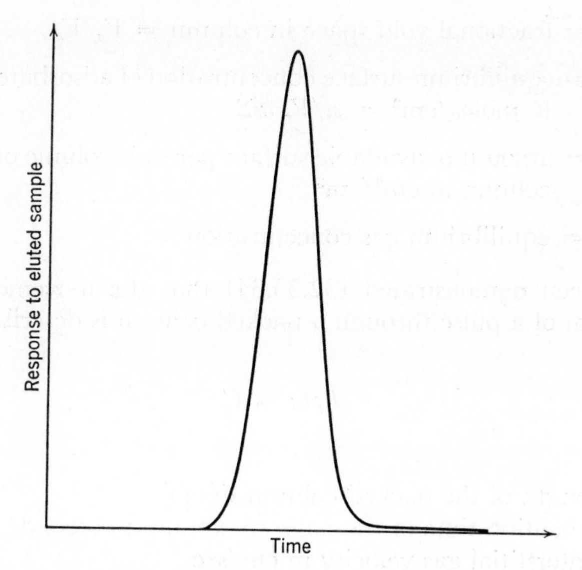

Fig. II-29. Typical elution peak of benzene desorbed from graphitized carbon black at 130°C.

to its limiting value as surface concentration tends to zero. The success of this precaution can be judged by the reproducibility of the measured retention times for a series of repeated experiments in which the sample gas is introduced at increasingly greater dilutions. Under the best conditions of low substrate heterogeneity and low equilibrium surface concentration, the elution peak will be nearly symmetrical. A good example is shown in Figure II-29, which is the elution response to benzene desorbed from graphitized carbon black at 130°C. Most elution peaks obtained in gas-solid chromatography are far from being so nearly symmetrical, chiefly because of the heterogeneous substrates usually employed.

2. *Theory*

Let ${}_gn$ be the total number of moles of gas in the column, in equilibrium with ${}_an$ moles of gas adsorbed; let V_g be the interstitial volume of the column and V_c the volume of the empty column; and let the column be filled with w grams per cm³ of packed adsorbent, having a specific surface of Σ cm²/gm: then

f = fractional void space in column $= V_g/V_c$ (II-42)

Γ = equilibrium surface concentration of adsorbate in moles/cm^2 $= {}_an/V_c\, w\Sigma$ (II-43)

$w\Sigma$ = amount of available surface per unit volume of column in cm^2/cm^3 (II-44)

${}_gn/V_g$ = equilibrium gas concentration (II-45)

It has been demonstrated (32,33,34) that the movement of the maximum of a pulse through a packed column is described by the relation

$$L/tu = C \qquad \text{(II-46)}$$

where

L = length of the packed column in cm
t = retention time of the pulse maximum in seconds
u = interstitial gas velocity in cm/sec

The quantity C, by means of equations II-42, 43, 44, 45, and 46, can be shown to be equivalent to

$$C = {}_gn/{}_an \qquad \text{(II-47)}$$

Let $\bar{p}$ = the average equilibrium pressure in the gas phase: then

$$\bar{p}\, V_g = {}_gnRT_c$$

where T_c is the temperature of the column: therefore,

$${}_gn = \bar{p}\, V_g/RT_c \qquad \text{(II-48)}$$

Let the average fraction of the adsorbent surface covered at pressure $\bar{p}$ be $\bar{\theta}$, and let us suppose a linear adsorption isotherm (Henry's law region), so that at equilibrium $\bar{p} = K\,\bar{\theta}$: then

$${}_an = (\bar{\theta}\, \Sigma\, w\, V_c/\beta N) \qquad \text{(II-49)}$$

where β is the two-dimensional van der Waals constant, which is assumed (35) to equal the limiting surface area of an adsorbed molecule at infinite compression. Combining equations II-47, 48, and 49 gives

$$C = (\beta N/\bar{\theta}\, \Sigma\, wV_c)\,(\bar{p}V_g/RT_c) \qquad \text{(II-50)}$$

Now let us introduce quantities that have been developed in the treatment of gas/solid adsorption:

$$V_\beta(\text{cm}^3 \text{ STP/g}) = (\Sigma/\beta N)\ 22{,}400 \qquad \text{(II-51)}$$

and

$$K\ (\text{mm Hg}) = \bar{p}/\theta \qquad \text{(II-52)}$$

This makes

$$C = (K/w\ V_\beta)\ (V_g/V_c)\ (22{,}400/RT_c) \qquad \text{(II-53)}$$

The practical application of equation II-46 requires that it be expressed in terms of quantities actually measured. The flow rate measured at room temperature (F_r) is used to calculate the actual flow rate (F_c) inside the column at temperature T_c:

$$F_c = F_r\ T_c/T_r \qquad \text{(II-54)}$$

The flow rate is sometimes further corrected to correspond to a zero pressure drop across the column, although for most work this correction is too small to be significant. The interstitial gas velocity is related to the volume flow rate by the expression

$$F_c = ua \qquad \text{(II-55)}$$

where a is the interstitial cross sectional area of the packed column. We then have also, of course, the equation

$$V_g = La \qquad \text{(II-56)}$$

Substituting equations II-54, 55, and 56 into equation II-46 gives

$$V_g T_r/t\ F_r T_c = C \qquad \text{(II-57)}$$

Introducing equation II-53 and rearranging gives

$$F_r\ t = (w\ V_c RT_r/22{,}400)\ (V_\beta/K) \qquad \text{(II-58)}$$

Note that the factor $w\ V_c\ V_\beta/22{,}400$ in equation II-58 has the significance of "total number of moles adsorbed by the packed column at infinite compression"; and that the slope of the adsorption isotherm is given by the factor K/V_β.

The parameters V_β and K in equation II-58 are significant for the description of the adsorption characteristics of the solid: V_β is

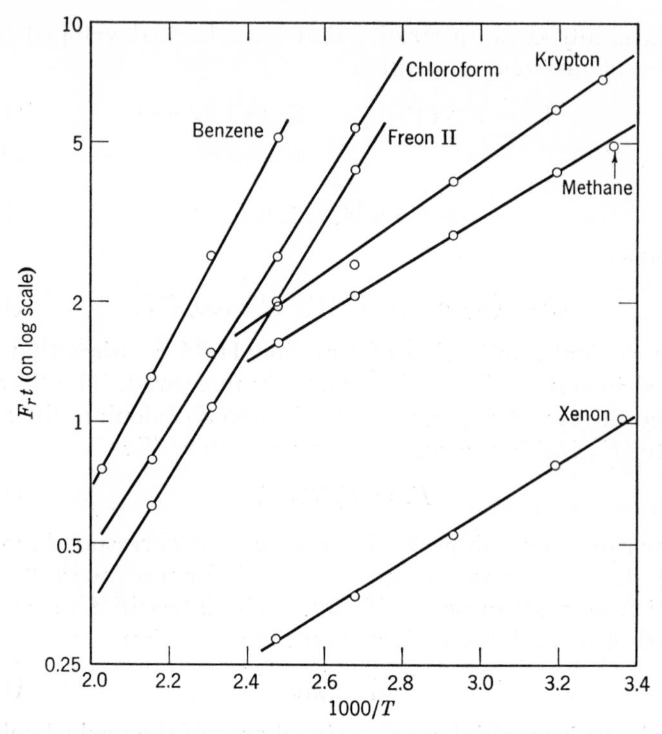

Fig. II-30. The variation of $\ln F_r t$ *vs* $1/T$ for a number of adsorbates on graphitized carbon black.

the specific monolayer capacity, hence, directly related to the specific surface area of the powder; and K is expressed by the relation

$$K = A^0 e^{-U_0/RT} \tag{II-59}$$

where A^0 and U_0 have been defined by equation I-32. The limiting isosteric heat of adsorption at infinitely dilute surface concentration can be obtained from the variation of K with temperature:

$$\lim_{\theta \to 0} q^{st} = RT^2 (d \ln K/dT) \tag{II-60}$$

or

$$\lim_{\theta \to 0} q^{st} = -R \, [d \ln K/d(1/T)]$$

hence, from equation II-58,

$$\lim_{\theta \to 0} q^{st} = R\,[d \ln (F_r\, t)/d(1/T)] \qquad \text{(II-61)}$$

The plot of $\ln(F_r\, t)$ *versus* $1/T$ should, therefore, be a straight line whose slope equals q^{st}/R. Previous attempts (36,37,38) to obtain $q^{st}\,(\theta \to 0)$ from chromatograms have not been accurate, as the presence of the factor T_c in the definition of C, equation II-57, was overlooked.

The variation of $\ln(F_r\, t)$ *versus* $1/T$, as determined experimentally (39) for a number of adsorbates on a graphitized carbon black, is shown in Figure II-30. The linear relation is clearly evident. The limiting heats of adsorption obtained from these data are reported in Tables VII-17 and VII-18, where they are compared with values derived from measurements of adsorption isotherms.

The chromatographic method has a particular advantage for the investigation of physical adsorption at high temperatures. In static systems, as contrasted with flow systems, the study is limited to those lower temperatures where decomposition of the adsorbate does not occur. In flow systems, however, the time of contact of the gas with the solid surface can be made quite short, thus minimizing or even eliminating decomposition reactions. Eberly (38) has shown, by this means, that physical adsorption of benzene and *n*-hexane still occurs even at temperatures well above their critical temperatures, which are 288.5 and 234.8 °C, respectively.

The chromatographic method, as here described, yields only the limiting isosteric heat, $\theta \to 0$. For an adsorbent that is nearly completely homotattic, then indeed we do have a measurement characteristic of the material, but such adsorbents unfortunately are exceedingly rare: the large number of adsorbents of current interest show surface heterogeneities that are usually quite pronounced. The limiting isosteric heat, $\theta \to 0$, may, under those conditions, be nothing more than the reflection of an adventitious surface condition of no particular significance beyond the specimen under examination. If the surface heterogeneity can be represented, however, by a continuous distribution of adsorptive potentials, U_0, that is symmetrical about a mean value, U_0', so that the distribution can be approximated by the Gaussian function,

$$\Phi(U_0) = (1/n)e^{-\gamma(U_0 - U_0')^2}$$

then it can be shown, Section V-4-B, that the limiting isosteric heat, $\theta \rightarrow 0$, is greater by an amount $1/(2\gamma RT)$ than the limiting isosteric heat of adsorption for a homotattic surface whose adsorptive potential is the same (U_0') as that of the average of the Gaussian distribution.

References

1. E. Wichers, *Rev. Sci. Instr.*, **13,** 502 (1942).
2. J. R. Partington, *An Advanced Treatise on Physical Chemistry*, Vol. 1, Longmans, Green, and Co., London, 1949, p. 267.
3. G. A. Miller, *J. Phys. Chem.*, **67,** 1359 (1963).
3a. Landolt-Bornstein, *Zahlenwerte und Funktionen*, Sections 13 241 and 13 242, Springer-Verlag, 1950–1951.
3b. H. H. Podgurski and F. N. Davis, *J. Phys. Chem.*, **65,** 1343 (1961).
4. G. Constabaris, J. H. Singleton, and G. D. Halsey, Jr., *J. Phys. Chem.*, **63,** 1350 (1959).
5. B. E. Blaisdell, *J. Math. Phys.*, **19,** 186 (1940); see also J. Kistemaker, *Physica*, **11,** 270 (1945).
6. B. B. Fisher and W. G. McMillan, *J. Chem. Phys.*, **28,** 549 (1958).
7. H. J. Hoge, *NBS-NACA Tables of Thermal Properties of Gases, Table 19.50*, July, 1950.
8. J. A. Morrison and D. M. Young, *Rev. Sci. Instr.*, **25,** 518 (1954).
9. G. Constabaris, J. H. Singleton, and G. D. Halsey, Jr., *J. Phys. Chem.*, **63,** 1350 (1959).
10. J. A. Hall in *Temperature, Its Measurement and Control in Science and Industry*, Vol. 2, Reinhold, New York, 1955, p. 113.
11. G. Constabaris, *Ph. D. thesis*, University of Washington, 1957.
12. G. W. Kanolt, *NBS Science Paper*, No. 520 (1926); also see R. E. Dodd and P. L. Robinson, *Experimental Inorganic Chemistry*, Elsevier, Amsterdam, 1954, p. 57.
13. S. Brunauer, P. H. Emmett, and E. Teller, *J. Am. Chem. Soc.*, **60,** 309 (1938).
14. P. H. Emmett in *Advances in Colloid Science*, Vol. 1, Interscience, New York, 1942, pp. 1–36.
15. P. H. Emmett and S. Brunauer, *J. Am. Chem. Soc.*, **59,** 1553 (1937).
16. L. A. Wooten and J. R. C. Brown, *J. Am. Chem. Soc.*, **65,** 113 (1943).
17. M. M. Dubinin, B. P. Bering, V. V. Serpinsky, and B. N. Vasil'ev in *Surface Phenomena in Chemistry and Biology*, Pergamon Press, New York, 1958, pp. 172–188.
18. T. N. Rhodin in *Advances in Catalysis*, Vol. 5, Academic Press, New York, 1953, p. 39.
19. S. J. Gregg, *J. Chem. Soc.*, **1955,** 1438; A. I. Sarkhov, *Bull. Acad. Sci. USSR, Div. Chem. Sci. (English Transl.)*, **1956,** 3.
20. H. B. Klevens, J. T. Carriel, R. J. Fries, and A. H. Peterson, *Proc. Second Intern. Congress Surface Activity*, Vol. 2, Academic Press, New York, 1957, p. 160.

21. W. Bushuk and C. A. Winkler, *Can. J. Chem.*, **33,** 1729 (1955).
22. R. A. Beebe in *Handbuch der Katalyse*, Vol. 4, G.-M. Schwab, ed., Springer-Verlag, Vienna, 1943, pp. 473–523.
23. G. H. Amberg, W. B. Spencer, and R. A. Beebe, *Can. J. Chem.*, **33,** 305 (1955).
24. R. A. Beebe and R. M. Dell, *J. Phys. Chem.*, **59,** 746 (1955).
25. I. M. Klotz, *Chemical Thermodynamics*, Prentice-Hall, New Jersey, 1950, p. 14.
26. J. A. Morrison and J. M. Los, *Faraday Soc. Discussions*, **1950,** 321.
27. J. J. Chessick and A. C. Zettlemoyer in *Advances in Catalysis*, Vol. 11, Academic Press, New York, 1959, pp. 263–299.
28. G. E. Boyd and W. D. Harkins, *J. Am. Chem. Soc.*, **64,** 1190 (1942).
29. A. C. Zettlemoyer, G. J. Young, J. J. Chessick, and F. H. Healey, *J. Phys. Chem.*, **57,** 649 (1953).
30. P. E. Berghausen in *Adhesion and Adhesives*, Clark, Rutzer, and Savage, eds., Wiley, New York, 1954, p. 225.
31. B. Millard, E. G. Caswell, E. E. Leger, and D. R. Mills, *J. Phys. Chem.*, **59,** 976 (195).
32. P. E. Eberly, Jr. and E. H. Spencer, *Trans. Faraday Soc.*, **57,** 289 (1961).
33. A. J. P. Martin and R. L. M. Synge, *Biochem. J.* (*London*), **35,** 1358 (1941).
34. S. A. Greene in R. L. Pecsok, ed., *Principles and Practice of Gas Chromatography*, Wiley, New York, 1959, pp. 29–30.
35. J. H. de Boer, *The Dynamical Character of Adsorption*, Clarendon Press, Oxford, 1953, pp. 172–4.
36. S. A. Greene and H. Pust, *J. Phys. Chem.*, **62,** 55 (1958).
37. J. F. Young in V. J. Coates, H. J. Noebels, and I. S. Fagerson, eds., *Gas Chromatography*, Academic Press, New York, 1958, pp. 15–23.
38. P. E. Eberly, Jr., *J. Phys. Chem.* **65,** 68 (1961).
39. S. Ross, J. K. Saelens, and J. P. Olivier, *J. Phys. Chem.* **66,** 696 (1962).

CHAPTER III

The Various Heats of Adsorption

We have considered in Chapter I the energetics of the adsorption process and described it in terms of energy changes, such as ${}_aE^{\text{vib}}$, ${}_gP^{\text{ads}}$, ΔE^{rot}, etc.; these terms give us a description of what is taking place, but they are not by themselves readily observable experimentally. The experimental determination of the energetics of adsorption usually proceeds by measuring a differential heat of adsorption, which may be the isosteric heat q^{st}, the adiabatic or isothermal calorimetric heats, or the integral heat of adsorption. These observed quantities necessarily lump together in different combinations that are peculiar to the experimental method all the individual potential and kinetic energy terms. Before we can hope to compare experimental observations with the conclusions drawn from any theory of adsorption regarding the energetics of the process, we must have firm definitions of the observed quantities in terms of the potential and kinetic energy changes that occur on adsorption.

1. The Integral and Differential Heats of Adsorption

Consider a process taking place in an isolated system, consisting of a container and a thermostat of infinite capacity in which ${}_an$ moles of an adsorbate originally at a pressure p_1 is transferred isothermally from the gas phase to the surface of an adsorbent that was previously in a vacuum. Let the new equilibrium pressure be p_2. The process is represented diagrammatically in Figure III-1. The initial and final conditions are as follows:

Initial	*Final*
Energy of gas $= {}_gE$ per mole	Energy of adsorbate $= {}_aE$ per mole
Energy of bath $= E_1$	Energy of bath $= E_2$
Temperature $= T$	Temperature $= T$
n moles of an ideal gas at $p = p_1$	$(n - {}_an)$ moles of an ideal gas at $p = p_2$
	${}_an$ moles adsorbed

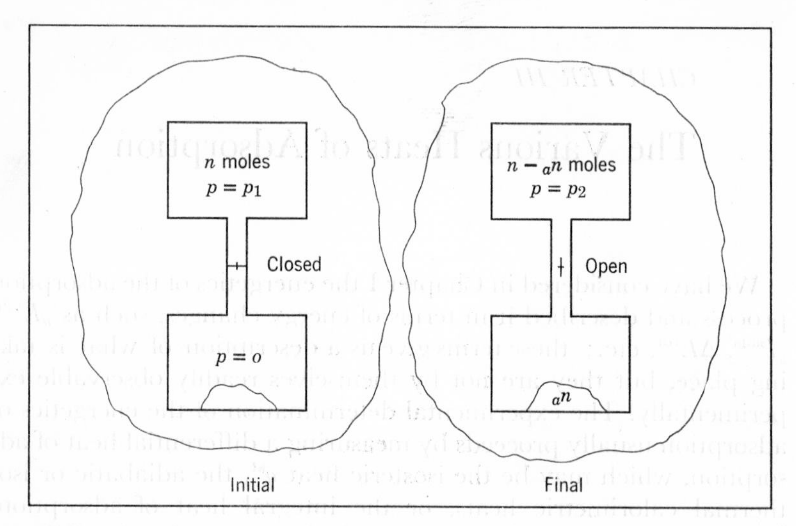

Fig. III-1. Diagram representing isothermal adsorption in an isolated system.

The integral heat of adsorption, q^{int}, is defined as the heat absorbed by the bath in the above process. Since the system is isolated, total energy is conserved. Therefore, the initial total energy of system equals the final total energy of system:

$$E_1 + (n)({}_gE) = E_2 + (n - {}_an)({}_gE) + ({}_an)({}_aE)$$

By the first law of thermodynamics,

$$q^{\text{int}} = (E_2 - E_1) + w$$

As no work is done, $w = 0$; therefore,

$$q^{\text{int}} = E_2 - E_1 = {}_an({}_gE - {}_aE) \qquad \text{(III-1)}$$

Since $\theta = {}_an/{}_an_m$ by equation II-31, equation III-1 becomes

$$q^{\text{int}} = ({}_an_m)\theta({}_gE - {}_aE) \qquad \text{(III-2)}$$

The differential heat of adsorption, q^{diff}, is defined as

$$q^{\text{diff}} = (\partial q^{\text{int}}/\partial_a n)_{T,\Sigma} = (1/{}_an_m)(\partial q^{\text{int}}/\partial\theta)_{T,\Sigma} \qquad \text{(III-3)}$$

hence,

$$q^{\text{diff}} = {}_gE - {}_aE - \theta(\partial_a E/\partial\theta)_{T,\Sigma} \qquad \text{(III-4)}$$

2. The Isothermal Heat of Adsorption

Imagine a system consisting of a thermostat bath of infinite capacity and a container, one side of which is a frictionless piston; inside the container is an ideal gas in equilibrium with an adsorbent. We are to consider the change in the energy of the bath that occurs when the volume of the container is decreased, by means of the piston, by an infinitesimal amount.

The initial and final conditions are as follows:

Initial	*Final*
Moles of gas $= {}_gn = (n - {}_an)$	Moles of gas $= n - {}_an - d({}_an)$
Moles adsorbed $= {}_an$	Moles adsorbed $= {}_an + d({}_an)$
Equilibrium pressure $= p$	Equilibrium pressure $= p + dp$
Volume $= V$	Volume $= V - d({}_gV)$
Total energy of bath $= E_1$	Total energy of bath $= E_1 + dE_1$
Energy of gas $= {}_gE$ per mole	Energy of gas $= {}_gE$ per mole
Energy of adsorbate $= {}_aE$ per mole	Energy of adsorbate $= {}_aE + d({}_aE)$ per mole

To carry out the process just described, external work must be done on the system; therefore, the total initial energy of system $+ dw$ equals the total final energy of system:

$$dE_{\text{total}} = -dw$$

$$E_1 + dE_1 + [n - {}_an - d({}_an)]({}_gE) + [{}_an + d({}_an)][{}_aE + d({}_aE)] - [E_1 + (n - {}_an)({}_gE) + ({}_an)({}_aE)] = -dw$$

$$dE_1 - ({}_gE)d({}_an) + ({}_aE)d({}_an) + ({}_an)d({}_aE) = -dw$$

The isothermal heat of adsorption, q^{th}, is defined (1) as

$$q^{\text{th}} = (\partial E_1/\partial({}_an))_{T,\Sigma} \tag{III-5}$$

Hence,

$$q^{\text{th}} = {}_gE - {}_aE - {}_an(\partial({}_aE)/\partial({}_an))_{T,\Sigma} - (\partial w/\partial({}_an))_{T,\Sigma}$$

or

$$q^{\text{th}} = {}_gE - {}_aE - \theta(\partial_a E/\partial\theta)_{T,\Sigma} - (\partial w/\partial({}_an))_{T,\Sigma} \tag{III-6}$$

Comparing equations III-4 and III-6 shows us that

$$q^{\text{th}} = q^{\text{diff}} - (\partial w/\partial({}_an))_{T,\Sigma} \tag{III-7}$$

The total work done on the system is the isothermal heat of compression of the gas in the calorimeter; it is $dw = pd({}_gV)$: for an ideal gas

$$pd({}_gV) + ({}_gV)dp = RTd({}_gn)$$

and since

$$d({}_gn) = -d({}_an)$$

then

$$dw = pd({}_gV) = -RTd({}_an) - ({}_gV)dp$$

or

$$dw/d({}_an) = -RT - ({}_gV)[dp/d({}_an)] \qquad \text{(III-8)}$$

The expression for the isothermal heat becomes, by substituting equation III-8 in equation III-7,

$$q^{\text{th}} = q^{\text{diff}} + RT + ({}_gV)[dp/d({}_an)] \qquad \text{(III-9)}$$

Equation III-9 is frequently but erroneously given as

$$q^{\text{th}} = q^{\text{diff}} + RT$$

The last term of equation III-9 is found experimentally as the slope of the adsorption isotherm multiplied by the volume of the gas in the calorimeter. The heat measured by an isothermal calorimeter is q^{th}, from which q^{diff} can be calculated by means of equation III-9.

3. The Adiabatic Heat of Adsorption

Imagine a system consisting of a perfectly insulated container, one side of which is a frictionless piston; inside the container is an ideal gas in equilibrium with the adsorbent. We are to consider the change of energy that occurs in the container when the volume is decreased, by means of the piston, by an infinitesimal amount. The argument is essentially that of Kington and Aston (1). The initial and final conditions are as follows:

Initial	*Final*
Moles of gas $= {}_g n = n - {}_a n$	Moles of gas $= n - {}_a n - d({}_a n)$
Moles adsorbed $= {}_a n$	Moles adsorbed $= {}_a n + d({}_a n)$
Equilibrium pressure $= p$	Equilibrium pressure $= p + dp$
Volume $= V$	Volume $= V - d({}_g V)$
Energy of gas $= {}_g E$ per mole	Energy of gas $= {}_g E$ per mole
Energy of adsorbate $= {}_a E$ per mole	Energy of adsorbate $= {}_a E + d_a E$ per mole
Total energy of system $= E$	Total energy of system $= E + dE$
Temperature $= T$	Temperature $= T + dT$

In the initial state the total energy, E, of the system is given by

$$E = {}_s E + (n - {}_a n)({}_g E) + ({}_a n)({}_a E) \qquad \text{(III-10)}$$

where ${}_s E$ is the energy of the solid phase, i.e., container plus the adsorbent; taking the derivative of equation III-10 gives

$$dE = d({}_s E) + (n - {}_a n)d({}_g E) - {}_g E\, d({}_a n) + {}_a n\, d({}_a E) + {}_a E\, d({}_a n) \qquad \text{(III-11)}$$

To evaluate the energy change given by equation III-11 we need expressions for $d({}_s E)$, $d({}_g E)$, and $d({}_a E)$. We shall assume that ${}_s E$ is a function of temperature only; then

$$d({}_s E) = {}_s C_v\, dT \qquad \text{(III-12)}$$

where ${}_s C_v$ is the heat capacity at constant volume of the container plus adsorbent: also

$$d({}_g E) = {}_g C_v\, dT \qquad \text{(III-13)}$$

and, since the energy of the adsorbed phase is a function of T and ${}_a n$,

$$d({}_a E) = [\partial({}_a E)/\partial T]_{an}\, dT + [\partial({}_a E)/\partial({}_a n)]_T\, d({}_a n)$$

or

$$d({}_a E) = {}_a C_v\, dT + [\partial({}_a E)/\partial({}_a n)]_T\, d({}_a n) \qquad \text{(III-14)}$$

Combining equations III-12, 13, and 14 with equation III-11:

$$dE = {}_s C_v\, dT + (n - {}_a n){}_g C_v\, dT - {}_g E\, d({}_a n) + {}_a n\, {}_a C_v\, dT + {}_a n\, [\partial({}_a E)/\partial({}_a n)]_T\, d({}_a n)$$

or

$$dE = [{}_sC_v + (n - {}_an)\,{}_gC_v + ({}_an)\,{}_aC_v]dT - {}_gE\,d({}_an) + {}_an[\partial({}_aE)/\partial({}_an)]_T\,d({}_an) + {}_aE\,d({}_an) \quad \text{(III-15)}$$

The adiabatic heat of adsorption is defined by the expression

$$q^{\text{ab}} = (\partial E/\partial T)_{an}[\partial T/\partial({}_an)]_S \quad \text{(III-16)}$$

From equation III-15,

$$(\partial E/\partial T)_{an} = {}_sC_v + (n - {}_an){}_gC_v + ({}_an){}_aC_v$$

and

$$(\partial E/\partial_a n)_S = q^{\text{ab}} - \{{}_gE - {}_aE - {}_an\,[\partial({}_aE)/\partial({}_an)]_T\} \quad \text{(III-17)}$$

Recalling equation III-4 allows us to express equation III-17 as follows:

$$(\partial E/\partial_a n)_S = q^{\text{ab}} - q^{\text{diff}} \quad \text{(III-18)}$$

The process described is adiabatic, therefore

$$dE = -p\,d({}_gV)$$

hence,

$$[\partial E/\partial({}_an)]_S = -p[\partial({}_gV)/\partial({}_an)]_S \quad \text{(III-19)}$$

Substituting equation III-19 in equation III-18 gives

$$q^{\text{ab}} = q^{\text{diff}} - p[\partial({}_gV)/\partial({}_an)]_S \quad \text{(III-20)}$$

Equation III-20 relates the measured adiabatic heat of adsorption to the differential heat of adsorption. The difference between the two is the adiabatic heat of compression, which must be used as a correction to the measured heat in order to produce the more basic quantity. To make this correction, the expression for the adiabatic heat of compression must be in terms that can be evaluated experimentally. This is done as follows. For an ideal gas,

$$p\left[\frac{\partial({}_gV)}{\partial({}_gn)}\right]_S = RT + {}_gnR\left[\frac{\partial T}{\partial({}_gn)}\right]_S - {}_gV\left[\frac{\partial p}{\partial({}_gn)}\right]_S$$

and, since $d({}_gn) = -d({}_an)$, then

$$-p\left[\frac{\partial({}_gV)}{\partial({}_an)}\right]_S = RT - {}_gnR\left[\frac{\partial T}{\partial({}_an)}\right]_S + {}_gV\left[\frac{\partial p}{\partial({}_an)}\right]_S \quad \text{(III-21)}$$

Now, since $p = p(T, {}_an)$,

$$\left[\frac{\partial p}{\partial({}_an)}\right]_S = \left(\frac{\partial p}{\partial T}\right)_{{}_an}\left[\frac{\partial T}{\partial({}_an)}\right]_S + \left[\frac{\partial p}{\partial({}_an)}\right]_T \quad \text{(III-22)}$$

Combining equations III-21 and 22 gives

$$-p\left[\frac{\partial({}_gV)}{\partial({}_an)}\right]_S = RT + {}_gV\left[\frac{\partial p}{\partial({}_an)}\right]_T + \left[{}_gV\left(\frac{\partial p}{\partial T}\right)_{{}_an} - {}_gnR\right]\left[\frac{\partial T}{\partial({}_an)}\right]_S \quad \text{(III-23)}$$

Equation III-23 provides us with an expression for the adiabatic heat of compression that can be evaluated experimentally from the slope of the adsorption isotherm, the slope of the adsorption isostere, and the temperature rise of the calorimeter vessel per mole adsorbed, all at the particular amount adsorbed to which q^{ab} refers.

4. The Isosteric Heat of Adsorption

The isosteric heat of adsorption is defined by the equation

$$(\partial \ln p/\partial T)_{{}_an} = [q^{st}/T({}_gV - {}_aV)] \quad \text{(III-24)}$$

where q^{st} is the isosteric heat of adsorption, ${}_gV$ and ${}_aV$ are the molar volumes of the gas phase and the adsorbed phase, respectively. With a view to our ultimate application of q^{st} we shall assume, following the argument given in Section II-2-A, that

$$(\partial \ln p/\partial T)_{{}_an} = (\partial \ln p/\partial T)_\theta$$

We shall evaluate q^{st} by deriving an exact thermodynamic expression for $(\partial \ln p/\partial T)_\theta$ following Hill (2).

At equilibrium between the gas and an adsorbed two-dimensional phase

$${}_g\mu = {}_a\mu$$

where ${}_g\mu$ and ${}_a\mu$ are the chemical potentials of the gas and adsorbed phases, respectively. If we change conditions infinitesimally to a new equilibrium state, then

$$d_g\mu = d_a\mu$$

hence:

$$\left[\frac{\partial({}_g\mu)}{\partial p}\right]_T dp + \left[\frac{\partial({}_g\mu)}{\partial T}\right]_p dT = \left[\frac{\partial({}_a\mu)}{\partial \theta}\right]_{p,T} d\theta + \left[\frac{\partial({}_a\mu)}{\partial p}\right]_{\theta,T} dp + \left[\frac{\partial({}_a\mu)}{\partial T}\right]_{\theta,p} dT$$

At constant θ, therefore:

$$\left(\frac{\partial p}{\partial T}\right)_\theta = \left[\left(\frac{\partial_a\mu}{\partial T}\right)_{\theta,p} - \left(\frac{\partial_g\mu}{\partial T}\right)_p\right] \Big/ \left\{\left[\frac{\partial({}_g\mu)}{\partial p}\right]_T - \left[\frac{\partial({}_a\mu)}{\partial p}\right]_{\theta,T}\right\} \tag{III-25}$$

Now

$$[\partial({}_g\mu)/\partial T]_p = ({}_g\mu - {}_gH)/T \tag{III-26}$$

and

$$[\partial({}_g\mu)/\partial p]_T = {}_gV \tag{III-27}$$

where ${}_gH$ is the molar enthalpy and ${}_gV$ is the molar volume. Considering ${}_a\mu = {}_a\mu(\Pi, p, T)$ and $\Pi = \Pi(\theta, p, T)$ we have

$$\left[\frac{\partial({}_a\mu)}{\partial T}\right]_{\theta,p} = \left[\frac{\partial({}_a\mu)}{\partial T}\right]_{p,\Pi} + \left[\frac{\partial({}_a\mu)}{\partial \Pi}\right]_{p,T}\left(\frac{\partial \Pi}{\partial T}\right)_{\theta,p}$$

and

$$\left[\frac{\partial({}_a\mu)}{\partial p}\right]_{\theta,T} = \left[\frac{\partial({}_a\mu)}{\partial p}\right]_{\Pi,T} + \left[\frac{\partial({}_a\mu)}{\partial \Pi}\right]_{p,T}\left(\frac{\partial \Pi}{\partial p}\right)_{\theta,T}$$

Using the following five relations:

$${}_a\mu = \left[\frac{\partial({}_aF)}{\partial({}_an)}\right]_{T,p,\Pi}; \quad d({}_aF) = ({}_aV)dp - ({}_aS)dT + (A)\,d\Pi\,;$$

$$\left[\frac{\partial({}_aF)}{\partial p}\right]_{T,\Pi} = {}_aV; \quad \left[\frac{\partial({}_aF)}{\partial T}\right]_{p,\Pi} = -\,{}_aS;$$

and ${}_aH = {}_aF + T({}_aS)$; the previous equations become

$$[\partial({}_a\mu)/\partial T]_{\theta,p} = [({}_a\mu - {}_aH)/T] + (\beta/\theta)(\partial\Pi/\partial T)_{\theta,p} \tag{III-28}$$

and

$$[\partial({}_a\mu)/\partial p]_{\theta,T} = {}_aV + (\beta/\theta)(\partial\Pi/\partial p)_{\theta,T} \qquad \text{(III-29)}$$

where β is the limiting area per mole of adsorbed phase.

Substituting equations III-26, 27, 28, and 29 in equation III-25 gives

$$\left(\frac{\partial p}{\partial T}\right)_\theta = \frac{\left(\frac{{}_a\mu - {}_aH}{T}\right) + \frac{\beta}{\theta}\left(\frac{\partial\Pi}{\partial T}\right)_{\theta,p} - \left(\frac{{}_g\mu - {}_gH}{T}\right)}{{}_gV - \left[{}_aV + \frac{\beta}{\theta}\left(\frac{\partial\Pi}{\partial p}\right)_{\theta,T}\right]}$$

Since ${}_g\mu = {}_a\mu$,

$$\left(\frac{\partial p}{\partial T}\right)_\theta = \frac{({}_gH - {}_aH) + \frac{T\beta}{\theta}\left(\frac{\partial\Pi}{\partial T}\right)_{\theta,p}}{T\left[{}_gV - {}_aV - \frac{\beta}{\theta}\left(\frac{\partial\Pi}{\partial p}\right)_{\theta,T}\right]} \qquad \text{(III-30)}$$

The above equation is exact; by ignoring pressure effects, dropping ${}_aV$ and assuming an ideal gas we obtain:

$$\left(\frac{\partial \ln p}{\partial T}\right)_\theta = \frac{1}{RT^2}\left[{}_gH - {}_aH + \frac{T\beta}{\theta}\left(\frac{\partial\Pi}{\partial T}\right)_\theta\right] = \frac{q^{st}}{RT^2}$$

therefore,

$$q^{st} = {}_gH - {}_aH + (T\beta/\theta)(\partial\Pi/\partial T)_\theta \qquad \text{(III-31)}$$

Equation III-31 is equivalent to equation II-30 previously derived.

5. The Relation between the Isosteric and Differential Heats of Adsorption

The isosteric heat of adsorption is given by equation III-31

$$q^{st} = {}_gH - {}_aH + (T\beta/\theta)(\partial\Pi/\partial T)_\theta \qquad \text{(III-31)}$$

For a perfect gas

$${}_gH = {}_gE + p\,{}_gV = {}_gE + RT \qquad \text{(III-32)}$$

and for the surface phase

$${}_aH = {}_aE + (\Pi\beta/\theta) \qquad \text{(III-33)}$$

Substituting equations III-32 and 33 in equation III-31 yields

$$q^{\text{st}} = {}_gE + RT - {}_aE - (\Pi\beta/\theta) + (T\beta/\theta)(\partial\Pi/\partial T)_\theta$$

or

$$q^{\text{st}} - RT = {}_gE - {}_aE - (\Pi\beta/\theta) + (T\beta/\theta)(\partial\Pi/\partial T)_\theta \quad \text{(III-34)}$$

The energy equation in two dimensions is

$$(\partial\,({}_aE)/\partial\sigma)_T = -\,\Pi + T(\partial\Pi/\partial T)_\sigma \quad \text{(III-35)}$$

using $\theta = \beta/\sigma$,

$$-\,\theta(\partial({}_aE)/\partial\theta)_{T,\Sigma} = -(\Pi\beta/\theta) + (T\beta/\theta)(\partial\Pi/\partial T)_\theta \quad \text{(III-36)}$$

Substituting equation III-36 in equation III-34 yields

$$q^{\text{st}} - RT = {}_gE - {}_aE - \theta(\partial({}_aE)/\partial\theta)_{T,\Sigma} = q^{\text{diff}} \quad \text{(III-37)}$$

6. The Relation of the Adsorptive Potential U_0 to the Differential Heat of Adsorption on a Homotattic Surface

The function U_0 has already been defined as equation I-30, the potential energy difference between the lowest energy state of a molecule in the gas phase and its lowest energy state in the adsorbed phase: i.e.,

$$U_0 = {}_gP^{\text{ads}} - {}_aE_0{}^{\text{vib}} \quad \text{(III-38)}$$

To develop the relation between U_0 and q^{diff} for a uniform surface, let the total energy of the gas phase at zero pressure consist of its kinetic energy plus the potential energy of adsorption, i.e.,

$${}_g^0E = {}_gP^{\text{ads}} + {}_gE^{\text{kin}}$$

where ${}_gE^{\text{kin}}$ is the total kinetic energy per mole of a molecule in a gas, which, for an ideal gas would be (3/2) RT.

Let ${}_aE^{\text{vib}}$ be the average vibrational energy per mole of a molecule in the adsorbed phase; then, for a mobile adsorbed film, the total energy (excluding interaction) per mole of a molecule in the surface film, ${}_a^0E$, is given by

$${}_a^0E = {}_aE^{\text{vib}} + {}_aE^{\text{kin}}$$

where ${}_aE^{\text{kin}}$ is the kinetic energy of translation and rotation of the adsorbed molecule. The total energy change on adsorption is

$$ {}^0_gE - {}^0_aE = {}_gP^{\text{ads}} - {}_aE^{\text{vib}} - \Delta E^{\text{kin}} \tag{III-39}$$

where

$$ -\Delta E^{\text{kin}} = {}_gE^{\text{kin}} - {}_aE^{\text{kin}} $$

then

$$ {}^0_gE - {}^0_aE = ({}_gP^{\text{ads}} - {}_aE_0^{\text{vib}}) - ({}_aE^{\text{vib}} - {}_aE_0^{\text{vib}}) - \Delta E^{\text{kin}} \tag{III-40}$$

where ${}_aE_0^{\text{vib}}$ is the zero-point vibrational energy of the adsorbed molecule, equivalent to $(^1/_2)Nh\nu$ where ν is the vibrational frequency of the adsorbed molecule with respect to the surface. Substituting equation III-38 in equation III-40 gives

$$ {}^0_gE - {}^0_aE = U_0 - ({}_aE^{\text{vib}} - {}_aE_0^{\text{vib}}) - \Delta E^{\text{kin}} \tag{III-41}$$

Let us recall equation III-4, which we can write as

$$ {}_gE - {}_aE = q^{\text{diff}} + {}_an[\partial({}_aE)/\partial({}_an)]_T \tag{III-42}$$

let us put

$$ {}^0_gE = {}_gE \tag{III-43}$$

which is to say that the gas phase is ideal; and, as at constant temperature ${}_aE$ is a function only of ${}_an$, then

$$ {}^0_aE = {}_aE - \int_0^{{}_an} [\partial({}_aE)/\partial({}_an)]_T \, d({}_an) \tag{III-44}$$

Combining equations III-41, 42, 43, and 44 gives

$$ q^{\text{diff}} = U_0 - ({}_aE^{\text{vib}} - {}_aE_0^{\text{vib}}) - \Delta E^{\text{kin}} $$
$$ - \int_0^{{}_an} \left(\frac{\partial({}_aE)}{\partial({}_an)}\right)_T d({}_an) - {}_an\left(\frac{\partial({}_aE)}{\partial({}_an)}\right)_T \tag{III-45}$$

Equation III-45 is the general expression for the relation between q^{diff} and U_0. Three special cases are of interest: (1) at low temperatures we can put ${}_aE^{\text{vib}} \simeq {}_aE_0^{\text{vib}}$; (2) as surface concentration approaches zero, the last two terms of equation III-45 approach zero; (3) when the surface phase behaves as a two-dimensional van der Waals gas (equation I-18). In the latter case,

$$ [\partial({}_aE)/\partial({}_an)]_T = -(\alpha/\beta)(1/{}_an_\beta) \tag{III-46}$$

and hence,

$$\int_0^{{}_a n} [\partial({}_aE)/\partial({}_an)]_T \, d({}_an) = -\alpha\theta/\beta \qquad \text{(III-47)}$$

and

$${}_an[\partial({}_aE)/\partial({}_an)]_T = -\alpha\theta/\beta \qquad \text{(III-48)}$$

combining equations III-45, 47, and 48 gives

$$q^{\text{diff}} = U_0 - ({}_aE^{\text{vib}} - {}_aE_0^{\text{vib}}) - \Delta E^{\text{kin}} + 2\alpha\theta/\beta \quad \text{(III-49)}$$

We have already shown, in equation I-41, that $2\alpha\theta/\beta = {}_aP^{\text{ia}}$, where ${}_aP^{\text{ia}}$ is the cooperative addition to the adsorptive potential, more commonly called the lateral interaction. In general, the last two terms of equation III-45 are always equal to ${}_aP^{\text{ia}}$, so that the general expression relating q^{diff} and U_0 can be written

$$q^{\text{diff}} = U_0 - ({}_aE^{\text{vib}} - {}_aE_0^{\text{vib}}) - \Delta E^{\text{kin}} + {}_aP^{\text{ia}} \quad \text{(III-50)}$$

The only term in equation III-50 that is dependent on θ is ${}_aP^{\text{ia}}$. We may write, therefore, that

$$q^{\text{diff}} = \lim_{\theta\to 0} q^{\text{diff}} + {}_aP^{\text{ia}} \qquad \text{(III-51)}$$

also

$$q^{\text{st}} = \lim_{\theta\to 0} q^{\text{st}} + {}_aP^{\text{ia}} \qquad \text{(III-52)}$$

The last two terms in equation III-45 can be expressed equally well in terms of θ where $\theta = {}_an/{}_an_\beta$:

$$\int_0^{{}_an} [\partial({}_aE)/\partial({}_an)]_T \, d({}_an) = \int_0^{\theta} [\partial({}_aE)/\partial\theta]_T \, d\theta$$

and

$${}_an[\partial({}_aE)/\partial({}_an)]_T = \theta[\partial({}_aE)/\partial\theta]_T$$

To prove that

$$-\theta[\partial({}_aE)/\partial\theta]_T - \int_0^{\theta} [\partial({}_aE)/\partial\theta]_T \, d\theta = {}_aP^{\text{ia}} \qquad \text{(III-53)}$$

assume the result; then

$$[\partial({}_aP^{\text{ia}})/\partial\theta]_T = -2[\partial({}_aE)/\partial\theta]_T - \theta[\partial^2({}_aE)/\partial\theta^2]_T$$

for the two-dimensional phase,

$$[\partial({}_aE)/\partial\sigma]_T = T(\partial\Pi/\partial T)_\sigma - \Pi$$

and

$$[\partial({}_aE)/\partial\sigma]_T = -(\beta/\sigma^2)\,[\partial({}_aE)/\partial\theta]_T = -(\theta^2/\beta)\,[\partial({}_aE)/\partial\theta]_T$$

since

$$\theta = \beta/\sigma$$

therefore,

$$[\partial({}_aE)/\partial\theta]_T = -(\beta T/\theta^2)(\partial\Pi/\partial T)_\theta + \beta\Pi/\theta^2 \qquad \text{(III-54)}$$

and

$$\left[\frac{\partial^2({}_aE)}{\partial\theta^2}\right]_T = \frac{2\beta T}{\theta^3}\left(\frac{\partial\Pi}{\partial T}\right)_\theta - \frac{\beta T}{\theta^2}\left(\frac{\partial^2\Pi}{\partial T\partial\theta}\right)_{\theta,T} + \frac{\beta}{\theta^2}\left(\frac{\partial\Pi}{\partial\,\theta}\right)_T - \frac{2\beta\Pi}{\theta^3}$$

therefore,

$$\left[\frac{\partial({}_aP)^{\text{ia}}}{\partial\theta}\right]_T = \frac{\beta T}{\theta}\left(\frac{\partial^2\Pi}{\partial T\partial\theta}\right)_{\theta,T} - \frac{\beta}{\theta}\left(\frac{\partial\Pi}{\partial\,\theta}\right)_T \qquad \text{(III-55)}$$

To evaluate the terms on the right-hand side, we write

$$[\partial\Pi/\partial\theta]_T = [\partial\Pi/\partial \ln p]_T\,[\partial \ln p/\partial\theta]_T$$

and use Gibbs' equation in the form $(\partial\Pi/\partial \ln p)_T = RT\theta/\beta$, for which see equation I-4; then

$$(\partial\Pi/\partial\theta)_T = (RT\theta/\beta)\,(\partial \ln p/\partial\theta)_T \qquad \text{(III-56)}$$

and

$$\left(\frac{\partial^2\Pi}{\partial T\partial\theta}\right)_{\theta,T} = \frac{R\theta}{\beta}\left(\frac{\partial p}{\partial\theta}\right)_T + \frac{RT\theta}{\beta}\left(\frac{\partial^2 \ln p}{\partial T\partial\theta}\right)_{\theta,T} \qquad \text{(III-57)}$$

Substitute equations III-56 and 57 in equation III-55; then

$$[\partial({}_aP^{\text{ia}})/\partial\theta]_T = RT^2\,[\partial^2 \ln p/\partial T\partial\theta]_{\theta,T} \qquad \text{(III-58)}$$

Equation III-58 is equivalent to equation I-39, by which ${}_aP^{\text{ia}}$ was originally defined: the assumption that equation III-53 is true is therefore justified.

The importance of equation III-50 is that it expresses the differential heat of adsorption in terms of a number of concepts that have a readily visualized physical basis; it reasserts and emphasizes that the differential heat of adsorption contains *inter alia* separate expressions for the adsorbate-adsorbent interaction and the adsorbate-adsorbate interaction; and since all the other experimentally determined heats of adsorption are related to q^{diff}, the same conclusion also holds true for them. The quantity U_0, expressing as it does the adsorbate-adsorbent interaction stripped of all other incidental energy changes such as lateral interaction, work terms, and kinetic and vibrational energy changes, is more suitable than any of the experimentally measured heats as an index of the fundamental "affinity" of a solid surface for adsorbing a particular gas molecule.

7. The Integral and Differential Enthalpy of Adsorption

The quantities previously described have been thermodynamically the heats or energies of adsorption, which are to be distinguished from the enthalpy change on adsorption. The total integral enthalpy of adsorption is the change of enthalpy on the isothermal adsorption of $_an$ moles of gas, initially in its standard state, by a bare surface to an equilibrium surface concentration θ and gas pressure p; the molar integral enthalpy, ΔH^{ads}, is defined as the total integral enthalpy divided by the number of moles adsorbed; the differential enthalpy of adsorption $\Delta \dot{H}^{\text{ads}}$, which also is a molar quantity, is the differential of the total enthalpy change with respect to the number of moles adsorbed, at a particular equilibrium pressure and surface concentration, i.e.,

$$\Delta \dot{H}^{\text{ads}} = d({_an}\,\Delta H^{\text{ads}})/d\,{_an} \tag{III-59}$$

All the thermodynamic functions can be expressed in the same way as total integral, molar integral, and molar differential quantities. The possibility of confusion is obvious. In this section we shall develop the relations of integral and differential enthalpies to quantities that can be determined experimentally.

The integral molar enthalpy change of adsorption is

$$\Delta H^{\text{ads}} = {_aH} - {_gH} \tag{III-60}$$

ΔH^{ads} is related to the isosteric heat by equation III-31, which can be rearranged as

$$\Delta H^{\text{ads}} = -q^{\text{st}} + (T\beta/\theta)(\partial\Pi/\partial T)_\theta \tag{III-61}$$

or

$$\Delta H^{\text{ads}} = -q^{\text{diff}} - RT + (T\beta/\theta)(\partial\Pi/\partial T)_\theta \tag{III-62}$$

By a convention of usage in this subject, the heat liberated on adsorption is taken as positive; the thermodynamic convention, by which heat *absorbed* by the system is positive, accounts for the difference in sign between q^{st} and ΔH^{ads}. The ΔH^{ads} in the above equations is an integral molar quantity; the differential molar enthalpy, $\Delta\dot{H}^{ads}$, is calculated from it by equation III-59. Applied to equation III-62 we get

$$\Delta\dot{H}^{ads} = -q^{diff} - {}_{a}n\left(\frac{\partial q^{diff}}{\partial_{a}n}\right)_T - RT + \beta T\left(\frac{\partial}{\partial\theta}\right)_T\left(\frac{\partial\Pi}{\partial T}\right)_\theta \qquad \text{(III-63)}$$

We shall evaluate both ΔH^{ads} and $\Delta\dot{H}^{ads}$ for each of the models of adsorption whose mathematical description is tabulated in Table I-1.

Case 1. The Two-Dimensional Ideal Gas. For this case, $(T\beta/\theta)$-$(\partial\Pi/\partial T)_\theta = RT$; hence, by equation III-62,

$$\Delta H^{ads} = -q^{diff} \qquad \text{(III-64)}$$

For the differential enthalpy we evaluate $(\partial q^{diff}/\partial_{a}n)_T = 0$ and $\beta T(\partial/\partial\theta)_T(\partial\Pi/\partial T)_\theta = RT$; hence,

$$\Delta\dot{H}^{ads} = -q^{diff} \qquad \text{(III-65)}$$

Case 2. The Volmer Equation. For this case, $(T\beta/\theta)(\partial\Pi/\partial T)_\theta = RT/[1 - \theta)$; hence, by equation III-62,

$$\Delta H^{ads} = -q^{diff} + RT[\theta/(1 - \theta)] \qquad \text{(III-66)}$$

To evaluate the differential enthalpy, $(\partial q^{diff}/\partial_{a}n)_T = 0$ and $\beta T(\partial/\partial\theta)_T(\partial\Pi/\partial T)_\theta = RT/(1 - \theta)^2$; hence,

$$\Delta\dot{H}^{ads} = -q^{diff} - RT + RT/(1 - \theta)^2 \qquad \text{(III-67)}$$

Case 3. The Two-Dimensional van der Waals Equation. For this case, $(T\beta/\theta)(\partial\Pi/\partial T)_\theta = RT/(1 - \theta)$; hence, by equation III-62,

$$\Delta H^{ads} = -q^{diff} + RT[\theta/(1 - \theta)] \qquad \text{(III-68)}$$

To evaluate the differential molar enthalpy, ${}_{a}n(\partial q^{diff}/\partial_{a}n)_T = 2\alpha\theta/\beta$, from equation III-49; and $\beta T(\partial/\partial\theta)_T(\partial\Pi/\partial T)_\theta = RT/(1 - \theta)^2$; hence,

$$\Delta\dot{H}^{ads} = -q^{diff} - 2\alpha\theta/\beta - RT + RT/(1 - \theta)^2 \qquad \text{(III-69)}$$

Case 4. The Langmuir Equation. For this case, $(T\beta/\theta)(\partial\Pi/\partial T)_\theta = (RT/\theta)\ln[1/(1 - \theta)]$; hence, by equation III-62,

$$\Delta H^{ads} = -q^{diff} - RT + (RT/\theta)\ ln\ [1/(1 - \theta)] \qquad \text{(III-70)}$$

To evaluate the differential molar enthalpy, ${}_an(\partial q^{\text{diff}}/\partial {}_an)_T = 0$ and $\beta T(\partial/\partial\theta)_T(\partial\Pi/\partial T)_\theta = RT/(1 - \theta)$; hence,

$$\Delta \dot{H}^{\text{ads}} = -q^{\text{diff}} + RT[\theta/(1 - \theta)] \qquad \text{(III-71)}$$

Case 5. The Fowler Equation. For this case, $(T\beta/\theta)(\partial\Pi/\partial T)_\theta = (RT/\theta)\ ln\ [1/(1 - \theta)]$; hence, by equation III-62,

$$\Delta H^{\text{ads}} = -q^{\text{diff}} - RT + (RT/\theta) \ln [1/(1 - \theta)] \qquad \text{(III-72)}$$

To evaluate the differential molar enthalpy, ${}_an(\partial q^{\text{diff}}/\partial {}_an)_T = 2\omega\theta$ and $\beta T(\partial/\partial\theta)_T(\partial\Pi/\partial T)_\theta = RT/(1 - \theta)$; hence,

$$\Delta \dot{H}^{\text{ads}} = -q^{\text{diff}} - 2\omega\theta + RT[\theta/(1 - \theta)] \qquad \text{(III-73)}$$

All the adsorption models considered above give $\Delta \dot{H}^{\text{ads}} = -q^{\text{diff}}$ for very low values of surface concentration; as θ tends closer to unity, $\Delta \dot{H}^{\text{ads}}$ approaches infinity. To obtain the standard differential enthalpy change, $\Delta \dot{H}_s^{\text{ads}}$, the values of θ in the above equations are replaced by θ_s and the value of q^{diff} corresponding to this (standard) concentration. De Boer (3) suggested that "we might introduce the standard state of the two-dimensional gas as that state where at 0°C the average distance between the molecules is the same as in the three-dimensional standard state (0°C, 760 mm)." His suggestion is now widely adopted: it yields the value $\sigma_s = 4.08\ T \times 10^{-16}$ cm² per molecule adsorbed. This surface concentration is low enough so that the two-dimensional ideal-gas law, case 1 above, is approximated at θ_s; hence,

$$\Delta \dot{H}_s^{\text{ads}} = -q_s^{\text{diff}} \qquad \text{(III-74)}$$

An ideal gas has a molar volume of 22,400 cm³ and therefore a molecular volume of $22{,}400/(6.02 \times 10^{23})$ cm³ or 3.72×10^{-20} cm³; the average distance apart is the cube root of 3.72×10^{-20} cm³ or 33.4×10^{-8} cm; the average molecular area is $(33.4 \times 10^{-8})^2$ cm² or 1114×10^{-16} cm². The standard two-dimensional pressure, Π_s, can then be calculated from $\Pi_s\sigma_s = kT$, where $\sigma_s = 1114 \times 10^{-16}$ cm² at $T = 273$°K; from which $\Pi_s = 0.338$ dynes/cm. The value of Π_s is invariant with temperature, but σ_s varies according to $\Pi_s\sigma_s = kT$, giving

$$\sigma_s = \frac{1.38 \times 10^{-16}}{0.338} T = (4.08 \times 10^{-16})\ T$$

TABLE III-1
Integral Molar and Differential Molar Enthalpies of Adsorption for Various Models of the Adsorbed Film

Isotherm equation	ΔH^{ads}	$\Delta \dot{H}^{ads}$
$p = K\theta$	$-q^{diff}$ (III-64)	ΔH^{ads} (III-65)
$p = K \frac{\theta}{1-\theta} e^{\theta/(1-\theta)}$	$-q^{diff} + RT \frac{\theta}{1-\theta}$ (III-66)	$\Delta H^{ads} + RT \frac{\theta}{(1-\theta)^2}$ (III-67)
$p = K \frac{\theta}{1-\theta} e^{\theta/(1-\theta)} e^{-2\alpha\theta/RT\beta}$	$-q^{diff} + RT \frac{\theta}{1-\theta}$ (III-68)	$\Delta H^{ads} + RT \frac{\theta}{(1-\theta)^2} - \frac{2\alpha\theta}{\beta}$ (III-69)
$p = K \frac{\theta}{1-\theta}$	$-q^{diff} + \frac{RT}{\theta} \ln \frac{1}{1-\theta} - RT$ (III-70)	$\Delta H^{ads} + RT \frac{1}{1-\theta} - \frac{RT}{\theta} \ln \frac{1}{1-\theta}$ (III-71)
$p = K \frac{\theta}{1-\theta} e^{-c\omega\theta/RT}$	$-q^{diff} + \frac{RT}{\theta} \ln \frac{1}{1-\theta} - RT$ (III-72)	$\Delta H^{ads} + RT \frac{1}{1-\theta} - \frac{RT}{\theta} \ln \frac{1}{1-\theta} - c\omega\theta$ (III-73)

TABLE III-2

Integral Molar and Differential Molar Free Energies of Adsorption for Various Models of the Adsorbed Film

Isotherm equation	ΔF^{ads}	$\Delta \dot{F}^{ads}$
$p = K\theta$	$RT \ln \theta + RT \ln \frac{K}{760}$ (III-77)	$\Delta F^{ads} + RT$ (III-78)
$p = K \frac{\theta}{1-\theta} e^{\theta/(1-\theta)}$	$RT \ln \frac{\theta}{1-\theta} + RT \frac{\theta}{1-\theta} + RT \ln \frac{K}{760}$ (III-79)	$\Delta F^{ads} + RT \frac{1}{(1-\theta)^2}$ (III-80)
$p = K \frac{\theta}{1-\theta} e^{\theta/(1-\theta)} e^{-2\alpha\theta/RT\beta}$	$RT \ln \frac{\theta}{1-\theta} + RT \frac{\theta}{1-\theta} + RT \ln \frac{K}{760} - \frac{2\alpha\theta}{\beta}$ (III-81)	$\Delta F^{ads} + RT \frac{1}{(1-\theta)^2} - \frac{2\alpha\theta}{\beta}$ (III-82)
$p = K \frac{\theta}{1-\theta}$	$RT \ln \frac{\theta}{1-\theta} + RT \ln \frac{K}{760}$ (III-83)	$\Delta F^{ads} + RT \frac{1}{1-\theta}$ (III-84)
$p = K \frac{\theta}{1-\theta} e^{-c\omega\theta/RT}$	$RT \ln \frac{\theta}{1-\theta} + RT \ln \frac{K}{760} - c\omega\theta$ (III-85)	$\Delta F^{ads} + RT \frac{1}{1-\theta} - c\omega\theta$ (III-86)

TABLE III-3

Integral Molar and Differential Molar Entropies of Adsorption for Various Models of the Adsorbed Film

Isotherm equation	ΔS^{ads}	$\Delta \dot{S}^{ads}$
$p = K\,\theta$	$-\dfrac{q^{diff}}{T} - R\ln\theta - R\ln\dfrac{K}{760}$ (III-87)	$\Delta S^{ads} - R$ (III-88)
$p = K\dfrac{\theta}{1-\theta}e^{\theta/(1-\theta)}$	$-\dfrac{q^{diff}}{T} - R\ln\dfrac{\theta}{1-\theta} - R\ln\dfrac{K}{760}$ (III-89)	$\Delta S^{ads} - R\dfrac{1}{1-\theta}$ (III-90)
$= K\dfrac{\theta}{1-\theta}e^{\theta/(1-\theta)}\,e^{-2\alpha\theta/RT\beta}$	$-\lim\limits_{\theta\to 0}\dfrac{q^{diff}}{T} - R\ln\dfrac{\theta}{1-\theta} - R\ln\dfrac{K}{760}$ (III-91)	$\Delta S^{ads} - R\dfrac{1}{1-\theta}$ (III-92)
$p = K\dfrac{\theta}{1-\theta}$	$-\dfrac{q^{diff}}{T} - R\ln\dfrac{\theta}{1-\theta} - \dfrac{R}{\theta}\ln(1-\theta)$ $- R - R\ln\dfrac{K}{760}$ (III-93)	$\Delta S^{ads} + \dfrac{R}{\theta}\ln(1-\theta)$ (III-94)
$p = K\dfrac{\theta}{1-\theta}e^{-c\omega\theta/RT}$	$-\lim\limits_{\theta\to 0}\dfrac{q^{diff}}{T} - R\ln\dfrac{\theta}{1-\theta} - \dfrac{R}{\theta}\ln(1-\theta)$ $- R - R\ln\dfrac{K}{760}$ (III-95)	$\Delta S^{ads} + \dfrac{R}{\theta}\ln(1-\theta)$ (III-96)

Equation III-62 is sufficiently general that it can be applied to experimental data to obtain the integral molar enthalpy change on adsorption. For this purpose, however, it is more convenient to use Hill's method (2) of determining ΔH^{ads}, in which he calculates Π as a function of p from the experimental data, and makes use of the relation

$$(\partial \ln p/\partial T)_{\Pi} = -\Delta H^{\text{ads}}/RT^2$$

To obtain $\Delta \dot{H}^{\text{ads}}$ from experimental data requires graphical differentiation of ${}_an\ \Delta H^{\text{ads}}$ with respect to ${}_an$.

The integral molar free energy (Gibbs function) is

$$\Delta F^{\text{ads}} = RT\ 1\text{n}(p/760) \tag{III-75}$$

The differential molar free energy is

$$\Delta \dot{F}^{\text{ads}} = \left(\frac{\partial}{\partial_a n}\right)_T \left({}_an\ RT \ln \frac{p}{760}\right)$$

$$= {}_an\ RT \left(\frac{\partial \ln p}{\partial_a n}\right)_T + RT \ln \frac{p}{760} \tag{III-76}$$

Equations III-75 and III-76 can be applied to the five cases of adsorption models considered above, by substituting for p the appropriate expressions derived from the adsorption isotherm equations listed in Table I-1.

The integral molar and differential molar entropies of adsorption can be calculated from the appropriate enthalpies and free energies by means of the following expressions:

$$\Delta S^{\text{ads}} = (\Delta H^{\text{ads}}/T) - (\Delta F^{\text{ads}}/T) \tag{III-77}$$

and

$$\Delta \dot{S}^{\text{ads}} = (\Delta \dot{H}^{\text{ads}}/T) - (\Delta \dot{F}^{\text{ads}}/T) \tag{III-78}$$

The tables preceding, Tables III-1, III-2, and III-3, are collections of the integral molar and differential molar thermodynamic functions, as evaluated for the various models of the adsorbed film listed in Table I-1.

References

(1). G. L. Kington and J. G. Aston, *J. Phys. Chem.*, **73,** 1929 (1951).
(2). T. L. Hill, *J. Chem. Phys.*, **17,** 520 (1949).
(3). J. H. de Boer, *The Dynamical Character of Adsorption*, Clarendon Press, Oxford, 1953, p. 112.

CHAPTER IV

Adsorption by a Heterogeneous Surface

1. Introduction

A heterogeneous surface is one described by a distribution of adsorptive potentials. Each "patch" of surface characterized by a given potential becomes the seat of an adsorbed phase, so that a number of surface phases coexist at a given equilibrium pressure, according to the phase rule principle discussed in Section I-1.

A crude preliminary calculation is sufficient to show that even relatively small variations of the adsorptive potential u_0 will greatly affect the behavior of the surface as an adsorbent: the adsorptive potential of argon on graphite is about 2 kcal/mole; the surface roughness or other causes could easily introduce a variation of 100 cal/mole; assuming $A^0 = 1 \times 10^6$ mm, then by equation I-32 the variation in K for this 5% variation of U_0 is 100% at 77°K, 30% at 200°K, and 17% at 300°K. Referring to the isotherm equations of Table I-1, the pressure required to adsorb like amounts would also vary by the same percentage factor. Relatively narrow distributions of adsorptive potential are, therefore, by no means insignificant, which goes far to explain the lack of success that has met all efforts to describe the real adsorbents by means of equations developed for the concept of a homotattic surface. But one can see that if the variation of the adsorptive potential were smaller, say 25 cal/mole, the heterogeneity would be significant at low temperatures, but would have little effect on the adsorption isotherm at high temperatures.

We have developed the quantity U_0 to serve as an index of a homotattic surface patch: consequently any equation, such as those of Section III-6, which contains this term as invariant with θ is restricted in its application to a homotattic patch. In particular, the isotherm equations of Table I-1, which contain U_0 implicitly in the constant K, are interconvertible with their corresponding two-dimensional equations of state only if the surface of the adsorbent is homotattic throughout (i.e., K is invariant with θ), and also if no change occurs in the *nature* of the adsorbate-adsorbate interaction as the surface concentration increases (i.e., α/β or $c\omega$ is independent of θ). An ideal solid surface of this sort is usually far from reality. An experimental approach to ideality can be made, nevertheless, by selecting as adsorbents materials with the same crystalline arrangement on all surfaces, such as cube crystals of sodium chloride, or materials with pronounced cleavage planes, such as mica or graphite. When such near-homotattic substrates are chosen, the adsorption isotherm shows many of the characteristics predicted by equations I-18b or I-20b of Table I-1. Thus, for example, first order phase changes and critical phenomena in the adsorbed film have been observed for methane, ethane, xenon, and diborane on sodium chloride (1), krypton on graphite (2), argon on calcium fluoride (3) and methane and krypton on sodium bromide (4), to mention only some well-authenticated results. The experimental data do not, however, conform exactly to the description afforded by equations I-18b or I-20b, a result that can well be attributed to a residual inhomogeneity of the substrate.

2. Introducing Surface Heterogeneity as a Variable

The foregoing account shows the necessity of finding an adequate treatment of the effects of surface heterogeneity. In recent years, the recognition of this need has resulted in numerous published efforts to interpret experimental results in terms of site distribution functions. The general method of all these treatments is to postulate that the surface is composed of infinitesimal patches of different energy that adsorb independently of one another. The distribution of adsorptive energies among these patches may be represented by a continuous distribution function, such as that shown in Figure IV-1, where f_i is the frequency of particular surface patches per unit energy interval; or

$$f_i = \frac{d\delta_i}{dU_0} = \Phi(U_0) \qquad \text{(IV-1)}$$

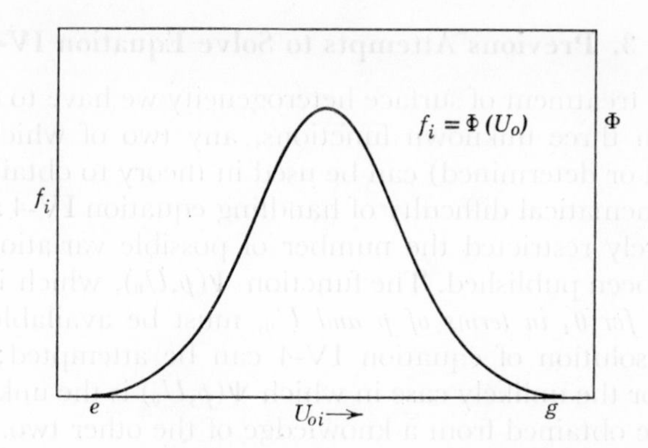

Fig. IV-1. Continuous distribution of adsorptive potential energies of a heterogeneous substrate.

where $d\delta_i$ is the fraction of surface having energies between U_{0i} and $U_{0i} + dU_0$. The summation of $d\delta_i$ over all permitted values of U_0 is equal to unity, i.e.,

$$\int_0^\theta d\delta_i = 1 \qquad \text{or} \qquad \int_e^g \Phi(U_0)\, dU_0 = 1 \qquad \text{(IV-2)}$$

Any distribution function that does not converge on unity is useless for this purpose, as was pointed out by Hill (5), since the actual surface has a definite extent.

At any equilibrium pressure, p, the fraction of the total surface covered with an adsorbed film, θ, is the sum of contributions from each patch of surface; i.e.,

$$\theta = \sum_i \Delta\delta_i \theta_i \qquad \text{or} \qquad \theta = \int_e^g \theta_i \Phi(U_0)\, dU_0 \qquad \text{(IV-3)}$$

The adsorption equilibrium at a given pressure on an individual patch can be represented functionally as $\theta_i = \Psi(p, U_0)$, where $\Psi(p, U_0)$ could be one of the isotherm equations in Table I-1, since each patch is a homotattic surface. The general form for any adsorption isotherm is therefore:

$$\theta = \sum \Delta\delta_i \Psi(p, U_0) \qquad \text{or} \qquad \theta = \int_e^g \Phi(U_0)\, \Psi(p, U_0)\, dU_0 \quad \text{(IV-4)}$$

The experimental data occur as a series of points on the unknown function $\theta = F(p)$, whose analytic form would be given by equation IV-4 when $\Psi(p, U_0)$ and $\Phi(U_0)$ are substituted and the integration carried out.

3. Previous Attempts to Solve Equation IV-4

In any treatment of surface heterogeneity we have to deal, therefore, with three unknown functions, any two of which if known (assumed or determined) can be used in theory to obtain the third. The mathematical difficulty of handling equation IV-4 analytically has severely restricted the number of possible variations that has hitherto been published. The function $\Psi(p,U_0)$, which is *an explicit expression for* θ_i *in terms of* p *and* U_0, must be available before an analytic solution of equation IV-4 can be attempted; except, of course, for the unlikely case in which $\Psi(p,U_0)$ is the unknown function to be obtained from a knowledge of the other two. The above restriction (*italicized*) on the form of the isotherm equation for the ideal surface, $\Psi(p,U_0)$, means that of all the isotherm equations listed in Table I-1, only the two-dimensional ideal gas (Henry's law equation) and the Langmuir equation are available for an analytical solution of equation IV-4.

The Henry's law equation is $\theta_i = p/K_i$, where K_i is a function of U_{0i} and T only. Substituting in equation IV-4 gives the adsorption isotherm

$$\theta = p \int_e^g \frac{\Phi(U_0)\, dU_0}{K_i} = Cp \tag{IV-5}$$

where C is a constant. This result shows that regardless of the form of the distribution function, Henry's law yields an adsorption isotherm that is a straight line through the origin. All equations for homotattic surfaces reduce to Henry's law at low surface concentration. Hence, every experimental adsorption isotherm, no matter how heterogeneous the substrate, reduces to equation IV-5 for low values of θ; although in practice, for surfaces of great heterogeneity, the pressures at which this occurs may be immeasurably small.

Langmuir himself was the first to propose the form that the Langmuir equation would take for a set of sites of different energies:

$$\theta = \sum_i \Delta\delta_i \theta_i = \sum_i \frac{\Delta\delta_i p}{K_i + p}$$

This equation is the summation form of equation IV-4. Zeldowitch (6) and later Roginsky (7) and Halsey and Taylor (8) replaced the concept of a set of sites by a continuous distribution of energies (our equation IV-1), thus permitting the summation to be replaced by an integral. By selecting an exponential distribution of the form $f_i = Ce^{-U_0/U_0'}$ over the whole range of U_0, both positive and negative, in combination with the Langmuir equation for $\Psi(p,U_0)$, these authors found the integration of equation IV-4 yielded the Freundlich adsorption equation in the form $\log \theta = (k'/U_0') \log p/p_0$. In these equations, U_0' is the average adsorptive energy.

An alternative way to use equation IV-4 is to assume functions for the adsorption isotherm, $F(p)$, and for $\Psi(p,U_0)$, and thence derive the distribution function $\Phi(U_0)$. That this can be done and that a unique solution exists for $\Phi(U_0)$ was demonstated rigorously by Sips (9). To describe the experimental isotherm, Sips chose for $F(p)$ an empirical equation of the form $\theta = Ap^c/(1 + Ap^c)$, and for $\Psi(p,U_0)$ took once more the Langmuir equation. He solved the integral equation by means of a Stieltjes transform and obtained for $\Phi(U_0)$ a unique distribution function which he showed to be nearly equivalent to a Gaussian distribution of adsorptive energies.

A creditable effort to evade the restrictions imposed by the Langmuir model was made by Harkins and Stearns (10), who sought to derive an adsorption isotherm for a mobile adsorbed film on a substrate having a heterogeneous distribution of adsorptive energies. Although these authors clearly outlined the requirements of an hypothesis of mobile adsorption, the simplifications introduced in deriving their explicit function $\Psi(p,U_0)$ effectively destroyed their model. With rare perception, Harkins and Stearns admitted that their equation had "no physical basis" but was really the result of a desire to solve explicitly for θ_i. Their paper represents the best effort yet made to solve the problem of mobile adsorption on a heterogeneous surface by an analytic method, and it failed to do so. The problem, stated in those terms, probably cannot be solved.

A more successful way to avoid introducing the Langmuir equation, with its mechanism of localized adsorption, was pointed out by Drain and Morrison (11). By differentiating equation IV-4 one obtains:

$$f_i = \frac{1}{\theta_i} \frac{d\theta}{dU_0} \tag{IV-6}$$

For the solution of this equation, a new set of experimental results are demanded, namely, q^{st} or q^{diff} as a function of θ. Equation IV-6 still contains the unknown funtion $\theta_i = \Psi(p,U_0)$. This function can, however, be eliminated when we know q^{st} as a function of θ at absolute zero. At that temperature, every surface patch is occupied in the serial order of its adsorptive potential beginning with the greatest: a surface patch is therefore either completely filled or completely empty, depending on how many adsorbed molecules are present on the total surface. As the molecules have no thermal energy, no adsorption equilibrium exists between patches of different energy on the surface: in mathematical language, $\theta_i = 1$ and hence $f_i = d\theta/dU_0$. To meet the physical requirements for this condition strenuous experimental efforts are required. Heats of adsorption and the heat capacity of the adsorbed film at a series of low temperatures must be measured as a function of surface concentration, and the zero degree heats of adsorption calculated therefrom. Only one adsorbent has yet been characterized by this method (see Figure VII-23).

In applying the above method Morrison and co-workers (11) used the experimentally obtained total differential heat of adsorption, q^{diff}, rather than U_0. At absolute zero the relation between q^{diff} and U_0 derived from equation III-50 is

$$q^{diff} = U_0 + {}_aP^{ia}$$

The additional term is not usually easily determined; the error introduced by disregarding it completely is, however, a blemish on an otherwise elegant method.

Summarizing all previously published attempts (12) to obtain a measure of surface heterogeneity, we find that the choice of model has almost invariably been a localized monolayer in which adsorbate-adsorbate interaction is ignored. While the neglect of this interaction always introduces a major error, the concept of localization can be justified for special cases: e.g., chemisorption. In general, however, the theoretical equations have been applied without due discrimination, their selection determined more by mathematical convenience than by a realistic estimate of the probable physical state of the adsorbed film. Mathematical simplicity is a dangerous criterion: "La nature," says Fresnel, "ne s'est pas embarassée des difficultés d'analyse; elle n'a évité que la complication des moyens." No analytic solution has yet been, or probably can be, based on the models of mobile or localized monolayers, taking into account both interaction and surface heterogeneity.

4. A Nonanalytic Solution of Equation IV-4

As we have seen, the analytic solution offers only intractable equations for the solution of θ_i, unless we introduce undesirable simplifications. In theory, the Hill-de Boer equation, equation I-18b, can be expanded by Taylor's theorem to an infinite series, and this series reverted to yield an explicit series for θ_i. In practice, however, this mathematical technique produces a series that, although ultimately convergent, was still found to be diverging at the thirteenth term. This solution is therefore not practicable. The some conclusion was reached with respect to the Fowler equation, equation I-20b.

The summation form of equation IV-4 suggests that the problem might be solved by use of a high-speed electronic computer. Such a solution would not yield an analytic expression for $\theta = F(p)$ but would give a series of computed isotherms (model isotherms) that would at least exhibit characteristics implicit in the model.

Two reasonable choices offer themselves for the function $\Psi(p,U_0)$: these are equations I-18b and I-20b, descriptive respectively of a mobile adsorbed film and of a localized adsorbed film. These equations are suitable because they include adsorbate-adsorbate interaction.

The surface of a solid is a nonequilibrium system with accidental energy variations resulting from cracks, edges, lattice defects, etc. A random distribution of adsorptive potentials, therefore, seems a reasonable supposition. The Gaussian probability function was selected:

$$f_i = \Phi(U_0) = \frac{1}{n} e^{-\gamma(U_0 - U_0')^2} \quad \text{(IV-7)}$$

but, instead of allowing U_0 to vary from plus to minus infinity, the function was confined to a range of 5 kcal/mole. To make the area under the curtailed distribution curve equal to unity, the normalizing factor, n, is given by:

$$n = \int_0^5 e^{-\gamma(U_0 - U_0')^2} \, dU_0 \quad \text{(IV-8)}$$

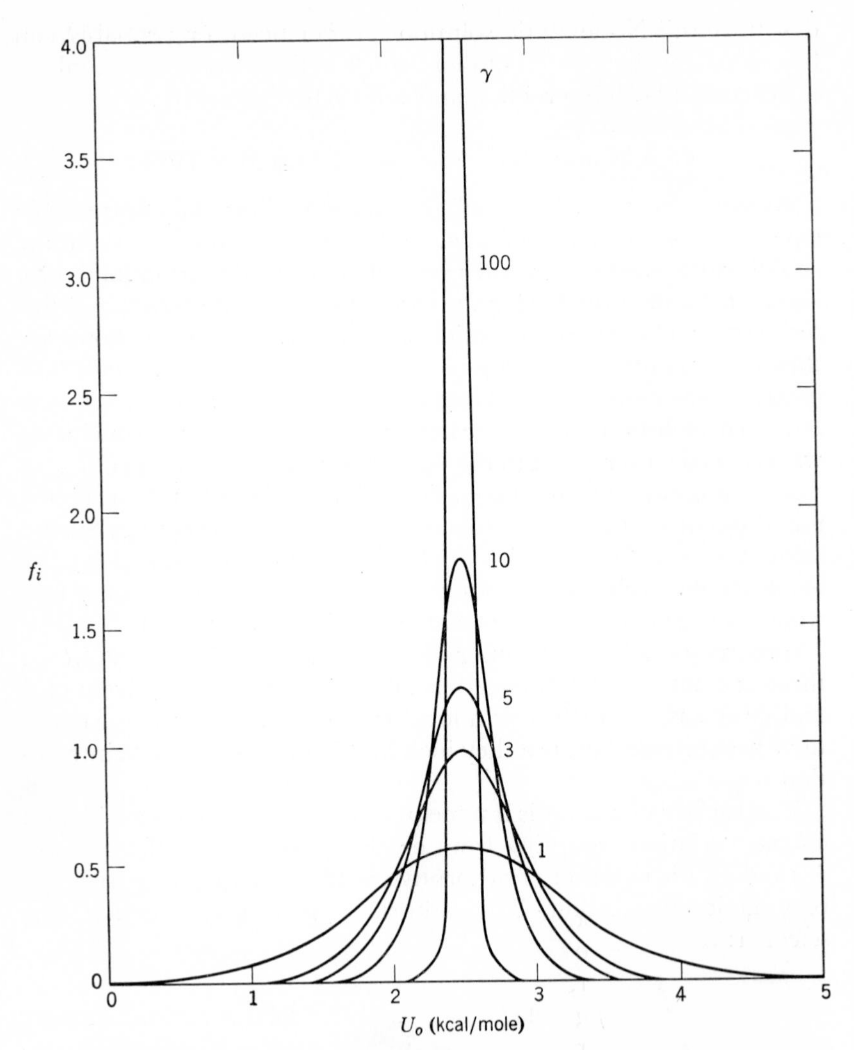

Fig. IV-2. Gaussian distributions of adsorptive potential energies for various values of γ in equation IV-7.

U_0' represents the average adsorptive energy of the surface, taken arbitrarily as 2.5 kcal/mole. The degree of heterogeneity of a surface can be expressed in terms of this Gaussian distribution by r, where $U_0' \pm r$ is the adsorptive energy range within which lies one half the total surface. The degree of heterogeneity is then expressed by $r = 0.4769/\sqrt{\gamma}$.

For use in the computer, integrations are replaced by summations. The definitions of the normalizing factor, n, becomes $\sum_i f_i \Delta U_0$, where ΔU_0 is taken as one three-hundredth of the range of U_0 from 0 to 5 kcal/mole, i.e., 1/60 kcal/mole. The computer calculates n for every γ value used.

The limits 0 and 5 were selected for the calculation of n because these limits are sufficient to include practically all the area under the complete Gaussian curve, for the largest width used in the present computations, namely $\gamma = 1$. The computer now has enough information to calculate f_i by equation IV-7 for any value of U_0 between 0 and 5 kcal/mole. The resulting distribution curves for several values of γ are shown in Figure IV-2.

For each distribution curve in Figure IV-2 there is a practical lower and upper limit of U_0 which we designate e and g, respectively, beyond which the values of f_i are insignificantly small. In the present computations the limits e and g were chosen as the points where $f_i = .001$. Within these limits the distribution curve contains an area that is at least 99.9% of the area that would result from the whole Gaussian curve. For each curve the range of U_0 between e and g is subdivided into 50 parts, thus effectively subdividing the total surface into 50 finite patches, each one of which is defined as that fraction $\Delta\delta_i$ of the surface having adsorptive energies between U_{0i} and $(U_{0i} + \Delta U_0)$, where $\Delta U_0 = (g - e)/50$: hence,

$$\Delta\delta_i = \int_{U_{0i}}^{U_{0i} + \Delta U_0} d\delta_i \qquad \text{(IV-9)}$$

In most cases the subdivision of the range of U_0 into 50 sections was found to be enough to free the resulting isotherm from irregularities; occasionally, however, 100 subdivisions had to be used.

To get the resulting isotherm, every finite surface patch is treated as an individual homotattic surface of energy U_{0i}, for which the amount adsorbed at a pressure p can be calculated by the Hill-de Boer equation, using

$$K_i = A^0 e^{-U_{0i}/RT} \qquad \text{(IV-10)}$$

where U_{0i} is the midpoint value for the ith patch of surface and A^0 has the significance described by equation I-33. Actually, A^0

need not be evaluated for the purpose of the computer: it is taken arbitrarily so as to make $K_i = 1$ when $U_{0i} = U_0' = 2.5$ kcal/mole and $T = 77.5$ °K in equation IV-10; under these conditions $A^0 = 3.37 \times 10^6$ mm. Equation IV-10 follows from equation I-32; the experimental evaluation of A^0 is discussed in Chapters VI and VII. The total coverage for the whole surface is then the

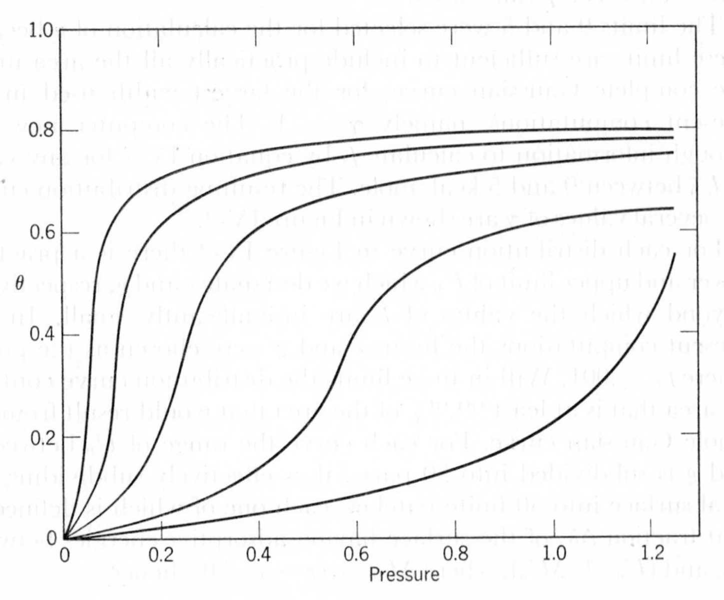

Fig. IV-3. Five typical adsorption isotherms for homotattic patches of different adsorptive potential energy.

summation of the 50 contributions, each one modified by the fraction that denotes its share of the whole surface. In mathematical shorthand:

$$\theta = \sum_{i=1}^{50} \Delta\delta_i \, \theta_i \qquad \text{(IV-11)}$$

The process just described is represented graphically in Figure IV-3, in which 5 (instead of 50) separate isotherms are shown for 5 homotattic patches of energy U_{01}, U_{02}, etc., composing a surface. At an equilibrium pressure p each patch is occupied to an extent

θ_1, θ_2, etc., as given by the Hill-de Boer isotherm, using K_1, K_2, etc. The total surface coverage, θ, is then given by the expression:

$$\theta = \Delta\delta_1\theta_1 + \Delta\delta_2\theta_2 + \Delta\delta_3\theta_3 + \Delta\delta_4\theta_4 + \Delta\delta_5\theta_5$$

where $\Delta\delta_1$, $\Delta\delta_2$, etc., represent the fractional part of the whole surface that pertains to each value of U_0.

The computed tables in the Appendix contain values of θ for values of p/K', where K' is the particular K value corresponding to the average, U_0', of the Gaussian distribution. When the computed isotherms are used for comparisons with experimental data, K' is a scale factor for the pressure axis, which determines U_0' for the experimental system. To make this more clear we can rewrite $\theta_i = \Psi(p,U_0)$ less generally as $\theta_i = \psi(p/K_i)$, since Table I-1 tells us that much about the form of the function. If now we multiply both p and K_i by the same factor, we leave θ_i unchanged, but we affect the isotherms by multiplying the pressure scale, and we shift the distribution curves of Figure IV-2 bodily along the U_0-axis. The value of U_0', which is merely a particular U_{0i}, is also shifted by the same amount. The factor K' needed to match a model isotherm with an experimental one determines U_0' for the experimental system, since in the model isotherms K' was arbitrarily chosen as unity.

5. The Integral and Differential Adsorptive Potentials for a Heterogeneous Surface

The computer can furnish us not only with θ as a function of p, but also with the integral adsorptive potential, $\mathbf{U}_0^{\text{int}}$, as a function of θ. The process for calculating the integral adsorptive potential can be illustrated simply by means of our previous 5 patch surface, where

$$\mathbf{U}_0^{\text{int}} = \Delta\delta_1\theta_1 U_{01} + \Delta\delta_2\theta_2 U_{02} + \Delta\delta_3\theta_3 U_{03} + \Delta\delta_4\theta_4 U_{04} + \Delta\delta_5\theta_5 U_{05}$$

In other words, the integral potential of each patch is the product of the area covered and adsorptive potential per unit area, or

$$\mathbf{U}_0^{\text{int}} = \sum_i \Delta\delta_i\theta_i U_{0i} \tag{IV-12}$$

From these integral adsorptive potentials, the computer can also obtain the differential adsorptive potential, $\mathbf{U}_0^{\text{diff}}$, defined as $d\mathbf{U}_0^{\text{int}}/d\theta$, which is actually computed as $\Delta\mathbf{U}_0^{\text{int}}/\Delta\theta$.

The quantity $\mathbf{U}_0^{\text{diff}}$ enters with the concept of heterogenous surface. It is not the adsorptive potential of any particular patch; it is the effective differential adsorptive potential for the *whole* surface at a given *total* surface concentration. A molecule in the gas phase at a fixed equilibrium pressure approaching a partially occupied heterogenous substrate is offered a range of adsorptive potentials; it will "see" crowded patches of high potential and sparsely populated patches of lower potential; the probability of its striking a bare spot on a particular patch is governed by the extent of the patch ($\Delta\delta_i$) and its degree of occupancy ($1 - \theta_i$). The molecule will linger longest on the patches of highest adsorptive potential, so that the probability, W_i, of finding it on the ith patch is:

$$W_i \propto \Delta\delta_i \, \tau_i \, (1 - \theta_i) \tag{IV-13}$$

where τ_i is the time of residence of a molecule on the ith patch and τ' is the corresponding lifetime on the patch of average adsorptive potential:

$$\tau_i/\tau' = e^{(U_{0i} - U_0')/RT} \tag{IV-14}$$

Consider the adsorption of one mole of gas by a surface of infinite extent at a fixed pressure and a fixed surface concentration: the fraction of the mole that will be adsorbed on the ith patch is proportional to W_i, and the corresponding contribution to the adsorptive potential is proportional to $W_i U_{0i}$. Therefore, the average (differential) adsorptive potential per mole, $\mathbf{U}_0^{\text{diff}}$, is given by

$$\mathbf{U}_0^{\text{diff}} = \sum_i W_i U_{0i} / \sum_i W_i \tag{IV-15}$$

Equation IV-15 is an alternative expression for $\mathbf{U}_0^{\text{diff}}$ to that given earlier, *viz.*,

$$\mathbf{U}_0^{\text{diff}} = \frac{d}{d\theta} (\mathbf{U}_0^{\text{int}}) \tag{IV-16}$$

Only for a homotattic surface is $\mathbf{U}_0^{\text{diff}}$ constant and independent of θ; for such a surface we could write

$$\mathbf{U}_0^{\text{diff}} = U_0' = U_0$$

6. The Relation of the Differential Adsorptive Potential $\mathbf{U}_0^{\text{diff}}$ to the Differential Heat of Adsorption for a Heterogeneous Surface

Equation III-50 shows the relation between q^{diff} and U_0 for a homotattic surface:

$$q^{\text{diff}} = U_0 - ({}_aE^{\text{vib}} - {}_aE_0{}^{\text{vib}}) - \Delta E^{\text{kin}} + {}_aP^{\text{ia}} \quad \text{(III-50)}$$

For a heterogeneous surface the energy terms of equation III-50 are to be replaced by operative average values: U_0 is replaced by $\mathbf{U}_0{}^{\text{diff}}$; ${}_aE^{\text{vib}}$ and ${}_aE_0{}^{\text{vib}}$ are replaced by ${}_a\mathbf{E}^{\text{vib}}$ and ${}_a\mathbf{E}_0{}^{\text{vib}}$, which are defined (by analogous reasoning to that leading to equation IV-15):

$${}_a\mathbf{E}^{\text{vib}} = \sum_i W_i\, {}_aE_i{}^{\text{vib}} / \sum_i W_i \quad \text{(IV-17)}$$

and

$${}_a\mathbf{E}_0{}^{\text{vib}} = \sum_i W_i\, {}_aE_{0i}{}^{\text{vib}} / \sum_i W_i \quad \text{(IV-18)}$$

ΔE^{kin} is assumed to be independent of the heterogeneity of the surface, since the absorbed phase on each patch is a two-dimensional gas; the over-all adsorbate-adsorbate interaction is described by

$${}_a\mathbf{P}^{\text{ia}} = \sum_i \Delta\delta_i\, {}_aP_i{}^{\text{ia}} \quad \text{(IV-19)}$$

If the molecular interaction constants of the adsorbate that determine ${}_aP_i^{\text{ia}}$ are the same for every patch, i.e., independent of U_{0i}, then

$${}_a\mathbf{P}^{\text{ia}} = {}_aP^{\text{ia}} \sum_i \Delta\delta_i = {}_aP^{\text{ia}} \quad \text{(IV-20)}$$

i.e., the lateral interaction potential for a heterogeneous surface is the same as that for a homotattic surface of the same total surface concentration, and can be calculated by equation III-51 or equation I-39 using either the equation of state or the isotherm equation for a homotattic surface.

The result of the above considerations is to give us an expression for the differential heat of adsorption by a heterogeneous surface:

$$\mathbf{q}^{\text{diff}} = \mathbf{U}_0{}^{\text{diff}} - ({}_a\mathbf{E}^{\text{vib}} - {}_a\mathbf{E}_0{}^{\text{vib}}) - \Delta E^{\text{kin}} + {}_aP^{\text{ia}} \quad \text{(IV-21)}$$

The various heats of adsorption can be obtained from $\mathbf{q}^{\text{diff}}$ by the equations developed in Chapter III.

References

1. S. Ross and H. Clark, *J. Am. Chem. Soc.*, **76,** 4291, 4297 (1954).
2. S. Ross and W. Winkler, *J. Colloid Sci.*, **10,** 330 (1955).
3. H. Edelhoch and H. S. Taylor, *J. Phys. Chem.*, **58,** 344 (1954).
4. B. B. Fisher and W. G. McMillan, *J. Chem. Phys.*, **28,** 549 (1958).
5. T. L. Hill, *J. Chem. Phys.*, **17,** 762 (1949).
6. J. Zeldowitch, *Acta Physicochim. U.R.S.S.*, **1,** 961 (1934).
7. S. A. Roginsky, *Compt. Rend. Acad. Sci. U.R.S.S.*, **45,** 61, 194 (1944); *Bull Acad. Sci. U.R.S.S.*, **14** (1945); *Acta Physicochim. U.R.S.S.*, **20,** 227 (1945); *ibid.*, **22,** 61 (1947).
8. G. D. Halsey, Jr. and H. S. Taylor, *J. Chem. Phys.*, **15,** 624 (1947); G. D. Halsey, Jr., *ibid.*, **16,** 931 (1948).
9. R. J. Sips, *J. Chem. Phys.*, **16,** 490 (1948); *ibid.*, **18,** 1024 (1950).
10. W. D. Harkins and R. S. Stearns, *J. Phys. Chem.*, **58,** 292 (1954).
11. J. A. Morrison, J. M. Los, and L. E. Drain, *Trans. Faraday Soc.*, **47,** 1023 (1951); L. E. Drain and J. A. Morrison, *Trans. Faraday Soc.*, **48,** 840 (1952); *ibid.*, **49,** 654 (1953).
12. As well as the references previously cited, 6–11, the following are included in our general summary: F. C. Tompkins, *Trans. Faraday Soc.*, **46,** 569 (1950); F. C. Tompkins and D. M. Young, *Trans. Faraday Soc.*, **47,** 77 (1951); G. D. Halsey, Jr., *J. Am. Chem. Soc.*, **73,** 2693 (1951); J. M. Honig and L. H. Reyerson, *J. Phys. Chem.*, **56,** 140 (1952); J. G. Aston, R. J. Tykodi, and W. A. Steele, *J. Phys. Chem.*, **59,** 1053 (1955). An excellent historical survey (with 155 references) of adsorbent-adsorbate interactions and surface heterogeneity in physical adsorption has been published by J. M. Honig, *Ann. N.Y. Acad. Sci.*, **58,** 741–797 (1954).

CHAPTER V

Implications of the Concept of Surface Heterogeneity

1. General Properties of the Isotherms

A. THE MOBILE ADSORBED FILM

A representative number of computed isotherms corresponding to argon adsorbed as a mobile film at 77.5 °K on a series of heterogeneous substrates, based on the distribution curves shown in Figure IV-2, are shown graphically in Figure V-1. The present model of mobile adsorption is seen to be capable of predicting adsorption isotherms that vary in shape from the sigmoidal form, already well known as characteristic of a near-homotattic surface, to isotherms that are concave throughout to the pressure axis. This feature of the model is certainly a requirement of any theory that pretends to offer a general description of adsorption phenomena. The isotherms for low values of γ resemble in form those that are described by the Langmuir equation; certainly had they occurred as experimental data the investigator would have attempted the customary p/V *vs* p plot that tests the fit of the Langmuir equation to experimental data. We have done so with these model isotherms, and in Figure V-2 we report that they do indeed follow the Langmuir equation in the range $\theta = 0.5$–0.8. The values of V_m, derived from the straight lines of Figure V-2 are, however, always too low, ranging from about 88% of the correct value for the less heterogeneous substrates down to 75% of the correct value for the most heterogeneous substrate. These findings are of value to future investigators as they reveal the disquieting fact that the Langmuir equation can be used to describe isotherms derived from a totally different model, and that the constants thereby determined, which include the important one of monolayer

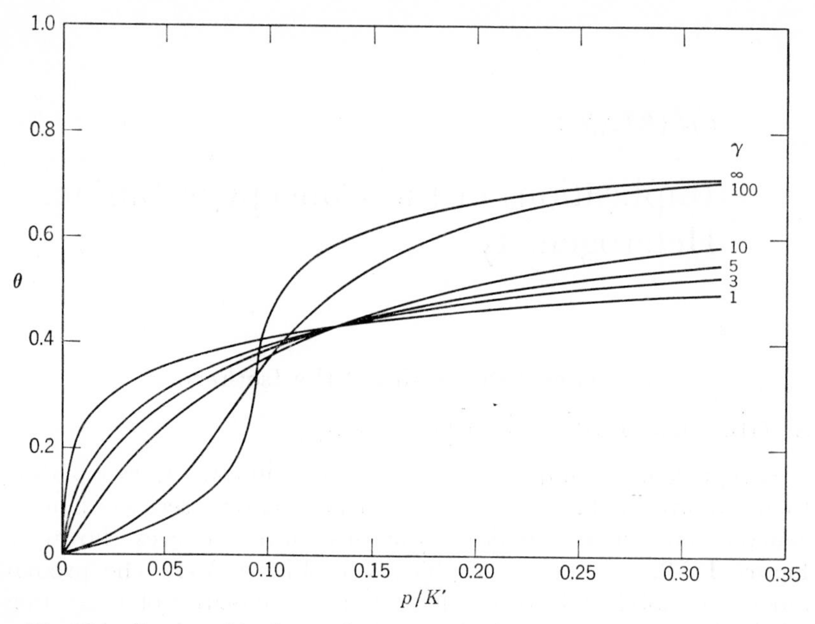

Fig. V-1. Computed isotherms for argon adsorbed as a mobile film at 77.5°K by substrates corresponding to absorptive potential distribution curves shown in Figure IV-2.

coverage, V_m, are always seriously in error. Nothing could be farther removed from the present model than the Langmuir model: the former model postulates interaction, heterogeneity of substrate and a mobile adsorbed film; whereas the latter model postulates no interaction, homogeneity of substrate and a localized adsorbed film. The straight lines of Figure V-2 give the clearest possible proof that merely to find data that fit the Langmuir equation cannot be accepted as any indication that the Langmuir model is a description of the physical situation. Certain clever hypotheses now on record would have been stifled at birth had this result been available earlier.

Let us turn our attention to the isotherms of sigmoidal form, which have actually been reproduced experimentally in studies of graphitized carbon (1) and other near-homotattic substrates. The usual interpretation that is made of experimental isotherms of this shape owes much to the influence of de Boer's exhaustive

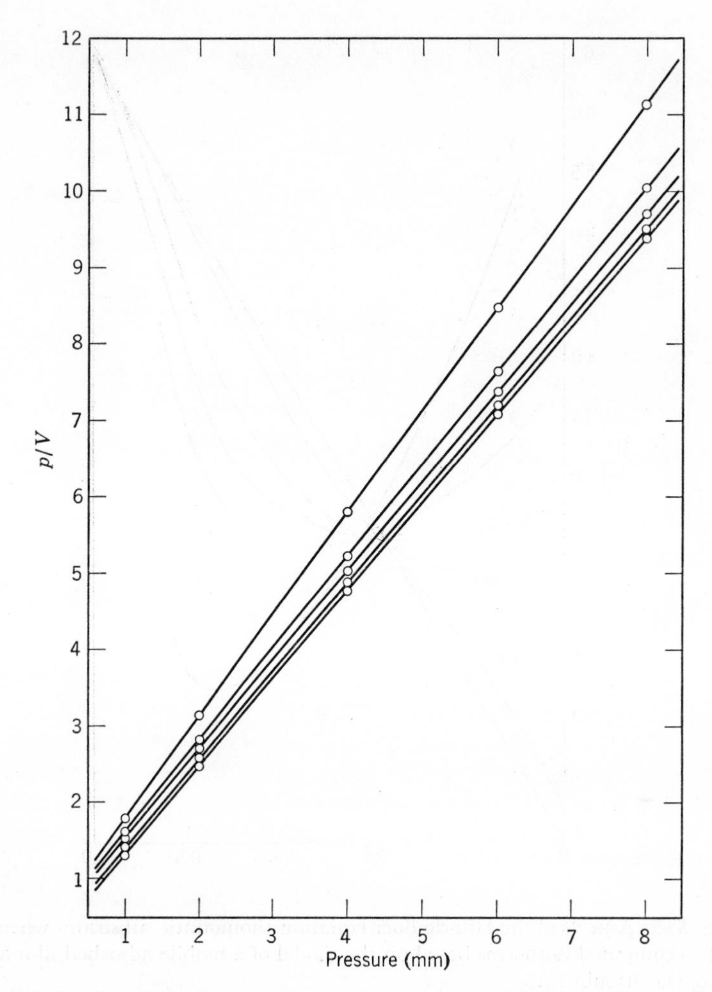

Fig. V-2. Test of the Langmuir equation as an empirical description of computed isotherms based on the model of a mobile adsorbed film and a heterogeneous substrate.

treatment of a nonideal, two-dimensional gas film on a homotattic surface. A tendency has developed to interpret such isotherms in terms of their deviation from the Hill-de Boer equation, and to ascribe the deviations to changes in adsorbate-adsorbate interaction, the surface being considered sufficiently close to homotattic

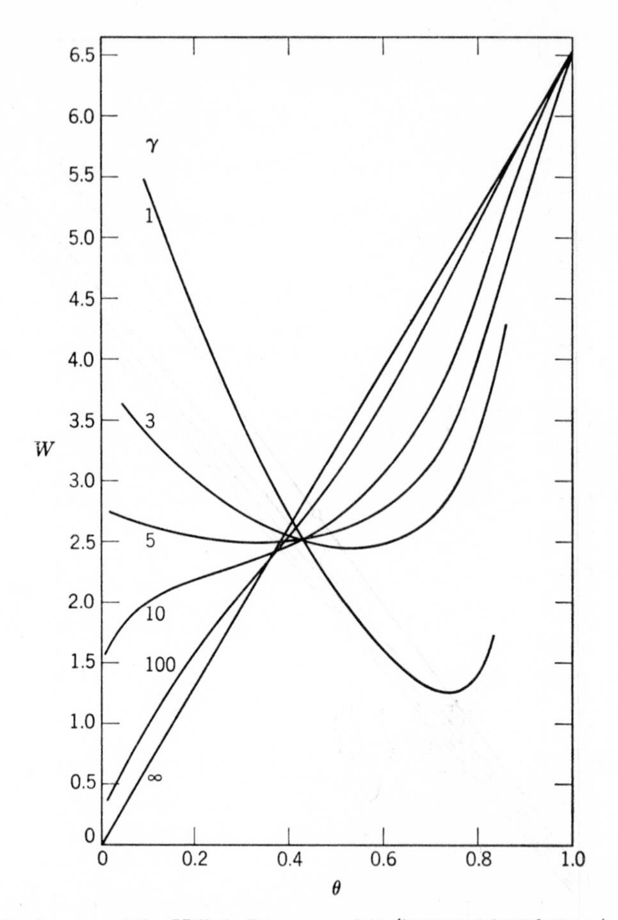

Fig. V-3. Aspects of the Hill-de Boer equation (homotattic substrate) when applied to computed isotherms based on the model of a mobile adsorbed film and a heterogeneous substrate.

that the effect of its heterogeneity is negligible (2,3,4). An examination of our model isotherms shows the likelihood of error in this assumption. Figure V-3 reports the deviations of a representative number of computed isotherms from the Hill-de Boer equation; in this diagram, the equation is plotted in its rectilinear form:

$$W = \ln \frac{\theta}{1 - \theta} + \frac{\theta}{1 - \theta} - \ln p = \frac{2\alpha\theta}{RT\beta} - \ln K \qquad \text{(V-1)}$$

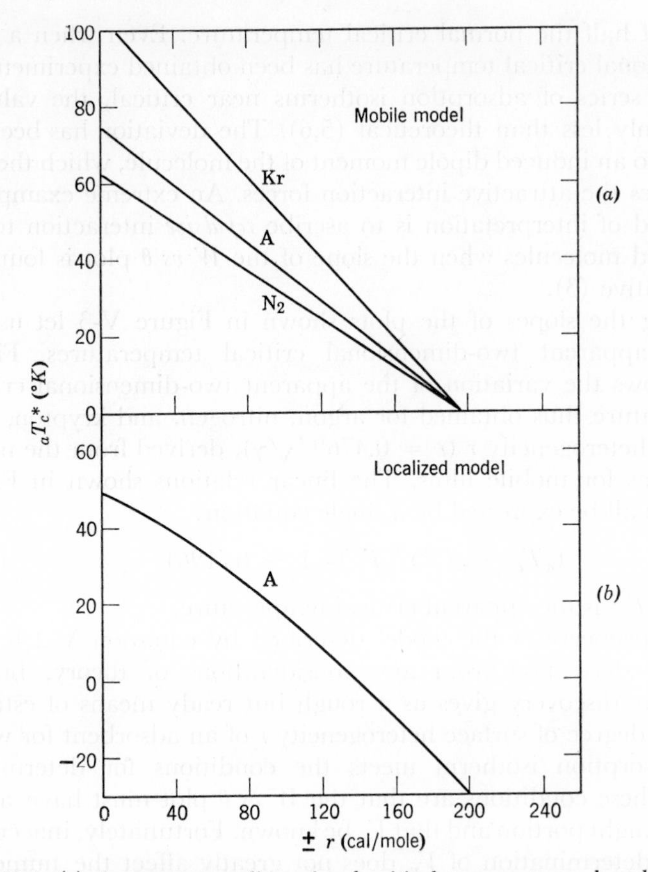

Fig. V-4. (a) Apparent two-dimensional critical temperatures, based on the slopes shown in Figure V-3. (b) Apparent two-dimensional critical temperatures, based on the application of the Fowler equation (homotattic substrate) to computed isotherms based on the model of a localized adsorbed film and a heterogeneous substrate.

If the curves of Figure V-3 had been obtained experimentally, one would measure the slope of the approximately straight portion between $\theta = 0.1$ and 0.5 and therefrom derive a value for $2\alpha/R\beta$, from which in turn an apparent critical temperature for two-dimensional condensation can be obtained by the relation $_aT_c = 8\alpha/27R\beta$. The two-dimensional critical temperature thus obtained has usually been found to be less than the theoretical value of half the normal critical temperature. Even when a two-dimensional critical temperature has been obtained experimentally from a series of adsorption isotherms near critical, the value is commonly less than theoretical (5,6). The deviation has been ascribed to an induced dipole moment of the molecule, which thereby decreases the attractive interaction forces. An extreme example of this kind of interpretation is to ascribe *repulsive* interaction to the adsorbed molecules when the slope of the W *vs* θ plot is found to be negative (3).

Using the slopes of the plots shown in Figure V-3 let us calculate apparent two-dimensional critical temperatures. Figure V-4 shows the variation of the apparent two-dimensional critical temperature thus obtained for argon, nitrogen, and krypton, with surface heterogeneity r ($r = 0.4769/\sqrt{\gamma}$), derived from the model isotherms for mobile films. The linear relations shown in Figure V-4 can all be expressed by a single equation:

$$({}_aT_c - {}_aT_c^*)/{}_aT_c = 1 - (r/196) \qquad \text{(V-2)}$$

where ${}_aT_c^*$ is the apparent critical temperature.

The property of the model described by equation V-2 has no obvious derivation from any considerations of theory, but its fortunate discovery gives us a rough but ready means of estimating the degree of surface heterogeneity r of an adsorbent for which the adsorption isotherm meets the conditions for determining ${}_aT_c^*$. These conditions are that the W *vs* θ plot must have a sensibly straight portion and that V_m be known. Fortunately, inaccuracy in our determination of V_m does not greatly affect the numerical value of ${}_aT_c^*$: e.g., a variation of 10–20% in V_m results in a variation of only 1 or 2 degrees in ${}_aT_c^*$. This fact greatly extends the range of possible practical application of equation V-2 as a method of estimating r, hence γ.

The curves of Figures V-3 and V-4 show that varieties of W *vs* θ plots and two-dimensional critical temperatures less than theoretical, all of which has been observed experimentally, can be described without postulating any induced variations (varying, that is, with the degree of surface covered) in the adsorbate-adsorbate lateral interaction forces, but by supposing instead that the substrate is heterogeneous. The present treatment of heterogeneity, described in Chapter IV, makes allowance for a uniform reduction of α due to the effect of a surface electric field, assumed—as far as its influence on the lateral interaction is concerned—to be constant at all portions of the surface. One of the implicit assumptions of the treatment, therefore, is that the operative value of α expresses an average for the lateral interaction, and that the use of this average does not vitiate the physical interpretation of the other adsorption parameters.

The alternative assumption—that the surface is completely homotattic, and that all the observed deviations from the Hill-de Boer equation are due to variations of α on different parts of the substrate—is less attractive. The proposition is indeed almost self-contradictory, as by postulating uniformity of the adsorbent one removes from consideration the primary mechanism that could be responsible for the variation of the lateral interactions. Moreover, any hypothesis that includes the assumption of complete surface uniformity of a solid is immediately suspect, unless there is some evidence to support it. In general, heterogeneity of a solid surface is the normal condition, and near-homotattic surfaces can be produced only by unusually favorable conditions.

B. THE LOCALIZED ADSORBED FILM

Fowler (7) has developed "a useful rough approximation to the accurate formulae" for localized films on a homotattic surface, taking into account adsorbate-adsorbate interaction. The adsorption isotherm thus derived is equation I-20 of Table I-1, which we have used to compute model isotherms for this type of film on surfaces of varying heterogeneity.

The method of obtaining the isotherms was exactly the same as that used for the previously described mobile model, including the same Gaussian distribution curves (Fig. IV-2).

As we mentioned previously the interaction constants $c\omega$ for a localized film take into account the crystalline lattice of the substrate, as that determines the arrangement of the adsorption sites on the surface. For our present model we have chosen an hexagonal layer-lattice structure, as many of the experimental data that we discuss were obtained with such adsorbents, e.g., graphitized carbon and boron nitride. For such a substrate Hill (8) has shown that the two-dimensional critical temperature of a localized film should be one third of the normal critical temperature of the

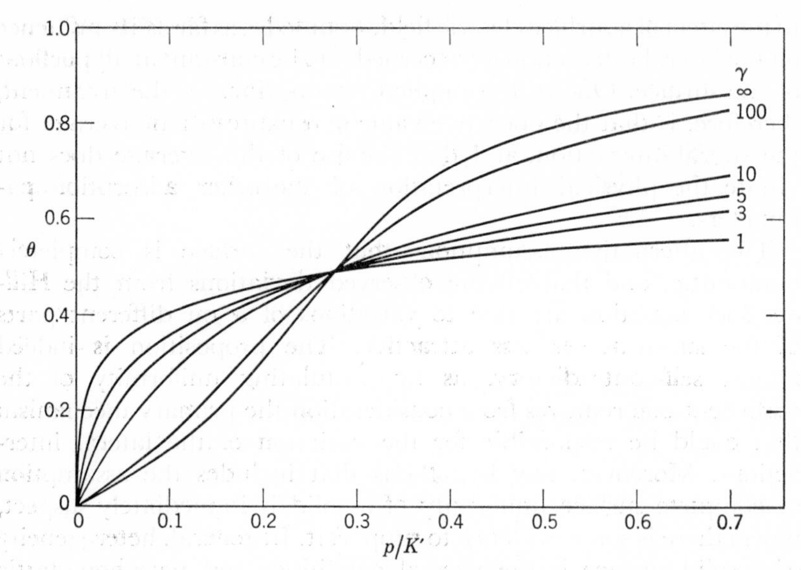

Fig. V-5. Computed isotherms for argon adsorbed as a localized film at 77.5°K by hexagonal substrates corresponding to adsorptive potential distribution curves shown in Figure IV-2.

adsorbate. This information gives us a means of estimating ω for such a substrate, assuming that the adsorbate-absorbate interaction in a localized film is of the same nature as that in a mobile film. At the critical temperature of a *mobile* adsorbed film

$$2\alpha/R\,{}_aT_c\,\beta = 4\alpha/R\,{}_gT_c\,\beta = 6.75 \tag{V-3}$$

since ${}_aT_c = 0.5\,{}_gT_c$. At the critical temperature of the *localized* adsorbed film specified by Hill,

$$c\omega/R\,{}_aT_c = 3c\,\omega/R\,{}_gT_c = 4.00 \tag{V-4}$$

By combining equations V-3 and V-4 we get $\omega = 1.185\,\alpha/\beta$, which allows us to estimate ω for an hexagonal substrate when the van der Waals constants of the adsorbate are known. (For definition of c, see equation I-43.)

For argon at 77.5°K, $c\omega/RT = 2.57$ and at 90.1°K, $c\omega/RT = 2.21$. These values of $c\omega/RT$ were used in computing a series of model isotherms for each of sixteen values of γ by the same method as that previously described.

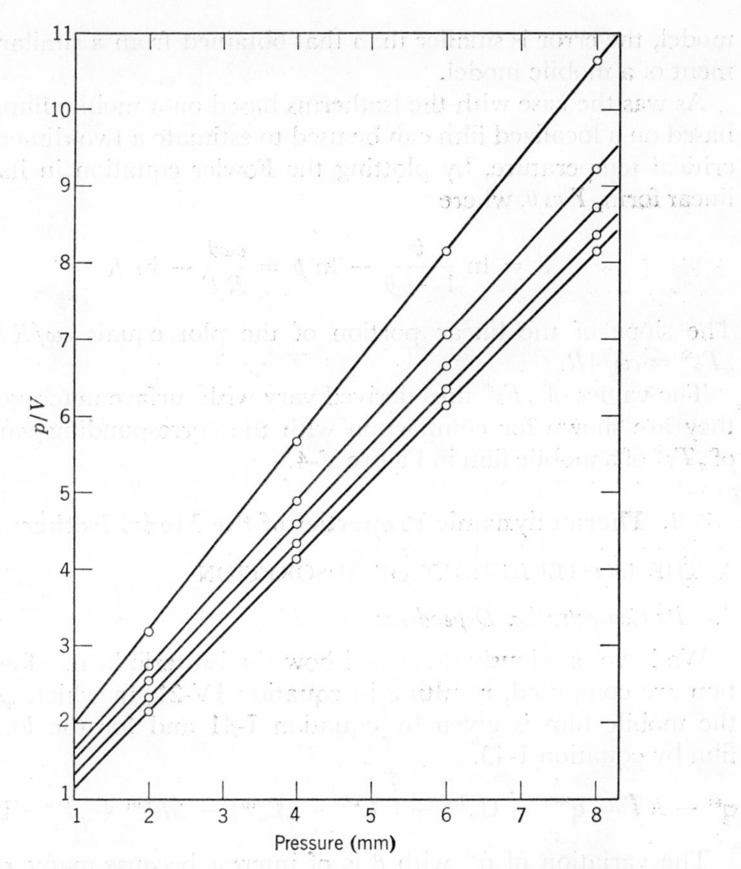

Fig. V-6. Test of the Langmuir equation as an empirical description of computed isotherms based on the model of a localized adsorbed film and a heterogeneous substrate.

For the model of a localized film a representative number of computed isotherms corresponding to argon adsorbed at 77.5 °K on a series of heterogeneous substrates, based on the distribution curves shown in Figure IV-2, are shown in Figure V-5. Here again the shapes of the isotherms vary from sigmoidal to concave. The latter isotherms are again reminiscent of the Langmuir equation and can indeed be approximated by that equation (see Fig. V-6), although the V_m thus derived is, as before, subject to error due to heterogeneity. As might perhaps be expected from a localized model, the error is smaller than that obtained from a similar treatment of a mobile model.

As was the case with the isotherms based on a mobile film, those based on a localized film can be used to estimate a two-dimensional critical temperature, by plotting the Fowler equation in its rectilinear form, Y *vs* θ, where

$$Y = \ln \frac{\theta}{1 - \theta} - \ln p = \frac{c\omega\theta}{RT} - \ln K \qquad \text{(V-5)}$$

The slope of the linear portion of the plot equals $c\omega/RT$ and ${}_aT_c{}^* = c\omega/4R$.

The values of ${}_aT_c{}^*$ thus derived vary with surface heterogeneity; they are shown for comparison with the corresponding variation of ${}_aT_c{}^*$ of a mobile film in Figure V-4.

2. Thermodynamic Properties of the Model Isotherms

A. THE ISOSTERIC HEAT OF ADSORPTION

1. *Its Concentration Dependence*

We have previously described how the isosteric heats of adsorption are computed, resulting in equation IV-21, in which ${}_aP^{\text{ia}}$ for the mobile film is given by equation I-41 and for the localized film by equation I-43.

$$\mathbf{q}^{\text{st}} - RT = \mathbf{q}^{\text{diff}} = \mathbf{U}_0{}^{\text{diff}} - ({}_a\mathbf{E}^{\text{vib}} - {}_a\mathbf{E}_0^{\text{vib}}) - \Delta E^{\text{kin}} + {}_aP^{\text{ia}} \quad \text{(IV-21)}$$

The variation of $\mathbf{q}^{\text{st}}$ with θ is of interest because many experimental data have been reported in this manner. Figure V-7 shows a representative number of model isosteric-heat curves for a mobile adsorbed film of argon on a series of heterogeneous surfaces, calculated by equation IV-21 from computed values of $\mathbf{U}_0{}^{\text{diff}}$ as defined by equation IV-16, corresponding to the distribution curves of Figure IV-2. These computed heat curves exhibit properties that are familiar to experimentalists. The initial decrease of isosteric heat with θ reflects the presence of high energy patches; at low surface concentrations the adsorbate-adsorbate interaction ${}_aP^{\text{ia}}$ does not yet play a prominent part. At higher surface concentrations the value of $\mathbf{U}_0{}^{\text{diff}}$ is reduced while that of ${}_aP^{\text{ia}}$ increases: the resultant isosteric heat may therefore go through a

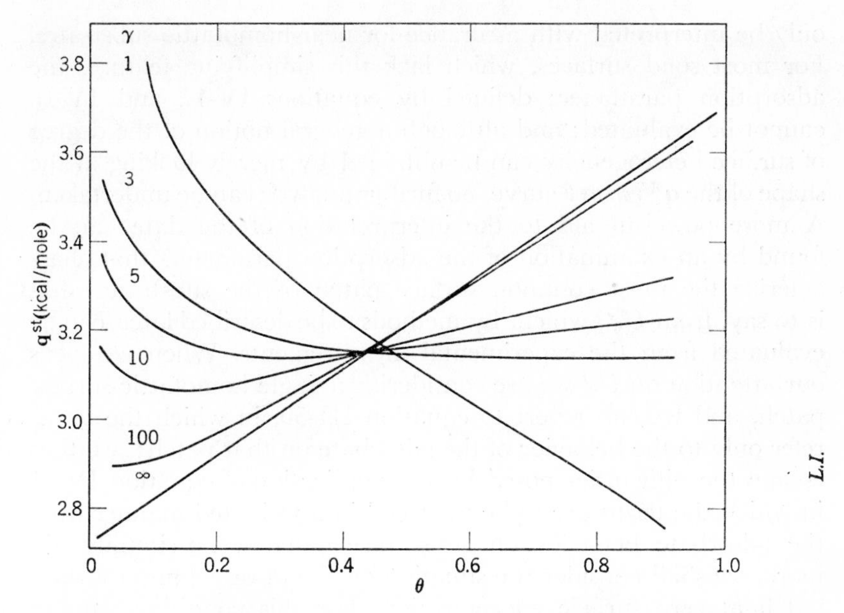

Fig. V-7. Computed isosteric heat curves as a function of θ, based on the model of a mobile adsorbed film on the series of heterogeneous substrates described in Figure IV-2.

minimum value. At still higher surface concentrations, particularly with the less heterogeneous substrates, $d({}_aP^{\text{ia}})/d\theta$ may be greater in absolute value than $d(\mathbf{U}_0^{\text{diff}})/d\theta$, with a resulting increase of $\mathbf{q}^{\text{st}}$ with θ. Finally, the falling off of $\mathbf{U}_0^{\text{diff}}$ to low values as the monolayer approaches completion causes $\mathbf{q}^{\text{st}}$ to decrease more sharply than ${}_aP^{\text{ia}}$ is increasing, which is the reason for the frequently observed maximum in $\mathbf{q}^{\text{st}}$ in this region. The model of a localized adsorbed film shows the same features as those illustrated in Figure V-7 for a mobile film. Experimental data in general show all the features represented in Figure V-7 and described above.

2. *Its Temperature Debendence*

The variation of $\mathbf{q}^{\text{st}}$ with θ, although it is readily enough determined experimentally either from adsorption isotherms at different temperatures or by direct calorimetric measurement, can only be interpreted with assurance for near-homotattic substrates. For most solid surfaces, which lack this simplifying feature, the adsorption parameters defined by equations IV-17 and IV-18 cannot be evaluated; and although a general notion of the degree of surface heterogeneity can be obtained by merely looking at the shape of the $\mathbf{q}^{st}$ *versus* θ curve, no further analysis can be undertaken. A more powerful aid to the interpretation of the data can be found by an examination of the adsorption parameters that characterize the most common surface patch of the substrate—that is to say, from U_0', which, by methods to be described later, can be evaluated from the experimental measurements. When we focus our attention on U_0' we are considering a single homotattic surface patch, and we can revert to equation III-50, in which the terms refer only to the behavior of the adsorbate on that patch; we thus escape the difficulties posed by the application of equation IV-21 in which the terms are to be averaged in a weighted manner from the adsorbate behavior on *all* surface patches. To simplify still more, we shall consider the situation on the most common surface patch at zero surface concentration. For this condition we can write:

$$\underset{\lim \theta \to 0}{q^{\text{diff}}} = U_0' - ({}_aE^{\text{vib}} - {}_aE_0{}^{\text{vib}}) + \Delta E^{\text{kin}} \qquad \text{(V-6)}$$

We now have a quantity of some fundamental interest, and one that we can often evaluate from the measurements, though with varying degrees of certainty, depending on the complexity of the observed system.

The nature of the model, i.e., mobile or localized adsorbed film, affects the vibrational and kinetic energy terms in equation V-6. The mobile adsorbed film, which we consider to behave as a two-dimensional gas, has one degree of translational freedom less than the gas phase with which it is in equilibrium. Therefore,

$$\Delta E^{\text{tr}} = -\ (^1/_2)\ RT \qquad \text{(V-7)}$$

We have defined ΔE^{kin} as

$$\Delta E^{\text{kin}} = \Delta E^{\text{tr}} + \Delta E^{\text{rot}} \qquad \text{(V-8)}$$

The value of ΔE^{rot} depends on the nature of the adsorbate molecule and the temperature of the adsorption: for monatomic molecules

ΔE^{rot} is always zero; polyatomic molecules are almost certainly not rotating freely in the adsorbed phase save at temperatures well above room temperature. No direct experimental observations of the rotational spectra of adsorbed molecules has yet been reported; the application of microwave spectroscopy to the investigation of adsorption systems could determine the rotational state of the adsorbate and would be of inestimable value, since thermal data alone provide only indirect and sometimes inconclusive evidence on this point.

The vibrational energy terms in equation V-6 do not refer to internal vibrations of the molecule, which one would not expect to be greatly affected by adsorption; they refer to the appearance of a new vibration of the adsorbed molecule as a whole with respect to the substrate. For the mobile film this vibration is normal to the surface and can be treated as a one-dimensional harmonic oscillator of frequency ν: therefore,

$$ {}_aE_0^{\text{vib}} = (^1/_2)\, Nh\nu \tag{V-9} $$

and

$$ {}_aE^{\text{vib}} = \frac{Nh\nu}{(e^{h\nu/kT} - 1)} + (^1/_2)\, Nh\nu \tag{V-10} $$

If the adsorbed film be localized a gas molecule loses all its translational energy on adsorption: i.e.,

$$ \Delta E^{\text{tr}} = -\, (^3/_2)\, RT \tag{V-11} $$

The remarks made above about ΔE^{rot} apply also to a localized adsorbed molecule. The vibrations of such a molecule consist of oscillations about its point of attachment to the surface; these oscillations are resolved along three mutually perpendicular axes, one of which is normal to the surface. Three distinct vibrational frequencies are possible, but the two that are parallel to the surface are probably equal to each other. We may, therefore, write

$$ {}_aE_0^{\text{vib}} = (^1/_2)\, Nh\nu^{\perp} + Nh\nu^{\parallel} \tag{V-12} $$

and

$$ {}_aE^{\text{vib}} = \frac{Nh\nu^{\perp}}{(e^{h\nu^{\perp}/kT} - 1)} + \frac{2Nh\nu^{\parallel}}{(e^{h\nu^{\parallel}/kT} - 1)} + (^1/_2)\, Nh\nu^{\perp} + Nh\nu^{\parallel} \tag{V-13} $$

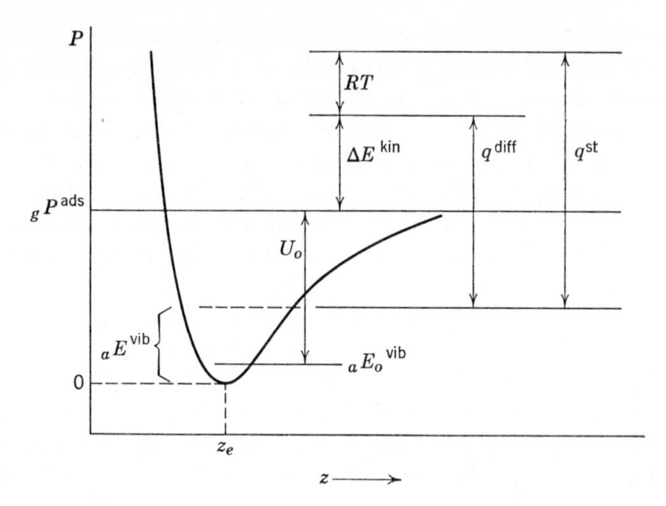

Fig. V-8. Potential energy well for adsorbate-adsorbent interaction, showing the relation of the various heats of adsorption.

The foregoing treatment shows that it is possible theoretically to distinguish between localized and mobile adsorbed films from the temperature dependence of U_0'. In practice, however, the distinction is not easily made because of the lack of independent information about the vibrational frequencies.

Figure V-8 is a schematic diagram of the energy differences between the gaseous and the adsorbed phases, which makes clear how the various potential and kinetic energies add up to give the various heats of adsorption. The quantity U_0 in this diagram is defined as

$$U_0 = {}_gP^{\text{ads}} - {}_aE_0^{\text{vib}} \tag{V-14}$$

B. THE ENTROPY OF ADSORPTION

1. *Its Temperature Dependence*

The entropy change on adsorption is the sum of changes in the translational, rotational, and vibrational entropies of the molecule:

$$\Delta S = \Delta S^{\text{tr}} + \Delta S^{\text{rot}} + {}_aS^{\text{vib}} \tag{V-15}$$

The *integral* translational entropy change per mole of a gas in its standard state (760 mm, temperature = T) to a mobile adsorbed film in its standard state (θ_s, T) is calculated, as shown below, to be:

$$ {}_gS_s^{\,\mathrm{tr}} - {}_aS_s^{\,\mathrm{tr}} = -\Delta S_s^{\,\mathrm{tr}} = \frac{R}{2}\ln M + \frac{R}{2}\ln T + 2.30 \qquad \text{(V-16)} $$

The corresponding *differential* translational entropy change is given by

$$ {}_gS_s^{\,\mathrm{tr}} - {}_a\dot{S}_s^{\,\mathrm{tr}} = -\Delta\dot{S}_s^{\,\mathrm{tr}} = \frac{R}{2}\ln M + \frac{R}{2}\ln T + 2.30 + R \qquad \text{(V-16a)} $$

In these equations, M is the molecular weight of the adsorbate. For the localized adsorbed film, the translational entropy change is simply the loss of all the translational entropy of the gas: i.e.,

$$ -\Delta S_s^{\,\mathrm{tr}} = -\Delta\dot{S}_s^{\,\mathrm{tr}} = {}_gS_s^{\,\mathrm{tr}} \qquad \text{(V-17)} $$

The translational partition function for n-dimensional space is

$$ f^{\mathrm{tr}} = \frac{(2\pi mkT)^{n/2}}{h^n}\, l^n \qquad \text{(V-18)} $$

for a gas in three dimensions, therefore, the translational partition function is

$$ {}_gf^{\mathrm{tr}} = \frac{(2\pi mkT)^{3/2}}{h^3}\, V \qquad \text{(V-19)} $$

and for a two-dimensional gas

$$ {}_af^{\mathrm{tr}} = \frac{(2\pi mkT)}{h^2}\, \sigma_s N \qquad \text{(V-20)} $$

In these equations, m is the mass of the molecule, V is the molar volume of an ideal gas at 760 mm, and $\sigma_s N$ is the molar area of an ideal two-dimensional gas in its standard state. The translational entropy is given in terms of the partition function by

$$ S^{\mathrm{tr}} = R\ln\frac{f^{\mathrm{tr}}}{N} + T\left(\frac{\partial \ln f^{\mathrm{tr}}}{\partial T}\right) + 1 \qquad \text{(V-21)} $$

for the gas phase, therefore,

$$ {}_gS_s^{\mathrm{tr}} = \frac{3}{2}R\ln M + \frac{5}{2}R\ln T - 2.31 \qquad \text{(V-22)} $$

and for the adsorbed phase

$$ {}_aS_s^{\mathrm{tr}} = R\ln M + 2\,R\ln T - 4.61 \qquad \text{(V-23)} $$

Equation V-22 refers to the gas in its standard state (760 mm); equation V-23 refers to the adsorbed phase in its standard state (σ_s = 4.08 T A²/molecule). Equation V-16 follows from these equations. The differential molar translational entropy of the adsorbed phase can be calculated from equation V-21 by means of equation III-88:

$$_a\dot{S}_s^{\text{tr}} = R \ln M + 2 R \ln T - 4.61 - R \qquad \text{(V-23a)}$$

Equation V-16a can then be derived.

The remarks previously made about the rotational states of adsorbed molecules indicated the present uncertain state of knowledge of that subject. There are no data on which to base calculations, other than *a priori* assumptions about how the molecule rotates in the adsorbed phase; the most probable model can sometimes be selected by considering the temperature dependence of the entropy of adsorption as we do in Chapter VII, or from other specific knowledge of the components of the system.

If the molecule be free to rotate in n independent ways the rotational partition function, as given by Halford (9) is

$$f^{\text{rot}} = \frac{1}{\pi\sigma}\left\{\frac{8\pi^3(I_A{}^a + I_B{}^b + \ldots I_G{}^g)^{1/n}kT}{h^2}\right\}^{n/2} \qquad \text{(V-24)}$$

where $a + b + \ldots\ldots g = n$; $I_A, I_B \ldots\ldots I_G$ are the moments of inertia and σ is the symmetry factor: hence,

$$S^{\text{rot}} = R\left[\ln \frac{1}{\pi\sigma}\left\{\frac{8\pi^3(I_A{}^a + I_B{}^b \ldots + I_G{}^g)^{1/n}kT}{h^2}\right\}^{n/2} + \frac{n}{2}\right] \qquad \text{(V-25)}$$

In these expressions, the degeneracy of the lowest energy state is omitted because it is not affected by the process of adsorption; the symmetry number is retained because it depends on the ability of the molecule to rotate and this may alter on adsorption.

The vibration of the adsorbed molecule with respect to the surface contributes to the entropy of adsorption: if the film be mobile and we assume the vibration to be that of an harmonic oscillator in one degree of freedom, for which the frequency does not change with temperature, the entropy contribution, $_aS^{\text{vib}}$, is given by

$$_aS^{\text{vib}} = \frac{_aE^{\text{vib}} - {}^1/_2\, Nh\nu}{T} - \frac{_aF^{\text{vib}} - {}^1/_2 Nh\nu}{T} \qquad \text{(V-26)}$$

or

$$_aS^{\text{vib}} = \frac{Nh\nu}{(e^{h\nu/kT} - 1)T} - R \ln (1 - e^{-h\nu/kT}) \qquad \text{(V-27)}$$

where ν is the frequency of the vibration.

If the adsorbed film be localized, and the model leading to equation V-13 is assumed, the entropy of vibration would be described by

$$_aS^{\text{vib}} = 2\left[\frac{Nh\nu^{\parallel}}{(e^{h\nu^{\parallel}/kT} - 1)T} - R \ln(1 - e^{-h\nu^{\parallel}/kT})\right]$$
$$+ \frac{Nh\nu^{\perp}}{(e^{h\nu^{\perp}/kT} - 1)T} - \text{R} \ln (1 - e^{-h\nu^{\perp}/kT}) \qquad \text{(V-28)}$$

The partition function for an harmonic oscillator is given by

$$f^{\text{vib}} = \prod_s \left(\frac{1}{1 - e^{-h\nu/kT}}\right) \qquad \text{(V-29)}$$

for the s modes of vibration. For the mobile adsorbed film with one mode of vibration, the partition function becomes

$$f^{\text{vib}} = \frac{1}{1 - e^{-h\nu/kT}} \qquad \text{(V-30)}$$

and for the localized film with two degenerate frequencies parallel to the surface and a third component normal to the surface, the partition function is

$$f^{\text{vib}} = \left(\frac{1}{1 - e^{-h\nu\|kT}}\right)^2\left(\frac{1}{1 - e^{-h\nu\perp kT}}\right) \qquad \text{(V-31)}$$

The vibrational entropy is obtained from the appropriate partition function by means of the expression

$$_aS^{\text{vib}} = R\left[\ln f^{\text{vib}} + T\,\frac{d \ln f^{\text{vib}}}{dT}\right] \qquad \text{(V-32)}$$

leading to equations V-27 and V-28.

2. *Its Concentration Dependence*

Of the three contributions to the entropy of adsorption—translational, rotational, and vibrational—only the translational entropy change is considered to be a function of θ, in determining the total entropy of adsorption as a function of θ. The rotational and vibrational entropies, while not inherently unaffected by surface concentration, may be held to introduce only "third order" effects with variations of θ.

The standard integral molar entropy change ΔS_s^{tr}, defined for a mobile film by equation V-16, is the translational entropy difference between a gas in its standard state and the adsorbed film in its standard state: the standard state of the gas is taken as 760 mm at the temperature of the experiment; the standard state of the surface phase was selected by de Boer and Kruyer (10) so that the average intermolecular separation would be the same in both phases at 0 °C. This corresponds to

$$\sigma_s = 4.08T \text{ A}^2/\text{molecule} \qquad \text{(V-33)}$$

therefore,

$$\theta_s = \frac{\beta}{\sigma_s} = \frac{\beta}{4.08T} \qquad \text{(V-34)}$$

For argon at 77.5 °K, $\sigma_s = 316.5$ A²/molecule and $\theta_s = 0.043$.

At any value of θ other than θ_s an entropy of compression (or expansion) of the two-dimensional gas, plus a partial molar entropy to account for the change in composition of the adsorbed film, must be added to the standard translational entropy change:

$$_aS^{gg} = \left(\frac{\partial \Delta S}{\partial \sigma}\right)_{an} d\sigma + \left(\frac{\partial \Delta S}{\partial_a n}\right)_{\sigma} d_a n \qquad \text{(V-35)}$$

where $_aS^{gg}$ designates an entropy term that we shall name *entropy of congregation*. For a two-dimensional van der Waals gas on a homotattic surface

$$_aS^{gg} = -R \ln \left[\frac{\theta}{1-\theta} \Big/ \frac{\theta_s}{1-\theta_s}\right] \qquad \text{(V-36)}$$

hence,

$$\Delta S^{ads} = \Delta S_s^{ads} - R \ln \left[\frac{\theta}{1-\theta} \Big/ \frac{\theta_s}{1-\theta_s}\right] \qquad \text{(V-37)}$$

The corresponding differential entropy difference for the same model is

$$\Delta \dot{S}^{ads} = \Delta \dot{S}_s^{ads} - R \ln \left[\frac{\theta}{1-\theta} \Big/ \frac{\theta_s}{1-\theta_s}\right] - R\left[\frac{1}{1-\theta} - \frac{1}{1-\theta_s}\right]$$

(V-37a)

The total entropy change on adsorption, from a gas at 760 mm to an adsorbed film at any surface concentration θ, is, therefore,

$$\Delta S^{\text{ads}} = \Delta S^{\text{tr}} + \Delta S^{\text{rot}} + {}_aS^{\text{vib}} \tag{V-38}$$

Equation V-37 may be derived as follows: the standard entropy change on adsorption is given by

$$\Delta S_s^{\text{ads}} = \frac{\Delta H_s^{\text{ads}}}{T} - \frac{\Delta F_s^{\text{ads}}}{T}$$

At any surface concentration θ other than θ_s,

$$\Delta S^{\text{ads}} = \frac{\Delta H^{\text{ads}}}{T} - \frac{\Delta F^{\text{ads}}}{T}$$

For a two-dimensional van der Waals gas on a homotattic surface:

$$\Delta H^{\text{ads}} = -q^{\text{diff}} + RT\frac{\theta}{1-\theta}, \qquad \Delta H_s^{\text{ads}} = -q_s^{\text{diff}} + RT\frac{\theta_s}{1-\theta_s} \tag{V-39}$$

and

$$\frac{q^{\text{diff}} - q_s^{\text{diff}}}{T} = \frac{2\alpha}{\beta T}(\theta - \theta_s) \tag{V-40}$$

Also

$$\frac{\Delta F^{\text{ads}} - \Delta F_s^{\text{ads}}}{T} = R\ln\frac{p}{p_s} \tag{V-41}$$

$$\left.\begin{aligned} \text{where } R\ln p &= R\ln K + R\ln\frac{\theta}{1-\theta} + R\frac{\theta}{1-\theta} - \frac{2\alpha\theta}{\beta T} \\ \text{and } R\ln p_s &= R\ln K + R\ln\frac{\theta_s}{1-\theta_s} + R\frac{\theta_s}{1-\theta_s} - \frac{2\alpha\theta_s}{\beta T} \end{aligned}\right\} \tag{V-42}$$

The substitutions give

$$\Delta S^{\text{ads}} - \Delta S_s^{\text{ads}} = -R\left[\ln\frac{\theta}{1-\theta}\bigg/\frac{\theta_s}{1-\theta_s}\right] \tag{V-43}$$

The expression for the differential entropy change, equation V-37a, can also be derived by an analogous method, using differential enthalpies and differential free energies of adsorption.

In a localized adsorbed film the adsorbate molecules have lost all three degrees of translational freedom, and perhaps rotational freedom as well, but have gained a certain amount of vibrational freedom. The thermal entropy of the surface phase, therefore, is made up of the rotational and vibrational entropies, neither of which, as we pointed out previously, are held to vary with θ. A new contribution to the total entropy of a localized film arises, however,

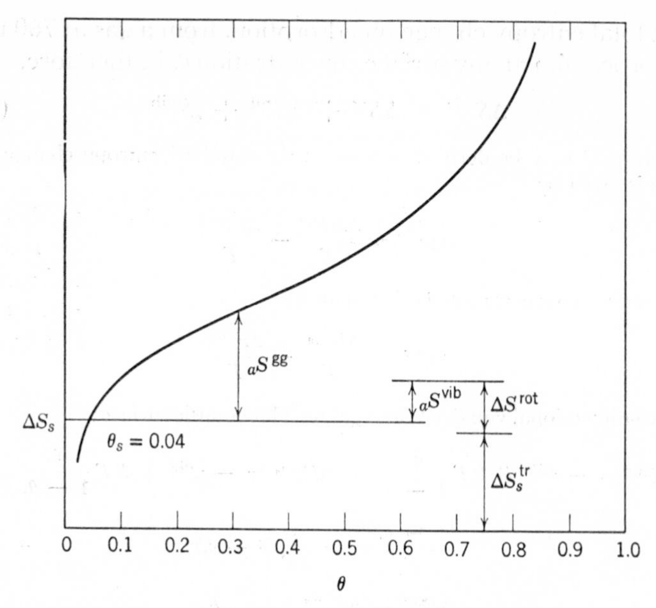

Fig. V-9. Differential molar entropies of adsorption for the case of a mobile adsorbed film on a homotattic surface.

from the number of ways of arranging the molecules among the surface sites (11): this is known as the configurational entropy of the system. For localized adsorbed films described either by equation I-19b or I-20b, the integral molar configurational entropy is given by

$$_aS^{\text{fig}} = -R\left[\ln\theta + \frac{1-\theta}{\theta}\ln(1-\theta)\right] \qquad \text{(V-44)}$$

The integral molar entropy, therefore, is

$$\Delta S^{\text{ads}} = \Delta S_s^{\text{ads}} - R\left[\ln\frac{\theta}{\theta_s} + \frac{1-\theta}{\theta}\ln(1-\theta) - \frac{1-\theta_s}{\theta_s}\ln(1-\theta_s)\right] \qquad \text{(V-45)}$$

where

$$\Delta S_s^{\text{ads}} = \Delta S_s^{\text{tr}} + \Delta S^{\text{rot}} + {}_aS^{\text{vib}} \qquad \text{(V-46)}$$

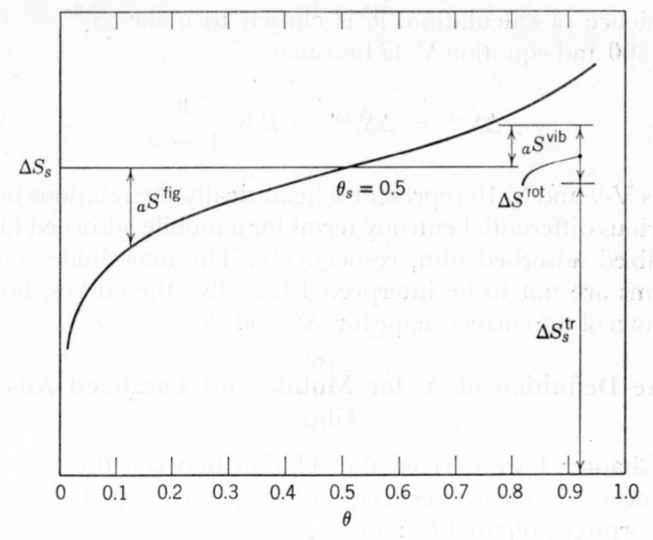

Fig. V-10. Differential molar entropies of adsorption for the case of a localized adsorbed film on a homotattic surface.

The corresponding differential molar entropies are

$$ {}_a\dot{S}^{\text{fig}} = -R \ln \frac{\theta}{1-\theta} $$

and

$$ \Delta\dot{S}^{\text{ads}} = \Delta\dot{S}_s^{\text{ads}} - R \ln \left[\frac{\theta}{1-\theta} \bigg/ \frac{\theta_s}{1-\theta_s} \right] \qquad \text{(V-47)} $$

where

$$ \Delta\dot{S}_s^{\text{ads}} = \Delta\dot{S}_s^{\text{tr}} + \Delta\dot{S}^{\text{rot}} + {}_a\dot{S}^{\text{vib}} $$

For the localized adsorbed film the standard state cannot be selected in the same way as that for the mobile adsorbed film: the distance of closest approach of two adsorbed molecules is determined by the spacing of the adsorption sites and not solely by the molecular dimensions of the adsorbate. Besides, there is no point in comparing what is essentially a two-dimensional solid with a three-dimensional gas. The standard state can therefore be selected for convenience of calculation. θ_s is chosen to make ${}_a\dot{S}_s^{\text{fig}} = 0$: i.e., $\theta_s = 0.500$ and equation V-47 becomes

$$\Delta\dot{S}^{\text{ads}} = \Delta\dot{S}_s^{\text{ads}} - R \ln \frac{\theta}{1-\theta} \tag{V-47a}$$

Figures V-9 and V-10 represent schematically the relations between the various differential entropy terms for a mobile adsorbed film and a localized adsorbed film, respectively. The magnitudes on these diagrams are not to be interpreted literally; the curves, however, are drawn of the correct shape for ${}_a\dot{S}^{\text{gg}}$ and ${}_a\dot{S}^{\text{fig}}$.

3. The Definition of A^0 for Mobile and Localized Adsorbed Films

In Chapter I we derived the relation between the constant K, which occurs in the isotherm equations appearing in Table I-1, and the adsorptive potential U_0: i.e.,

$$K = A^0 e^{-U_0/RT} \tag{I-32}$$

where

$$\ln A^0 = -\frac{\Delta\dot{S}^{\text{ads}}}{R} + ({}_aE^{\text{vib}} - {}_aE_0^{\text{vib}})/RT + \Delta E^{\text{kin}}/RT$$
$$- {}_aP^{\text{ia}}/RT - \ln g(\theta, T) + \ln 760 - 1 + \frac{\beta}{R\theta}\left(\frac{\partial \Pi}{\partial T}\right)_\theta \tag{I-33}$$

Equation I-33 is completely general; but we now have defined models of mobile and localized adsorbed films in sufficient detail to enable us to express all the quantities in equation I-33 in terms of these models.

For the mobile adsorbed film that is described by the Hill-de Boer isotherm, equation I-18b,

$$\ln g(\theta, T) = \ln \frac{\theta}{1-\theta} + \frac{\theta}{1-\theta} - \frac{2\alpha\theta}{RT\beta}$$

We evaluate

$$\frac{\beta}{R\theta}\left(\frac{\partial \Pi}{\partial T}\right)_\theta - 1 = \frac{\theta}{1-\theta}$$

by means of equation I-18a: $g(\theta, T)$ is defined by equation I-24: and from equation I-41,

$$\frac{{}_aP^{\text{ia}}}{RT} = \frac{2\alpha\theta}{RT\beta} \tag{V-49}$$

Substituting these equations in equation I-33 and using equations V-38 and V-43 gives a particular expression for ln A^0, evaluated for $\theta = \theta_s$, relevant to this model:

$$\ln A^0 = -\left(\frac{\Delta S_s^{\text{tr}}}{R} + \frac{\Delta S^{\text{rot}}}{R} + \frac{{}_aS^{\text{vib}}}{R}\right) + ({}_aE^{\text{vib}} - {}_aE_0^{\text{vib}})/RT + \frac{\Delta E^{\text{kin}}}{RT} - \ln\frac{\theta_s}{1-\theta_s} + \ln 760 \tag{V-50}$$

The terms on the right-hand side of equation V-50 that depend on the vibrational frequency ν are not fixed by the model. The frequency-dependent terms can be consolidated by invoking equation V-26:

$$\ln A^0 = -\left(\frac{\Delta S_s^{\text{tr}}}{R} + \frac{\Delta S^{\text{rot}}}{R}\right) + ({}_aF^{\text{vib}} - {}_aE_0^{\text{vib}})/RT + \frac{\Delta E^{\text{kin}}}{RT} - \ln\frac{\theta_s}{1-\theta_s} + \ln 760 \tag{V-51}$$

For the localized adsorbed film that is described by the Fowler isotherm, equation I-20b,

$$\ln g(\theta, T) = \ln\left(\frac{\theta}{1-\theta}\right) - \frac{c\omega\theta}{RT} \tag{V-52}$$

$$\frac{\beta}{R\theta}\left(\frac{\partial \Pi}{\partial T}\right)_\theta = \frac{1}{\theta}\ln\left(\frac{1}{1-\theta}\right)$$

and from equation I-43,

$$_aP^{\text{ia}}/RT = \frac{c\omega\theta}{RT} \tag{V-53}$$

Substituting these equations, and equation V-45 for ΔS^{ads}, in equation I-33, gives us a particular expression for ln A^0 relevant to the localized adsorbed film:

$$\ln A^0 = -\left(\frac{\Delta S_s^{\,\mathrm{tr}}}{R} + \frac{\Delta S^{\mathrm{rot}}}{R}\right) + ({}_aF^{\mathrm{vib}} - {}_aE_0{}^{\mathrm{vib}})/RT + \frac{\Delta E^{\mathrm{kin}}}{RT}$$

$$- \ln \theta_s - 1 - \left(\frac{1 - \theta_s}{\theta_s}\right) \ln (1 - \theta_s) + \ln 760 \quad \text{(V-54)}$$

4. A Limiting Case of the Models: Low Surface Concentration

A. THE INITIAL SLOPE OF THE ISOTHERM

We have shown previously in Section IV-3 that when the adsorption isotherm for each surface patch i has the form $p = K_i\theta_i$, the resulting isotherm for the heterogeneous surface, regardless of the form of the distribution function, is

$$p = \theta/C \quad \text{(IV-5)}$$

The Henry's Law isotherm, $p = K_i\theta_i$, is true of an ideal two-dimensional gas; all other types of adsorbed film reduce to nearly ideal behavior at sufficiently low surface concentrations (about $\theta_i < 0.02$). Provided that all, or nearly all, surface patches are occupied to an extent less than 2% of the monolayer, the adsorption isotherm for any system will have the linear form of equation IV-5. A practical limitation of this condition is that the surface must not have too wide a range of adsorptive potentials, otherwise the linear portion of the isotherm will occur at total surface concentrations too small to be observed. Some experimental systems meet the necessary conditions and it therefore becomes important to evaluate the constant in equation IV-5 in terms of the Gaussian distribution of adsorptive potentials.

The equation for the Gaussian distribution of adsorptive potentials is

$$\Phi(U_0)dU_0 = \frac{1}{n} e^{-\gamma(U_0 - U_0')^2}\, dU_0 \quad \text{(IV-7)}$$

The limiting isotherm for a heterogeneous surface is

$$\theta = p \int_e^g \frac{\Phi(U_0)dU_0}{K_i} \quad \text{(IV-5)}$$

where

$$K_i = A^0\, e^{-U_{0i}/RT} \quad \text{(IV-10)}$$

Combining the above equations:

$$\theta = \frac{p}{nA^0} \int_e^g e^{-\gamma(U_0 - U_0')^2 + (U_0/RT)} \, dU_0 \tag{V-55}$$

Let $X = U_0 - U_0'$; then,

$$\theta = \frac{p}{nA^0} e^{-U_0/RT} \int_{U_0'-e}^{g-U_0'} e^{-\gamma X^2 + (X/RT)} \, dX \tag{V-56}$$

For values of γ greater than about 1, the integrand becomes negligible within the range of the limits of the integral in equation V-56: we can without error, therefore, set the limits of integration as $-\infty$ to $+\infty$; the normalizing factor n then becomes $\sqrt{\pi/\gamma}$. The value of the definite integral is given (12) as

$$\int_{-\infty}^{+\infty} e^{-\gamma X^2 + (X/RT)} \, dX = \sqrt{\frac{\pi}{\gamma}} \, e^{1/4\gamma(RT)^2} \tag{V-57}$$

Equation V-56 becomes

$$p = K'\theta e^{-1/4\gamma(RT)^2} \tag{V-58}$$

where, from equation IV-10

$$K' = A^0 e^{-U_0'/RT} \tag{V-59}$$

The constant C in equation IV-5 is then given by

$$C = \frac{1}{A^0} e^{U_0'/RT + 1/4\gamma(RT)^2} \tag{V-60}$$

Let the slope of the experimental isotherm $(\partial V/\partial p)_{T,\Sigma}$, equal Z; then

$$Z = CV_\beta \tag{V-61}$$

hence,

$$\ln Z + \ln A^0 = \frac{U_0'}{RT} + \ln V_\beta + \frac{1}{4\gamma(RT)^2} \tag{V-62}$$

The method of applying equation V-62 to experimental data is discussed in the next chapter.

B. THE HEATS OF ADSORPTION AT ZERO COVERAGE

The isosteric heat of adsorption is defined by equation I-34

$$q^{st} = RT^2 \left(\frac{\partial \ln p}{\partial T}\right)_\theta \tag{I-34}$$

Applied to the limiting form of the adsorption isotherm for a heterogeneous surface, equation V-58, this equation gives

$$\lim_{\theta \to 0} \mathbf{q}^{st} = RT^2 \frac{d \ln K'}{dT} + \frac{1}{2\gamma RT} \tag{V-63}$$

For the average homotattic patch of this surface (considered independently as the whole adsorbent) the limiting adsorption isotherm is $p = K'\theta$ and the application of equation I-34 gives

$$\lim_{\theta \to 0} q^{st} = RT^2 \frac{d \ln K'}{dT} \tag{V-64}$$

These equations show that the limiting isosteric heat of adsorption for a heterogeneous surface is greater by an amount $1/2\,\gamma RT$ than

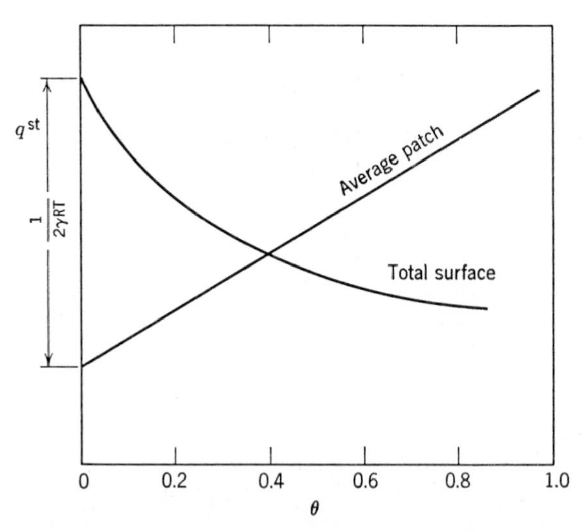

Fig. V-11. Variation of the isosteric heat on the whole heterogeneous surface and on its average patch as a function of surface concentration, showing the relation between the two heats at zero surface concentration.

the limiting isosteric heat of adsorption for a homotattic surface whose adsorptive potential is the same as that of the average of the Gaussian distribution. This relation is illustrated schematically in Figure V-11. The equations derived in this section for the limiting values of the isosteric heats are all equally applicable to mobile and to localized adsorbed films: the distinction between them arises only in the A^0 term, where the appropriate equations for the two models are equations V-50 and V-54, respectively. In Figure V-11 the difference between the two models, other than the curves themselves, lies in the points of intersection, which for the mobile adsorbed film is at $\theta \simeq 0.4$, and for the localized adsorbed film at $\theta = 0.5$.

From the discussion in Section IV-6 and equation IV-21 it is apparent that at zero coverage

$$\lim_{\theta \to 0} \mathbf{q}^{st} = \mathbf{U}_0^{diff} - ({}_a\mathbf{E}^{vib} - {}_a\mathbf{E}_0^{vib}) - \Delta E^{kin} + RT \quad \text{(V-65)}$$

hence,

$$\lim_{\theta \to 0} \mathbf{U}_0^{diff} = U_0' + \frac{1}{2\gamma RT} \quad \text{(V-66)}$$

if we assume that the average vibrational frequency of the adsorbed atoms, as defined in equations IV-17 and IV-18, at the limit of zero surface concentration, is the same as the frequency of the atoms on the average patch of the surface; the assumption cannot be greatly in error save for extremely heterogeneous substrates.

For some actual surfaces the distribution of the highest energy sites may not be described accurately by the Gaussian formula; equation V-63 above, which is based on the validity of the Gaussian distribution, can be used to indicate whether a significant deviation exists from the extremity of this distribution.

References

1. S. Ross and W. Winkler, *J. Colloid Sci.*, **10,** 319 (1955).
2. H. Clark and S. Ross, *J. Am. Chem. Soc.*, **75,** 6081 (1953).
3. J. H. de Boer and S. Kruyer, *Trans. Faraday Soc.*, **54,** 540 (1958).
4. J. L. Shereshefsky and B. R. Mazumder, *J. Phys. Chem.*, **63,** 1630 (1959); see also Discussion of this paper by S. Ross, *ibid.*, **63,** 1638 (1959).
5. S. Ross and H. Clark, *J. Am. Chem. Soc.*, **76,** 4291, 4297 (1954).
6. B. B. Fisher and W. G. McMillan, *J. Chem. Phys.*, **28,** 549, 555, 563 (1958).

7. R. H. Fowler and E. A. Guggenheim, *Statistical Thermodynamics*, Cambridge University Press, Cambridge, 1949, p. 431, equation 1007, 10.
8. T. L. Hill, *J. Phys. Chem.*, **59,** 1065 (1955).
9. J. O. Halford, *J. Chem. Phys.*, **2,** 694 (1934).
10. J. H. de Boer and S. Kruyer, *Koninkl. Ned. Akad. Wetenschap. Proc.*, **55B,** 451 (1952).
11. D. H. Everett, *Proc. Chem. Soc. (London)*, **1957,** 38.
12. D. B. de Haan, *Nouvelles Tables d'Integrales Definies*, G. E. Stechert, New York, 1939.

CHAPTER VI

Methods of Evaluating Parameters of the Theory from Observed Isotherms

To have a complete quantitative description of the adsorbed state we should know the extent of the surface under investigation; how the adsorbed molecules interact with the surface and with each other; the area occupied by an adsorbed molecule; the translational, rotational, and vibrational degrees of freedom of molecules on the surface; and all of these as functions of temperature and concentration. No adsorption system can possibly be measured at present in such detail, but techniques other than the measurement of adsorption isotherms can give us information about some of these conditions: e.g., electron microscopy, field emission microscopy, calorimetry, infrared and microwave absorption spectroscopy, and rates of diffusion through porous solids. The full application of these techniques still lies in the future; at present we must rely almost entirely on the measurement of adsorption isotherms, which require for their analysis the creation of a feasible model and a number of supporting assumptions. In the previous chapters we have developed models for the adsorption process. We now propose to bring these models to the test of experience by evaluating for specific physical systems the following characteristics: the distribution of adsorptive potentials of a surface in terms of the Gaussian parameters γ and U_0'; the mutual interaction of the adsorbate, $2\alpha/\beta$; the limiting area of the adsorbed molecule, β; the monolayer capacity of the surface, V_β; the changes of translational and rotational energies and entropies on adsorption; and the vibrational frequency, ν.

1. Method of Relating Experimental Isotherms to Computed Model Isotherms

The tables of computed isotherms in the Appendix differ from experimentally obtained isotherms in that they record the variation of θ with p/K' rather than the experimentally observed V *versus* p. On comparing experimentally observed data with the computed isotherms, V is related to θ through the constant V_β (defined in Section I-2), which appears graphically as a scale factor for the ordinate axis; K', which is defined by equation V-59, is a scale factor for the pressure axis (Section IV-4). The constants V_β and K' are therefore determined when the experimental and computed isotherms are brought into coincidence. The remaining two parameters, γ and $2\alpha/\beta$, determine the shape of the isotherm. For some simple adsorbates, e.g., the inert gases, the values of $2\alpha/\beta$ can be determined *as a first approximation* without reference to adsorption data by means of equation I-42; but even for argon, krypton, and xenon appreciable deviations from the calculated values may be introduced by the presence of a sufficiently strong

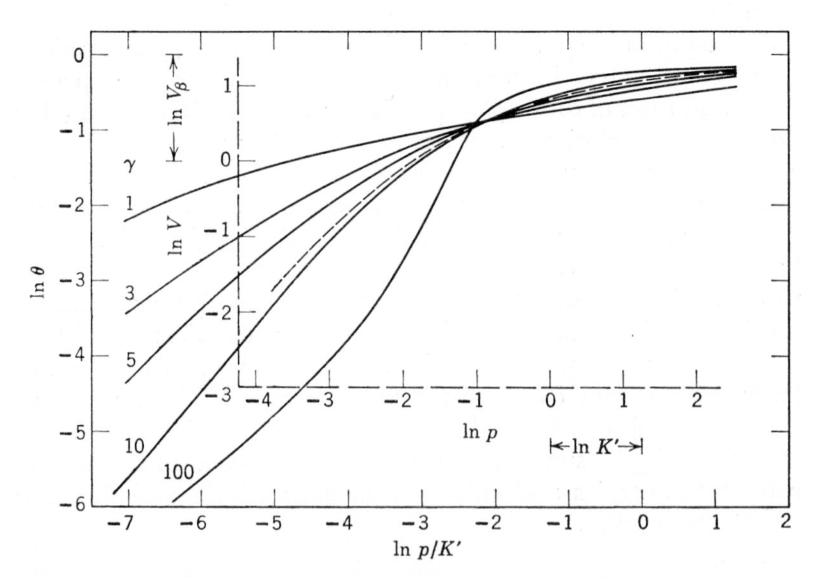

Fig. VI-1. Model isotherms (solid lines) and a superposed experimental isotherm (dashed line) in its proper interpolated position.

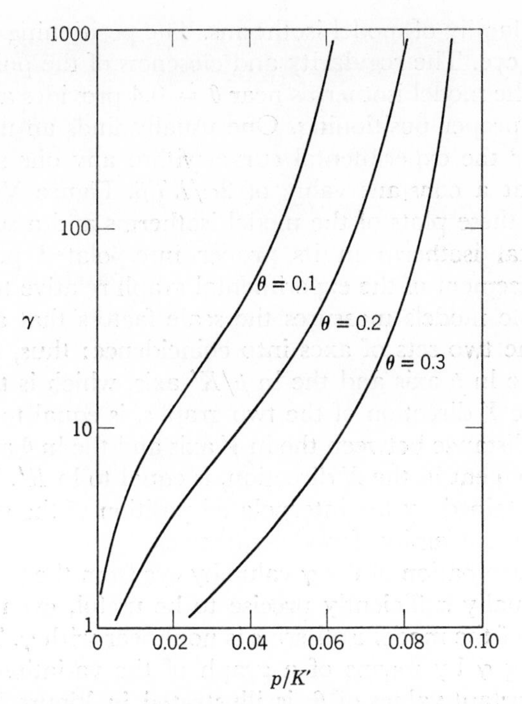

Fig. VI-2. Graphical aid for the interpretation of γ. Values of p/K' and V_β can be obtained with more certainty than the value of γ; hence, p/K' at $\theta = 0.1$, 0.2, and 0.3 are obtained and the three corresponding values of γ read from this diagram.

electric field emanating from the surface. Assuming for the moment, however, that a value of $2\alpha/\beta$ has been obtained for the adsorbate with the required accuracy by equation I-42 or by methods discussed later in this chapter, the final parameter γ can then be found by matching the shape of the experimental isotherm against those of model isotherms computed for the given value of $2\alpha/RT\beta$.

The model isotherms corresponding to the value of $2\alpha/RT\beta$ chosen for the adsorbate and temperature required are plotted on a large sheet of rectangular coordinate paper as $\ln \theta$ *versus* $\ln p/K'$ for a series of γ-values. The experimental isotherm is plotted on a separate sheet of paper to the same scale as $\ln V$ *versus* $\ln p$. The two sheets of graph paper are superimposed so that the axes are parallel and the curve for the experimental points is interpolated within the family of model isotherms. The positioning of the curve is done by eye. The regularity and closeness of the points of intersection of the model isotherms near $\theta = 0.4$ provides an additional aid to the proper positioning. One usually finds an unambiguous location for the experimental curve within any one set of model isotherms at a constant value of $2\alpha/RT\beta$. Figure VI-1 shows a number of these plots of the model isotherms and a superimposed experimental isotherm in its proper interpolated position. The X-Y displacement of the experimental graph relative to that of the graph of the models measures the scale factors that are required to bring the two sets of axes into coincidence: thus, the distance between the $\ln p$ axis and the $\ln p/K'$ axis, which is the displacement in the Y direction of the two graphs, is equal to $\ln V_\beta$; similarly, the distance between the $\ln V$ axis and the $\ln \theta$ axis, which is the displacement in the X direction, is equal to $\ln K'$. The parameter γ is obtained by the interpolated position of the experimental curve within the family of model isotherms.

The determination of the γ value by eye from the position of the curve is usually sufficiently precise to be useful, even though the separation of the model isotherms is nonlinear with γ. The method of obtaining γ by means of a graph of the variations of γ with p/K' at constant values of θ, is illustrated in Figure VI-2, which is based on the computed tables for $2\alpha/RT\beta = 6.51$. The values of p/K' at $\theta = 0.1$, 0.2, and 0.3 are obtained from the experimental data and applied to Figure VI-2, from which three estimates of γ can be read off and averaged.

A circumstance that must be observed is that the model isotherms were computed for $T = 77.5\,°\mathrm{K}$ in equation IV-10

$$K_i = A^0 e^{-U_{0i}/RT} \tag{IV-10}$$

The computed isotherms can be adapted to match adsorption data obtained at a temperature other than 77.5 °K. Suppose a computed isotherm of a certain shape is selected as a match to the experimental data; the question arises, is the value of γ attached to this computed isotherm still meaningful at a temperature other than 77.5 °K? In other words, what value of γ would need to be used to produce a computed isotherm of the same shape if the experimental temperature T had been used in equation IV-10? The shape of the computed isotherm is determined by the set of K_i and

f_i values associated with it; to produce isotherms of the same shape at two different temperatures the necessary conditions are:

$$K_{i,\,77.5} = K_{i,\,T}$$

and

$$f_{i,\,77.5} = f_{i,\,T}$$

hence,

$$U_{0i,\,77.5}/77.5 = U_{0i,\,T}/T \qquad \text{(VI-1)}$$

and

$$\gamma_{77.5}(U_{0i} - U_0')^2{}_{77.5} = \gamma_T(U_{0i} - U_0')^2{}_T \qquad \text{(VI-2)}$$

Substituting equation VI-1 in VI-2 gives

$$\gamma_T = \gamma_{77.5}\left(\frac{77.5}{T}\right)^2 \qquad \text{(VI-3)}$$

The answer to the question posed earlier is: γ_T as given by equation VI-3 is the value of γ required to produce a computed isotherm of the desired shape at the temperature of the experiment.

A numerical example will help to make clear the manner in which this correction is applied. An experimental adsorption isotherm was measured at 273°K; this isotherm was matched against the computed isotherms for the appropriate value of $2\alpha/RT\beta$ with T equal to 273°K, yielding values of V_β, K', and an apparent value of $\gamma(\gamma_{77.5})$ of, say, 120. The true value of $\gamma(\gamma_T)$ that actually represents the distribution of adsorptive potentials needed to make the model correspond to the data is given by $\gamma = 120\ (77.5/273)^2 = 9.7$.

The reader must not suppose from the foregoing that γ is in any way assumed to be dependent on temperature. The correction that we have just introduced arises from using a set of tables for temperatures other than the one for which they were computed; the advantage of this procedure is that we avoid having to compute additional sets of tables for every possible experimental temperature. References to γ, when they are unqualified, that are made throughout this book are to the true value of γ.

An important practical corollary of equation VI-3 is that the adsorption isotherm of a heterogeneous substrate approaches that of a completely homotattic substrate as the temperature of the adsorption is increased: thus, for example, an adsorbent for which $\gamma = 100$ would, at a temperature of 310°K (4 × 77.5°), produce an isotherm with an apparent γ of 1600, which is hardly to be distinguished from that corresponding to $\gamma = \infty$. In practice, therefore, the effect of a small degree of heterogeneity is better observed in the isotherm at a low temperature, and conversely, the isotherm can be made to simulate more and more closely the ideal model described by equation I-18 as the temperature is raised.

2. Perturbations of Lateral Interaction Potentials by the Substrate

A. ORIENTATION OF THE ADSORBATE

In Section I-7 we discussed the lateral interaction potential of an adsorbed phase described by the two-dimensional van der Waals equation of state and found

$$ {}_aP^{\mathrm{ia}} = \frac{2\alpha\theta}{\beta} \qquad \text{(I-41)} $$

For the simple and ideal situation the adsorbed molecules, as far as their mutual interaction is concerned, are identical with those in the gas phase: this situation permits us to calculate the two-dimensional van der Waals constants α and β from the familiar (three-dimensional) constants a and b, as follows:

$$ \alpha^{id} = a\left(\frac{9\pi}{256\,b}\right)^{1/3} \qquad \text{(VI-4)} $$

and

$$ \beta^{id} = 2b\left(\frac{9\pi}{256\,b}\right)^{1/3} \qquad \text{(VI-5)} $$

hence, equation I-42 follows immediately: $\alpha^{id}/\beta^{id} = a/2b$. We now find it advantageous to distinguish the ideal values of α and β, calculated by equations VI-4 and VI-5, by the symbols α^{id} and β^{id}. When the symbols α and β are not thus differentiated they will refer to values of the van der Waals constants that are actually operative in the description of the isotherm. Equations VI-4 and VI-5 are not true when the adsorbate is either oriented or polarized by the substrate; nevertheless, the values of α^{id} and β^{id} may be corrected for effects due to orientation and polarization. In Table VI-1 we report, for a number of gases and vapors, the three-dimensional van der Waals constants, a and b (molar values), and the two-dimensional constants, α^{id} and β^{id} (molecular values), derived therefrom. The critical temperature of each gas and its ideal two-dimensional critical temperature is also included, as well as the average diameter calculated from b.

An adsorbate molecule can be oriented at the surface if, for a given orientation, the corresponding adsorptive potential is a maximum, and if the energy required to displace it from this orientation is several times the thermal energy, kT. In the absence of an electric field, this condition could be realized by an asymmetric polyatomic molecule for which the dispersion forces of attraction to the substrate are markedly greater for a particular orientation. In the presence of an electric field, an adsorbate with a permanent dipole moment would tend to have its dipole aligned with the field, and this effect may overbalance the tendency to assume the orientation associated with maximum dispersion forces. The two effects may, however, also cooperate in favor of the same orientation. Molecules whose polarization ellipsoid is asymmetric, whether or not there is a permanent dipole, will experience an orienting moment when in an electric field, which may counteract or add to the orienting moments due to other effects.

Orientation, no matter by what mechanism it is produced, affects the lateral interaction potential: de Boer (1) has derived the following expression for α.

$$\alpha = 3\pi\, \xi^2 E/16d^4 \qquad \text{(VI-6)}$$

in which ξ is the effective polarizability of the adsorbate molecules in the plane of the adsorbed film, E is their "characteristic energy" related to the ionization potential, and d is their effective van der Waals diameter: when the molecules are oriented the values of ξ and d are no longer the average values appropriate for the freely rotating molecules. These average values may be used in equation VI-6 to obtain α^{id}, also defined by equation VI-4. For the freely rotating molecule the average diameter is related to the van der Waals b and hence to β^{id}:

$$\beta^{id} = \pi(d^{\text{average}})^2/2 \qquad \text{(VI-7)}$$

TABLE VI-1

Adsorbate Interaction Parameters and Critical Temperatures

	Molar		Molecular				
	a ergs cm^3 mole^{-2} $\times 10^{-12}$	b cm^3 mole^{-1}	d cm $\times 10^8$	α^{id} ergs cm^2 molecule^{-2} $\times 10^{30}$	β^{id} cm^2 molecule^{-1} $\times 10^{16}$	${}_gT_c$, °K	${}_aT_c^{id}$, °K
Helium	0.034	23.7	2.66	1.32	11.1	5.2	2.6
Hydrogen	0.245	26.8	2.77	9.15	12.1	33.2	16.6
Neon	0.212	17.1	2.39	9.20	8.95	44.4	22.2
Argon	1.35	32.2	2.95	47.4	13.6	151	75.5
Krypton	2.33	39.8	3.17	76.0	15.7	210	105
Xenon	4.34	51.0	3.43	131	18.5	290	145
Nitrogen	1.39	39.2	3.15	45.7	15.5	126	63
Oxygen	1.36	31.8	2.93	48.0	13.5	154	77
Chlorine	6.53	56.2	3.54	191	19.7	417	208
Carbon monoxide	1.49	39.8	3.17	48.6	15.7	134	67
Carbon dioxide	3.61	42.7	3.23	115	16.4	304	152
Nitrous oxide	3.80	44.2	3.27	120	16.8	310	155

Sulfur dioxide	6.75	56.4	3.55	196	19.8	430	215
Hydrogen sulfide	4.45	42.9	3.24	142	16.4	373	187
Carbon disulfide	11.7	76.8	3.94	310	24.3	546	273
Ammonia	4.2	37.0	3.08	141	15.0	405	203
Methane	2.26	42.8	3.24	75.0	16.4	191	95.5
Ethane	5.49	63.8	3.72	153	21.7	305	153
Ethylene	4.48	57.1	3.57	132	20.0	283	141
Acetylene	4.41	51.3	3.44	134	18.6	309	155
Propane	8.70	84.4	4.05	222	25.8	369	189
Cyclohexane	22.8	142.4	4.83	488	36.6	554	277
Benzene	18.1	115.4	4.51	427	32.7	562	281
Methyl chloride	7.52	64.8	3.72	210	21.7	416	208
Chloroform	15.2	102.2	4.33	368	29.4	536	268
Carbon tetrachloride	20.4	138.3	4.78	441	35.9	556	278
Chlorobenzene	25.5	145.3	4.86	542	37.1	632	316
Dimethyl ether	8.11	72.5	3.86	217	23.4	399	200
Diethyl ether	17.5	134.4	4.72	380	34.9	467	234
Acetone	14.0	99.4	4.28	337	28.9	508	254
Water	5.46	30.5	2.89	196	13.1	647	323

For the oriented molecule, d is related to the experimentally observed value of β by an analogous equation:

$$\beta = \pi(d^{\text{oriented}})^2/2 \qquad \text{(VI-8)}$$

For the freely rotating molecule the time-average polarizability is independent of direction and is related to the polarizabilities along the three principal axes of the molecule by the expression

$$\xi = (\xi_1 + \xi_2 + \xi_3)/3; \qquad \text{(VI-9)}$$

whereas the oriented molecule has a polarizability in the plane parallel to the surface given by

$$\xi = (\xi_2 + \xi_3)/2, \qquad \text{(VI-10)}$$

if we consider the ξ_1 axis to be normal to the surface. In Table VI – 2 we quote from Landolt-Bornstein (2) a compilation of electron polarizabilities along the three principal axes of a number of substances commonly used as adsorbates.

The correction of α^{id} for the effect of orientation can be made by means of a factor defined as

$$\omega = (d^4/\xi^2)^{id}(\xi^2/d^4)^{op} \qquad \text{(VI-11)}$$

where the superscripts *id* and *op* refer to values of d and ξ for the ideal (freely rotating) and for the operative or actual two-dimensional phase, respectively; hence,

$$\alpha = \omega\alpha^{id} \qquad \text{(VI-12)}$$

In addition a second-order correction due to the Keesom alignment effect should be applied when the adsorbate molecules possess a large (>1 debye) permanent dipole moment: the dipoles of such molecules are strong enough to slightly restrict their rotation in the gas phase, thus leading to a contribution to their mutual attraction. If the molecules are now oriented in the adsorbed phase the correction of α^{id} is completely given by the factor ω; the Keesom alignment effect must therefore be eliminated from the calculation in order to avoid an excess correction. Let the van der Waals a be considered as the sum of two terms a^{id}, and a_μ, where a^{id} refers to a condition of unrestricted free rotation and a_μ is the result of Keesom alignment: then

$$\alpha = \omega\left(1 - \frac{a_\mu}{a}\right)\alpha^{id} \qquad \text{(VI-13)}$$

where a_μ is given approximately by

$$a_\mu = 4\pi \mu^4/9RTd^3 \qquad \text{(VI-14)}$$

In equation VI-13 the factor a_μ/a is small, being about 0.10 for $\mu = 1.7D$ and 0.01 for $\mu = 1D$.

B. POLARIZATION OF THE ADSORBATE

The direction of the electric field at a solid surface, at least where its effect has been directly measured by means of contact potentials, is such that it tends to orient the dipoles, whether induced or permanent, in the adsorbed layer, with their positive ends farther away from the surface. This results in the positive surface potential that has been found experimentally by Mignolet (3) and others. As well as tending to orient the adsorbed molecules in the way we have just described, the presence of an electric field at the interface has also another important effect on the adsorbate: namely, the induction of a dipole moment in the adsorbed molecules. The existence of parallel-oriented dipoles in the adsorbed phase, which on many grounds may reasonably be inferred, introduces mutual repulsion that partially counteracts the normal dispersion forces of attraction between the adsorbate molecules. De Boer (4) has derived an expression for the reduction of the two-dimensional van der Waals constant α resulting from this effect:

$$\lambda = -{}_a\mu^2 \left(\frac{\pi}{d^{\text{oriented}}}\right) \qquad \text{(VI-15)}$$

where λ is the change in α, ${}_a\mu$ is the total effective dipole moment of the adsorbed molecules normal to the surface and d^{oriented} is the diameter of the molecules as obtained from equation VI-8. For a molecule without a permanent dipole

$${}_a\mu = \mu^{\text{ind}} \qquad \text{(VI-16)}$$

where μ^{ind} is the dipole moment induced by the electric field F, according to the following relation:

$$\mu^{\text{ind}} = F\,\xi \qquad \text{(VI-17)}$$

In equation VI-17, the polarizability ξ is, for a freely rotating molecule, the average polarizability given by equation VI-9; and, for an

TABLE VI-2
Average and Principal Polarizabilities of Molecules (2)

Name	Formula	$\xi^{av} \times 10^{24}$, cm^3	$\xi_1 \times 10^{24}$, cm^3	$\xi_2 \times 10^{24}$, cm^3	$\xi_3 \times 10^{24}$, cm^3	Comments
Hydrogen	H_2	0.79	0.934	0.718	0.718	ξ_1 Symmetry axis
Neon	Ne	0.39	0.39	0.39	0.39	Sphere
Argon	A	1.63	1.63	1.63	1.63	Sphere
Krypton	Kr	2.46	2.46	2.46	2.46	Sphere
Xenon	Xe	4.00	4.00	4.00	4.00	Sphere
Nitrogen	N_2	1.76	2.38	1.45	1.45	ξ_1 Symmetry axis
Oxygen	O_2	1.60	2.35	1.21	1.21	ξ_1 Symmetry axis
Chlorine	Cl_2	4.61	6.60	3.62	3.62	ξ_1 Symmetry axis
Carbon monoxide	CO	1.95	2.60	1.625	1.625	ξ_1 Symmetry axis
Nitrous oxide	N_2O	3.00	4.86	2.07	2.07	ξ_1 Symmetry axis
Carbon dioxide	CO_2	2.65	4.01	1.97	1.97	ξ_1 Symmetry axis
Sulfur dioxide	SO_2	3.72	5.49	2.72	3.49	ξ_3 Symmetry axis $\xi_2 \perp$ plane OSO
Hydrogen sulfide	H_2S	3.78	4.04	3.44	4.01	ξ_3 Symmetry axis $\xi_2 \perp$ plane HSH
Carbon disulfide	CS_2	8.74	15.14	5.54	5.54	ξ_1 Symmetry axis
Ammonia	NH_3	2.26	2.42	2.18	2.18	ξ_1 Symmetry axis
Methane	CH_4	2.60	2.60	2.60	2.60	Regular tetrahedron

Ethane	C_2H_6	4.47	5.48	3.97	3.97	ξ_1 Symmetry axis
Ethylene	C_2H_4	4.26	5.61	3.59	3.59	ξ_1 Symmetry axis
Acetylene	C_2H_2	3.33	5.12	2.43	2.43	ξ_1 Symmetry axis
Propane	C_3H_8	6.29	5.01	6.93	6.93	$\xi_1 \perp$ Plane CCC
Cyclohexane	C_6H_{12}	10.87	9.25	11.68	11.68	$\xi_1 \perp$ Plane of ring
Benzene	C_6H_6	10.32	6.35	12.31	12.31	$\xi_1 \perp$ Plane of ring
Methyl chloride	CH_3Cl	4.56	5.42	4.14	4.14	ξ_1 Symmetry axis
Methyl bromide	CH_3Br	5.55	6.85	4.9	4.9	ξ_1 Symmetry axis
Dichloromethane	CH_2Cl_2	6.48	5.02	8.47	5.96	$\xi_1 \perp$ Plane ClCCl $\xi_2 \perp$ Plane HCH
Chloroform	$CHCl_3$	8.23	6.68	9.01	9.01	ξ_1 Symmetry axis
Carbon tetrachloride	CCl_4	10.5	10.5	10.5	10.5	Regular tetrahedron
Ethyl chloride	C_2H_5Cl	6.4	6.60	5.01	7.59	
Chlorobenzene	C_6H_5Cl	12.25	13.24	7.58	15.93	$\xi_2 \perp$ Plane ring $\xi_3 \parallel$ CCl bond
Dimethyl ether	CH_3OCH_3	5.16	6.30	4.31	4.86	$\xi_2 \perp$ Plane COC ξ_3 Symmetry axis
Diethyl ether	$C_2H_5OC_2H_5$	8.73	11.26	7.07	7.87	$\xi_2 \perp$ Symmetry axis ξ_3 Symmetry axis
Acetone	CH_3COCH_3	6.33	7.10	4.82	7.08	$\xi_2 \perp$ Symmetry axis ξ_3 Symmetry axis
Pyridine	C_6H_5N	9.5	11.88	5.78	10.84	$\xi_2 \perp$ Plane ring ξ_3 through N

oriented molecule, is its polarizability in a direction parallel to the electric field. For a molecule with a permanent dipole,

$$ {}_a\mu = \mu^{\text{ind}} + \bar{\mu} \qquad \text{(VI-18)} $$

where $\bar{\mu}$ is the component of the permanent dipole μ in the direction of the electric field: $\bar{\mu}$ can be calculated by means of the Langevin function, if the electric field produces the major part of the orienting force:

$$ \bar{\mu} = \mu \left(\coth \frac{\mu F}{kT} - \frac{kT}{\mu F} \right) \qquad \text{(VI-19)} $$

Combining all the corrections to α^{id} that we have discussed into one general equation gives

$$ \alpha = \omega \left(1 - \frac{a_\mu}{a} \right) \alpha^{id} + \lambda \qquad \text{(VI-20)} $$

which, since the Keesom effect is significant only for some special cases, can be reduced to

$$ \alpha = \omega \alpha^{id} + \lambda \qquad \text{(VI-21)} $$

The necessity of introducing these corrections of α^{id} and the possibility that orientation has an effect on β^{id} brings some uncertainty into the process of matching isotherms. Because the correct value of $2\alpha/RT\beta$ is itself uncertain, we are obliged to seek a match for the experimental isotherm out of several families of computed isotherms, each family comprising a range of γ values for a given value of $2\alpha/RT\beta$. In the process of matching, it becomes evident that a range of values of $2\alpha/\beta$ exists within which range any value will yield a value of γ capable of bringing experimental and theoretical isotherms into coincidence, as the effect on the shape of the isotherm of decreasing $2\alpha/\beta$ is effectively offset by increasing the value of γ. The range of ambiguity is larger for the more heterogeneous substrates and becomes progressively narrower for more and more homogeneous substrates. For a completely homotattic substrate, there can be no ambiguity as γ is fixed (i.e., equal to infinity) and $2\alpha/\beta$ can be determined from equation I-18 in the form given by equation V-1. For practical purposes, near-homotattic substrates, which may be defined as those with γ values, either true or apparent, greater than 1000, have adsorption isotherms that are hardly to be distinguished from those of γ equal to infinity: the range of uncertainty is therefore reduced for such substrates to negligible proportions. A very heterogeneous surface, on the other hand, with a low value of γ, can hardly be incorporated into any adsorption system or measured in any practicable set of conditions that would raise its apparent value of γ sufficiently high to take advantage of this fact, even though the apparent γ increases with T^2. Moderately uniform substrates, however, with a true γ of about 100 or greater, can be investigated at room temperatures, which will raise the apparent γ value by a factor of 10 to 20, thus allowing $2\alpha/\beta$ to be determined with some confidence, by treating the surface as completely homotattic.

If the adsorbate has a rigid molecular structure and some elements of molecular symmetry, the discrepancy between the experimentally observed and ideal values of α can be used to estimate the electric field F at the interface, on the basis of the polarization theory. The way in which these calculations is made is illustrated in Chapter VII for several experimental systems. A value of the surface field having been found, the information can be used to calculate the correction factors ω and λ in equation VI-21 for many other adsorbates on the substrate in question, assuming that the surface field is not itself affected more by one adsorbate than by another.

The procedures just outlined meet, at least partially, the dilemma, which is already well known in this field of study, that the observations can be accounted for in more than one way, by allowing the proportion of the effects to be ascribed to the heterogeneity of the substrate and to the lateral interaction terms to vary. An alternative procedure, which would in most cases be less in error, is to minimize, by appropriate variation of the conditions of the experiment, the perturbations due to the lateral interaction potential, in order to observe the effect of the degree of heterogeneity more clearly. This can be accomplished, for example, by selecting as an adsorbate a substance that is not subject to orientation and whose interaction with an electric field is as small as possible. The obvious choice lies among the least polarizable of the inert gases, i.e., neon and argon; helium is a less practicable choice because of the extremely low temperatures required and its possible anomalous behavior as an adsorbed phase. With neon and argon we may take $2\alpha/\beta$ as equal to $2\alpha^{id}/\beta^{id}$, at least as a first approximation. With this value in hand, γ can be obtained, again to a first approximation, by the usual method of fitting isotherms. If γ is found to be about 100 or more, and if one wished to pursue the investigation, the surface field can be estimated by means of isotherms at room temperature, the value of $2\alpha/\beta$ for argon suitably corrected by equation VI-21, and a more accurate value of γ determined from the low temperature isotherms.

3. The Analysis of K'

The parameter K' is defined by equation V-59:

$$K' = A^0 \, e^{-U_0'/RT} \tag{V-59}$$

where, for the mobile model,

$$\ln A^0 = -\left[\frac{\Delta S_s^{\mathrm{tr}} + \Delta S^{\mathrm{rot}}}{R}\right] + \frac{{}_aF^{\mathrm{vib}} - {}_aE_0^{\mathrm{vib}}}{RT} + \frac{\Delta E^{\mathrm{kin}}}{RT} - \ln\frac{\theta_s}{1-\theta_s} + \ln 760 \tag{V-51}$$

The analysis of K' can be undertaken with more or less rigor depending on how many data are available at different temperatures and on the complexity of the adsorbate molecule. We shall describe how this analysis is performed for three successively more complex types of adsorption system.

Case 1. A Monatomic Gas Adsorbed at a Low Temperature. For the adsorption of a monatomic gas $\Delta S^{\mathrm{rot}} = 0$, ΔS_s^{tr} is given by equation V-16, and $\Delta E^{\mathrm{kin}} = -{}^1/_2 RT$. The frequency dependent term is given (see equations V-26 and V-27) by:

$${}_aF^{\mathrm{vib}} - {}_aE_0^{\mathrm{vib}} = RT \ln\left(1 - e^{-h\nu/kT}\right) \tag{VI-22}$$

At low temperatures this reduces to

$${}_aF^{\mathrm{vib}} - {}_aE_0^{\mathrm{vib}} \simeq 0 \tag{VI-23}$$

Under these conditions equation V-51 becomes

$$\ln A^0 = -\frac{\Delta S_s^{\mathrm{tr}}}{R} - \tfrac{1}{2} - \ln\frac{\theta}{1-\theta_s} + \ln 760 \tag{VI-24}$$

All the terms comprising $\ln A^0$ are known for any specific monatomic gas as a function of temperature; hence, from experimental values of K', U_0' is obtained. If K' is available from isotherms at several temperatures, the better procedure would be to plot the logarithmic form of equation V-59,

$$\ln K' - \ln A^0 = -U_0'/RT \tag{VI-25}$$

as ($\ln K' - \ln A^0$) *versus* $1/T$. If the simplifying assumptions are valid, the straight line obtained must pass through the origin; if it does so, the slope of the line equals $-U_0'/R$. If the line does not pass through the origin, the assumption leading to equation VI-23; i.e., that the temperature of the experiment is low enough to confine the adsorbed molecules to their ground state of vibration, is not valid, and the analysis must proceed according to the next section.

Case 2. A Monatomic Gas Adsorbed at a Higher Temperature. We have two unknowns, U_0' and ν; it is therefore necessary to have experimentally determined values of K' at several temperatures. From these data by means of equation V-64, which can be written in the form

$$\lim_{\theta \to 0} q^{\mathrm{st}} = RT^2 \frac{d \ln K'}{dT} = -R \frac{d \ln K'}{d(1/T)} \tag{V-64}$$

one can obtain, by plotting $\ln K'$ *versus* $1/T$, the limiting value at zero surface concentration of q^{st}; thence,

$$\lim_{\theta \to 0} q^{\mathrm{diff}} = \lim_{\theta \to 0} q^{\mathrm{st}} - RT$$

The standard integral molar entropy of adsorption can now be calculated by means of equation III-91:

$$\Delta S_s{}^{\mathrm{ads}} = \lim_{\theta \to 0} \frac{q^{\mathrm{diff}}}{T} - R \ln \frac{\theta_s}{1 - \theta_s} - R \ln \frac{K'}{760} \tag{III-91}$$

This entropy term contains both translational and vibrational entropy changes ($\Delta S^{\mathrm{rot}} = 0$ for the case under discussion); the translational change of entropy contribution can be calculated independently by the Sackur-Tetrode equation, equation V-16; thence,

$$_aS^{\mathrm{vib}} = \Delta S_s{}^{\mathrm{ads}} - \Delta S_s{}^{\mathrm{tr}} \tag{VI-26}$$

From the value of the vibrational entropy thus derived the vibrational frequency ν can be obtained by the use of equation V-27. Finally U_0' can be calculated from the limiting value of q^{diff} and $({}_aE^{\text{vib}} - {}_aE_0^{\text{vib}})$ by means of equation V-6:

$$U_0' = \lim_{\theta \to 0} q^{\text{diff}} + ({}_aE^{\text{vib}} - {}_aE_0^{\text{vib}}) + (^1/_2)RT$$

Case 3. Polyatomic Molecules as Adsorbates. This complex case requires the use of the completely general equation V-51 for its complete description, and so introduces two more unknown quantities, *viz.*, the changes of rotational entropy and rotational energy on adsorption. No completely unambiguous solution can be obtained from adsorption data alone—the best that can be done is to construct a number of possible models for the behavior of the adsorbed molecules and judge the degree of probability of each alternative. Any additional information, other than adsorption data, about the character of the adsorbate or of the adsorbent may then become of value in helping to decide which is the most reasonable model.

One may first calculate the standard integral molar entropy change on adsorption, as described in case 2, using equation III-91. This entropy change includes translational, rotational, and vibrational changes of entropy on adsorption. Various trial models can now be assumed as descriptions of the adsorbed molecule: namely, the loss of none, one, two, or three degrees of rotational freedom. Each of these postulated models will have its particular value of ΔS^{rot}, which can be calculated by equation V-25. If the sum of ΔS_s^{tr} and ΔS^{rot} is greater than ΔS_s^{ads}, the difference can be attributed to vibrational entropy, ${}_aS^{\text{vib}}$, whence ν can be calculated. One cannot expect, therefore, that these entropy calculations can help us decide which of the postulated models is more feasible. We illustrate this situation in Chapter VII, where we consider the adsorption of benzene by graphite. For this case, fortunately, we are able to call upon our previous knowledge of the magnitude of the surface electric field of graphite to help us pick out the most likely orientation of the adsorbate molecule. In general, however, one could not count upon having this useful additional knowledge.

Recently certain investigations of high precision have been reported (5) pertaining to the Henry's-law region of the adsorption isotherm. The analysis of such data can be made by means of equation V-62:

$$\ln Z + \ln A^0 - \frac{1}{4\gamma(RT)^2} = \frac{U_0'}{RT} + \ln V_\beta \qquad \text{(V-62)}$$

It is necessary, however, that values of V_β and γ be known, presumably from an adsorption isotherm in the region of higher coverage that has been analyzed in terms of the present theory. Equation V-62 can then be solved by processes similar to those described above in cases 1 or 2, where the condition $\Delta S^{\text{rot}} = 0$ can be reasonably assumed.

References

1. J. H. de Boer, *The Dynamical Character of Adsorption*, Clarendon Press, Oxford, 1953, pp. 146–148.
2. Landolt-Bornstein, *Zahlenwerte und Funktionen aus Physik, Chemie, Astronomie, Geophysik, und Technik, 6th edition*, Vol. 1, Part 3, Springer-Verlag, Berlin, 1951, pp. 509–512.
3. J. C. P. Mignolet, *Discussions Faraday Soc.*, **8,** 105 (1950); idem, in W. E. Garner, ed., *Chemisorption* Butterworths, London, 1957, p. 118.
4. J. H. de Boer, *op. cit.*, p. 155.
5. See references 21 and 22, Chapter VII.

CHAPTER VII

Results Obtained from Analyses of Experimental Data

The theory developed in the previous chapters has been presented with a view to its ultimate application to experimental observations, so that it can be tested by evaluating and verifying the fundamental parameters of adsorption systems that are evoked by its use. The experimental data at our disposal consist largely of adsorption isotherms determined with graphitized carbon blacks; some of the adsorption systems have been measured at a number of different temperatures, others only at two temperatures. The extent to which the analysis can be carried depends on the quantity of data available, their accuracy, and the complexity of the adsorption system. The examples cited in this chapter are each analyzed as far as the limits imposed by these conditions allow.

In the first part of this chapter we discuss the adsorbents whose heterogeneity is described adequately by a Gaussian distribution of adsorptive potentials (Gaussian Adsorbents); the second part contains some examples of adsorbents for which such a distribution is inadequate, and for which a multimodal distribution of adsorptive potentials is a satisfactory postulate.

In every case the model of mobile adsorption has successfully produced descriptions of the experimental data that are consistent at two or more temperatures. The model isotherms based on the Fowler equation could be made to correspond with any of the experimental isotherms, but the values of V_β, γ, and U_0' thereby derived do not agree with one another for the same system at two temperatures; the model based on the Hill-de Boer equation on the other hand gives values for these parameters that do so agree. The failure of the model of localized adsorption to describe these experimental data cannot be ascribed to an inaccurate selection of the numerical value of $c\omega$ in equation I-20b, as the success of the alternative model, using equation I-18b, excludes the possibility that a mere change of a constant in equation I-20b would suffice to make that equation effective. These equations are the functions $\Psi(p, U_0)$ in equation IV-4 and they are not at all alike; if one of them is adequate, the other is necessarily inadequate.

The first conclusion, therefore, that we obtain by an application of our computations is that temperatures below that of liquid nitrogen are required to cause a physically adsorbed film to become localized. But while we may have disposed of the applicability of the Fowler equation to these data, there may be other theories of localized adsorption that would not be so easily eliminated. Nevertheless, on the basis of some calculations by Hill, it seems more reasonable to expect that temperatures below that of liquid nitrogen would be required to cause a physically adsorbed mobile film to become localized. Hill compared the mobile → localized transition to that of hindered rotation in molecules such as ethane. The potential energy barrier to the transition is small so that mobility is achieved at quite low temperatures.

1. Gaussian Adsorbents

A. A NUMBER OF ADSORPTION SYSTEMS MEASURED AT LOW TEMPERATURES (*CA.* 80°K)

In this section we plan to demonstrate that the foregoing theory can be used effectively to analyze single adsorption isotherms measured over the monolayer range of surface concentration with a variety of substrates, yielding information that, although not of the highest accuracy because of the paucity of the data, could be of practical utility. In later sections we shall include examples of how a more complete application of fundamental theory, made possible by the availability of the appropriate data, leads to a more detailed and more accurate description of adsorption systems.

When the isotherm data available are limited to a single temperature, the factors that determine the temperature dependence (i.e., rotational and vibrational changes of the adsorbate molecule on adsorption) cannot always be evaluated; nevertheless, such data are useful enough, provided that they refer to conditions where simplifying assumptions would not introduce much error. For example, with the inert gases, no change in molecular rotation occurs on adsorption; again, if temperatures are low enough, the adsorbed molecules will be mostly in their ground state of vibration,

TABLE VII-1

Parameters Derived from Comparisons of Experimental Data with the Simplified Model of a Mobile Adsorbed Film

Adsorbent	Adsorbate	Temp, °K	V_β cm³ (STP)/g	γ	K'	U_0' Calcd.	U_0' Obsvd.
MT (3100)	Argon	77.5	2.3	240	1.89	2.12	2.02
MT (3100)	Argon	90.1	2.2	240	14.8	2.14	2.02
P-33 (2700)	Argon	77.5	3.68	170	1.90	2.12	2.01
P-33 (2700)	Argon	90.1	3.68	170	14.8	2.14	2.01
P-33 (2700)	Nitrogen	77.5	3.15	120	1.30	2.13	2.03
P-33 (2700)	Nitrogen	90.1	3.15	120	10.3	2.15	2.03
P-33 (2700)	Krypton	90.1	3.41	330	0.45	2.82	3.50
P-33 (2000)	Argon	77.5	3.65	100	1.92	2.12	2.13
P-33 (2000)	Argon	90.1	3.70	100	15.7	2.13	2.13
P-33 (2000)	Nitrogen	77.5	3.10	70	1.13	2.15	2.09
P-33 (2000)	Nitrogen	90.1	3.05	70	9.62	2.17	2.09
P-33 (1500)	Argon	77.5	3.7	30	2.28	2.09	2.05
P-33 (1500)	Argon	90.1	3.6	30	17.8	2.11	2.05
P-33 (1500)	Nitrogen	77.5	3.1	20	1.39	2.12	2.07
P-33 (1500)	Nitrogen	90.1	2.9	20	10.2	2.15	2.07
P-33 (1000)	Argon	77.5	3.60	8.5	2.88	2.06	2.13
P-33 (1000)	Argon	90.1	3.60	8.5	25.0	2.05	2.13
P-33 (1000)	Nitrogen	77.5	3.15	6.0	1.96	2.06	2.12
P-33 (1000)	Nitrogen	90.1	3.15	6.0	16.9	2.06	2.12
P-33 (1000)	Krypton	90.1	3.12	16	0.50	2.80	—
BN (Pultz)	Argon	77.5	5.86	310	8.86	1.88	1.85
BN (Pultz)	Argon	90.1	5.11	310	60.0	1.89	1.85
BN (Pultz)	Nitrogen	77.5	4.85	220	7.18	1.86	1.84
BN (Pultz)	Nitrogen	90.1	4.70	220	48.0	1.87	1.84
BN (Winkler)	Nitrogen	77.5	3.12	70	7.18	1.86	1.84
BN (Winkler)	Nitrogen	90.1	3.12	70	48.0	1.87	1.84
Linde 13X	Argon	77.5	196	22	.223	2.46	2.24
Linde 13X	Argon	90 1	196	22	2.18	2.48	2.24
Diamond	Argon	77.2	0.95	3	330	1.32	—
Diamond	Argon	90.0	1.00	3	1500	1.32	—
Diamond	Nitrogen	77.2	0.86	2	480	1.21	—
Diamond	Nitrogen	90.0	0.86	2	1850	1.22	—
Rutile	Argon	85	33.3	2	63.5	1.76	1.72
Rutile	Oxygen	100	34.1	2	325	1.77	1.91
Anatase	Argon	77.5	4.16	2	33.0	1.68	—

and so this factor would also not contribute significantly to the temperature dependence of the adsorption.

A number of isotherms that would seem to fit this category have been collected. Lacking information about the nature of the electric

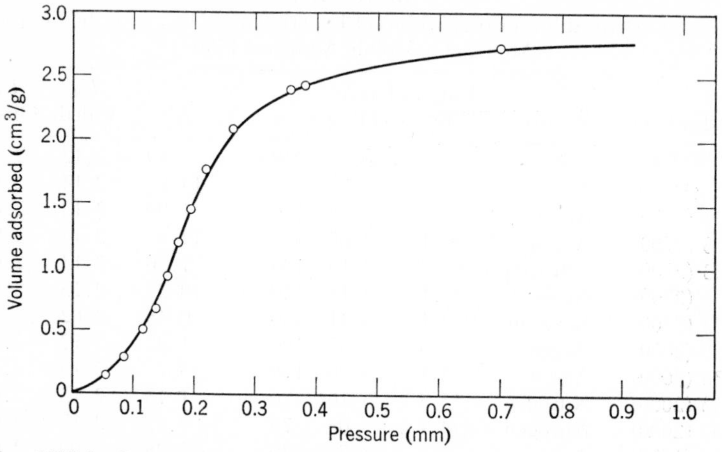

Fig. VII-1. Comparison of experimental adsorption isotherms (individual points) with the theoretical description of a mobile adsorbed film (solid line). The parameters used for the calculated curve are reported in Table VII-1. The data refer to argon adsorbed by P-33 (2700°C) at 77.5°K (2).

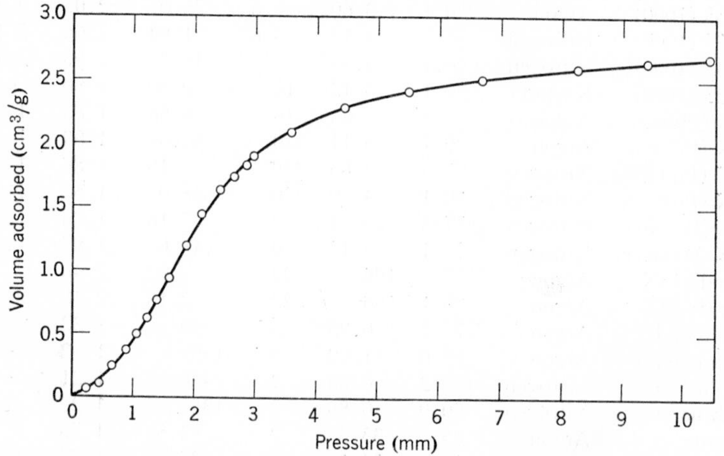

Fig. VII-2. Same as Figure VII-1. The data refer to argon adsorbed by P-33 (2700°C) at 90.1°K (2).

field at the interface we are, for the present, obliged to ignore it: that is, to assume that it would not polarize the adsorbate sufficiently to affect the determination of the parameters, γ, V_β and K' that would be derived by taking the ideal values of α/β as given by equa-

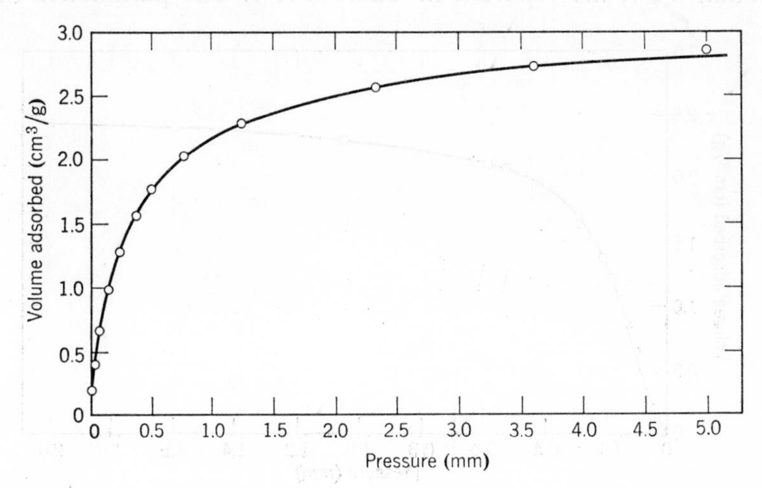

Fig. VII-3. Same as Figure VII-1. The data refer to argon adsorbed by P-33 (1000°C) at 77.5°K (4).

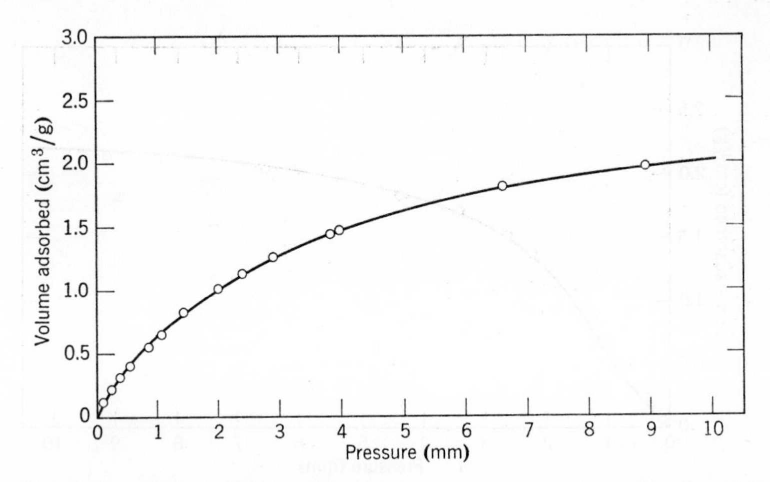

Fig. VII-4. Same as Figure VII-1. The data refer to argon adsorbed by P-33 (1000°C) at 90.1°K (4).

tion I-42. This assumption is not correct and we estimate in later sections, for some of these adsorbents, the size of the contribution that we neglect at present: generally speaking, it is about 5%. The results obtained by treating the data by the procedure given in section VI-1, are reported in Table VII-1. The parameters thus

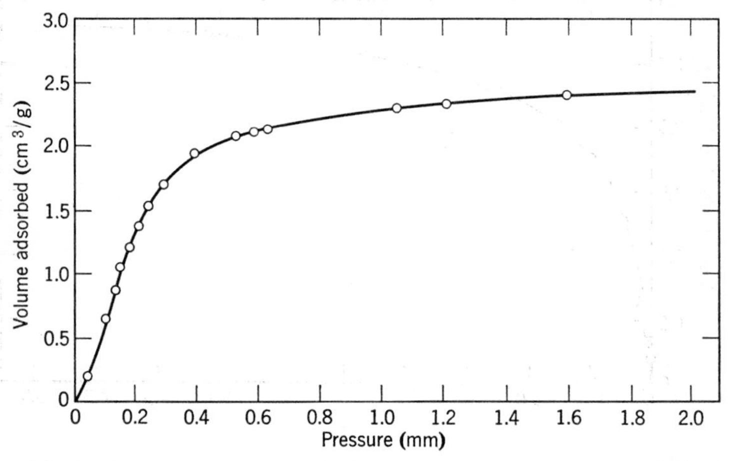

Fig. VII-5. Same as Figure VII-1. The data refer to nitrogen adsorbed by P-33 (2700°C) at 77.5°K (2).

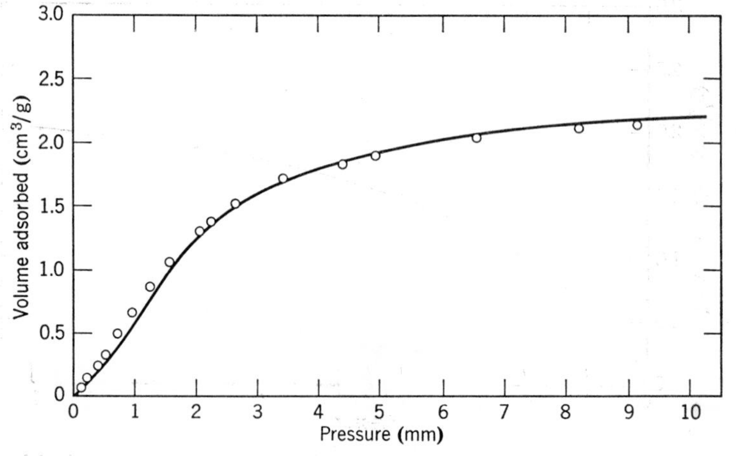

Fig. VII-6. Same as Figure VII-1. The data refer to nitrogen adsorbed by P-33 (2700°C) at 90.1°K (2).

derived are from the best fits that can be obtained within the family of model isotherms defined by the appropriate value of $2\alpha^{id}/RT\beta^{id}$.

Illustrations of the precision of the fitting are shown in Figures VII-1–VII-19, in which the experimental data, shown by circles, are compared with model isotherms computed specially for the

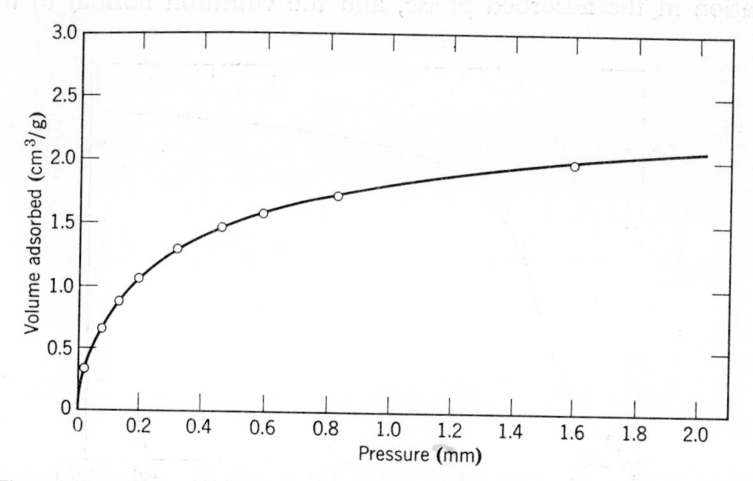

Fig. VII-7. Same as Figure VII-1. The data refer to nitrogen adsorbed by P-33 (1000°C) at 77.5°K (4).

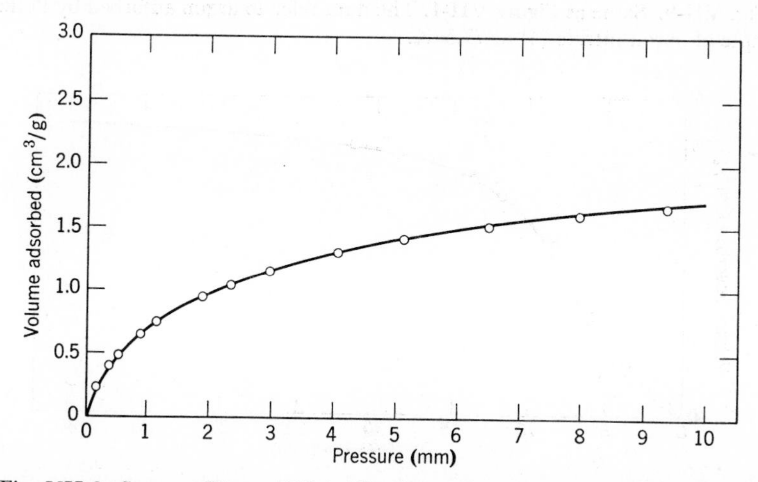

Fig. VII-8. Same as Figure VII-1. The data refer to nitrogen adsorbed by P-33 (1000°C) at 90.1°K (4).

values of γ, V_β, and K' that had been obtained by the graphical interpolation method described in Chapter VI; the solid line in these diagrams represents the computed isotherm.

The interpretation of the values of K' was made in accordance with the assumptions described under Section VI-3 (*Case 1*): i.e., free rotation in the adsorbed phase, and the vibration normal to the

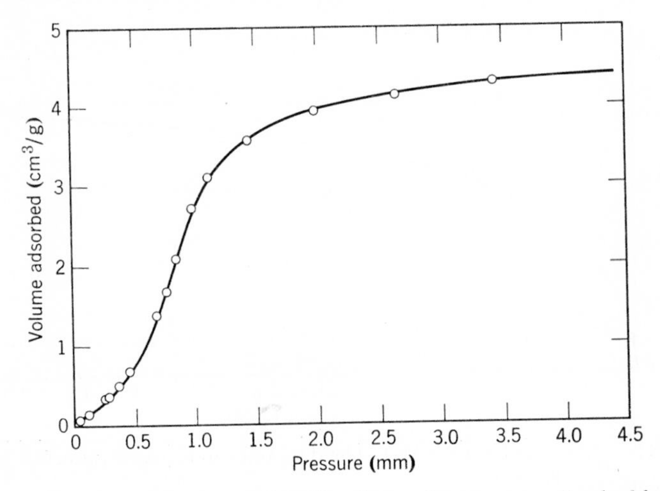

Fig. VII-9. Same as Figure VII-1. The data refer to argon adsorbed by Pultz's sample of boron nitride (4) at 77.5°K.

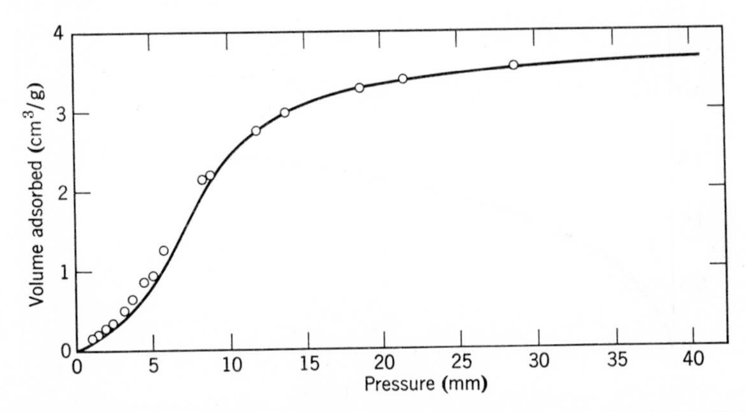

Fig. VII-10. Same as Figure VII-1. The data refer to argon adsorbed by Pultz's sample of boron nitride (4) at 90.1°K.

surface restricted to the ground state. The calculation of U_0' is then made by equations VI-24 and VI-25.

Although isotherms at two temperatures were available for most of the systems reported in Table VII-1, the evaluation of U_0' was made independently for each isotherm, rather than from the temperature dependence of ln K'. The invariance of γ, V_β, and

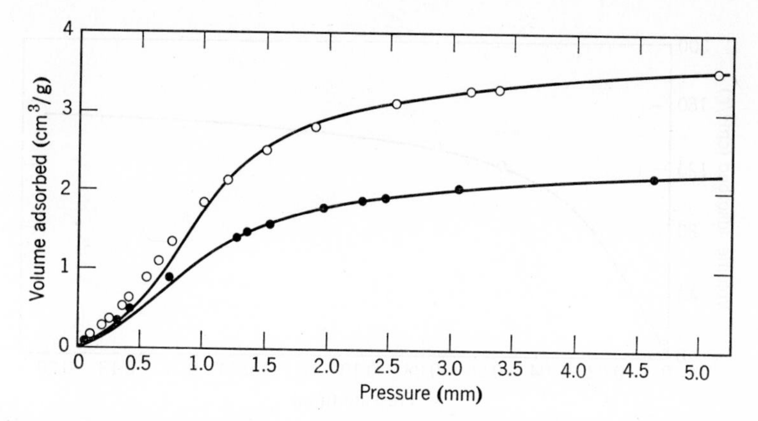

Fig. VII-11. Same as Figure VII-1. The data refer to nitrogen adsorbed by Pultz's sample of boron nitride (4) at 77.5°K (upper curve), and nitrogen adsorbed by Winkler's sample of boron nitride (5) at 77.5°K (lower curve).

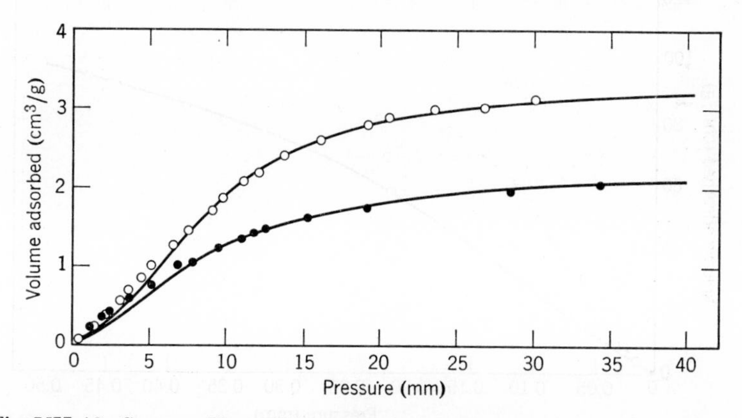

Fig. VII-12. Same as Figure VII-1. The data refer to the same systems reported by Figure VII-11, upper and lower curves corresponding respectively; the temperature of the measurements = 90.1°K.

U_0' over a small temperature range is a requirement of the theory and of the present simplifying assumptions, and when actually confirmed by comparison of the experimental and computed isotherms, creates confidence in the model and in the physical significance of the values obtained. The agreement of these constants at each temperature is seen to be excellent.

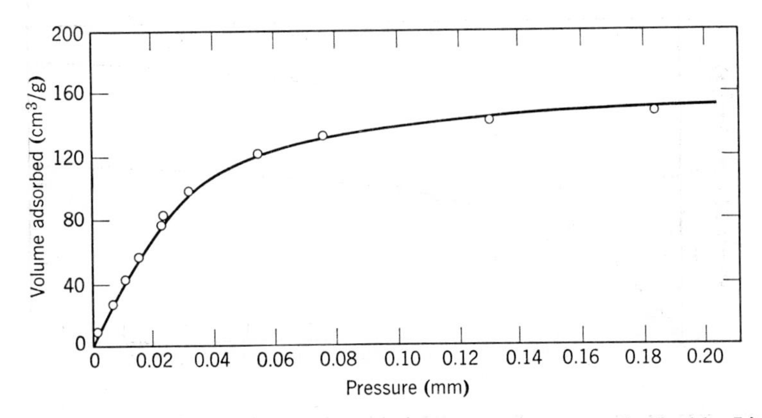

Fig. VII-13. Same as Figure VII-1. The data refer to argon adsorbed by Linde sieve 13X at 77.5°K (7).

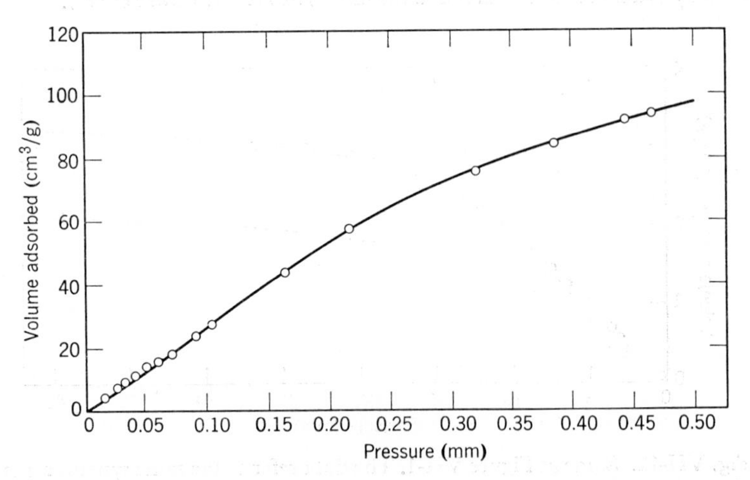

Fig. VII-14. Same as Figure VII-1. The data refer to argon adsorbed by Linde sieve 13X at 90.1°K (8).

Also included in Table VII-1 are a few isotherms that are of interest chiefly because they refer to adsorbents that have received attention in important papers by other workers, although data are not available at two temperatures.

The values of V_β for the various adsorbents that are reported in Table VII-1 usually differ by less than 20% from estimates of

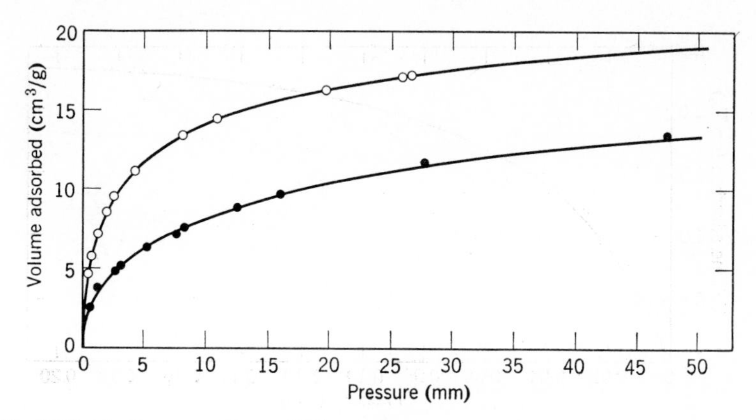

Fig. VII-15. Same as Figure VII-1. The data refer to argon adsorbed by rutile (10) at 85.0°K (upper curve); and oxygen adsorbed by rutile (10) at 100°K (lower curve).

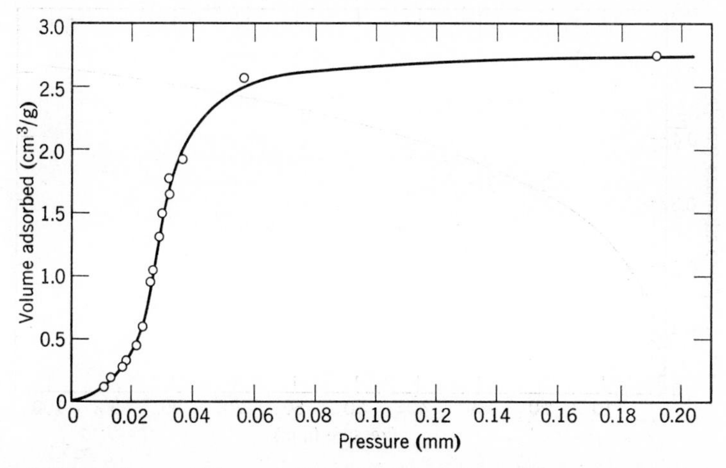

Fig. VII-16. Same as Figure VII-1. The data refer to krypton adsorbed by P-33 (2700°C) at 90.1°K (2).

"V_m" that have been derived for the same samples by other methods, such as the point B method (11) or the B.E.T. equation (12). This comparison is included in Table VII-2. The results of the present method of deriving V_β do not show the rather erratic changes with the tempetature of adsorption that are occasionally so distressing in the determination of "V_m". Both V_β and "V_m" meas-

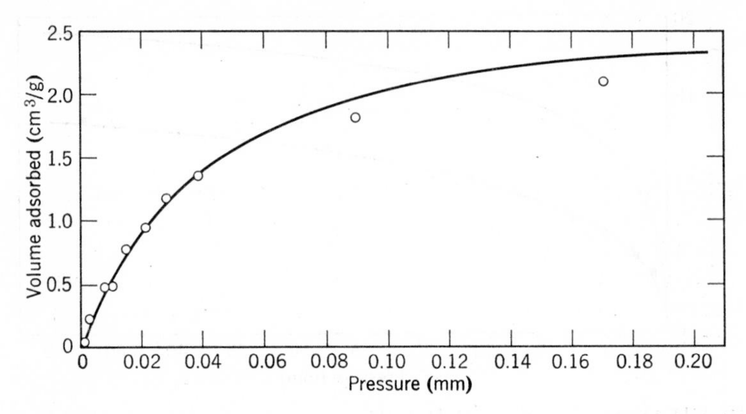

Fig. VII-17. Same as Figure VII-1. The data refer to krypton adsorbed by P-33 (1000°C) at 90.1°K (4).

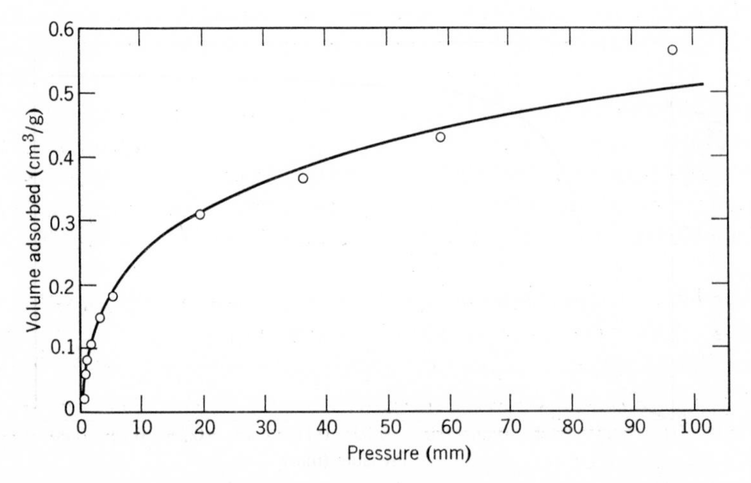

Fig. VII-18. Same as Figure VII-1. The data refer to argon adsorbed by diamond dust at 77.2°K (9).

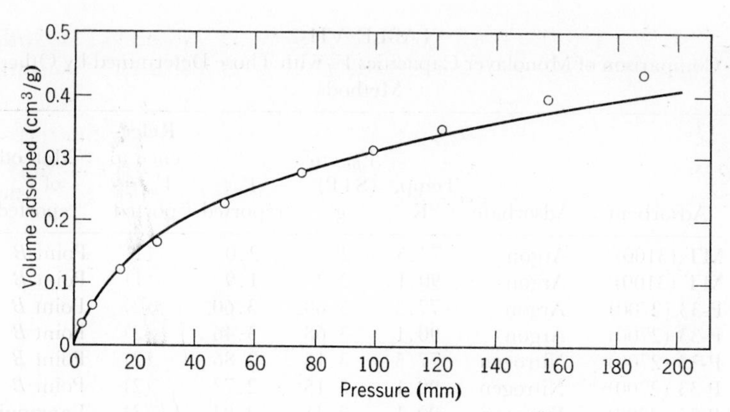

Fig. VII-19. Same as Figure VII-1. The data refer to argon adsorbed by diamond dust at 90.1° (9).

ure the surface area of an adsorbent; their use to determine this quantity requires different values for the cross sectional area of the adsorbed molecules: thus, for example, for argon $\beta^{id} = 13.6$ A^2; σ_0, as commonly used for the B.E.T. measurement, is 14.6 A^2 at 77°K and 15.5 A^2 at 90°K.

The monolayer capacity of a solid surface, as determined by the present theory, would not be expected to show a large temperature coefficient since it is dependent on the two-dimensional van der Waals β, hence, on the collision diameters of the adsorbate molecules. In the temperature interval of 12° between 78 and 90°K, the change in V_β (probably less than 1%) would be immeasurable. Nothing in the present theory links V_β with the liquid density; the temperature dependence of the distance of closest approach of adsorbed molecules at infinite pressure is much less than would be inferred from the temperature dependence of the density of the liquid.

We have already demonstrated, by using the model isotherms as though they represented actual experimental data, that an application of the Langmuir equation yields a value of "V_m" that is smaller than the value of V_β used to generate the curves, and that the discrepancy becomes greater as the surface is more heterogeneous (i.e., lower values of γ). Since the B.E.T. theory is based on the Langmuir equation, we expect to find, as indeed in all instances we do find, that the B.E.T.-derived value of "V_m" is lower than our

TABLE VII-2
Comparison of Monolayer Capacities V_β with Those Determined by Other Methods

Adsorbent	Adsorbate	Temp, °K	V_β, cm³ (STP)/g	V_m, reported	Reference to V_m reported	Method of V_m reported
MT (3100)	Argon	77.5	2.3	2.0	(1)	Point *B*
MT (3100)	Argon	90.1	2.2	1.9	(1)	Point *B*
P-33 (2700)	Argon	77.5	3.68	3.60	(2)	Point *B*
P-33 (2700)	Argon	90.1	3.68	3.46	(2)	Point *B*
P-33 (2700)	Nitrogen	77.5	3.15	2.86	(2)	Point *B*
P-33 (2700)	Nitrogen	90.1	3.15	2.72	(2)	Point *B*
P-33 (2700)	Krypton	90.1	3.41	2.94	(3)	Langmuir
P-33 (2000)	Argon	77.5	3.65	3.60	(4)	Point *B*
P-33 (2000)	Argon	90.1	3.70	—	—	—
P-33 (2000)	Nitrogen	77.5	3.10	2.90	(4)	Point *B*
P-33 (2000)	Nitrogen	90.1	3.05	—	—	—
P-33 (1500)	Argon	77.5	3.7	3.50	(4)	Point *B*
P-33 (1500)	Argon	90.1	3.6	—	—	—
P-33 (1500)	Nitrogen	77.5	3.1	2.95	(4)	Point *B*
P-33 (1500)	Nitrogen	90.1	2.9	—	—	—
P-33 (1000)	Argon	77.5	3.60	3.50	(4)	Point *B*
P-33 (1000)	Argon	90.1	3.60	3.36	(4)	Point *B*
P-33 (1000)	Nitrogen	77.5	3.15	3.00	(4)	Point *B*
P-33 (1000)	Nitrogen	90.1	3.15	2.85	(4)	Point *B*
P-33 (1000)	Krypton	90.1	3.12	2.83	(4)	Point *B*
BN (Pultz)	Argon	77.5	5.86	5.06	(4)	Langmuir
BN (Pultz)	Argon	90.1	5.11	4.26	(4)	Langmuir
BN (Pultz)	Nitrogen	77.5	4.85	4.23	(4)	Langmuir
BN (Pultz)	Nitrogen	90.1	4.70	3.70	(4)	Langmuir
BN (Winkler)	Nitrogen	77.5	3.12	2.78	(5)	Langmuir
BN (Winkler)	Nitrogen	90.1	3.12	—	—	—
Linde 13X	Argon	77.5	196	185	(6)	B.E.T.
Linde 13X	Argon	90.1	196	—	—	—
Diamond	Argon	77.2	0.95	.327	(9)	B.E.T.
Diamond	Argon	90.0	1.00	—	—	—
Diamond	Nitrogen	77.2	0.86	.350	(9)	B.E.T.
Diamond	Nitrogen	90.0	0.86	—	—	—
Rutile	Argon	85	33.3	19.4	(10)	B.E.T.
Rutile	Oxygen	100	34.1	20.6	(10)	B.E.T.

V_β, and that this discrepancy is the greater with the more heterogeneous substrates.

We cannot offer any external or independent comparison of the values of γ assigned to the adsorbents in Table VII-1, since this is the first time that surface heterogeneity has been measured in terms

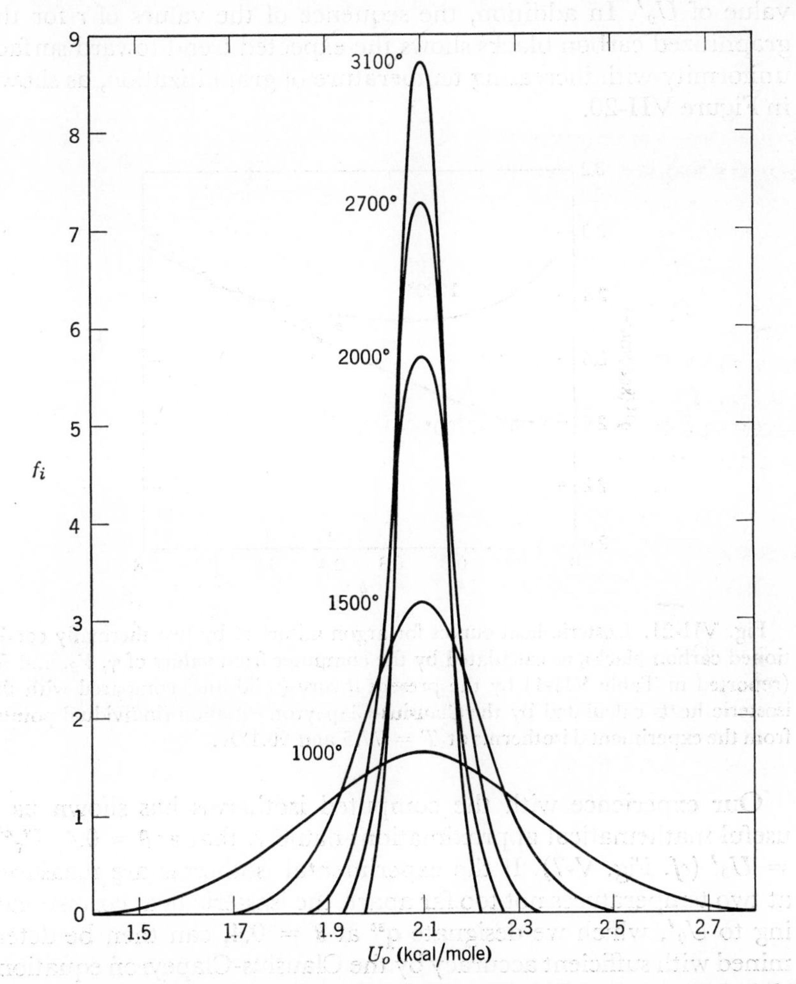

Fig. VII-20. Absorptive energy distribution curves for argon adsorbed by carbon blacks that have been thermally conditioned (graphitized) at different temperatures.

of a Gaussian distribution; nevertheless, the values of γ give reasonable widths to the distribution curves as measured by the relation

$$r = 477\sqrt{1/\gamma}$$

where one half of the surface is within $\pm r$ calories of the mean value of U_0'. In addition, the sequence of the values of r for the graphitized carbon blacks shows the expected trend toward surface uniformity with increasing temperature of graphitization, as shown in Figure VII-20.

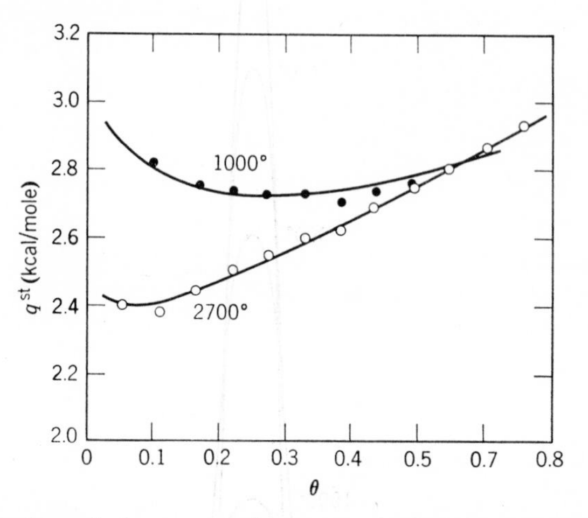

Fig. VII-21. Isosteric heat curves for argon adsorbed by two thermally conditioned carbon blacks, as calculated by the computer from values of γ, V_β, and K' (reported in Table VII-1) by the present theory (solid line) compared with the isosteric heats calculated by the Clausius-Clapeyron equation (individual points) from the experimental isotherms at $T = 77.5$ and 90.1°K.

Our experience with the computed isotherms has shown us a useful mathematical approximation: namely, that at $\theta = 0.4$, $\mathbf{U_0}^{\text{diff}} = U_0'$ (*cf.* Fig. V-7). If the experimental isotherms are measured at two temperatures not too far apart, the isosteric heat corresponding to U_0', which we designate $\mathbf{q}^{\text{st}}$ at $\theta = 0.4$, can then be determined with sufficient accuracy by the Clausius-Clapeyron equation:

$$\mathbf{q}^{\text{st}} = \left[R \left(\frac{T_1 T_2}{T_2 - T_1} \right) \ln \frac{p_2}{p_1} \right]_{\theta = 0.4} \qquad \text{(VII-1)}$$

Then, by equation III-49,

$$U_0' = \mathbf{q}^{\mathrm{st}}{}_{\theta = 0.4} - R\left(\frac{2T_1T_2}{T_2 + T_1}\right) + ({}_aE^{\mathrm{vib}} - {}_aE_0{}^{\mathrm{vib}}) + \Delta E^{\mathrm{kin}} - \frac{0.8\alpha^{id}}{\beta^{id}} \quad \text{(VII-2)}$$

We do not at present, for the adsorption systems listed in Table VII-1, have sufficient information to calculate U_0' accurately by the above equation. For our present purpose, therefore, we shall make the following assumptions: (a) at the low temperatures referred to in Table VII-1 the adsorbed molecules are mostly in their ground state of vibration; hence

$${}_aE^{\mathrm{vib}} \simeq {}_aE_0{}^{\mathrm{vib}}$$

(b) the translational energy per mole of a molecule in the gas phase is given by $^3/_2\,RT$, and in the adsorbed phase by $^2/_2\,RT$: that is, the adsorbed phase behaves as a two-dimensional gas; further, for argon, krypton, nitrogen, and oxygen we assume that no change takes place in the rotational energy on adsorption; therefore,

$$\Delta E^{\mathrm{kin}} = -\,(^1/_2)RT$$

Equation VII-2 then becomes:

$$U_0' = \mathbf{q}^{\mathrm{st}}_{\theta = 0.4} - 3R\left(\frac{T_1T_2}{T_1 + T_2}\right) - \frac{0.8\alpha^{id}}{\beta^{id}} \quad \text{(VII-3)}$$

The approximate values of U_0' calculated by equation VII-3, are listed in Table VII-1 (where available) as U_0' (observed).

Using the values of γ, V_β, and U_0' (observed) determined for argon and nitrogen on P-33 (2700°) and P-33 (1000°), reported in Table VII-1, the computer gives us $\mathbf{U}_0^{\mathrm{int}}$,* and hence, by equations IV-16, IV-21, and I-41, gives us $\mathbf{q}^{\mathrm{st}}$ as a function of θ. These computed isosteric heats are shown in Figures VII-21 and VII-22 by the solid lines; the observed isosteric heats determined from the

* Computed values of $\mathbf{U}_0^{\mathrm{int}}$ were obtained but, apart from the present application, were not further used by us. They have not been included in the tables (Appendix) because their limited usefulness did not seem to justify the additional expense.

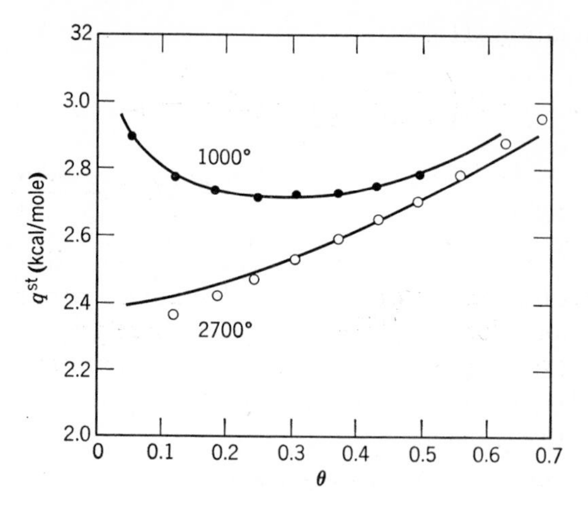

Fig. VII-22. Same as Figure VII-21, but referring to nitrogen as adsorbate.

adsorption isotherms are shown in the same diagram as individual points. In spite of the approximations involved, the agreement between the heat curves predicted by our model and those observed is striking; and makes it more probable that the parameters obtained by our method of curve-fitting are not merely empirical quantities but quantitative measurements of physical constants of the system.

We shall now discuss the individual adsorbents in more detail.

1. Graphitized Carbon Blacks

The P-33 series of graphitized carbon blacks appears, by the present method, to undergo little change in V_β during the process of graphitizing: the point *B* method gives a set of lower values of V_m that change somewhat irregularly with the graphitizing process. For these carbon blacks the B.E.T. plots of the data are not linear and so B.E.T. values cannot be introduced for this comparison. A more extensive series of such comparisons for various adsorbents would no doubt be of interest but the appropriate data are not yet available.

The values of γ reported in Table VII-1 for the graphitized carbon blacks describe quantitatively the decrease of surface heterogeneity that had previously been demonstrated qualitatively, by means of X-ray diffraction and electron micrographs (see frontispiece), to occur with the graphitizing process (13). The decrease of heterogeneity is found to be associated with a slight increase in the average adsorption energy; we therefore deduce that the "perfect" graphite surface has an adsorptive energy slightly greater than that of MT (3100°). The increase of the adsorptive energy with the process of graphitizing a carbon black may be traced to the superior interaction taking place between an adsorbate molecule and carbon atoms that are in a denser and more regular array on the more crystalline substrate. The increase of density on graphitizing, accompanied by a decrease in the interlammellar spacing, has also been observed experimentally (10). In Figure VII-20 are depicted the distribution curves of carbon blacks of varying degrees of graphitizing as reported in Table VII-1. The blacks referred to as P-33 are all derived from the same parent material; the Sterling MT (3100°) has a different origin; nevertheless, the latter carbon black fits smoothly and regularly into the series.

2. *Boron Nitride*

Boron nitride has a hexagonal-layer-lattice structure that is isoelectronic and isometric with graphite, although unlike graphite it is not an electrical conductor; comparisons of the two are therefore of interest. Two different specimens of boron nitride have been reported above as adsorbents: these differ both in specific surface area and degree of heterogeneity; nevertheless, they agree in their average adsorptive energies. They both have near-homotattic surfaces and we conclude that the value of U_0' reported corresponds closely to that for the perfect basal cleavage plane of boron nitride.

The sample of boron nitride measured by Pultz, with a γ value of 310, has a more uniform surface than any of the graphitized carbon blacks. As reported by Ross and Pultz (4), however, the apparent ${}_aT_c$ of this sample would give it, by equation V-2, a degree of heterogeneity corresponding to a γ value of about 65. The discrepancy in this case can be traced to a slight contamination of the surface, which renders invalid the direct application of equation V-2 for the determination of γ. See Section VII-2 for a discussion of the correct way to apply equation V-2 to nonGaussian adsorbents.

A slight asymmetry toward the high-energy end of the distribution curve, due to the presence of traces of a surface boric oxide, is indicated: the experimental adsorption isotherm at low coverage shows more adsorption than can be accounted for by the best matching with the models. The magnitude of the effect can be seen in Figures VII-9–VII-12. Analyses, by Mr. F. R. Klebacher (14) in this laboratory, of the sulfuric acid washings from these and other samples of boron nitride have shown the presence of a soluble boron compound. Klebacher reported 0.08% of a soluble boron compound in the specimen used by Winkler, and 1 ppm in that used by Pultz.

The average adsorptive energies of both argon and nitrogen on nonpolar adsorbents should be very little different, as the polarizabilities of these two adsorbates are about the same. The carbon black series and the boron nitride samples show themselves to be nonpolar by this criterion: the operating adsorptive forces are only dispersion forces.

3. Diamond Dust

The data for the adsorption of argon and nitrogen on diamond dust (9) were supplied by courtesy of Dr. Victor R. Deitz of the National Bureau of Standards. The extreme heterogeneity of the surface is not surprising in view of the character of the crystal, whose cleavage planes are not pronounced, and the nature of the sample, which is the dust resulting from the splitting of gems. The most outstanding feature of the adsorption of argon or nitrogen by this surface is the unusually low values of their average adsorptive potentials, which are lower by one third than the corresponding potentials on a graphite surface. Graham (15) has calculated the ratio of the adsorptive potentials, based on dispersion energies alone, for the adsorption of a gas on the surfaces of graphite and of diamond: his calculated result is 1.44; our results give a ratio of 1.5 for argon on the two surfaces and 1.6 for nitrogen. The immediate practical result of the low adsorptive potentials is that the specific area of the dust determined by the B.E.T. method is grossly in error. The part of the surface having an adsorptive potential less than that of the liquid adsorbate remains sparsely occupied, even as multilayers build up over the more energetic patches of the surface; this effect, combined with that produced by a low value of γ (for which see Table VII-1), leads to a value of V_m by the B.E.T. method that is about one third of the value of V_β.

4. *Linde 13X Synthetic Zeolite*

The Linde sieve 13X is a synthetic zeolite based on the faujasite structural framework (16). It is supplied in its ion exchange form Na_2X and has a structure consisting of open cages of about 12 A diameter; these pores can admit about 20 argon or nitrogen molecules. The adsorption characteristics of this material have been reported by Barrer and Stuart (17), who deduced that this synthetic zeolite is an energetically homogeneous adsorbent for argon but is heterogeneous for nitrogen. Their conclusion is based on the shapes of the $\mathbf{q}^{st}$ *versus* V curves, which are a horizontal straight line for the argon adsorption and a curve of negative slope for the nitrogen adsorption. We have already referred to the serious error that is introduced by neglecting adsorbate-adsorbate interaction; the misinterpretation of a horizontal heat curve as evidence of an energetically homogeneous substrate is an example of this error.

The adsorption of argon at 77.5 and 90.1 °K by Linde 13X, as measured by Messrs. Chen and Albers in our laboratory, has been used to obtain V_β, U_0', and γ by our present method (Figs. VII-13 and VII-14); the values of V_β and U_0' are in reasonable agreement with those reported by Barrer and Stuart (17); according to our interpretation, however, this adsorbent, with a γ value of 22, cannot be considered as energetically homogeneous for the adsorption of argon. Barrer and Rees (18) have proposed an equation of state for a homotattic surface, which they have applied to the observed argon isotherms with this adsorbent; such a procedure is of course unjustified if the surface is heterogeneous.

5. *Rutile and Anatase*

The sample of rutile that has been the subject of extensive adsorption and calorimetric measurements by Morrison and co-workers (10) is a classic adsorbent because of their extensive and careful measurements. We have used the published data for the adsorption of argon at 85 °K and of oxygen at 100 °K to obtain the parameters reported in Table VII-1; the description of the experimental isotherm by the model is shown graphically in Figure VII-15. Morrison and Drain have also given an adsorptive energy distribution for the argon/rutile system by the method described in Section IV-3 and we have included their diagram for comparison with our adsorptive energy distribution (Fig. VII-23). The higher adsorptive energies reported by Morrison and Drain result from their neglect of adsorbate-adsorbate interaction; a correction for this effect

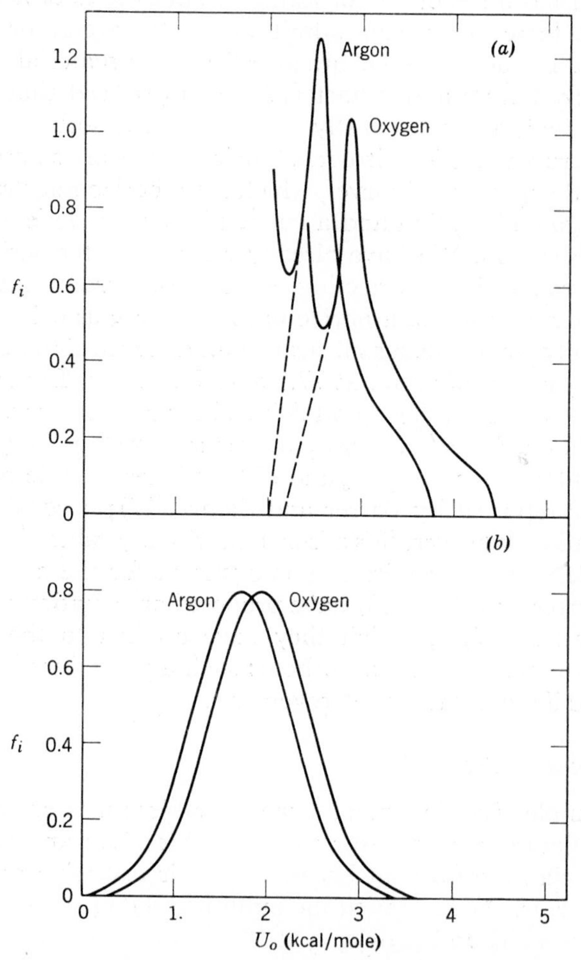

Fig. VII-23. Adsorptive energy distributions of argon and oxygen on the surface of rutile: (a) according to Morrison *et al.* (10), (*b*) by the present theory.

could well bring the two distribution curves into general agreement.

The adsorption of argon at 77.5 °K by a sample of anatase has been determined for comparison. Although the specific surface area of the anatase is considerably less than that of the rutile, the average adsorptive energies are nearly the same; both samples also show great heterogeneity at the surface. For these adsorbents at least the difference in the natures of their crystalline lattices had no great influence in determining the character of the surface.

B. THE KINETIC-MOLECULAR BEHAVIOR OF CHLOROFORM ADSORBED BY GRAPHITE

A number of adsorption isotherms of chloroform on graphitized carbon black P-33 (2700°) in the temperature range 250–350 °K, and in the monolayer region of adsorption, have been reported by Machin and Ross (19). The adsorbent used has been shown to be near-homotattic by the evidence of the argon and nitrogen adsorption isotherms reported and discussed in Section VII-1-A; and in the range of temperatures at which the chloroform adsorption was carried out, the apparent value of γ is sufficiently large that the isotherms can be treated as if γ were equal to infinity. That this conclusion is valid is demonstrated by the observation that the isotherms at the lowest temperatures display unmistakably the vertical discontinuities that are interpreted as two-dimensional phase changes: such vertical discontinuities in the isotherm cannot occur

TABLE VII-3
Adsorption Parameters for $CHCl_3$ on a Homotattic Graphite Surface

T, °K	${}_{a}n_{\beta}$, μ mole/g	$2\alpha/\beta$, kcal/mole	K', mm Hg
323.2	81	2.63	90.1
297.4	81	2.62	24.6
285.3	81	2.89	14.6
268.2	81	2.88	5.75
248.7	81	2.84	1.46
		average = 2.77 +.12	

$\alpha^{id} = 368 \times 10^{-30}$ erg cm² molecule^{-2}
$\alpha = 283 \times 10^{-30}$ (observed)
$\beta^{id} = \beta = 29.4$ A² molecule^{-1}

with surfaces having other than extremely narrow distributions of adsorptive potentials, no matter how low the temperature. This statement is based on the characteristics of the computed subcritical isotherms of heterogeneous substrates. The observation of a vertical discontinuity in the isotherm is, therefore, proof that the surface

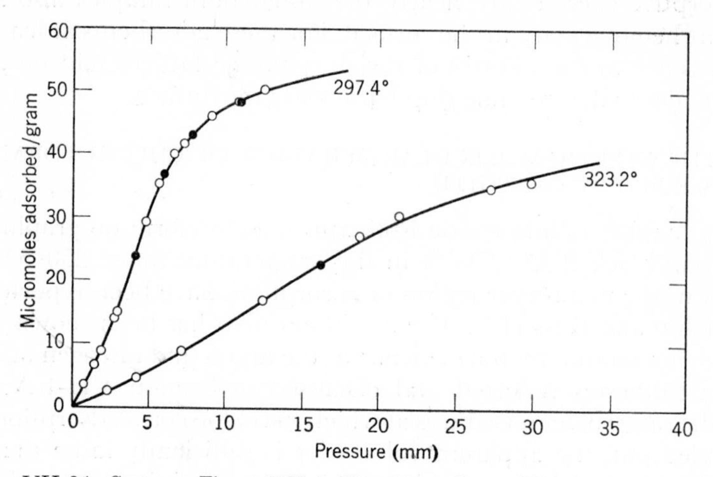

Fig. VII-24. Same as Figure VII-1. The data refer to $CHCl_3$, adsorbed by P-33 (2700°C) at 297.4 and 323.2°K (19).

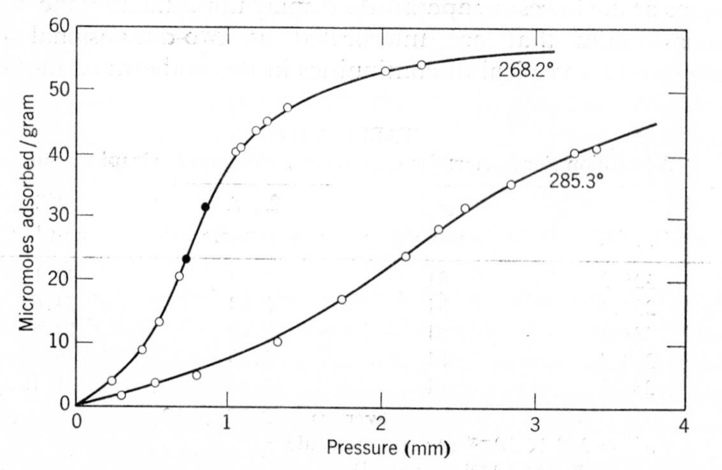

Fig. VII-25. Same as Figure VII-1. The data refer to $CHCl_3$ adsorbed by P-33 (2700°C) at 268.2 and 285.3°K (19).

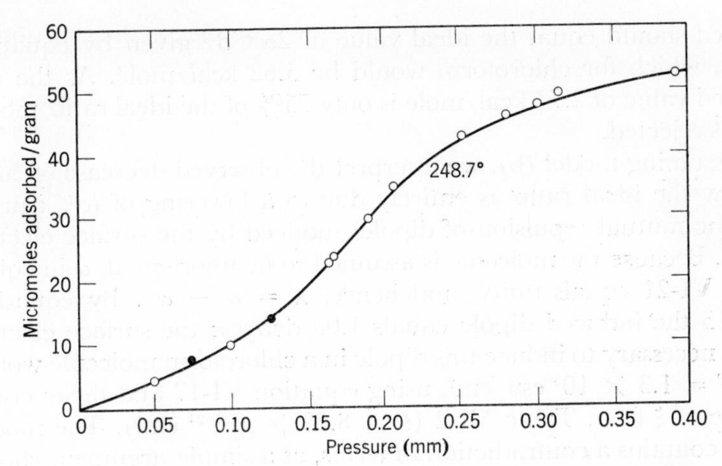

Fig. VII-26. Same as Figure VII-1. The data refer to $CHCl_3$ adsorbed by P-33 (2700°C) at 248.7°K (19).

may be treated as ideally uniform, particularly so at higher temperatures because of the increase in the apparent value of γ with temperature.

The experimental isotherms were superimposed on a set of theoretical isotherms computed according to equation I-18b, i.e., $\gamma = \infty$, for various values of $2\alpha/RT\beta$, using the method previously described in Section VI-1. In this way, the values of the parameters V_β, K', and $2\alpha/\beta$ were obtained. As a cross-check the data were also plotted according to equation V-1 as W *versus* θ, using the value of V_β that had been found by fitting; this method is a more sensitive test of the validity of equation I-18b, particularly at high or low surface concentration. Values of the parameters that satisfied both methods are reported in Table VII-3, and graphical representations of the agreement of the experimental data with the computed isotherms are reported in Figures VII-24–VII-26.

Some possible models of the kinetic-molecular behavior of the adsorbed chloroform are: (a) free rotation, insignificant surface electric field; (b) free rotation, large surface electric field; (c) orientation along symmetry axis due to dispersion forces alone; (d) orientation along symmetry axis due primarily to a surface electric field.

Model (a) implies that the adsorbed molecule is essentially in the same condition as in the gas phase; hence, the value of $2\alpha/\beta$ observed should equal the ideal value of $2\alpha^{id}/\beta^{id}$ given by equation I-42, which for chloroform would be 3.62 kcal/mole. As the observed value of 2.77 kcal/mole is only 75% of the ideal ratio, model (a) is rejected.

Assuming model (b), we interpret the observed decrease of $2\alpha/\beta$ below the ideal ratio as entirely due to a lowering of α^{id}, caused by the mutual repulsion of dipoles induced by the surface electric field. Because the molecule is assumed to be unoriented, ω in equation VI-21 equals unity, and hence, $\lambda = \alpha - \alpha^{id}$. By equation VI-15 the induced dipole equals 1.08 debyes; the surface electric field necessary to induce this dipole in a chloroform molecule would be $F = 1.3 \times 10^5$ esu/cm², using equation VI-17 and the average value of ξ from Table VI-2 ($\xi = 8.23 \times 10^{-24}$ cm³). The model now contains a contradiction in terms, as a simple argument shows that a field of this magnitude would, by its interaction with the permanent dipole possessed by chloroform, prevent the molecule from rotating freely. The energy required to rotate a dipole μ in an electric field F is equal to $2\mu F$; hence, the fraction of molecules able to rotate is given by $e^{-2\mu F/kT}$. As μ is 1.05 debyes for chloroform, the fraction of molecules able to rotate could be only 7 in 10,000 at 0°C. This conclusion is contrary to the initial assumption of the model, which must, therefore, also be rejected.

Model (c) supposes that there is no surface electric field; hence, the lowering of α^{id} could be caused only by the alignment of the permanent dipole plus the anisotropy of the polarizability. The chloroform molecule would be expected to orient itself, in response to the dispersion force interaction, with the three chlorine atoms closest to the surface and the permanent dipole of the molecule normal to the surface. The alignment of the permanent dipole would then reduce α^{id} by introducing a mutual repulsion term λ; the lateral polarizability of the oriented molecule, however, is greater than the average polarizability of the freely-rotating molecule, and this factor, therefore, would have the effect of increasing α^{id}: i.e., ω in equation VI-21 is greater than unity. For the purpose of establishing the model we can put β (oriented) $= \beta^{id}$ for this almost symmetrical molecule; then taking the values of the lateral interaction constants of chloroform from Table VII-3 and the polarizabilities from Table VI-1 we find $\omega = 1.2$ and $\lambda = -80 \times 10^{-30}$ erg cm² molecule^{-2}. Substituting these values in equation VI-21 gives α

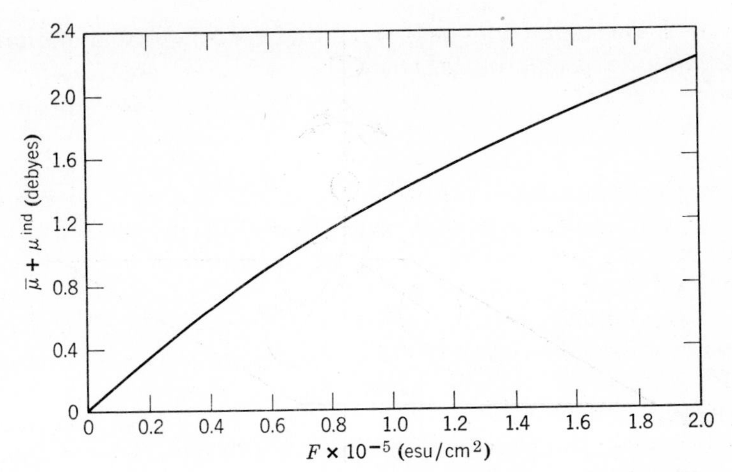

Fig. VII-27. Graphical aid for the solution of F based on the experimentally obtained value of $_a\mu$.

$= 361 \times 10^{-30}$ erg cm^2 molecule^{-2}, which is only slightly less than $\alpha^{id} = 368 \times 10^{-30}$. The observed value of α is 283×10^{-30}, however, which is much less than the value predicted by this model.

Model (d), as we shall show, avoids the self-contradictions that are inherent in the previous models. As we have seen, the molecules must be oriented and, in the absence of an electric field, we cannot account for the low value of α that was observed: the calculations perfomed for model (c) indicated that λ must equal -158×10^{-30} in order to satisfy equation VI-21; the total effective dipole moment normal to the surface can then be calculated by equation VI-15, giving $_a\mu = 1.74$ debye. By equation VI-18

$$_a\mu = \mu^{\text{ind}} + \bar{\mu} \tag{VI-18}$$

Both μ^{ind} and $\bar{\mu}$ can be calculated as functions of the electric field: μ^{ind} is given by equation VI-17:

$$\mu^{\text{ind}} = F\,\xi_1 \tag{VI-17}$$

where $\xi_1 = 6.68 \times 10^{-24}$ cm^3 for chloroform; $\bar{\mu}$ can be calculated by means of the Langevin function, equation VI-19, if we assume that the major part of the orienting force derives from the electric field. By plotting $_a\mu$ for various values of F, as shown in Figure VII-27, the value of the field required to give the observed value of $_a\mu$

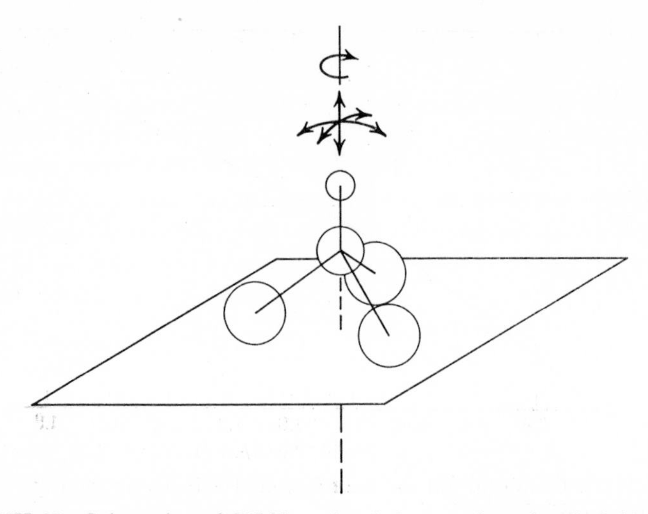

Fig. VII-28. Orientation of $CHCl_3$ molecule by a surface electric field, showing rotation only about the vertical axis and harmonic oscillations resolved in three mutually perpendicular planes.

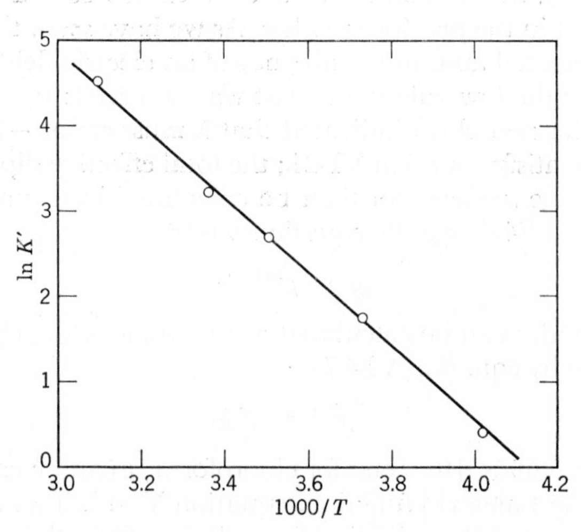

Fig. VII-29. Experimentally determined values of K' for chloroform on a graphite surface plotted as a function of T. This type of plot is used to determine the limiting isosteric heat of adsorption at zero surface concentration.

may be interpolated, yielding $F = 1.12 \times 10^5$ esu/cm², or 3.36 $\times 10^7$ volts/cm.

Accepting tentatively as a working hypothesis that the chloroform molecules are oriented at the surface of graphite in the manner illustrated by Figure VII-28, we can use the observed values of K' at different temperatures to arrive at a statistical-mechanical description of the adsorption process. The values of K' as a function of T are first used to solve equation V-64 graphically as shown in Figure VII-29:

$$\lim_{\theta \to 0} q^{\mathrm{st}} = -R \frac{d \ln K'}{d(1/T)} \tag{V-64}$$

thence,

$$\lim_{\theta \to 0} q^{\mathrm{diff}} = \lim_{\theta \to 0} q^{\mathrm{st}} - RT$$

The standard molar entropy of adsorption at each temperature T can now be calculated by equation III-91:

$$\Delta S_s^{\mathrm{ads}} = \lim_{\theta \to 0} \frac{q^{\mathrm{diff}}}{T} - \mathrm{R} \ln \frac{\theta_s}{1 - \theta_s} - R \ln \frac{K'}{760} \tag{III-91}$$

This entropy term contains translational, rotational, and vibrational entropy changes on adsorption. For our present model of adsorbed chloroform: ΔS_s^{tr} is given by equation V-16, ΔS^{rot} is given by equation V-25, using $\sigma = 3$ for both the gaseous and adsorbed states; $\theta_s = \beta/4.08T$. The remaining entropy term is ${}_aS^{\mathrm{vib}}$, which is obtained by subtraction;

$${}_aS^{\mathrm{vib}} = \Delta S_s^{\mathrm{ads}} - \Delta S_s^{\mathrm{tr}} + \Delta S^{\mathrm{rot}} \tag{V-15}$$

The adsorbed chloroform molecule has lost the freedom to rotate about the two principal axes that are parallel to the surface, the motion with respect to these axes being restricted to oscillations of frequency ν_1 and ν_2; the molecule retains its rotation about the symmetry axis; in addition, the center of mass of the molecule vibrates with respect to the surface, with a frequency ν_3. The vibrational frequencies ν_1 and ν_2 are clearly equal to one another by virtue of symmetry. We need a relation between ν_1, ν_2, and ν_3 in order to carry out subsequent calculations: as it would seem likely that some coupling exists between the three vibrations, we may reasonably assume that $\nu_1 = \nu_2 = \nu_3 = \nu$. The total vibrational energy increase on adsorption is therefore to be divided equally between the three vibrations that have made their appearance. The frequency ν can now be calculated by means of equation V-27 in the form

$$\frac{{}_aS^{\text{vib}}}{3R} = \frac{h\nu/kT}{(e^{h\nu/kT} - 1)} - \ln\,(1 - e^{-h\nu/kT}) \qquad \text{(VII-4)}$$

The above equation can be solved graphically with the aid of a plot of $h\nu/kT$ *versus* ${}_aS^{\text{vib}}/3R$ in the appropriate range of values. A collection of the quantities involved in these calculations is reported in Table VII-4.

TABLE VII-4
Quantities Used in Calculating Entropy Changes and Adsorptive Potential of Chloroform with Respect to a Graphite Surface

	T_1	T_2	T_3	T_4	T_5	T_6	T_7
Temperature, °K	323.2	297.4	285.3	268.2	248.7	231.2	223.2
$\lim_{\theta \to 0} q^{\text{diff}}$ (cal/mole)	7430	7480	7500	7540	7580	7610	7630
$-\Delta S_s^{\text{ads}}$	11.18	10.93	11.14	11.21	11.01	11.54	11.64
$-\Delta S_s^{\text{tr}}$	12.80	12.72	12.68	12.62	12.54	12.46	12.42
$-\Delta S^{\text{rot}}$	13.09	12.92	12.84	12.72	12.56	12.42	12.36
${}_aS^{\text{vib}}/3$	4.90	4.90	4.79	4.71	4.70	4.45	4.38
$h\nu/kT$	0.231	0.231	0.244	0.255	0.256	0.291	0.301
$\nu \times 10^{-12}$ (sec^{-1})	1.55	1.43	1.45	1.42	1.33	1.40	1.40
U_0' (kcal/mole)	8.18	8.17	8.15	8.14	8.14	8.11	8.12

The average adsorptive potential, U_0', can be calculated from the values of ν and the limiting differential heat, q^{diff}, using equation III-49:

$$U_0' = \lim_{\theta \to 0} q^{\text{diff}} + ({}_aE^{\text{vib}} - {}_aE_0^{\text{vib}}) + \Delta E^{\text{kin}} \qquad \text{(III-49)}$$

which for the present case can be written

$$U_0' = \lim_{\theta \to 0} q^{\text{diff}} + \frac{3Nh\nu}{(e^{h\nu/kT} - 1)} - (1/2)RT \qquad \text{(VII-5)}$$

Values of U_0' thus obtained are tabulated in Table VII-4.

The nature of the above calculation endows a fictitious variation with T on ν and U_0': actually we have artificially smoothed out the real variation of q^{st} with T by taking the slope of the curve shown in Figure VII-29 as constant; the real variation of q^{st} within the temperature range of the experimental data is about 1%, which is probably less than the precision with which K' can be determined by the present method of matching isotherms. Had this variation been determined, the values of ν and U_0' would have shown a random scatter about a mean. Since we know that ν and U_0' are invariant with T we obtain a good estimate of them from a simple arithmetic mean of the values reported in Table VII-4, which gives $\nu = 1.43 \times 10^{12}\ \text{sec}^{-1}$ and $U_0' = 8.14$ kcal/mole.

An alternative method of evaluating these quantities would be to use equations V-59 and V-51:

$$K' = A^0 e^{-U_0'/RT} \tag{V-59}$$

and

$$\ln A^0 = -\left(\frac{\Delta S_s^{\text{tr}}}{R} + \frac{\Delta S^{\text{rot}}}{R}\right) + \left(\frac{{}_aF^{\text{vib}} - {}_aE_0^{\text{vib}}}{RT}\right) + \frac{\Delta E^{\text{kin}}}{RT} - \ln\frac{\theta_s}{1-\theta_s} + \ln 760 \tag{V-51}$$

The solution of these equations is done graphically by trial and error: a value of ν is selected, $\ln A^0$ is calculated by equation V-51 for each temperature; and the function $(\ln K' - \ln A^0)$ plotted *versus* $1/T$. That value of ν that produces a straight line through the origin for this plot is the desired solution; the slope of the line yields the value of U_0'/R. For our present model of adsorbed chloroform, ΔS_s^{tr} is given by equation V-16; ΔS^{rot} is given by equation V-25 using $\sigma = 3$ for both the gaseous and adsorbed states;

$$\Delta E^{\text{kin}} = -(3/2)RT$$

$${}_aF^{\text{vib}} - {}_aE_0^{\text{vib}} = 3RT \ln\left(1 - e^{-h\nu/kT}\right) \tag{V-27}$$

$$\theta_s = \beta/(4.08\ T) \tag{V-34}$$

Using the experimental values of K' reported in Table VII-3, a value of ν was found that satisfied equation V-59. The frequency thus derived is

$$\nu = \nu_1 = \nu_2 = \nu_3 = 1.35 \times 10^{12}\ \text{sec}^{-1}$$

and

$$U_0' = 8.05 \text{ kcal/mole}$$

The latter calculation conforms more with the theory than the former, since the slope of the plot of ($\ln K' - \ln A^0$) *versus* $1/T$ is correctly treated as constant, i.e., U_0' is invariant with temperature; whereas it is less correct to treat q^{st} as invariant with temperature. Nevertheless, the more precise treatment is tiresome in practice as it involves a process of trial and error. Both treatments give substantially the same answers, and can be used (as here) to get independent checks of the correctness of the calculations.

C. THE KINETIC-MOLECULAR BEHAVIOR OF TRICHLOROFLUOROMETHANE ADSORBED BY GRAPHITE

A number of adsorption isotherms of trichlorofluoromethane (Freon-11) on the same graphitized carbon black in the temperature range 220–290°K and in the monolayer region of adsorption have been reported by W. D. Machin and S. Ross (19). The arguments that were employed in discussing the adsorption of chloroform by the same graphite adsorbent apply *mutatis mutandis* to the adsorption of Freon-11. The adsorption parameters, determ'ned as described in the previous section, are reported in Table VII-5, and the graphical representations of the agreement between experimental data and computed isotherms are reported in Figures VII-30, VII-31, VII-32, and VII-33.

TABLE VII-5

Adsorption Parameters for $CFCl_3$ on a Homotattic Graphite Surface

T, °K	${}_a n_\beta$, μ mole/g	$2\alpha/\beta$, kcal/mole	K', mm Hg
286.2	82	2.63	28.8
273.2	82	2.74	15.2
260.2	82	2.68	7.03
248.2	82	2.77	3.46
238.2	82	2.68	1.75
223.2	82	2.68	0.522
		average = 2.70 ± .03	

$\alpha^{id} = 343 \times 10^{-30}$ erg cm² molecule⁻¹

$\alpha = 297 \times 10^{-30}$ (observed)

$\beta^{id} = 31.2$ A² molecule⁻¹

The same models of the kinetic-molecular behavior of adsorbed chloroform are possible for adsorbed Freon-11, and the same arguments apply by which all but one of these models were found to involve contradictions. We must, therefore, asume as before that

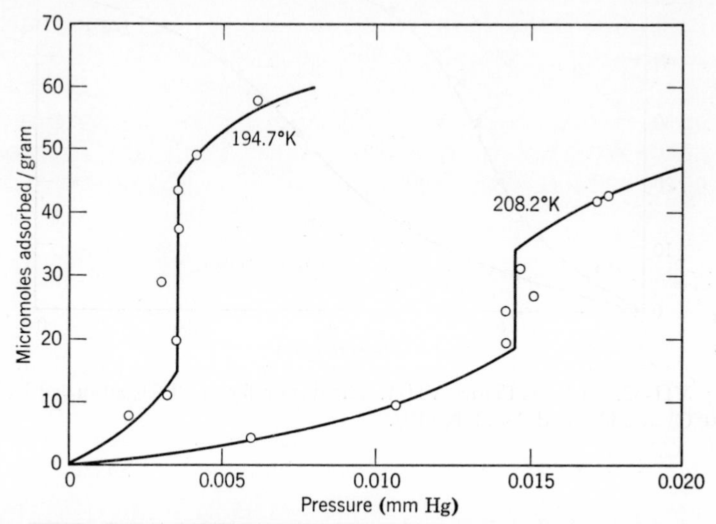

Fig. VII-30. Same as Figure VII-1. The data refer to $CFCl_3$ adsorbed by P-33 (2700°C) at 194.7 and 208.2°K (19).

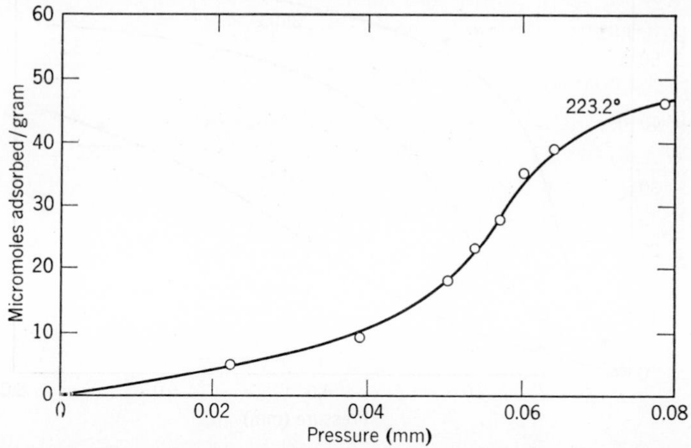

Fig. VII-31. Same as Figure VII-1. The data refer to $CFCl_3$ adsorbed by P-33 (2700°C) at 223.2°K (19).

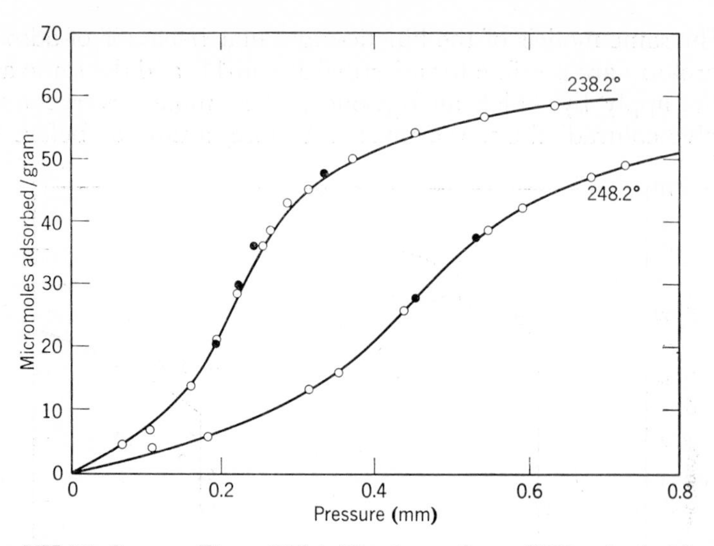

Fig. VII-32. Same as Figure VII-1. The data refer to $CFCl_3$ adsorbed by P-33 (2700°C) at 238.2 and 248.2°K (19).

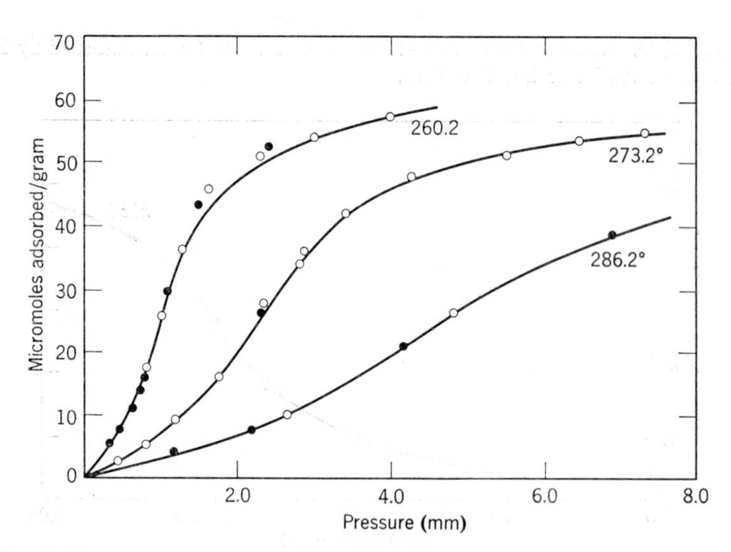

Fig. VII-33. Same as Figure VII-1. The data refer to $CFCl_3$ adsorbed by P-33 (2700°C) at 260.2, 273.2, and 286.2°K (19).

the molecule is oriented about its symmetry axis by the influence of a surface electric field emanating from the substrate. To calculate the strength of the electric field from the present data we require the values of the polarizabilities along each of the three principal axes of the molecule; unfortunately these have not been reported for Freon-11. The average polarizability is 8.45×10^{-24} cm^3, from bond refractivities; and as an approximation we take ξ_2 and ξ_3 to have the same values for Freon-11 as for the corresponding axes of chloroform; namely, $\xi_2 = \xi_3 = 9.01 \times 10^{-24}$ cm^3; from which, by equation VI-9, we get $\xi_1 = 7.35 \times 10^{-24}$ cm^3. Taking the values of the lateral interaction constants from Table VII-5, and the polarizabilities as approximated above, we find $\omega = 1.14$ and $\lambda = -96.1 \times 10^{-30}$ erg cm^2/molecule^2. The total effective dipole normal to the surface is then, by equation VI-15, ${}_a\mu = 1.17$ debyes. The method outlined in the previous section can now be used to obtain the value of the field required to produce the observed value of ${}_a\mu$, yielding $F = 1.27 \times 10^5$ esu/cm^2. This value of the field for graphite is in good agreement with the value of 1.12×10^5 obtained from the chloroform adsorption data, especially in view of the approximation made in obtaining the directional polarizabilities of Freon-11.

An adsorbed molecule of Freon-11 seems to have the same kinetic behavior as the adsorbed chloroform molecule shown in the diagram of Figure VII-28. The calculation of ν and U_0' is done exactly as for chloroform. By the direct calculation of ΔS_s^{ads}, equation III-91, for which the quantities involved are in Table VII-6, the results are:

$$\nu = \nu_1 = \nu_2 = \nu_3 = 0.69 \times 10^{12}\ \text{sec}^{-1}$$

$$U_0' = 8.05\ \text{kcal/mole}$$

By the alternative method of trial and error and a linear plot of equation V-51, the results are:

$$\nu = \nu_1 = \nu_2 = \nu_3 = 0.76 \times 10^{12}\ \text{sec}^{-1}$$

$$U_0' = 8.17\ \text{kcal/mole}$$

These results agree as closely as the data and methods of treatment permit.

TABLE VII-6

Quantities Used in Calculating the Adsorptive Potential of Freon-11 with Respect to a Graphite Surface

	T_1	T_2	T_3	T_4	T_5	T_6	T_7	T_8
Temperature, °K	286.2	273.2	260.2	248.2	238.2	223.2	208.2	194.7
$\lim_{\theta \to 0} q^{\text{diff}}$ (cal/mole)	7340	7370	7390	7420	7440	7470	7500	7520
$-\Delta S_s^{\text{ads}}$	11.80	11.94	11.95	12.11	12.17	12.13	12.60	12.65
$-\Delta S_s^{\text{tr}}$	12.82	12.78	12.72	12.68	12.64	12.58	12.50	12.43
$-\Delta S^{\text{rot}}$	17.66	17.58	17.49	17.39	17.29	17.19	17.04	16.91
${}_aS^{\text{vib}}/3$	6.23	6.14	6.09	5.99	5.92	5.88	5.65	5.56
$h\nu/kT$	0.118	0.124	0.127	0.134	0.138	0.141	0.158	0.165
$\nu \times 10^{-12}$ (sec^{-1})	0.706	0.704	0.689	0.691	0.686	0.655	0.687	0.671
U_0' (kcal/mole)	8.09	8.08	8.07	8.06	8.05	8.04	8.02	8.01

D. THE KINETIC-MOLECULAR BEHAVIOR OF TRICHLOROFLUOROMETHANE ADSORBED BY BORON NITRIDE

Boron nitride is a chemically inert, refractory white solid. The crystal has an hexagonal layer structure similar to that of graphite. According to Moeller (20): "In solid boron nitride, the unit cell is an hexagonal unit of alternating boron and nitrogen atoms, B_3N_3, the constants of which are a = 2.5038 ± 0.0001 A and c = 6.660 ± 0.001 A. The structure has been considered to be that of graphite, the B—N bond distance of 1.45 A being comparable to the C—C bond distance of 1.42 A in that substance and lying between the theoretical single and double B—N bond distances of 1.54 and 1.36 A, respectively. Indeed, the compound has been called 'inorganic graphite.' According to Pease, however, the hexagonal rings are packed directly on top of each other, with the positions of the boron and nitrogen atoms being interchanged in adjacent layers."

The sample of boron nitride used for the measurements of the adsorption of Freon-11 was the same sample as that previously mentioned in Section VII-1-A-(2), in which the presence of traces of a surface boric oxide had been detected. We have subsequently found, however, that some of this impurity disappears on heating

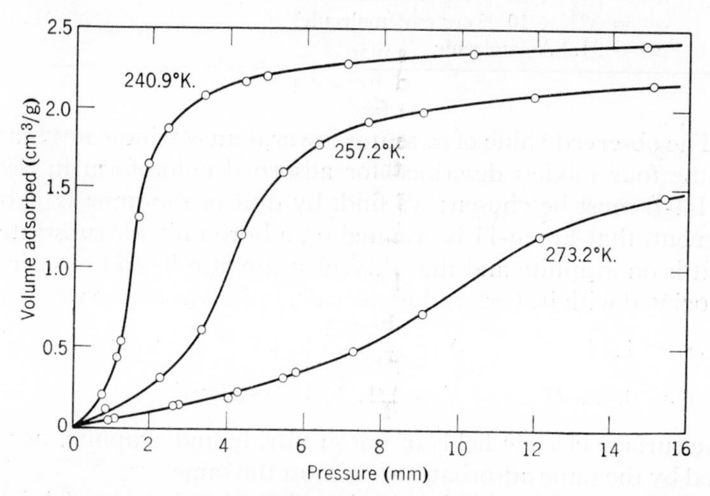

Fig. VII-34. Same as Figure VII-1. The data refer to $CFCl_3$ adsorbed by boron nitride at 240.9, 257.2, and 273.2°K.

the sample to 400–500 °C *in vacuo* for several hours. The sample thus treated then appears completely homotattic with respect to adsorbed Freon-11 at room temperatures, although the argon adsorption isotherm at 77.5 °K still shows signs of the presence of impurity.

A number of adsorption isotherms of Freon-11 on a sample of boron nitride treated as described above were measured by one of us (J. P. O.); these isotherms are fitted to equation I-18b, as illustrated in Figure VII-34, and the adsorption parameters derived thereby are reported in Table VII-7.

TABLE VII-7

Adsorption Parameters for $CFCl_3$ on a Homotattic Boron Nitride Surface

T, °K	V_β, cm^3 STP/g	$2\alpha/\beta$, kcal/mole	K', mm Hg
273.2	3.05	2.87	72.3
257.2	3.05	2.92	33.1
240.9	3.05	3.00	14.2
233.2	3.05	3.01	8.85
	average =	2.95 ± .06	

$\alpha^{id} = 343 \times 10^{-30}$ erg cm²/molecule²
$\alpha = 321 \times 10^{-30}$ erg cm²/molecule²
$\beta^{id} = 31.2$ A²/molecule

The observed value of α, as it is lower than α^{id}, indicates that one of the four models described for adsorbed chloroform in Section VII-1-B must be chosen; we find, by dint of the same type of argument, that Freon-11 is oriented on a boron nitride substrate just as it is on graphite and that this substrate also has an electric field associated with it. Performing the calculations as before we obtain:

$$\omega = 1.14 \qquad \lambda = -68.5 \times 10^{-30} \text{ erg cm}^2/\text{molecule}^2$$

$$_a\mu = 0.986\ D \qquad F = 1.1. \times 10^5 \text{ esu/cm}^2$$

The surface electric fields of boron nitride and graphite, as measured by the same adsorbate, are almost the same.

The calculation of U_0' and ν, based on the above model, is done exactly as before (Section VII-1-B). By the direct calculation of

ΔS_s^{ads}, equation III-91, for which the quantities involved are in Table VII-8, the results are:

$$\nu = \nu_1 = \nu_2 = \nu_3 = 0.53 \times 10^{12} \text{ sec}^{-1}$$

$$U_0' = 6.7 \text{ kcal/mole}$$

TABLE VII-8

Quantities Used in Calculating the Adsorptive Potential of Freon-11 with Respect to a Boron Nitride Surface

	T_1	T_2	T_3	T_4
Temperature, °K	273.2	257.2	240.9	233.2
$\lim_{\theta \to 0} q^{diff}$ (cal/mole)	6090	6120	6150	6170
$-\Delta S_s^{ads}$	10.37	10.46	10.65	10.69
$-\Delta S_s^{tr}$	12.78	12.70	12.64	12.62
$-\Delta S^{rot}$	17.58	17.47	17.33	17.25
${}_aS^{vib}/3$	6.66	6.57	6.44	6.39
$h\nu/kT$	0.095	0.099	0.106	0.109
$\nu \times 10^{-12}$ (sec^{-1})	0.541	0.533	0.534	0.529
U_0' (kcal/mole)	6.82	6.81	6.79	6.79

By the alternative method of trial and error and a linear plot of equation V-59, the results obtained are:

$$\nu = \nu_1 = \nu_2 = \nu_3 = 0.50 \times 10^{12} \text{ sec}^{-1}$$

$$U_0' = 6.7 \text{ kcal/mole}$$

These results agree satisfactorily.

E. A SECOND LOOK AT ADSORBED ARGON AND NITROGEN ON GRAPHITE AND BORON NITRIDE

In Section VII-1-A we used adsorption isotherms of argon or nitrogen on various solids measured at 77.5 and 90.1 °K to evaluate the adsorption parameters V_β, γ, and K' for the systems. That the values obtained were only approximate was due to the introduction of a simplifying assumption, namely, that no corrections need be made to α^{id} and β^{id}. This assumption implies, first, that the adsorbate molecules are not restricted in rotation and second, that the adsorbate is not appreciably polarized by the surface. For two of the adsorbents that were included in that discussion we now have positive evidence for the presence of a surface electric field, and can, therefore, re-evaluate the data to obtain more accurate results. In Section VI-2-B we pointed out that the effect of a surface electric field is generally to decrease lateral interactions, as measured by α; and that when isotherm matches are made with a decreased value of $2\alpha/RT\beta$, a higher value of γ is obtained. This result is illustrated in this section; it shows that the effect of an electric field at the surface can be confused with that of surface heterogeneity, at least within a limited range of possible values. We emphasize once again the desirability of determining one of these properties of the substrate by an independent method.

1. Argon

Taking the surface field of graphite as 1.2×10^5 esu/cm², as determined in the preceeding sections, the dipole induced thereby in argon is, according to equation VI-17, equal to 0.195 debyes. The correction term λ to be added to α^{id} is, according to equation VI-15,

$$\lambda = -{}_a\mu^2 \pi/d \qquad \text{(VI-15)}$$

from which $\lambda = -4.06 \times 10^{-30}$ erg cm²/molecule². The operative value of α is given by equation VI-21.

$$\alpha = \omega \alpha^{id} + \lambda \qquad \text{(VI-21)}$$

Since the argon molecule is isotropic, the orientation factor ω is equal to unity; with $\alpha^{id} = 47.4 \times 10^{-30}$ erg cm²/molecule², α becomes 43.3×10^{-30} erg cm²/molecule². The value of β for argon is not affected by the surface field, hence $\beta = \beta^{id} = 13.6$ A²/molecule and $2\alpha/\beta = 637 \times 10^{-16}$ ergs/molecule or 918 cal/mole. For comparison the value of $2\alpha^{id}/\beta^{id}$ is 1000 cal/mole. Adsorption isotherms of argon on graphite at 77.5 and 90.1 °K ought to be matched against model isotherms computed for $2\alpha/RT\beta$ equal to 6.0 and 5.2, respectively. The corresponding idealized values of $2\alpha/RT\beta$, which were actually applied in Section VII-1-A as a first approximation indiscriminately to all adsorbents, are 6.5 and 5.6. In Table VII-9 we report new sets of adsorption parameters for argon adsorbed by two of the graphites listed in Table VII-1. The substrates are the same or similar to the ones for which the surface field was evaluated in Sections VII-1-B and C. In ad-

TABLE VII-9
Parameters for Adsorbed Argon on Substrates of Known Surface Electric Field

Adsorbent	Temp., °K	$2\alpha/RT\beta$	V_β, cm³ STP/g	γ	K'	$\lim_{\theta\to 0} q^{st}$, kcal/mole	$-\Delta S_s^{ads}$, eu	$\nu \times 10^{-12}$, sec⁻¹	U_0', kcal/mole
MT (3100°C)	77.5	6.0	2.35	700	1.66	2.27	9.99	2.5	2.10
MT (3100°C)	90.1	5.2	2.25	700	13.0		8.67	2.2	2.10
P-33 (2700°C)	77.5	6.0	3.83	400	1.67	2.28	9.13	2.7	2.11
P-33 (2700°C)	90.1	5.2	3.75	400	13.2		8.81	2.3	2.10
BN (Pultz)	77.5	5.6	5.90	∞	7.45	—	—	—	—
BN (400°C)	77.5	5.6	6.30	∞	7.6	—	—	—	—
P-33 (1000°C)	77.5	6.0	3.85	8	2.77	2.46	12.42	1.98	2.16
P-33 (1000°C)	90.1	5.2	3.85	8	25.7		12.10	1.90	2.15

dition, we include in Table VII-9 a heterogeneous carbon black for which we assume that the surface field is the same as that of graphite. For a molecule of such relatively low polarizability as argon the effect of a distribution of intensity of the surface field over the surface can probably be taken as unimportant: the observed heterogeneity is chiefly the result of a distribution of dispersion forces, as tacitly assumed by the theory underlying the computation of the model isotherms.

For the boron nitride adsorbent we are able to assign a value of γ equal to infinity, at least for the major part of the surface, on the basis of a recent determination by R. Cripps of this laboratory of a krypton adsorption isotherm at 77.5 °K, in which the vertical discontinuity caused by two-dimensional condensation was clearly evident. The computed tables for model substrates of slight heterogeneity reveal that this mathematical discontinuity disappears as soon as the surface departs from homotaxis, and is replaced by the appearance of a slope less than 90°. The presence of the vertical discontinuity in an observed isotherm enables us, therefore, to assert confidently that a major portion of the substrate is composed of a homotattic surface plane. In re-fitting the argon adsorption data for the boron nitride we have put γ equal to infinity, and thus readily evaluated $2\alpha/RT\beta$. The observed value, reported in Table VII-9, leads to F for boron nitride of 1.5×10^5 esu/cm², which is a rather higher result than that derived from the Freon-11 adsorption isotherm ($F = 1.1 \times 10^5$ esu/cm²).

When values of K' are given for more than one temperature, one can go on to calculate ν and U_0' by the procedure described in Section VI-3 (*Case 2*). The values of q^{st}, ΔS_s^{ads}, ν, and U_0' thus obtained are reported in Table VII-9. As only two temperatures are available the result for ν, which is derived by a calculation that involves a subtraction and a long extrapolation, is not as accurate as one would wish. With more precise data, which became available later, better accuracy was obtained (see Section VII-1-F): we find that about 100% error in ν can be expected when using data conventionally obtained at 77.5 and 90.1 °K. The other quantities derived from the data are not so sensitive to the precision of measurement.

Comparing the new adsorption parameters with those obtained when the surface field is completely neglected we find that differences exist that would be significant for some purposes, such as an effort to describe the adsorption system in terms of its kinetic-molecular behavior by an analysis of K'; these differences would not be important, however, for some practical purposes such as the determination of specific surface area, where only a 3% error would occur on neglecting the effect of the surface field. The values of U_0', calculated from K', differ by only 1% from the values obtained previously; the estimation of the vibrational frequency would, however, be much more affected by neglecting the surface field. We conclude, for the adsorption of argon, that by disregarding the surface field of the adsorbent no serious errors are introduced that would prevent the practical application of the method of fitting for the determination of specific surface areas, adsorptive potentials, and for a rough estimate of the degree of heterogeneity of the surface. Inasmuch as we have only limited information about the magnitude of surface fields of solids, the approximate method is usually the only method applicable; argon is the preferred adsorbate for this purpose; polyatomic adsorbates, which are more sensitive to the surface field, are apt to introduce more serious errors for that reason.

2. Nitrogen on P-33 (2700°)

The nitrogen molecule is anisotropic with respect to shape and polarizability: this fact, which introduces the possibility of orientation on adsorption, makes impossible a calculation of α from α^{id} by taking into account the surface field, because of the uncertainty about the correct model, i.e., whether the nitrogen is oriented or not. We are obliged, therefore, to find the operative values of α and β by obtaining the best match between the experimental and model isotherms, guided by knowledge previously acquired about the substrate. The surface is known to be nearly homotattic, so much so that at room temperatures it can be considered completely homotattic; but that it is not actually so, however, is shown by the true γ value of 400 obtained with argon at liquid air temperatures; we can assume that for nitrogen as adsorbate, since nitrogen is a substance of about the same molecular size, polarizability, and heat of adsorption as argon, the true value of γ will again be about 400. With this as a guide we were able to obtain reliable values of the adsorption parameters for nitrogen

TABLE VII-10
Parameters of Adsorbed Nitrogen on P-33 (2700°) Graphite

T, °K	$2\alpha/RT\beta$	V_β, cm³ STP/g	lim q^{st}, $\theta \to 0$ kcal/mole	$-\Delta S_s^{ads}$	γ	K'	$\nu \times 10^{-12}$	U_0', kcal/mole
77.5	4.1	3.30 }	2.31	9.93	300	0.78	1.22 }	2.16 }
90.1	3.5	3.28 }		10.1	300	6.35	0.98 }	2.15 }

on graphite at 77.5 and 90.1 °K, by the method of matching against model isotherms, using the same data cited in Table VII-1. These parameters are reported in Table VII-10, and graphical comparisons of model and experimental isotherms are shown in Figures VII-35 and VII-36.

The average value of $2\alpha/\beta$ is 631 cal/mole. We obtain a value of β from the ratio of the monolayer capacities (V_β) found for argon and nitrogen, assuming that for argon $\beta = \beta^{id} = 13.6$ A², by the following equation:

$$\beta_{N_2} = \beta_A \left(\frac{V_{\beta,A}}{V_{\beta,N_2}}\right) \tag{VII-6}$$

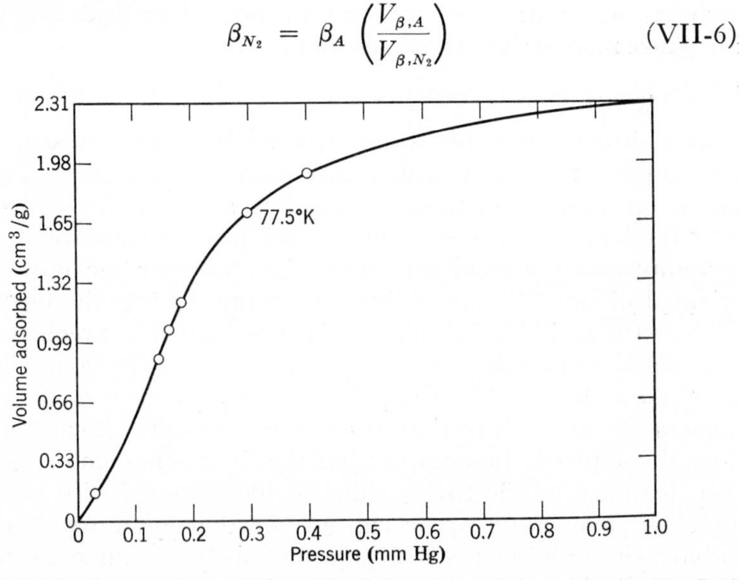

Fig. VII-35. Same as Figure VII-1. The data refer to nitrogen adsorbed by P-33 (2700°C) at 77.5°K. The theoretical description is now more precise for the low-pressure data than that given in Figure VII-5.

from which $\beta = 15.6$ A^2 for nitrogen. This value of β is practically the same as β^{id}, *viz.*, 15.55 A^2. The operative value of α is then found to be 34.2×10^{-30} erg cm^2/molecule^2; whereas α^{id} is 45.7×10^{-30}.

The effectiveness of a surface field in inducing a dipole in a nitrogen molecule depends on whether the molecules are rotating freely, standing on end, or lying flat. If the molecules are rotating freely, a field of 1.2×10^5 esu/cm^2, which we have found for graphite, would induce a dipole of 0.21 debyes, resulting in an operative

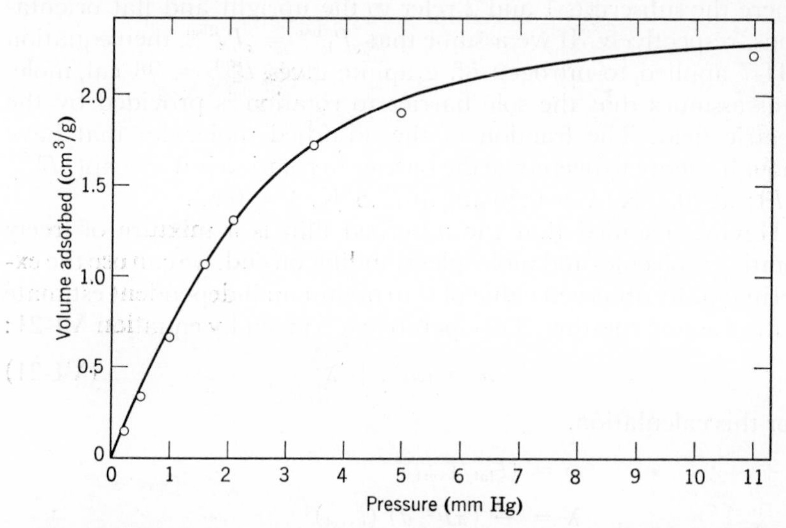

Fig. VII-36. Same as Figure VII-1. The data refer to nitrogen adsorbed by P-33 (2700°C) at 90.1°K. The theoretical description is now more precise for the low-pressure data than that given in Figure VII-6.

value of α of 41.3×10^{-30} erg cm^2/molecule^2. If the molecules stand on end, this surface field, because of the greater polarizability of nitrogen along the long axis, would induce a dipole of 0.29 debyes, with a resulting operative value of α, calculated by equation VI-21, of 22.8×10^{-30} erg cm^2/molecule^2. If the molecules lie flat on the surface, the corresponding induced dipole is 0.17 debyes and the operative value of α is 50.8×10^{-30}. We see, therefore, that none of these models yields the observed value of α; the true situation must correspond to an intermediate condition.

The factors that could influence the orientation of nitrogen on graphite are the dispersion forces and the electric field. The presence of the surface field, since it tends to orient the molecules with their axis of greatest polarizability colinear with the field, introduces a barrier to the rotation of that axis out of that alignment. The total energy barrier to free rotation is given by

$$B^{rot} = ({}_gP_1{}^{ads} - {}_gP_2{}^{ads})$$
$$= ({}_gP_1{}^{disp} - {}_gP_2{}^{disp}) + (F^2/2)\,(\xi_1 - \xi_2) \qquad \text{(VII-7)}$$

where the subscripts 1 and 2 refer to the upright and flat orientations, respectively. If we assume that ${}_gP_1{}^{disp} = {}_gP_2{}^{disp}$, then equation VII-7 applied to nitrogen on graphite gives $B^{rot} = 98$ cal/mole. This assumes that the sole barrier to rotation is provided by the electric field. The fraction of the adsorbed molecules that have enough energy to overcome the barrier to rotation is $X = \exp(-B^{rot}/RT)$; at 90.1 °K, $X = 0.58$ and at 77.5 °K, $X = 0.53$.

Having assumed that the adsorbed film is a mixture of freely rotating molecules and molecules standing on end, we can use the experimentally observed value of α to obtain an independent estimate of the fraction rotating. The operative α is given by equation VI-21:

$$\alpha = \omega\alpha^{id} + \lambda \qquad \text{(VI-21)}$$

For this calculation,

$$\omega = (\xi_{lat}/\xi_{vert})^2$$
$$\lambda = -(\pi F^2/d)\,(\xi_{vert})^2$$

where ξ_{lat} and ξ_{vert} refer to the weighted average, or effective, polarizability of a molecule in the film, parallel to and normal to the surface. Let X be the fraction of the molecules rotating, then

$$\xi_{lat} = X\xi_{ave} + (1 - X)\xi_2$$

and

$$\xi_{vert} = X\xi_{ave} + (1 - X)\xi_1$$

where the values of ξ_1, ξ_2, and ξ_{ave} for nitrogen are those given in Table VI-2. Solving the above equations for X yields $X = 0.64$, referring to an average temperature of 84 °K, since the observed value of α is an average. The result, based on the observed value of α, is close to the estimate obtained from equation VII-7, when we assumed that the surface field created the sole barrier to rotation. We conclude, therefore, that the dispersion forces for all orientations are nearly equal; a difference of only 25 cal/mole for the flat orientation *versus* the upright orientation is enough for equation VII-7 to yield $X = 0.64$. The internal consistency of these calculated results supports the hypothesis that nitrogen adsorbed by graphite at these temperatures is significantly restricted in rotation due to interaction with the electric field. In Section VIII-4 we make use of the hypothesis that the dispersion potential between nitrogen and graphite is independent of the orientation of the nitrogen; the energy of interaction thus calculated is then in satisfactory agreement with the experimental observations.

F. A DETAILED INVESTIGATION OF THE INERT GASES ADSORBED BY GRAPHITE

The excellent measurements of the adsorption of neon, argon, krypton, and xenon by the graphite P-33(2700°), obtained by Sams, Constabaris, and Halsey (21), were made available to us prior to their publication by courtesy of Professor G. D. Halsey, Jr. These data are of inestimable value to produce a more rigorous test of our model and, from the results thus derived, to extend our quantitative understanding of physical adsorption.

The data all refer to the "Henry's-law region" of the adsorption isotherm, for the treatment of which we have previously developed equation V-62:

$$\ln Z + \ln A^0 = \frac{U_0'}{RT} + \ln V_\beta + \frac{1}{4\gamma(RT)^2} \qquad \text{(V-62)}$$

where

$$Z = \lim_{V \to 0} \left(\frac{\partial V}{\partial p}\right)_T$$

The values of both V_β and γ for the adsorption of argon by this adsorbent have been reported in Table VII-9: $V_\beta = 3.83$ cm^3 STP/g and $\gamma = 400$ kcal^{-2} mole^2. Values of γ for the other inert gases on this adsorbent would be of the same order of magnitude, and consequently make the last term of equation V-62 negligible compared to the other terms. Thus, for example, by taking into account the term in γ when making the calculations from the data for argon, the value of U_0' is diminished by 0.1% and the value of ν is diminished by 0.2%; these corrections are just about the same magnitude as the precision of the result. But although we may disregard the γ term in the calculation, it must not be inferred that such low surface heterogeneity is without effect on the shape of the isotherm, but only that it is without appreciable effect on the initial slope at low values of θ. The isotherms computed for $\gamma = \infty$ and $\gamma = 100$ in Figure V-1 show clearly the small but distinct differences to which we refer. Equation V-62 now reduces to:

$$\ln Z + \ln A^0 = (U_0'/RT) + \ln V_\beta \qquad \text{(VII-8)}$$

The relatively high temperatures of these experiments demand that we follow equation V-51 for the evaluation of $\ln A^0$. For inert gas adsorbates other than argon the values of V_β were determined from the argon value by the relation (see equation VII-6)

$$V_\beta = \frac{3.83\ (13.6)}{\beta^{id}} = \frac{52.1}{\beta^{id}} \qquad \text{(VII-9)}$$

in which β^{id} is in A^2 per molecule.

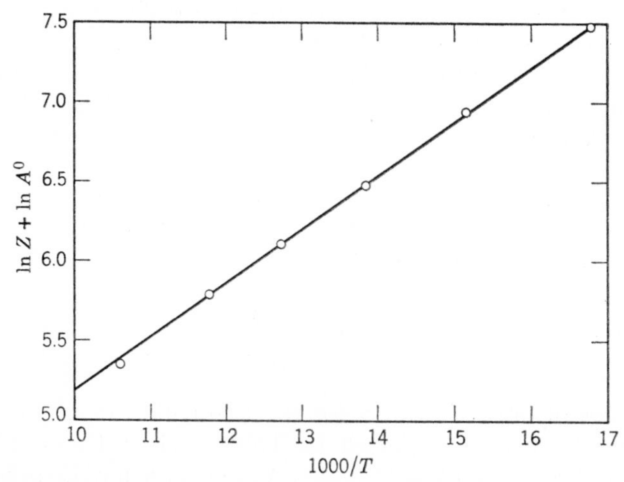

Fig. VII-37. The adsorption data of Constabaris *et al.* (21) for neon adsorbed by graphitized carbon black, P-33 (2700°C) at several temperatures, plotted according to equation VII-8 for evaluation of U_0' and ν.

The alternative method of trial and error, described in Section VII-1-B, was followed for the determination of U_0' and ν for each of the inert gases. The trial slope U_0'/R, and the trial intercept, $\ln V_\beta$, were obtained by the method of least squares for each value

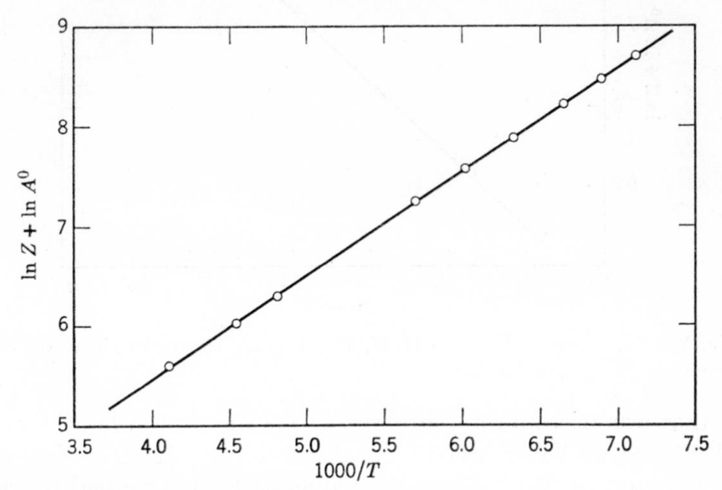

Fig. VII-38. Same as Figure VII-37. The data refer to argon, instead of neon, adsorbed at the same substrate.

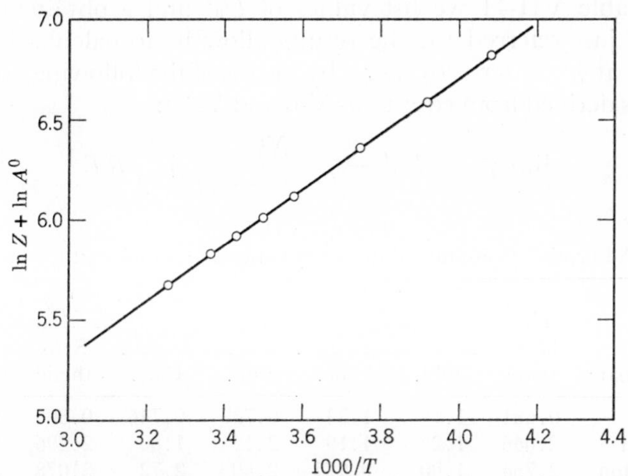

Fig. VII-39. Same as Figure VII-37. The data refer to krypton, instead of neon, adsorbed at the same substrate.

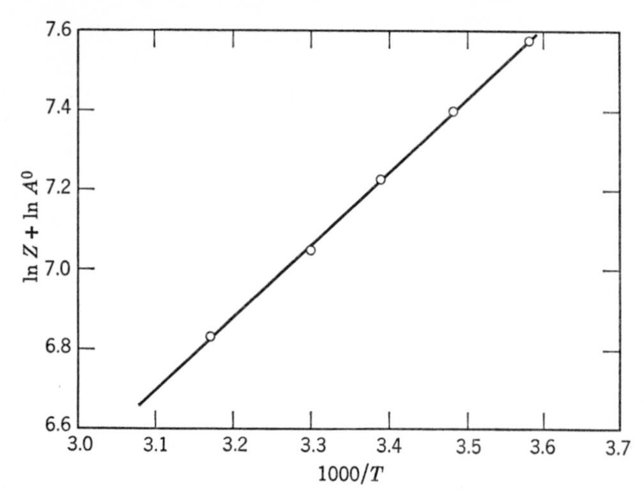

Fig. VII-40. Same as Figure VII-37. The data refer to xenon, instead of neon, adsorbed at the same substrate.

of ν that was tried. The results reported in Figures VII-37 through VII-40 and in Table VII-11 are derived from the graphical solution of equation V-62, using the values of V_β determined by equation VII-9. The quantities involved in the calculations are reported separately for each adsorbate in Table VII-12.

In Table VII-11 we list values of U_0' and ν obtained by the method just referred to; the results allow us to calculate the isosteric heat q^{st}, at zero coverage, by means of the following equation, which is derived from equations V-6 and V-10:

$$\lim_{\theta \to 0} q^{st} = U_0' - \frac{Nh\nu}{(e^{h\nu/kT} - 1)} + \frac{3}{2} RT \qquad \text{(VII-10)}$$

TABLE VII-11
Adsorption Constants of the Inert Gases on Graphitized Carbon

Adsorbate	U_0', kcal/mole	$\nu \times 10^{-12}$ sec^{-1}		${}_gP^{ads}$ kcal/mole		$\lim_{\theta \to 0} q^{st}$ kcal/mole	
		Obs.	Calc.	Obs.	Calc.	Synthetic	Thermo.
Neon	0.681	1.19	1.23	0.737	0.746	0.809	0.820
Argon	2.066	1.28	1.19	2.12	1.80	2.296	2.310
Krypton	2.754	1.00	0.975	2.80	2.72	3.078	3.075
Xenon	3.654	0.850	0.805	3.69	3.37	3.987	3.974

TABLE VII-12
Parameters Associated with the Adsorption of the Inert Gases by Graphite

T, °K	Z, cm³/mm $\times 10^4$	$-\Delta S_s^{\mathrm{tr}}$	$-\left[\dfrac{{}_aF^{\mathrm{vib}} - (^1/_2)Nh\nu}{RT}\right]$
A.—Neon on Graphite			
59.5	21.96	9.34	0.49
65.9	11.43	9.45	0.55
72.3	6.696	9.54	0.61
78.6	4.243	9.62	0.67
84.8	2.877	9.70	0.72
94.3	1.733	9.81	0.80
B—Argon on Graphite			
140.6	37.86	10.88	1.05
145.1	29.37	10.92	1.08
150.1	22.55	10.94	1.11
158.1	15.28	11.00	1.15
166.1	10.68	11.06	1.18
175.1	7.500	11.10	1.23
207.8	2.614	11.28	1.38
220.4	1.890	11 34	1.43
240.0	1.22	11.42	1.51
C—Krypton on Graphite			
245.2	3.875	12.16	1.73
255.4	3.021	12.20	1.76
266.9	2.337	12.25	1.80
279.5	1.778	12.29	1.85
285.1	1.595	12.31	1.87
291.2	1.431	12.34	1.88
297.2	1.288	12.35	1.90
307.2	1.08	12.39	1.94
D—Xenon on Graphite			
279.2	8.428	12.74	2.00
287.0	6.959	12.77	2.02
295.0	5.745	12.79	2.05
303.2	4.744	12.82	2.07
315.0	3.727	12.86	2.11

The isosteric heats of adsorption synthesized in this way by equation VII-10 are based directly on the model of a mobile adsorbed film that we have developed; they can be compared with values of q^{st} obtained by the use of the thermodynamic definition, which has no reference to any model, *viz.*,

$$q^{st} = RT^2 \left(\frac{\partial \ln p}{\partial T} \right)_\theta$$

At zero coverage this definition becomes

$$\lim_{\theta \to 0} q^{st} = -RT^2 \left(\frac{d \ln Z}{dT} \right) \qquad \text{(VII-11)}$$

The values of q^{st} calculated by equation VII-10 and those by equation VII-11 are in excellent agreement for all four adsorbates (Table VII-11) at the temperatures indicated, thus demonstrating that the consequences of our model of the process is quantitatively in agreement with the thermodynamics of the data. The present model, however, is not necessarily unique in this respect.

G. THE ADSORPTION OF THE ISOTOPIC PAIRS H_2-D_2 AND CH_4-CD_4 BY GRAPHITE

Constabaris, Sams, and Halsey (22) have measured the adsorption of H_2, D_2, CH_4, and CD_4 by the graphitized carbon black P-33 (2700°) in the very dilute range (less than about 10% of the monolayer). These adsorbates behave just as did the inert gases with the same adsorbent: the isotherms all have the limiting form at low pressures of straight lines through the origin. Our method of treating the experimental data is the same as that described in Section VII-1-F, making use of equations VII-8 and VII-9. The plots of equation VII-8 for these applications are exemplified (for methane) in Figure VII-41. In Table VII-13 we report the results of the analysis in terms of U_0' and ν for the four adsorbates. The comparison of the limiting isosteric heats obtained from the theoretical model and from the experimental data is also included in this table, and the two are shown to be in good agreement.

A comparison of the vibrational frequencies of each of the isotopic pairs shows the expected decrease for the species of greater mass, although the ratios are not exactly those of the square roots of the molecular weights that would result from harmonic vibrations.

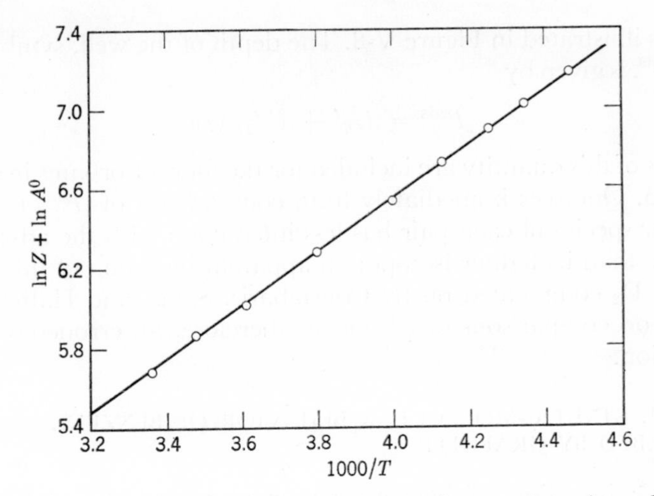

Fig. VII-41. Same as Figure VII-37. The data refer to methane, instead of neon adsorbed at the same substrate.

The interaction energies between adsorbate and adsorbent are not accurately reflected by the limiting isosteric heats that are reported in Table VII-13, although prior to the present method of treating the data such an interpretation had been universally made. A better index of the interaction is to be had from the depth of the potential well, or the maximum loss of potential energy of adsorp-

TABLE VII-13

Adsorption Parameters of the Isotopic Pairs H_2-D_2 and CH_4-CD_4

	H_2	D_2	CH_4	CD_4
β (A^2/molecule)	12.1	12.1	16.4	16.4
V_β (cm^3 STP/g)	4.31	4.31	3.17	3.17
U_0' (kcal/mole)	0.956	1.032	2.678	2.644
$\nu \times 10^{-12}$ (sec^{-1})	10	7	2.48	2.04
q^{st} (thermo.)[a]	1.293	1.337	3.032	3.005
q^{st} (synthetic)[a]	1.281	1.334	3.046	2.994
${}_gP^{ads}$ kcal/mole	1.436	1.366	2.796	2.741
${}_gP^{elec}$ (kcal/mole)	0.082	0.082	0.270	0.270
${}_gP^{disp}$ (kcal/mole)	1.354	1.284	2.526	2.471

[a] q^{st} at 114°K for H_2-D_2; q^{st} at 260.6°K for CH_4-CD_4.

tion as illustrated in Figure V-8. The depth of the well, symbolized by ${}_{g}P^{\text{ads}}$, is given by

$$ {}_{g}P^{\text{ads}} = U_0' + (1/2)Nh\nu $$

Values of this quantity are included for the four adsorbates in Table VII-13. One sees immediately from comparisons of ${}_{g}P^{\text{ads}}$ that the heavier species of each pair has less interaction with the adsorbent surface than its lighter isotope: an apparent inversion of this effect for H_2-D_2 commented on by Constabaris, Sams, and Halsey, was based on comparisons of q^{st} and is, therefore, an erroneous interpretation.

H. THE KINETIC-MOLECULAR BEHAVIOR OF BENZENE ADSORBED BY GRAPHITE

The adsorption of benzene vapor at room temperatures by graphitized carbon P-33 (2700°) has been measured in the monolayer region by one of us (J. P. O.). The determination of V_β and K' were made by fitting the experimental isotherms to equation I-18b. At these temperatures this surface may be treated as com-

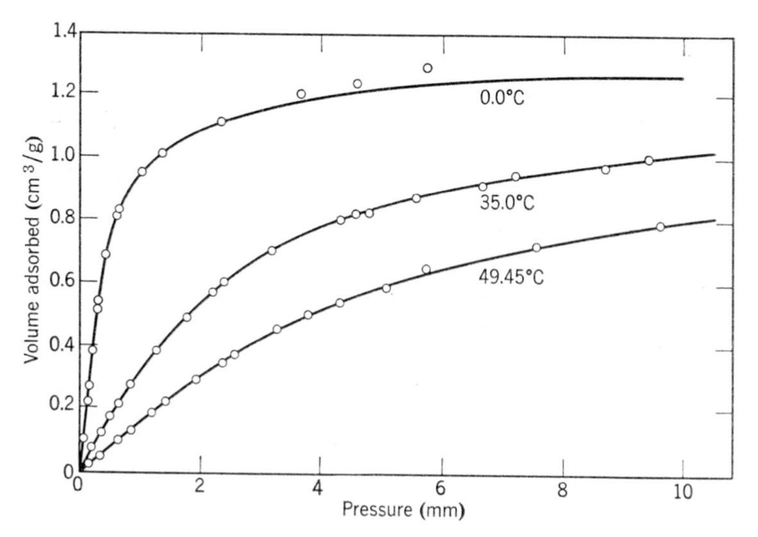

Fig. VII-42. Same as Figure VII-1. The data refer to benzene adsorbed by graphitized carbon black, P-33 (2700°C) at 0.0, 35.0, and 49.45°C.

TABLE VII-14
Adsorption Parameters for C_6H_6 on a Homotattic Graphite Surface

T, °K	V_β, cm³ STP/g	$2\alpha/\beta$, kcal/mole	K', mm Hg
322.6	1.70	1.04	9.93
308.2	1.70	1.10	4.75
273.2	1.70	1.06	0.705
	average =	1.07 ±0.02	

$\alpha^{id} = 427 \times 10^{-30}$ erg cm² molecule^{-2}
$\alpha = 112 \times 10^{-30}$ erg cm² molecule^{-2}
$\beta^{id} = 32.7$ A²/molecule
$\beta = 30.6$ A²/molecule

pletely homotattic. The graphical representations of the quality of the matches is shown in Figure VII-42, and the values of the parameters obtained are reported in Table VII-14.

Compared to the adsorbates that we have considered hitherto, the benzene molecule is markedly anisotropic both in its configuration and in its polarizability. In considering possible orientations of adsorbed benzene, other than a freely rotating molecule, one thinks first of the molecule lying flat on the surface, as this position brings all six carbon atoms close to the surface; this orientation has indeed been postulated by Kemball (23) for benzene adsorbed on mercury, and by Kiselev (24) for benzene adsorbed on graphite. On the other hand, an orientation in which the plane of the ring is normal to the surface cannot be discounted, since the anisotropy of polarizability would favor this position in the presence of a strong electric field and, as can be shown by a calculation using the approximate method developed in Chapter VIII, the dispersion forces calculated for the two orientations are actually nearly equal.

The benzene molecule in its freely rotating state has rotations about three principal axes: one is the cyclic rotation of the ring; the other two are rotations about axes in the plane of the ring. The latter two rotations are equivalent, so that, if rotational restrictions are to be introduced on adsorption of the molecule, these rotations are either both restricted or they are both unrestricted. At the same time the energy differences between each of the six equivalent positions of the ring during the cycling rotation are so small that there is no reason to consider the restriction of that rotation on adsorption, except at temperatures very close to absolute zero. As possible models, therefore, for the kinetic-molecular behavior of adsorbed benzene we have only to consider:

(a) Free rotation,
(b) Orientation with the plane of the ring parallel to the surface; two degrees of rotation lost,
(c) Orientation with the plane of the ring normal to the surface; two degrees of rotation lost.

Each of the above models is to be thought of as in the presence of a strong electric field, which we have already shown to exist at a graphite surface. For all three models the usual "adsorption vibration" normal to the surface is introduced; in addition, each degree of rotation that is lost is replaced by a torsional vibration about the former axis of rotation. For simplicity, as was done for chloroform and Freon-11, we suppose that the torsional and the "adsorption" vibrations have the same frequency.

The free-rotation model requires that the observed lowering of

TABLE VII-15
Quantities Used in Calculating the Adsorptive Potential of Benzene with Respect to a Graphite Surface

	T_1	T_2	T_3
	Free Rotation		
Temperature, °K	322.6	308.2	273.2
$\lim_{\theta \to 0} q^{diff}$ (cal/mole)	8610	8640	8710
$-\Delta S_s^{ads}$	10.56	10.53	10.81
$-\Delta S_s^{tr}$	13.37	13.31	13.19
$-\Delta S^{rot}$	0	0	0
$_aS^{vib}$	2.81	2.78	2.38
$h\nu/kT$	0.673	0.683	0.844
$\nu \times 10^{-12}$ (sec^{-1})	4.5	4.4	4.8
U_0' (kcal/mole)	9.00	9.00	8.93
	Two Degrees of Rotation Lost		
$-\Delta S^{rot}$	16.95	16.85	16.61
$_aS^{vib}/3$	6.59	6.54	6.33
$h\nu/kT$	0.099	0.101	0.120
$\nu \times 10^{-12}$ (sec^{-1})	0.66	0.65	0.64
U_0' (kcal/mole)	9.48	9.47	9.43

α^{id} be due to parallel dipoles induced by the electric field at the interface. According to equation VI-15 the induced dipole would be 2.08 debyes; the electric field necessary to cause this effect would, by equation VI-17, be 2.02×10^5 esu/cm^2. This value for the field at a graphite surface is rather larger than we found previously, i.e., 1.2×10^5 esu/cm^2, but still sufficiently close to warrant our investigating the temperature dependence of K' on the basis of this model, using the procedure described in Section VII-1-B. The quantities involved in the calculation are reported in Table VII-15; the results obtained are $U_0' = 8.97$ kcal/mole and $\nu = 4.6 \times 10^{12}$ sec^{-1}.

The orientation of adsorbed benzene with the plane of the ring parallel to the surface means that ω in equation VI-21 is not equal to unity, but must include terms that take into account both the increased cross-sectional area and the increased polarizability that are associated with the "flat" orientation; hence,

$$\omega = \left(\frac{\xi_2 + \xi_3}{2\xi_{av}}\right)^2 \left(\frac{\beta^{id}}{\beta^{flat}}\right)^2 = 0.954$$

and

$$\lambda = \alpha - \omega\alpha^{id} = -294 \times 10^{-30} \text{ erg cm}^2/\text{molecule}^2$$

The induced dipole, calculated by equation VI-15 equals 2.02 debyes, and the requisite field would then be $F = 3.18 \times 10^5$ esu/cm^2. This value is so much greater than our previous estimates that we must regard this orientation as improbable. Further evidence against this model is given by the observed value of β, which is much less than the *ca.* 40 A^2/molecule required for the flat orientation. The temperature dependence of K' was analyzed for this model, and the quantities involved in the calculation are shown in Table VII-15. The results obtained are: $U_0' = 9.46$ kcal/mole and $\nu = \nu_1 = \nu_2 = \nu_3 = 0.65 \times 10^{12}$ sec^{-1}.

The third model of adsorbed benzene is kinetically identical with the above model, and differs only in having the plane of the ring oriented normal to, rather than parallel with, the surface. The direction of greater polarizability of the molecule is now aligned with the field and the directions of lesser polarizability determine the lateral interactions; these effects combine in such a way that

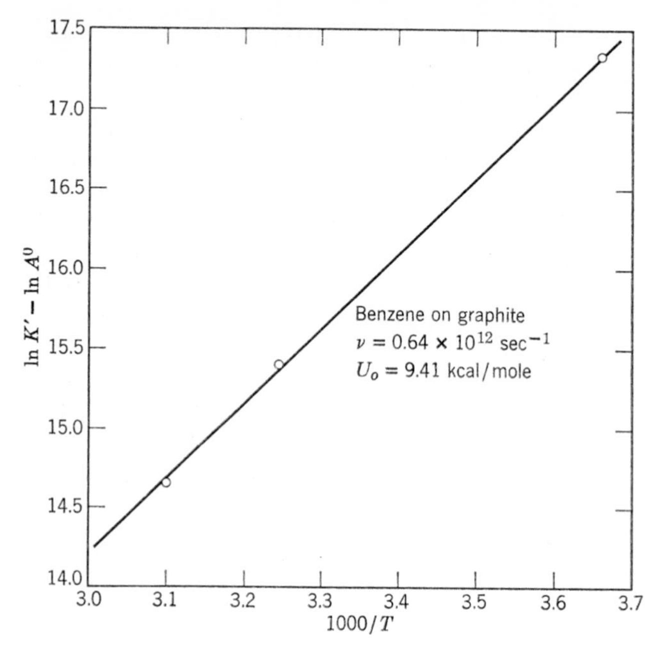

Fig. VII-43. The adsorption data reported in Figure VII-42 plotted according to equation V-59 for evaluation of U_0' and ν.

a much lower field strength is adequate to cause the observed lowering of α^{id}: i.e.,

$$\omega = \left(\frac{\xi_1 + \xi_2}{2\xi_{av}}\right)^2 = 0.817$$

and $\lambda = \alpha - \omega\alpha^{id} = -237 \times 10^{-30}$ erg cm²/molecule². In this calculation we assume that the collision cross section of the oriented molecule is the same as when it is freely rotating, i.e., $\beta^{perp} = \beta^{id}$. The induced dipole, calculated by equation VI-15, is 1.80 debyes, which leads to a field strength of 1.46×10^5 esu/cm². This value for the field for graphite is in closer agreement with those values obtained from chloroform and Freon-11 adsorption, and so makes this model, on this basis alone, more probable than any of the others. Additional evidence to the same effect is given by the experimentally observed value of β (30.6 A²), which is actually a little less than the theoretical value for a freely rotating molecule (i.e., $\beta^{id} = 32.7$ A²).

TABLE VII-16

Parameters Associated with Assumed Models of the Kinetic-Molecular Behavior of Adsorbed Benzene on Graphite

	Units	Observed	Model (a)	Model (b)	Model (c)
$\lim_{\theta \to 0} q^{st}$	kcal/mole	9.25	9.25	9.25	9.25
$-\Delta S_s^{ads}$	eu	10.6	10.6	10.6	10.6
U_0'	kcal/mole	—	8.97	9.46	9.46
α	erg cm²/molecule²	112×10^{-30}	112×10^{-30}	112×10^{-30}	112×10^{-30}
ν	sec^{-1}	—	4.4×10^{12}	0.64×10^{12}	0.64×10^{12}
α^{id} (molecule)		—	427×10^{-30}	407×10^{-30}	349×10^{-30}
β	A²/molecule	30.6	32.7	40	32.7
μ^{ind}	debyes	—	2.08	2.02	1.80
F	esu/cm²	1.2×10^{5} [a]	2.02×10^{5}	3.18×10^{5}	1.46×10^{5}

[a] As reported in Sections VII-1-B and C.

The investigation of the temperature dependence of K' for this model is the same as for the previous model, and the quantities involved are reported in Table VII-15. The rotational symmetry numbers used in equation V-25 to calculate ΔS^{rot} are 12 for the molecule in the gas phase and 6 for the adsorbed molecule. As a check, the graphical solution of equation V-59 is shown in Figure VII-43, from which the numerical results are:

$$U_0' = 9.41 \text{ kcal/mole and } \nu = \nu_1 = \nu_2 = \nu_3 = 0.64 \times 10^{12} \text{ sec}^{-1}$$

Comparisons of the calculated quantities according to each model discussed are included in Table VII-16. As the table shows, the thermodynamic quantities do not serve to distinguish between the assumed models. We are, however, enabled to find the most probable model because we have previous knowledge of the electric field associated with the graphite surface; we are aided in this search by the corroborative evidence of the observed low collision cross section.

J. LIMITING HEATS OF ADSORPTION ON GRAPHITIZED CARBON BLACK DETERMINED CHROMATOGRAPHICALLY

The measurement of the heat of adsorption by means of gas-solid chromatography (Section II-2-C) promises to offer an important aid to the present method of analysing adsorption isotherms. In matching experimental and computed isotherms there is always a certain latitude in selecting the parameters V_β, γ, K', and $2\alpha/\beta$ for the best description. As might well be expected with such a number of variables, internal adjustments and compensations make it possible to find various sets of parameters that seem to be equally good as answers. The chromatographic method of measuring the limiting heat of adsorption as θ tends to zero, is an independent technique for determining one of the quantities characteristic of the adsorption system, thereby helping to reduce the uncertainty of an uncorroborated result.

For a few adsorbates the values of U_0' and ν relative to a graphite substrate have been obtained, so that the limiting isosteric heat can be calculated for any temperature, either by equations VII-10 or VII-5. In Table VII-17, the values of U_0' and ν are taken from the preceding sections of this book; the limiting isosteric heat

TABLE VII-17
Comparison of Limiting Isosteric Heats, kcal/mole, of Adsorbates on Graphite

	U_0'	$\nu \times 10^{-12}$ sec^{-1}	Adsorption isotherm T, °K	Adsorption isotherm q^{st}	Chromatograph Temp. range, °K	Chromatograph q^{st}
Krypton	2.754	1.00	352	3.15	302–403	3.20
Xenon	3.654	0.85	350	4.03	297–403	3.94
Methane	2.678	2.48	351	3.13	299.5–403	2.90
Freon-11	8.17	0.76	418	7.9	372–463	7.86
Chloroform	8.05	1.35	418	8.1	372–463	8.00
Benzene	9.41	.64	448	9.1	403–493	9.18

(column 5) is calculated for the temperature designated T, using equation VII-10 for the adsorbates krypton, xenon, and methane, and equation VII-5 for the adsorbates Freon-11, chloroform, and benzene. For comparison, the limiting isosteric heats obtained from chromatograms in the temperature range indicated are given (see Figure II-30).

In the next set of comparisons, Table VII-18, the limiting isosteric heats derived from adsorption isotherms, by equation II-60, are not recalculated to correspond to the temperature range of the chromatograms. Either inadequate data or the complexity of the

TABLE VII-18
Comparisons of Limiting Isosteric Heats, kcal/mole, of Adsorbates on Graphite (not Temperature Corrected)

	Adsorption isotherm temp. range, °K	q^{st}	Chromatograph temp. range, °K	q^{st}
Ethane	173.1–296.7	4	299.5–372	3.78
Propane	258.2–296.7	6.5	296.5–372	6.26
Butane	258.2–296.7	8	341–433	8.10
Pentane	—	—	372–463	8.61
Hexane	—	—	372–463	9.83
Ethylene	229.1–266.2	4.5	313–403	4.40
Butadiene	—	—	372–463	7.65
Cyclopropane	244.2–296.7	5.5	313–403	5.98
Dimethyl-ether	258.2–296.7	6	341–462	5.50
Carbon tetrachloride	231.2–323.2	8.5	372–463	8.35

required analyses have hitherto prevented the evaluation of the necessary quantities, U_0' and ν. The temperature corrections obtained in Table VII-17 show that less than 100 cal/mole is usually involved.

2. Non-Gaussian Adsorbents

The distribution of adsorptive potential energies may not have a form that is symmetrical about a mean; it could possess enough asymmetry so that the adsorption isotherm could not be described by a model that assumed a Gaussian distribution. Imagine, for example, an otherwise uniform surface that is contaminated by a small amount of nonvolatile impurity of higher adsorptive potential: the adsorption isotherms of a gas on this substrate would show a small "knee" at the low-pressure end of the isotherm, thereby indicating the presence of the impurity. The shape of the isotherm would be the same whether we consider the impurity as a coating on a part of the surface of the major constituent or as a mechanically separate ingredient of the mixture: we could, therefore, account for such an isotherm by assigning separately to each of the surfaces its own values of the adsorption parameters V_β, γ, and K'. By doing so, we are actually describing substrates of non-Gaussian adsorptive energy distributions in terms of a sum of Gaussian distributions. Ross, Olivier and Hinchen (26a) have published some examples of adsorption isotherms that they interpreted quantitatively on the supposition of a dual distribution of adsorptive potentials (see below). Although not described quite in this way, a number of adsorbents have been previously reported that are effectively mixtures of two surfaces. The most clear-cut example is a slightly oxidized graphite surface for which the adsorption isotherm of water vapor shows a pronounced knee whose height varies with the degree of oxidation of the surface; reducing the surface with hydrogen eliminates the knee (25). The great dissimilarity of the adsorptive potentials of the two surfaces for the polar water molecule makes it obvious to the eye that two distinct adsorbing surfaces are present. Molybdenum disulfide, most samples of which have a partially oxidized surface, is another example (26). The adsorption isotherms of argon and nitrogen on boron nitride show a similar knee, though to a much less marked extent; it is, in fact, only visible when the low pressure portion is plotted on an expanded scale as was done by Winkler (5).

The presence of traces of boron oxide in the sample was detected by wet analysis; the prominence of the initial concavity of the isotherm to the pressure axis was found to relate to the amount of oxide contaminant in the boron nitride.

The surface of a solid is rarely smooth but is interrupted by cracks, crevices, capillaries, cavities, corners, and edges. Even on a molecular scale roughness is frequently introduced by lattice disorder or spiral dislocations. The utmost effort to obtain completely homotattic substrates has not yet succeeded in avoiding residual inhomogeneities. When graphite has a chemically pure adsorbing surface, the inhomogeneity has the form of a symmetrical distribution about a mean; the mean is representative of the basal plane of the graphite lattice. This type of random distribution presumably arises from the factors that disturb the geometrical smoothness of the surface. For oxidized graphite, molybdenum disulfide, and boron nitride the asymmetry of the distribution curve has been traced to the simultaneous presence of two chemically distinct adsorbing surfaces, each presumably with its own random distribution of adsorptive energies. In general, we may consider each chemically distinct adsorbing surface that is present in a mass of adsorbent to have its individual random distribution of adsorptive energies. This applies whether the chemical difference arises from different surface planes of the same crystal; surface contamination; oxide, hydroxide, or other surface compounds; or even gross admixture of other chemical substances.

A. CADMIUM BROMIDE

The adsorption isotherm of argon at 77.5 °K on cadmium bromide (26a) is plotted in Figure VII-44. The shape of this adsorption isotherm has characteristics that are indicative of both low and high degrees of heterogeneity: the latter is indicated by the concave shape of the isotherm with respect to the pressure axis and the former by the sigmoidal shape at higher pressures. No one of the model adsorption isotherms, such as those shown in Figure V-1, possesses both those features.

Using the tables of model isotherms and a process of trial and error, the experimental isotherm shown in Figure VII-44 can be described quantitatively by a dual distribution of adsorptive energies. We have no information concerning the surface electric field

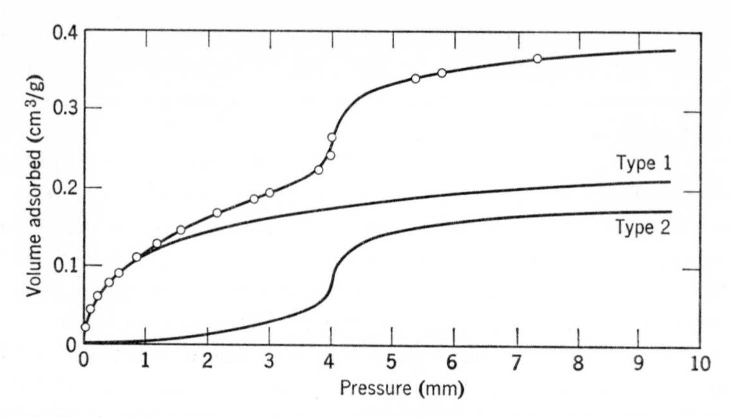

Fig. VII-44. Adsorption isotherm of argon on cadmium bromide at 77.5°K. The two lower curves are hypothetical isotherms for two surface constituents; the uppermost curve is the sum of the two hypothetical curves, and the circles indicate experimental measurements (26a).

of cadmium bromide and therefore used the tables calculated for $2\alpha^{id}/RT\beta^{id} = 6.5$. The two model isotherms ultimately obtained are referred to as type 1 and type 2 in the diagram; their sum provides a satisfactory match with the experimental points, which are indicated by circles. The adsorption parameters for each of the two surfaces are reported in Table VII-19; the numerical values would be slightly different were the magnitude of the surface field

TABLE VII-19
Adsorption Parameters of Argon Adsorbed at 77.5°K by the Dual Surface of Cadmium Bromide

	V_β, cm³ STP/g	γ	K'	U_0' (calculated), kcal/mole
Type 1	0.37	3	20.6	1.76
Type 2	0.26	∞	42.1	1.65

known and taken into account. The two surface constituents are: first, a relatively heterogeneous surface ($\gamma = 3$) with an average adsorptive energy, U_0' (calculated), of 1.76 kcal/mole, which is present as 59% of the total surface; second, an essentially homotattic surface with an adsorptive energy, U_0' (calculated), of 1.65 kcal/mole, present as 41% of the total surface.

The adsorbent was chosen purposely as one likely to give a near-homotattic surface, because of the hexagonal layer-lattice structure of the crystal; the surface constituent of low heterogeneity is therefore identified with the basal plane of cadmium bromide, consisting of close-packed bromine atoms. The cadmium bromide was prepared by dehydration of the tetrahydrate and the second surface constituent could perhaps be cadmium hydroxide, formed by surface hydrolysis. To test this supposition, J. J. Hinchen in our laboratory prepared a sample of finely divided cadmium bromide under anhydrous conditions, by subliming the material in a stream of dry argon and collecting the particles on a microporous filter. Care was taken to preserve the sample from contact with anything but argon. Detailed information about the method of preparation will be published elsewhere. The adsorption isotherm of argon at 77.5 °K on this substrate then showed only the characteristics (i.e., the same value of K') of the type 2 surface; evidence of type 1 was almost entirely lacking. This result confirms the truth of our analysis of the surface into two distinct distributions, one of which, the more heterogeneous, is probably the result of surface hydrolysis.

B. SODIUM BROMIDE

Cube crystals of sodium bromide have the same lattice plane, i.e., {100}, on all faces. Such crystals have often been favorite candidates for the preparation of a completely homotattic adsorbent, and the assumption has always been made that the heterogeneity of the resulting surface is negligible. No definite criteria have hitherto been available to check this assumption. In Figure VII-45 we report the argon adsorption at 77.5 °K on a specimen of sodium bromide (26a), prepared according to the directions given by Fisher and McMillan (27). The shape of the isotherm is itself sufficient in the light of the present reasoning to indicate a dual distribution of adsorptive energies. A quantitative description of the isotherm in these terms has been obtained by a process of trial and error, using $2\alpha^{id}/RT\beta^{id} = 6.5$, and is shown in Figure VII-45; the adsorption parameters are reported in Table VII-20. The surface has two constituents, one of high and the other of low degree of heterogeneity, as was the case with the cadmium bromide described previously. Of the total surface, 75% consists of a near-homotattic constituent ($\gamma = 200$), which presumably derives from the {100} planes of the crystal; this portion of the surface has a relatively low average-argon-adsorptive energy U_0' (calculated), of 1.49 kcal/

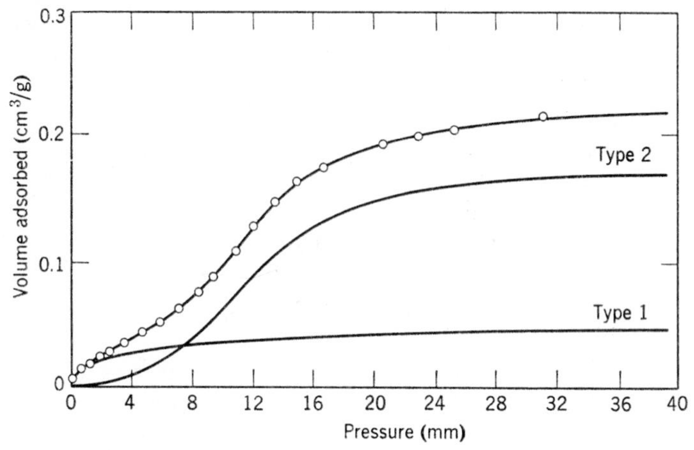

Fig. VII-45. Same as Figure VII-44. The data refer to the adsorption of argon on sodium bromide at 77.5°K (26a).

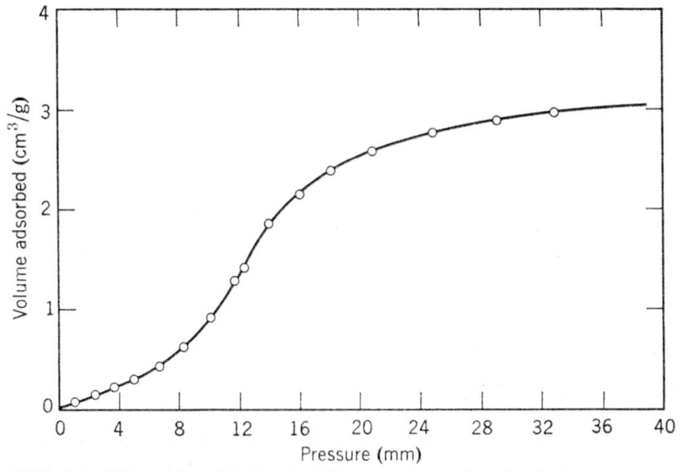

Fig. VII-46. Adsorption isotherm of argon on an anhydrous preparation of sodium bromide at 77.5°K (26a). The computed isotherm corresponding closely to the experimental data is now the same as that labeled Type 2 in Figure VII-45.

mole. The remaining 25% of the surface has a wide distribution ($\gamma = 3$) of adsorptive energies, with the average U_0' (calculated), at 1.59 kcal/mole. Once again, chemisorbed water may well be the source of the second surface constituent.

A verification of this interpretation of the surface of sodium bromide, when prepared under usual conditions, as a dual distribution is again provided by a new preparation (by J. J. Hinchen) under anhydrous conditions of a sample of the sublimed crystals, which when used as an adsorbent shows evidence of type 2 surface only. The adsorption of argon at 77.5 °K by these crystals is shown in Figure VII-46: the whole isotherm is described by the same values of K' and γ that are reported in Table VII-20 for the type 2 surface of sodium bromide. The surface area of the new preparation is much greater than that of the sample reported in Table VII-20.

TABLE VII-20
Adsorption Parameters of Argon Adsorbed at 77.5° K by the Dual Surface of Sodium Bromide

	V_β, cm^3 STP/g	γ	K'	U_0' (calculated), kcal/mole
Type 1	0.080	3	61.6	1.59
Type 2	0.24	200	120	1.49

C. OTHER EXAMPLES FROM PREVIOUSLY REPORTED DATA

The shape of the isotherm in Figure VII-44 resembles all those reported for the alkali halides as adsorbents, as well as for other inorganic salts that were prepared in attempts to procure near-homotattic substrates: examples are the sodium chloride of Orr (28) and those of Ross and co-workers (29,30,31); the potassium chloride of H. Clark (32); the calcium fluorides of Taylor and Edelhoch (33) and Ross and Winkler (31), and the sodium bromide of Fisher and McMillan (27). A part of the surface of each of the adsorbents (type 2) was sufficiently homotattic to allow two-dimensional phase transitions to become manifest as near-discontinuities in the isotherms on lowering the temperature of adsorption; if these portions of the surface had any other than a very narrow

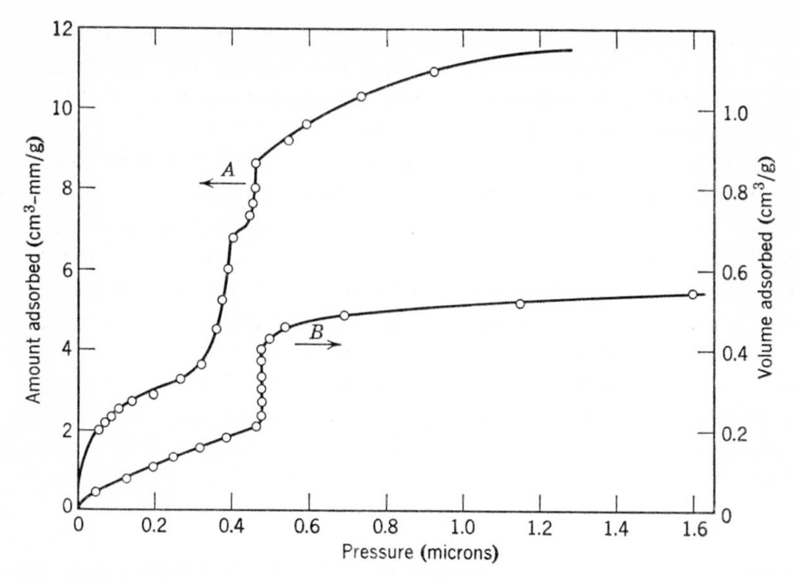

Fig. VII-47. Adsorption isotherms of ethane on sodium chloride at 90.1°K: *A*, adsorbent as prepared displayed both {100} and {111} surfaces—data of Ross and Boyd (29); *B*, adsorbent as prepared displayed only {100} surface—data of Ross and Winkler (31).

range of adsorptive potentials, the phase transitions would occur over too wide a range of pressure to be identified as such. On the type 1 part of these surfaces, for example, phase transitions cannot be identified because of the great heterogeneity of this part of the substrate.

Ross and Boyd (29) prepared crystals of sodium chloride in which both the {100} and the {111} surfaces were developed; the adsorption isotherm of ethane at 90.1 °K gave evidence of two type 2—i.e., near-homotattic—surfaces, as well as showing the initial knee, which is evidence of a type 1 surface. This isotherm is shown in Figure VII-47, which also includes for comparison an ethane isotherm measured at the same temperature for a sodium chloride adsorbent that had only {100} faces developed; the pressure characteristic of the phase transition of ethane on the homotattic {100} surface of sodium chloride at 90.1 °K shows on both isotherms as the location of a discontinuity at $p = 4.5 \times 10^{-3}$ mm; the homotattic {111} portion of the surface is responsible for the convex shape (with respect to the pressure axis) of the isotherm beyond the "knee," though the rise is not sufficiently pronounced to be described as another discontinuity. The evidence is almost enough to determine relative magnitudes of the surface fields and the adsorptive potentials of the {111} and {100} crystal faces. The absence of a phase change on the {111} faces is probably genuine; though enough heterogeneity in the characteristic distribution associated with those faces could, even if a phase change were to occur, dissolve the evidence. An examination of the crystals with the microscope indicated, however, that the two crystal faces were equally well developed, in which case the two-dimensional phase change of ethane on the {100} and its absence on the {111} faces shows that the former substrate has the weaker surface field. Furthermore, the relative pressures at which the two crystal faces make themselves manifest in the adsorption isotherm—i.e., lower pressure for {111} than for {100}—is evidence of a smaller adsorptive potential for the {100} faces.

D. SURFACE ELECTRIC FIELD

An estimate of the surface field of a near-homotattic surface can be made from the observed value of the two-dimensional critical temperature ${}_aT_c$. According to de Boer (34):

$$\alpha/\alpha^{id} = {}_aT_c/{}_aT_c^{\,id} \qquad \text{(VII-12)}$$

where α and ${}_aT_c$ refer to the observed two-dimensional van der Waals interaction constant and the observed two-dimensional critical temperature, respectively; the same symbols with the superscript *id* refer to the ideal values of these quantities; and

$$\lambda = \alpha - \alpha^{id} = -\pi\mu^2/d$$

where μ is the dipole induced on the adsorbate by the electric field F of the substrate, and d is the diameter of the adsorbed molecule. The induced dipole is related to the surface field by

$$\mu = F\xi$$

where ξ is the polarizability of the adsorbate. These equations are written for an adsorbate that has no permanent dipole moment and that is not oriented by the substrate; they are approximations otherwise.

Hence,

$$F = \frac{1}{\xi}\left[\frac{\alpha^{id}d}{\pi}\left(1 - \frac{{}_aT_c}{{}_aT_c^{\,id}}\right)\right] \qquad \text{(VII-13)}$$

In spite of the presence of the type 1 part of the surface, one can frequently estimate the two-dimensional critical temperature of the phase transition, provided that the type 2 part of the surface is uniform enough to make the transition manifest: thus, we can use this procedure, for example, with the supra- and sub-critical isotherms of methane, ethane, and xenon adsorbed by the {100} face of sodium chloride, published by Ross and Clark (30). Using estimates of ${}_aT_c$ from these data in equation VII-13 yields a value for the surface field of $F = 1.5 \pm 0.5 \times 10^5$ esu/cm²; the poor precision of the result is due to the difficulty of interpolating the temperature of the critical isotherm by eye. The surface field of the {100} face of sodium bromide can be estimated in a similar way, using the data of Fisher and McMillan (27) for the adsorption of methane and krypton. The result for the surface field is again between 1 and 2×10^5 esu/cm².

A better estimate of the surface field of the {100} face of sodium bromide can be made from the argon isotherm reported in Figure VII-45 by assuming that the type 2 part of the surface is actually much closer to the homotattic than would be indicated by a γ value of 200; since this value had been assigned on the basis of no surface field, the assumption that it is actually higher is well justified. If we now put $\gamma > 1000$ for the type 2 surface we find on refitting it to the model isotherms that $2\alpha/RT\beta = 5.7$, rather than the previously assumed ideal ratio of 6.5. Assuming $\beta = \beta^{id}$ for argon, we calculate the induced dipole to be 0.24 debye and the surface field to be 1.4×10^5 esu/cm². The same result is derived from the argon adsorption isotherm for the sodium bromide prepared under anhydrous conditions (Fig. VII-46).

For potassium chloride, H. Clark (32) observed phase transitions to take place at temperatures lower than on a sodium chloride substrate: we deduce that the surface field of potassium chloride is greater than that of sodium chloride or sodium bromide.

The configuration of ionic surfaces at which both positive and negative ions are present does not lead to the concept of a uniform surface field expressible by a single number, but rather to a field fluctuating from point to point along the surface. This varying field induces varying dipoles in the adsorbed molecules and the average component of these dipoles normal to the surface is the quantity that we calculate from the observed lowering of α^{id} or ${}_aT_c{}^{id}$. When we use equation VII-13 for such a surface we are calculating the equivalent uniform surface field that, if it were actually present, would cause the observed effects.

References

1. P. Y. Hsieh, *Ph.D. thesis, Rensselaer Polytechnic Institute*, 1959.
2. S. Ross and W. Winkler, *J. Colloid Sci.*, **10,** 319 (1955).
3. S. Ross and W. Winkler, *J. Colloid Sci.*, **10,** 330 (1955).
4. W. W. Pultz, *Ph.D. thesis, Rensselaer Polytechnic Institute*, 1958; S. Ross and W. W. Pultz, *J. Colloid Sci.*, **13,** 397 (1958).
5. W. Winkler, *Ph.D. thesis, Rensselaer Polytechnic Institute*, 1955.
6. R. M. Barrer and W. I. Stuart, *Proc. Roy. Soc. (London)*, **249A,** 464,484 (1959).
7. E. S. Chen, unpublished results, Rensselaer Polytechnic Institute laboratory.
8. E. W. Albers, *Ph.D. thesis, Rensselaer Polytechnic Institute*, 1961.
9. J. de D. Lopez-Gonzales, F. G. Carpenter, and V. R. Deitz, *J. Phys. Chem.*, **65,** 1112 (1961).
10. J. A. Morrison, J. M. Los, and L. E. Drain, *Trans. Faraday Soc.*, **47,** 1023 (1951), L. E. Drain and J. A. Morrison, *Trans. Faraday Soc.*, **48,** 840 (1952), *ibid.*, **49,** 654 (1953).
11. S. Brunauer and P. H. Emmett, *J. Am. Chem. Soc.*, **57,** 1754 (1935), P. H. Emmett and S. Brunauer, *ibid.*, **59,** 1553 (1937).
12. S. Brunauer, P. H. Emmett, and E. Teller, *J. Am. Chem. Soc.*, **60,** 309 (1938).
13. W. D. Schaeffer, W. R. Smith, and M. H. Polley, *Ind. Eng. Chem.*, **45,** 1721 (1954), M. H. Polley, W. D. Schaeffer, and W. R. Smith, *J. Phys. Chem.*, **57,** 469 (1953).
14. F. R. Klebacher, *B.S. thesis, Rensselaer Polytechnic Institute*, 1957.
15. D. Graham, *J. Phys. Chem.*, **64,** 1089 (1960).
16. R. M. Barrer, *Nature*, **178,** 1410 (1957); idem, *Colston Papers*, Vol. X, Butterworth's, London, 1958, pp. 6–34.
17. R. M. Barrer and W. I. Stuart, *Proc. Roy. Soc. (London)*, **249A,** 464,484 (1959).
18. R. M. Barrer and L. C. V. Rees, *Trans. Faraday Soc.*, **55,** 992 (1959).
19. W. D. Machin and S. Ross, *Proc. Roy. Soc. (London)*, **265A,** 455 (1962).
20. T. Moeller, *Inorganic Chemistry*, Wiley, New York, 1952, pp. 757–8.
21. G. Constabaris and G. D. Halsey, Jr., *J. Chem. Phys.*, **27,** 1433 (1957); G. Constabaris, J. H. Singleton, and G. D. Halsey, Jr., *J. Phys. Chem.*, **63,** 1350 (1959); J. R. Sams, Jr., G. Constabaris, and G. D. Halsey, Jr., *J. Phys. Chem.*, **64,** 1689 (1960).
22. G. Constabaris, J. R. Sams, Jr., and G. D. Halsey, Jr., *J. Phys. Chem.*, **65,** 367 (1961).
23. C. Kemball, *Proc. Roy. Soc. (London)*, **187A,** 73 (1946).

24. N. N. Avgul, A. V. Kiselev, I. A. Lygina, and D. P. Poschkus, *Izvestia Akad. Nauk. S.S.S.R., Otd. Khim. Nauk* 1196 (1959); *Bull. Acad. Sci. USSR, Div. Chem. Sci.* (English transl.) 1155 (1959).
25. C. Pierce, R. N. Smith, J. W. Wiley, and H. Cordes, *J. Am. Chem. Soc.*, **73,** 4551 (1951); B. Millard, E. G. Caswell, E. E. Leger, and D. R. Mills, *J. Phys. Chem.*, **59,** 976 (1955).
26. E. V. Ballou and S. Ross, *J. Phys. Chem.*, **57,** 653 (1953); P. Cannon, *ibid.*, **64,** 858 (1960).
26a. S. Ross, J. P. Olivier, and J. J. Hinchen, *Advances in Chemistry Series,* **33,** 317 (1961).
27. B. B. Fisher and W. G. McMillan, *J. Chem. Phys.*, **28,** 549,555,563 (1958).
28. J. W. Orr, *Proc. Roy. Soc. (London)*, **173A,** 349 (1939).
29. S. Ross and G. E. Boyd, *U.S.A.E.C. Report MDDC 864*, (1947).
30. S. Ross and H. Clark, *J. Am. Chem. Soc.*, **76,** 4291,4297 (1954).
31. S. Ross and W. Winkler, *J. Am. Chem. Soc.*, **76,** 2637 (1954).
32. H. Clark, *Ph.D. thesis*, *Rensselaer Polytechnic Institute*, 1954.
33. H. Edelhoch and H. S. Taylor, *J. Phys. Chem.*, **58,** 344 (1954).
34. J. H. de Boer, *The Dynamical Character of Adsorption*, Clarendon Press, Oxford, 1953, p. 155.

CHAPTER VIII

The Nature of the Adsorptive Forces

Observations of a large number of adsorption systems disclose that the heat evolved during adsorption may be anywhere between a few hundred calories to over a hundred kilocalories per mole of adsorbate. Early workers in the field used a value of 20 kcal/mole to define the upper limit of physical adsorption, believing that heats of adsorption above this value could arise only from a chemical bonding between the adsorbate and the surface. We are now aware that there is a wide range of heats of adsorption that could arise from any one of four adsorption mechanisms or combinations thereof; and, therefore, that a knowledge of the heat of adsorption is not enough, save in extreme cases, to distinguish between adsorption mechanisms.

1. Origins of the Adsorptive Potential

Four general types of molecular interaction lead to a potential energy for adsorption:

(a) *London-type dispersion forces* resulting from induced-dipole/induced-dipole and multi-polar attractions,

(b) *Induction forces* brought about by the operation of a surface electric field on induced or permanent dipoles of resident molecules,

(c) *Charge transfer* (1) between the adsorbed molecule and the surface resulting in a no-bond resonance state,

(d) *Dative bonding* resulting from a chemical reaction between the adsorbate and surface atoms.

Of these four types of interaction the first two certainly lead to physical adsorption, the fourth equally certainly results in chemisorption, the third, which is a postulated mechanism whose actual occurrence has not yet been adequately demonstrated, exists in a twilight zone that defies classification as either physical or chemical adsorption. This book has heretofore not been concerned with the source of the adsorptive potential, but has postulated only that the adsorbed film be mobile; its considerations could apply, therefore, to all adsorption situations other than (d) above. The present chapter is devoted to a consideration of the adsorptive potential arising from (a), (b), and (c); that is, excluding chemisorption properly so called. This general topic has been discussed by de Boer (2) for a number of specific cases. The theory and techniques that we have developed in this book enable us to derive from experimental data a quantity, ${}_{0}P^{ads}$, the depth of the adsorptive potential well (see Fig. V-8), which quantity can also be calculated *a priori* for some simple systems from their physical properties; we are therefore better able for such systems to make comparisons of theory and observation and so have an independent test of conclusions reached in previous sections.

2. Expressions for the Dispersion Potential

A. THE DIRECT LATTICE SUMMATION

Dispersion forces between two atoms can be represented by a potential function expressed in terms containing inverse powers of the internuclear separation s. The simplest function of this sort includes a potential energy of attraction proportional to the inverse sixth power of the separation and a repulsion that is zero at distances of separation greater than a particular value s_e and infinite at separations less than s_e; this is the so-called hard sphere or van der Waals model. Such an approximate potential function can be improved in two respects: investigations of the second virial coefficient have revealed that the potential energy of repulsion is best described as proportional to the inverse twelfth power of the separation; and the term in s^{-6}, which accounts for the greater part of the total potential, and which is due to the attraction of mutually induced dipoles, should have added to it the dipole-quadrupole and quadrupole-quadrupole attraction, expressed as terms in s^{-8} and s^{-10}, respectively. A more complete potential function for the forces between two atoms is, therefore:

$$P_{ij} = C_1 s_{ij}^{-6} + C_2 s_{ij}^{-8} + C_3 s_{ij}^{-10} - \mathcal{R} s_{ij}^{-12} \quad \text{(VIII-1)}$$

where P_{ij} is the potential energy lost by two particles, i and j, on approaching each other from an infinite separation to a separation s. C_1, C_2, and C_3 are dipole-dipole, dipole-quadrupole, and quadrupole-quadrupole coefficients, respectively; $\mathcal{R}$ is a repulsion constant. To calculate the potential energy lost by a gas molecule on approaching a surface, the potential difference P_{ij} has to be summed for the simultaneous interactions of the gas molecule (i) with each of the atoms (j) of the adsorbent.

$$\sum_j P_{ij} = {}_gP = C_1\sum_j s_{ij}^{-6} + C_2\sum_j s_{ij}^{-8} + C_3\sum_j s_{ij}^{-10} - \mathcal{R}\sum_j s_{ij}^{-12} \quad \text{(VIII-2)}$$

For a polyatomic adsorbate it may be necessary to sum the interactions over each atom of the adsorbate molecule as well. The various values of s_{ij} in equation VIII-2 can be calculated for a particular adsorbent of known crystal lattice in terms of a single distance z, which is defined as the distance between the center of an adsorbate atom and the mathematical plane in which lie the centers of the surface atoms of the adsorbent. For a layer-lattice structure, z_e is the sum of half the interlamellar spacing and the radius r_e of the adsorbate molecule.

The constants C_1, C_2, and C_3 in equation VIII-2 can be evaluated by means of the following approximate quantum mechanical equations (3,4):

$$C_1 = 6\,mc^2\,\xi_i\xi_j\,\frac{1}{\dfrac{\xi_i}{\chi_i} + \dfrac{\xi_j}{\chi_j}} \quad \text{(VIII-3)}$$

$$C_2 = \frac{45h^2}{32\pi^2 m}\,\xi_i\xi_j\left\{\frac{1}{2\left(\dfrac{\xi_j}{\chi_j}\Big/\dfrac{\xi_i}{\chi_i} + 1\right)} + \frac{1}{2\left(\dfrac{\xi_i}{\chi_i}\Big/\dfrac{\xi_j}{\chi_j} + 1\right)}\right\} \quad \text{(VIII-4)}$$

$$C_3 = \frac{105h^4}{256\pi^4 m^3 c^2}\,\xi_i\xi_j\left\{\frac{\xi_i/\chi_i}{3\left(\dfrac{\xi_j}{\chi_j}\Big/\dfrac{\xi_i}{\chi_i}\right) + 1} + \frac{3}{4\left(\dfrac{\chi_i}{\xi_i} + \dfrac{\chi_j}{\xi_j}\right)} + \frac{\xi_j/\chi_j}{3\left(\dfrac{\xi_i}{\chi_i}\Big/\dfrac{\xi_j}{\chi_j}\right) + 1}\right\} \quad \text{(VIII-5)}$$

In these equations ξ represents polarizability and χ represents diamagnetic susceptibility: the subscripts i and j refer to the adsorbate and the adsorbent, respectively.

No theoretical expression for the repulsion constant $\mathcal{R}$ has been developed; it can, however be evaluated in terms of the equilibrium separation, $z = z_e$, where the potential function has its maximum value. At $z = z_e$, $d_g P/dz = 0$: hence,

$$\mathcal{R} = \frac{C_1 \frac{d}{dz}\left(\sum_j s_{ij}^{-6}\right)_{z \to z_e} + C_2 \frac{d}{dz}\left(\sum_j s_{ij}^{-8}\right)_{z \to z_e} + C_3 \frac{d}{dz}\left(\sum_j s_{ij}^{-10}\right)_{z \to z_e}}{\frac{d}{dz}\left(\sum_j s_{ij}^{-12}\right)_{z \to z_e}} \quad \text{(VIII-6)}$$

The derivatives in equation VIII-6 can be obtained in two ways: the summations may be plotted *versus* z in the range bracketing z_e and the slopes found graphically, or these plots may be expressed by an empirical equation and the slopes obtained analytically. The latter method is more accurate, as an equation of the form

$$\sum_j s_{ij}^{-n} = az^{-b} \quad \text{(VIII-7)}$$

has been found to hold (4).

The method described above has been used by Kiselev *et al.* (4) to calculate the dispersion interactions of a number of adsorbates with the cleavage surface of graphite. A prime disadvantage of the method of direct lattice sums, apart from the laborious calculations, is the dependence of the answer on the lattice position above which the adsorbate molecule is assumed to be situated (see Fig. VIII-1). For a mobile adsorbed film the proper procedure would be to find the weighted average interaction of all possible positions on the surface. In practice, only a few principal positions are ever worked out and a straight arithmetic average of the interactions for these positions would not be characteristic of the mobile molecule.

Figure VIII-1 is a schematic diagram of an argon atom on the three principal positions of the basal plane of graphite: on top of a carbon atom, between two carbon atoms, and in the center of a ring. The results calculated by Kiselev *et al.* are reported for neon, argon, and krypton in Table VIII-1. We do not include a number of other calculations made by the same authors for polyatomic adsorbates, as the writers did not take into account the anisotropies

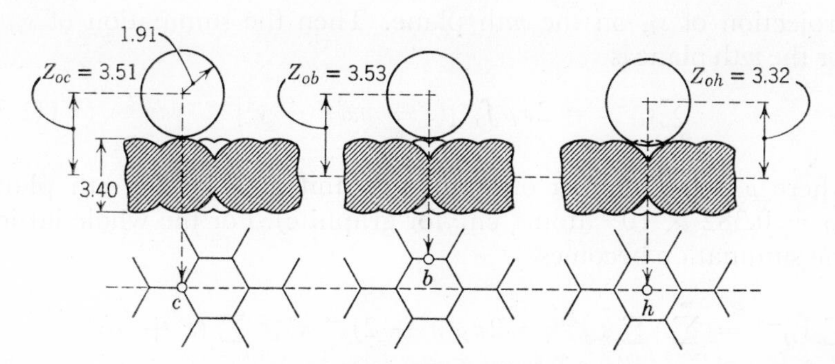

Fig. VIII-1. Argon atoms in different positions relative to a graphite substrate, drawn to scale; from reference (4).

TABLE VIII-1
Calculated Values of ${}_gP^{\text{disp}}$ (kcal/mole) by Equation VIII-2 for Inert Gas Molecules Situated at Different Parts of the Graphite Substrate

Adsorbent	Above C atom	Between two C atoms	Center of the ring	Average
Neon	0.77	0.84	1.11	0.91
Argon	1.95	2.07	2.64	2.22
Krypton	2.59	2.78	3.47	2.95

of polarizability and magnetic susceptibility when calculating dispersion interactions for oriented molecules.

B. CROWELL'S EXPRESSION FOR THE LATTICE SUMS

1. The Original Application

Crowell (5) has shown how to evaluate the lattice sums, $\sum_j s_{ij}^{-n}$, for a layer-lattice structure by an analytical method. The lattice is approximated by a set of layer planes, each with a uniform distribution of matter, and separated by the interplanar distance d. The lattice sum is approximated by integrating over the planes and forming the sum of the resulting terms. Let the adsorbate molecule, at a distance z from the surface plane, be separated by a distance s_{ij} from any point in the mth plane below the surface: the vertical distance of the adsorbate from this plane is $(z + md)$. Let y be the projection of s_{ij} on the mth plane. Then the summation of s_{ij}^{-n} for the mth plane is

$$\sum_{j(m)} s_{ij}^{-n} = 2\pi\rho \int_0^\infty [(z + md)^2 + y^2]^{-n/2} y\,dy \qquad \text{(VIII-8)}$$

where ρ is the number of atoms per unit area in the mth plane ($\rho = 0.382 \times 10^{16}$ atoms/cm² for graphite). For the whole lattice the summation becomes

$$\sum_j s_{ij}^{-n} = \sum_{m=0}^{\infty} \sum_{j(m)} s_{ij}^{-n} = 2\pi\rho(n-2)^{-1}\, d^{2-n} \sum_{m=0}^{\infty} (x + m)^{2-n} \qquad \text{(VIII-9)}$$

where $x = z/d$. The summation of $(x + m)^{2-n}$ is the generalized Riemann zeta function of x and $n - 2$. These functions have been tabulated (6) as derivatives of $\Psi(x) = (d/dx)\ln(x!)$. The fundamental expansion of this expression for the kth derivative of $\Psi(x)$ is

$$\Psi^k(x) = (-1)^{k+1}\, k! \sum_{m=0}^{\infty} (x + m)^{-k-1} \qquad \text{(VIII-10)}$$

Equation VIII-9 can now be written as

$$\sum_j s_{ij}^{-n} = \frac{2\pi\rho(-1)^n}{(n-2)d^{n-2}(n-3)!} \Psi^{n-3}(x) \qquad \text{(VIII-11)}$$

The functions $\Psi^k(x)$ have been tabulated for values of k up to 4. For higher values of k the function must be calculated by equation VIII-10. Fortunately one can neglect all but the first term of the series for $k >$ *ca.* 4, provided $x < 2$ as is commonly the case in adsorption calculations.

The concept, incorporated in this approximation, that matter is distributed continuously in each layer plane, while it might not be a permissible assumption for a localized adsorbed film, is actually close to the truth for a mobile adsorbed film, in which the rapid translation of the molecules along the surface prevents their responding to its fine structure. This approximation produces just the sort of average that is desired and which is so difficult to obtain from the direct lattice sum.

Crowell (7) has used his lattice sum expression, equation VIII-11, to calculate the interaction of inert gases with graphite, assuming that the dispersion energy of attraction could be adequately accounted for by the dipole-dipole interaction only: i.e.,

$$ {}_gP = C_1 \sum_j s_{ij}^{-6} - \mathcal{R} \sum_j s_{ij}^{-12} \qquad \text{(VIII-12)} $$

Substituting equation VIII-11 for the lattice sums gives

$$ {}_gP = \frac{C_1 \pi \rho}{12 d^4} \Psi^3(x) - \frac{\mathcal{R} \pi \rho}{5 z^{10}} \qquad \text{(VIII-13)} $$

where, according to equation VIII-6,

$$ \mathcal{R} = -\frac{C_1 d^6 x_e^{11}}{24} \Psi^4(x_e) \text{ and } x_e = \frac{z_e}{d} $$

Therefore,

$$ {}_gP = \frac{C_1 \pi \rho}{12 d^4} \left[\Psi^3(x) + \left(\frac{z_e}{z} \right)^{10} \frac{x_e}{10} \Psi^4(x_e) \right] \qquad \text{(VIII-14)} $$

Equation VIII-14 is the analytical description of the potential well for this model. The maximum potential energy loss, or the depth of the well, is that value of ${}_gP$ (which we term ${}_gP^{\text{disp}}$) at which $z = z_e$: i.e.,

$$ {}_gP^{\text{disp}} = \frac{C_1 \pi \rho}{12 d^4} [\Psi^3(x_e) + \frac{x_e}{10} \Psi^4 (x_e)] \qquad \text{(VIII-15)} $$

An advantage of having an analytic expression such as equation VIII-14 for the potential well, is that it can be used to estimate the vibrational frequency of the adsorbed molecule with respect to the surface. For vibrations of small amplitudes the system behaves as an harmonic oscillator for which the frequency is given by

$$ \nu = (1/2\pi)\sqrt{\kappa/m} \qquad \text{(VIII-16)} $$

where m is the mass of the adsorbate molecule and κ is the force constant defined by

$$ \kappa = -\left(\frac{d^2{}_gP}{dz^2} \right)_{z \to z_e} \text{ or } \kappa = \frac{1}{d^2} \left(\frac{d^2{}_gP}{dx^2} \right)_{x \to x_e} \qquad \text{(VIII-17)} $$

Applied to equation VIII-14,

$$ \kappa = \frac{C_1 \pi \rho}{12 d^6} \left[\frac{120}{x_e^6} + \frac{11 \Psi^4(x_e)}{x_e} \right] \qquad \text{(VIII-18)} $$

2. *A Refined Application*

Kiselev *et al.* (4) have shown that the dipole-quadrupole and quadrupole-quadrupole terms in equation VIII-2 are not negligible, as was assumed by Crowell, nor are they cancelled by the repulsion term, as has been stated by de Boer (2). A more accurate result could, therefore, be obtained by using all the terms in equation VIII-2 in conjunction with Crowell's expression for the lattice sums, equation VIII-11. Instead of equation VIII-13 we would now have:

$$ {}_gP = \frac{C_1\pi\rho}{12d^4}\,\Psi^3(x) + \frac{C_2\pi\rho}{3z^6} + \frac{C_3\pi\rho}{4z^8} - \frac{\mathcal{R}\pi\rho}{5z^{10}} \qquad \text{(VIII-19)} $$

where

$$ \mathcal{R} = -\frac{C_1 d^6}{24}\,x_e^{\,11}\Psi^4(x_e) + C_2 z_e^{\,4} + C_3 z_e^{\,2} $$

hence,

$$ {}_gP = \frac{C_1\pi\rho}{12d^4}\left[\Psi^3(x) + \left(\frac{z_e}{z}\right)^{10}\frac{x_e}{10}\,\Psi^4(x_e)\right] + \frac{C_2\pi\rho}{z^6}\left[\frac{1}{3} - \frac{1}{5}\left(\frac{z_e}{z}\right)^4\right] + \frac{C_3\pi\rho}{z^8}\left[\frac{1}{4} - \frac{1}{5}\left(\frac{z_e}{z}\right)^2\right] \qquad \text{(VIII-20)} $$

The expression for ${}_gP^{\text{disp}}$ is found to be:

$$ {}_gP^{\text{disp}} = \frac{C_1\pi\rho}{12d^4}\left[\Psi^3(x_e) + \frac{x_e}{10}\,\Psi^4(x_e)\right] + \frac{2}{15}\,\frac{C_2\pi\rho}{z_e^{\,6}} + \frac{1}{20}\,\frac{C_3\pi\rho}{z_e^{\,8}} \qquad \text{(VIII-21)} $$

As before, the vibrational frequency of the adsorbed molecule with respect to the surface can be obtained by means of equation VIII-16. Applied to equation VIII-19,

$$ \kappa = -\frac{C_1\pi\rho}{12d^6}\left[\frac{120}{x_e^{\,6}} + \frac{11\Psi^4(x_e)}{x_e}\right] + \frac{7C_2\pi\rho}{z_e^{\,8}} + \frac{4C_3\pi\rho}{z_e^{\,10}} \qquad \text{(VIII-22)} $$

C. THE CONTRIBUTION OF THE SURFACE LAYER OF THE ADSORBENT TO THE DISPERSION FORCES

It would be interesting to know how much of the total dispersion forces between the adsorbate and the whole lattice of the adsorbent is contributed by interactions with only the surface atoms. This can be readily calculated by use of the previous equations. Again assuming that the dispersion interaction is adequately accounted for by the 6-8-10-12 potential function (equation VIII-2), we evaluate the lattice sums for the surface plane only: i.e., put $m = 0$ in equation VIII-10; hence,

$$\sum_j s_{ij}^{-6} = \frac{\pi\rho}{2} z^{-4}; \qquad \sum_j s_{ij}^{-8} = \frac{\pi\rho}{3} z^{-6}; \qquad \sum_j s_{ij}^{-10} = \frac{\pi\rho}{4} z^{-8}$$

and

$$\sum_j s_{ij}^{-12} = \frac{\pi\rho}{5} z^{-10}$$

from which

$$_gP = \frac{C_1\pi\rho}{2} z^{-4} + \frac{C_2\pi\rho}{3} z^{-6} + \frac{C_3\pi\rho}{4} z^{-8} - \frac{\pi\rho}{5} (C_1 z_e^6 + C_2 z_e^4 + C_3 z_e^2) z^{-10} \quad \text{(VIII-23)}$$

which was obtained by evaluating the repulsion coefficient, as before, from the condition that at $z = z_e$, $d_gP/dz = 0$. The expression for $_gP^{\text{disp}}$ is, therefore,

$$_gP^{\text{disp}} = \frac{3C_1\pi\rho}{10} z_e^{-4} + \frac{2C_2\pi\rho}{15} z_e^{-6} + \frac{C_3\pi\rho}{20} z_e^{-8} \quad \text{(VIII-24)}$$

Equation VIII-24 represents the contributions of the surface layer of the adsorbent to the total dispersion interaction. We can also calculate the contribution made to the vibrational force-constant by the first layer. According to equations VIII-17 and VIII-23,

$$\kappa = 12C_1\pi\rho z_e^{-6} - 8C_2\pi\rho z_e^{-8} + 4C_3\pi\rho z_e^{-10} \quad \text{(VIII-25)}$$

The vibrational frequency that would result from the force constant κ contributed by the surface layer only of the adsorbent is given (very nearly) by

$$\nu = (\sqrt{10}/\pi z_e)/\sqrt{_gP^{\text{disp}}/m} \quad \text{(VIII-26)}$$

D. THE LATTICE TREATED AS A CONTINUUM

A rough approximation of the lattice sums is made by treating the lattice as a continuum by integrating, rather than summing, the s_{ij}^{-n} terms in equation VIII-1: i.e., by writing

$$_gP = \rho_v \iiint P_{ij}\, dV \tag{VIII-27}$$

where the integration is over a semi-infinite volume and ρ_v is the number of adsorbent atoms per cm³ of solid. Using the terms in s_{ij}^{-6} and s_{ij}^{-12} only, since we are already approximating so much, we get

$$_gP = \frac{C_1 \pi \rho_v}{6 z_e^3} \left[\left(\frac{z_e}{z}\right)^3 - \frac{1}{3}\left(\frac{z_e}{z}\right)^9 \right] \tag{VIII-28}$$

which was obtained by evaluating the repulsion coefficient as before from the condition that at $z = z_e$, $d_gP/dz = 0$. The expression for $_gP^{\text{disp}}$ for this model is

$$_gP^{\text{disp}} = C_1 \pi \rho_v / 9 z_e^3 \tag{VIII-29}$$

For evaluating the vibrational frequency of the adsorbed molecule with respect to the surface, this model yields, by equations VIII-17 and VIII-16:

$$\kappa = 3 C_1 \pi \rho_v / z_e^5 \tag{VIII-30}$$

hence

$$\nu = (3\sqrt{3}/2\pi z_e)\sqrt{_gP^{\text{disp}}/m} \tag{VIII-31}$$

3. Calculations of Dispersion Potentials by Various Models for the Inert Gases Adsorbed by Graphite

We have developed general expressions in the preceding section for the calculation of the dispersion potential contribution to the potential energy for adsorption and for the vibrational frequency of the adsorbed molecule, using various expressions for the lattice sums. We now propose to apply these equations, for the purpose of comparison with experiment, to the adsorption of the inert gases by graphite. In Section VII-1-F we report the observed values of $_gP^{\text{ads}}$ and ν for neon, argon, krypton, and xenon on P-33(2700°) graphite. The experimental value of $_gP^{\text{ads}}$ contains contributions from both the dispersion interactions and a polarization potential due to interaction with the surface electric field: i.e.,

$$_{g}P^{\text{ads}} = {}_{g}P^{\text{disp}} + {}_{g}P^{\text{elec}} \qquad \text{(VIII-32)}$$

where

$$_{g}P^{\text{elec}} = (1/2)F^2 \xi \qquad \text{(VIII-33)}$$

F and ξ are the surface field of the adsorbent and the polarizability of the adsorbate, respectively. The surface field of graphite is about 1.2×10^5 esu/cm^2, as determined in the previous chapter; using this value for F, we calculate $_{g}P^{\text{elec}}$ for each of the inert gases: the results are reported in Table VIII-3. The quantities involved in the calculations of the dispersion potentials of the inert gases on graphite are reported in Table VIII-2.

The observed values of $_{g}P^{\text{disp}}$ and ν calculated by equation VIII-32 are compared in Table VIII-3 with the results calculated using the various expressions for dispersion potentials and frequencies that we have discussed above. The best agreement (and an excellent one) between observation and theory is given by the use of equations VIII-21 and VIII-22, based on the layerwise integration of the 6-8-10-12 potential function. If the multipolar attractions are neglected in the calculation, the values of $_{g}P^{\text{disp}}$ are from 3 to 5% less; the vibrational frequencies calculated on this basis are also reduced, though to a slightly smaller extent. We can also see from Table VIII-3 that if the lattice is treated as a continuum the answers are considerably smaller: the calculated dispersion potentials are

TABLE VIII-2
Parameters Involved in the Calculation of the Dispersion Potentials of the Inert Gases on Graphite

	Carbon	Neon	Argon	Krypton	Xenon
r_e (A)	1.70	1.59	1.91	2.01	2.25
z_e (A)	—	3.29	3.61	3.71	3.95
$-\chi \times 10^{30}$ (cm^3)	10.5	12.0	32.2	46.5	71.5
$\xi \times 10^{24}$ (cm^3)	0.937	0.398	1.63	2.48	4.00
$C_1 \times 10^{45}$ (kcal mole^{-1} cm^6)	—	0.214	0.77	1.15	1.82
$C_2 \times 10^{60}$ (kcal mole^{-1} cm^6)	—	0.026	0.104	0.157	0.252
$C_3 \times 10^{76}$ (kcal mole^{-1} cm^{10})	—	0.04	0.17	0.31	0.43

TABLE VIII-3

Observed and Calculated Values of the Dispersion Potentials and of the Vibrational Frequencies for the Inert Gases Adsorbed by Graphite

Quantity	Source	Ne	A	Kr	Xe
	Observed values				
${}_gP^{ads}$ kcal/mole	Table VII-11	0.737	2.12	2.80	3.69
${}_gP^{elec}$ kcal/mole	Equation VIII-33	0.043	0.180	0.256	0.415
${}_gP^{disp}$ kcal/mole	Equation VIII-32	0.694	1.94	2.54	3.28
$\nu \times 10^{-12}$ sec^{-1}	Table VII-11	1.19	1.28	1.00	0.85
	Calculated values				
${}_gP^{disp}$	Layerwise integration of 6-8-10-12 potential function	0.764	1.92	2.59	3.22
$\nu \times 10^{-12}$		1.20	1.23	0.96	0.80
${}_gP^{disp}$	Layerwise integration of 6-12 potential function	0.729	1.84	2.48	3.11
$\nu \times 10^{-12}$		1.16	1.20	0.93	0.78
${}_gP^{disp}$	Integration of 6-8-10-12 potential function for surface layer only	0.693	1.71	2.29	2.81
$\nu \times 10^{-12}$		1.16	1.18	0.92	0.76
${}_gP^{disp}$	Integration of 6-12 potential function for a continuum	0.23	0.64	0.88	1.2
$\nu \times 10^{-12}$		0.55	0.60	0.47	0.40
${}_gP^{disp}$	Direct lattice sums (4) from Table VIII-1	0.91	2.22	2.95	—

only about one third of the observed values, and the calculated frequencies are about half the observed values. For a crystalline lattice more closely packed than that of graphite, this particular approximation would not be so much in error. By treating the lattice as a continuum we discount the true density of the surface plane, which is the source of the major part of the dispersion interaction. The contribution of the surface layer of the adsorbent to the total dispersion forces, as calculated by equation VIII-24, is shown in Table VIII-3 to amount to 90%. Thus, the vibrational restoring force is almost completely provided by the surface layer, and the vibrational frequency is therefore quite well approximated by equation VIII-26, which was derived by considering the surface layer only.

The variation of ν with ${}_gP^{disp}$ has the general form described by the following equation:

$$\nu \propto (1/z_e)\sqrt{{}_gP^{disp}/m}$$

This relation is explicit, for example, in equations VIII-26 and VIII-31; for the models that involve layerwise integration the relation is implicit, but can be shown to hold by plotting ν *versus* $(1/z_e)\sqrt{{}_gP^{\text{disp}}/m}$. Plots of this type are shown in Figure VIII-2, based on the data of Table VIII-3 for the various models. The straight lines in the diagram show the relation predicted by each model.

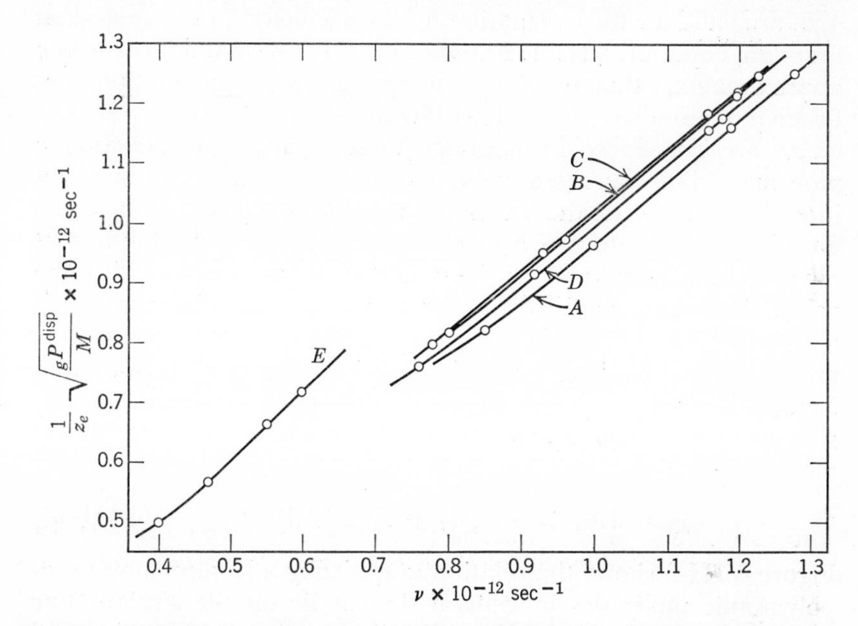

Fig. VIII-2. The variation of the vibrational frequency of adsorbed atoms as a function of their dispersion interaction with the substrate, according to experimental measurements and to various models of the interaction (see Table VIII-3). *A*, observed values; *B*, layerwise integration of 6-8-10-12 potential function; *C*, layerwise integration of 6-12 potential function; *D*, integration of 6-8-10-12 potential function for surface layer only; *E*, integration of 6-12 potential function for a continuum.

The observed values of ${}_gP^{\text{disp}}$ and ν are also used, along with the values of z_e given in Table VIII-2, to show, by means of the circled points, the actual relation found (Curve *A*). On this basis it is clear that the 3-9 potential function, in which the lattice is treated as a continuum, equations VIII-29 and VIII-31, is definitely excluded as a valid model (Curve *E*).

4. Calculations of the Dispersion Interaction of Polyatomic Molecules with Graphite

The excellent agreement, demonstrated in the previous section, of the observed and theoretical values of the dispersion interactions between the inert gases and a graphite surface is evidence that the approximate quantum mechanical expressions, equations VIII-3, 4, 5, are valid for the calculations of the coefficients in the potential function, equation VIII-1. Equations VIII-3, 4, and 5 incorporate approximations that restrict their application to interactions between monatomic particles, and Pitzer (8) has warned that serious errors are introduced by applying these equations to polyatomic molecules. Presumably a valid approach would be to sum all interactions between the individual atoms of the adsorbate and the adsorbent. Each atom of a polyatomic adsorbate molecule has its own values of ξ_i, χ_i, and z_e; the uncertainties associated with these values destroy the usefulness of the answer becase of its large probable error. An alternative method might be to convert polyatomic groups into an "equivalent atom" for which equations VIII-3, 4, and 5 would be valid. As a first step in this direction we note that ξ_i and χ_i for the inert gas atoms have a linear relation to each other; namely,

$$\chi_i = 1.65 \times 10^{-5}\, \xi_i + 5.42 \times 10^{-30} \qquad \text{(VIII-34)}$$

Figure VIII-3 shows the relation graphically and also shows that polyatomic molecules in general do not lie on the straight line described by equation VIII-34. By disregarding the experimental values of the diamagnetic susceptibility χ_i of a polyatomic molecule and substituting for if the value given by equation VIII-34, we have found that equations VIII-3, 4, and 5 now give coefficients that yield dispersion energies in agreement with observation. What has been done is, in effect, to endow the polyatomic molecule with a diamagnetic susceptibility equal to that of an hypothetical inert gas atom of the same polarizability. It seems that equation VIII-34 represents a necessary condition for the successful application of equations VIII-3, 4, and 5.

As examples of this empirical, or perhaps semi-theoretical, procedure we include the calculations for the interaction with graphite of hydrogen, methane, ethane, and nitrogen. Table VIII-4 con-

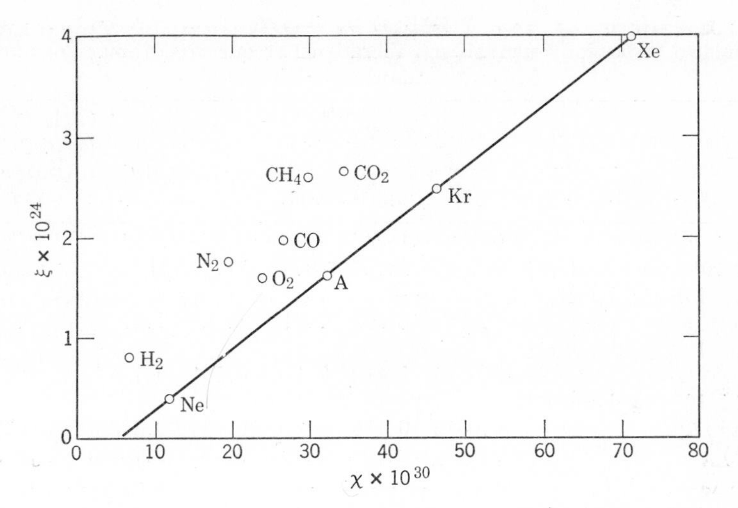

Fig. VIII-3. Relation between diamagnetic susceptibility and polarizability of the inert gas atoms, also showing that polyatomic molecules have lower diamagnetic susceptibilities than hypothetical monatomic molecules of the same polarizability.

tains the values of ξ_i (taken from Table VI-2), χ_i calculated by equation VIII-34, C_1, C_2, and C_3 calculated by equations VIII-3, 4, and 5, the equilibrium separation z_e, and, finally, the calculated values of ${}_gP^{\text{disp}}$ and ν obtained from equations VIII-21 and VIII-22.

The equilibrium separation z_e is, as with the inert gases, the sum of half the interlamellar spacing of graphite and the radius, r_e, of the adsorbate molecule. The calculated dispersion energy is sensitive to slight variations in the numerical value selected for r_e; with the inert gases the appropriate value of r_e is obtained from the second virial coefficient of the gas, which thus refers to a separation corresponding to a minimum potential energy between two atoms and is, therefore, analogous to the condition for minimum potential energy of adsorption. Following the same reasoning we might have chosen the radius determined from the structure of the crystalline state, which bears the same analogy to the adsorption situation; with the inert gases, in fact, either method gives the same radius within 0.5%. A radius based on collision properties, as *e.g.*, from the gas viscosity or the van der Waals b, would be inappropriate in this context.

TABLE VIII-4
Calculated Dispersion Potentials and Vibrational Frequencies of Some Polyatomic Molecules Adsorbed by Graphite

					C_2H_6 perp.		
	H_2	N_2	CH_4	C_2H_6 Flat	1st seg.		2nd seg.
$\xi_i \times 10^{24}$ (cm^3)	0.79	1.76	2.60	3.97	2.74		2.74
$\chi_i \times 10^3$ (cm^3) calculated	18.5	34.5	48.5	71.0	50.6		50.6
$C_1 \times 10^{45}$ (kcal mole^{-1} cm^6)	0.398	0.833	1.21	1.82	1.27		1.27
$C_2 \times 10^{60}$ (kcal mole^{-1} cm^8)	0.0513	0.112	0.165	0.251	0.174		0.174
$C_3 \times 10^{76}$ (kcal mole^{-1} cm^{10})	0.0815	0.185	0.276	0.424	0.292		0.292
z_e (A)	3.34	3.78	3.78	3.78	3.78		5.32
${}_gP^{disp}$ (kcal/mole)	1.345	1.734	2.520	3.783	—	3.384	—
$\nu \times 10^{-12}$ (sec^{-1})	5	1.4	2.2	1.9	—		—

We assume hydrogen to be rotating freely on the graphite surface at 114°K and, therefore, the value of r_e was obtained from second virial coefficients (9). The radius of nitrogen adsorbed at 84°K was obtained in the same way (9). The radius of methane was obtained from crystal structure data (10). For ethane, which might be oriented on adsorption, we take the smaller dimension of the molecule as equal to the diameter of methane; the longer dimension is greater than the smaller by the length of the carbon-carbon bond (1.54 A). To obtain ${}_gP^{disp}$ for ethane, we calculate the separate contributions from the two CH_3-segments, taking their centers as the carbon atom and sharing the polarizability of the molecule equally between the segments.

The calculated values of ${}_gP^{disp}$ in Table VIII-4 are to be compared with values obtained from experiment. The experimental values for hydrogen and methane are reported in Table VII-13. For hydrogen, ${}_gP^{disp} = 1.354$ kcal/mole and $\nu = 10 \times 10^{12}$ sec^{-1}. The calculated value of ${}_gP^{disp}$ is in excellent agreement; the calculated vibrational frequency, however, is low. The larger observed value of ν could well be due to the interaction of the high frequency lattice vibrations with an adsorbate of such small mass. For methane, the experimental values, ${}_gP^{\text{disp}} = 2.526$ kcal/mole and $\nu = 2.48 \times 10^{12}$ sec^{-1}, are both in excellent agreement with the calculated values reported above.

We do not have experimental values of ${}_gP^{\text{disp}}$ or ν for nitrogen, but the values of K' at 77.5 and 90.1°K reported in Table VII-1 allow us to calculate $\lim_{\theta \to 0} q^{\text{st}}$ from equation V-64, giving a value of 2.12 kcal/mole at $T = 84$°K. From our calculated value of ${}_gP^{\text{disp}}$ and ν we can synthesize a value of $\lim_{\theta \to 0} q^{\text{st}}$ by equation VII-10, which gives 2.00 kcal/mole for the freely rotating molecule. If, as we suggested in Section VII-E-2, the dispersion interaction of nitrogen with graphite is nearly independent of the orientation of the nitrogen molecule and about 40% of the adsorbed nitrogen is oriented with the long axis normal to the surface, we should add about 30 cal/mole to the calculated value of ${}_gP^{\text{elec}}$. In addition, the quadrupole of nitrogen will interact with the gradient of the surface electric field and so contribute a quantity ${}_gP^{\text{quad}}$ toward q^{st} (11):

$$ {}_gP^{\text{quad}} = (Q/4)\,(dF/dz) \qquad \text{(VIII-35)} $$

where Q is the quadrupole moment, which for nitrogen is 1.5×10^{-26} esu/cm^2. In our previous considerations we have not had to consider dF/dz as long as it is small; for a molecule with a permanent quadrupole moment, however, even small field gradients are significant. Thus, for example, a value of dF/dz equal to 0.17×10^5 esu/cm^2/A which we could otherwise ignore, would yield ${}_gP^{\text{quad}} = 90$ cal/mole by interaction with nitrogen. By taking these factors into account, therefore, we see that the calculated dispersion potential of nitrogen on graphite must be close to the correct value.

The dispersion potential calculation for ethane adsorbed by graphite gives a pronounced preference of 400 cal/mole for the flat orientation; the anisotropy of the polarizability, however, reduces this preference to about 240 cal/mole, as ${}_gP^{\text{elec}}$ for the flat orientation is 410 against 570 cal/mole for the upright position. At 260°K and with a barrier to rotation of 240 cal/mole, about 60% of the adsorbed molecules have enough energy to rotate. We can synthesize $\lim_{\theta \to 0} q^{\text{st}}$ approximately by means of a straight average of ${}_gP^{\text{ads}}$ for the two orientations, *viz.*, 4.07 kcal/mole, from which we obtain $\lim_{\theta \to 0} q^{\text{st}} = 4.3$ kcal/mole: the experimental value of $\lim_{\theta \to 0} q^{\text{st}}$ for ethane, as determined by Albers (12), is 4 kcal/mole.

The general good agreement between the observed values and the results of these calculations shows that the procedure on which equation VIII-34 is based might be applicable to a wide range of polyatomic adsorbates, particularly diatomic molecules and hydrocarbons. Molecules or segments of molecules containing several highly polarizable atoms probably could not be treated as a whole, but ought to be summed as contributions from each atom; this warning would apply to substances such as halogenated hydrocarbons. The chief obstacle in the way of obtaining trustworthy results from these calculations is the lack of information about the appropriate molecular dimensions of all but the simplest molecules. For instance, too few measurements of the second virial coefficient of carbon tetrachloride as a function of temperature are on record to allow the accurate determination of its intermolecular separation.

5. A Simple Approximation for the Calculation of Dispersion Interactions with Graphite

A few approximations are all that are required to reduce the rather complex equations we have been using to a much simpler form. Figure VIII-3 shows that the polarizability and diamagnetic susceptibility of inert-gas-type molecules are very nearly directly proportional to one another. Using this approximation, the co efficients C_1, C_2, and C_3 defined by equations VIII-3, 4, and 5 all become directly proportional to ξ_i. In addition one can see from Table VIII-3 that the multipolar contributions are nearly a constant proportion of the total interaction potential, and that the contribution of the surface layer of the adsorbent is also a nearly constant fraction of the total interaction potential. We could expect, therefore, that ${}_gP^{\text{disp}}$ for a graphite surface would be given fairly accurately by a relation of the form

$$ {}_gP^{\text{disp}} = C\, \xi_i\, z_e^{-4} \qquad \text{(VIII-36)} $$

Using the results reported in Tables VIII-2, VIII-3, and VIII-4 for ${}_gP^{\text{disp}}$ and z_e, we have tested equation VIII-36 and found that

C is a constant for these adsorbates; the value of C is 198 ± 5 kcal A/mole, when z_e is in A and ξ_i is in A^3. With this value of C, equation VIII-36 reproduces the experimental values of $_gP^{disp}$ to within 2%.

References

1. P. M. Gundry and F. C. Tompkins, *Trans. Faraday Soc.*, **56,** 846 (1960).
2. J. H. de Boer, *Advances in Colloid Science*, vol. 3, Interscience, New York, 1950, pp. 1–66.
3. J. E. Lennard-Jones and B. N. Dent, *Trans. Faraday Soc.*, **24,** 92 (1928).
4. N. N. Avgul, A. V. Kiselev, I. A. Lygina, and D. P. Poschkus, *Izvestia Akad. Nauk S.S.S.R., Otd. Khim. Nauk* 1196 (1959); *Bull. Acad. Sci. USSR, Div. Chem. Sci.* (English transl.) 1155 (1959).
5. A. D. Crowell, *J. Chem. Phys.*, **22,** 1397 (1954).
6. H. T. Davis, *Tables of Higher Mathematical Functions,* vol. 2, Principia Press, Bloomington, Ind., 1933.
7. A. D. Crowell, *J. Chem. Phys.*, **26,** 1407 (1957).
8. K. S. Pitzer, *Advances in Chemical Physics*, vol. 2, Interscience, New York, 1959, p. 68.
9. J. E. Lennard-Jones, *Proc. Phys. Soc.*, **43,** 461 (1931).
10. R. W. G. Wyckoff, *Crystal Structures*, vol. 4, Interscience, New York, 1960, p. 25.
11. G. L. Kington and A. C. Macleod, *Trans. Faraday Soc.*, **55,** 1799 (1959).
12. E. W. Albers, *Ph.D. thesis, Rensselaer Polytechnic Institute*, 1961.

*ADDENDUM**

Dr. W. A. Steele has recently published a paper on monolayer adsorption with lateral interaction on heterogeneous surfaces (1), in which he is occupied with the same problem that we discuss in this book. We shall demonstrate herewith that Steele's theory of physical adsorption based on statistical mechanics leads to the same result as that previously derived by us from a kinetic treatment and published in scientific journals in 1961 and 1962 (2,3,4,5). Dr. Steele was apparently not aware that the final reduction to practice of his most general formulation invoked just those simplifying assumptions that would cause his model to become identical with that of one of our earlier discussions of the subject. Nowadays nobody is surprised to find that a given physical model would yield the same result, whether formulated by kinetics or by statistical mechanics. The mathematical symbolism introduced, however, although necessary, often obscures the mental picture of the physical model. One makes a contribution merely by disclosing the innate similarity of two treatments that appear to have no connection.

The most general form of the adsorption-isotherm equation for a heterogeneous substrate is, as we pointed out (2) in 1961, the solution of the equation

$$\theta = \int_{e}^{g} \Phi\,(U_0)\,\Psi(p, U_0)\,dU_0 \tag{IV-4}$$

where θ is the relative surface concentration of adsorbate; $\Psi(U_0)$ dU_0 is the distribution of adsorptive energies that exists for a given adsorbate-adsorbent pair; and $\Psi(p, U_0)$ describes the adsorption isotherm for a homotattic patch of adsorptive energy U_{0i}. In this general form equation (IV-4) includes all types of adsorption from completely localized to completely mobile, depending on the function chosen for $\Psi(p, U_0)$; it also includes any distribution of energies U_0 that can be expressed by a mathematical function, $\Psi(U_0)$ dU_0.

* Added in proof.

Dr. Steele produces the equivalent of equation (IV-4) in terms of statistical mechanics, in which the grand partition function provides for W different kinds of sites of atomic dimension on the substrate, and the sum is made of all possible assignments of N molecules to these different groups of sites. The variation of potential energy over the site can also be adjusted to take into account any desired degree of surface mobility of the adsorbed atoms.

The resemblance between the two general formulations is all the more pronounced when one recalls that all the common functions for $\Psi(p, U_0)$, such as the Langmuir equation, the Fowler-Guggenheim equation, the Hill-de Boer equation, and the two-dimensional virial equation, have at one time or another been derived by the use of statistical mechanics. Equation (IV-4) can therefore describe a model of adsorption in statistical-mechanical terms and with no less detail than any application of Steele's most general grand partition function. Both equation (IV-4) and Steele's most general form for the grand partition function are quite useless, however, until specific functions are fed into them. This is the point where one has to introduce the "restricted models" and "highly simplified assumptions" that Dr. Steele so deprecates when indulged in by others; but in fact we differ only from him in having reached that point, via equation (IV-4), much sooner.

We now assumed (Section V-4) that at very low surface concentrations the adsorbate would behave as a two-dimensional ideal gas; we also supposed that the substrate could be described by a Gaussian distribution of adsorptive sites. These assumptions gave us the forms for the functions $\Psi(p, U_0)$ and $\Phi(U_0)\,dU_0$, respectively. The adsorption isotherm that results has the form of Henry's law,

$$V^{\text{ads}} = Z\,p \tag{Ad-1}$$

where V^{ads} is the amount adsorbed in cc. (STP)/g. and p is the pressure in mm. The variation of Z with temperature is given, (3,4), by

$$\ln Z + \ln A^\circ = \frac{U_0'}{RT} + \ln V_\beta + \frac{1}{4\gamma\,(RT)^2} \tag{V-62}$$

where U_0' is the midpoint of the Gaussian distribution as expressed by the equation

$$\Phi(U_0)\,dU_0 = \sqrt{\frac{\gamma}{\pi}}\exp - \gamma(U_0 - U_0')^2\,dU_0 \qquad \text{(Ad-2)}$$

and ln A^0 is a function of the changes in translation, rotation, and vibration that occur on adsorption. For a mobile adsorbed film ln A^0 is given by the equation*

$$\ln A^0 = -\left[\frac{\Delta S_s^{\text{tr}}}{R} + \frac{\Delta S^{\text{rot}}}{R}\right] + ({}_aF^{\text{vib}} - {}_aE_0^{\text{vib}})/RT + \frac{\Delta E^{\text{kin}}}{RT} - \ln\frac{\theta_s}{1-\theta_s} + \ln 760 \qquad \text{(V-51)}$$

where the terms given refer to changes of entropy, free energy, and kinetic energy on adsorption, and θ_s is the standard concentration of the adsorbed phase.

The practical test of his theory offered by Steele is the analysis of the adsorption isotherms of neon and argon on a heterogeneous carbon black sample known as "Black Pearls No. 71," using also the adsorption data for the near-homotattic graphite adsorbent known as P-33(2700). The original data were reported by Halsey and his co-workers (6,7,8). We, too, have used those portions of the same data that referred to the near-homotattic graphite, for which we have published (3) our analysis in terms of equation (V-62). We quote the following results that we obtained for neon and argon adsorbed by P-33(2700).

TABLE Ad-I
Adsorption Parameters for P-33(2700) Carbon Black

	V_β cc. (STP)/g.	U_0' kcal/mole
Neon	5.82	0.681
Argon	3.83	2.066

The value of γ, equation (Ad-2), for both neon and argon on this near-homotattic adsorbent is so large (> 400 kcal^{-2}/mole^{-2})

* Our first attempt to derive this parameter was wrong, and so it is printed incorrectly in reference (3), equation (24), and in reference (4), equation (26): the extra term given in these two places, $-[\theta_s/(1-\theta_s)]$, is fortunately very small and made no practical difference to the results. The correct derivation is given on pp. 158–159.

that the last term of equation (V-62) is negligible. For the heterogeneous carbon black B.P., however, this is by no means true. One would therefore write equation (V-62) for each of these adsorbents as follows:

$$\ln (Z)_{\text{P-33}} + \ln (A^0)_{\text{P-33}} = (U_0')_{\text{P-33}}/RT + \ln (V_\beta)_{\text{P-33}} \quad \text{(Ad-3)}$$

and

$$\ln (Z)_{\text{B.P.}} + \ln (A^0)_{\text{B.P.}} = (U_0')_{\text{B.P.}}/RT + \ln (V_\beta)_{\text{B.P.}} + 1/(4\gamma_{\text{B.P.}} R^2T^2) \quad \text{(Ad-4)}$$

We shall now follow Steele's reasoning step by step, using our own symbolism. First, he assumes that "the differences in adsorption properties of the two carbon blacks (per unit area at fixed T, p) are primarily due to differences in site distribution functions." In our notation this means that we can write $(U_0')_{\text{P-33}} = (U_0')_{\text{B.P.}}$ and $(A^0)_{\text{P-33}} = (A^0)_{\text{B.P.}}$

The first of these equations we ourselves have actually verified experimentally for a series of carbon blacks that had been thermally treated at 1000°, 1500°, 2000°, and 2700° for a number of hours. The surface heterogeneity of each one was well described by a Gaussian distribution of adsorptive energies; the distribution curve became more and more narrow the higher the temperature of thermal treatment, but the midpoint of the distribution did not vary (see Fig. VII-20). If the B.P. carbon black is like a member of this series, then it would indeed be correct to write $(U_0')_{\text{B.P.}} = (U_0')_{\text{P-33}}$.

The invariance of ln A^0 means, however, that each patch of the heterogeneous surface adsorbs molecules with the same average loss of translational, rotational, and vibrational energies. For a substrate of great heterogeneity this does not seem too probable; but that, nevertheless, is the implication of Steele's assumption that "the variation of energy within a single site $(\epsilon_{\text{I}}^{(s)}(\mathbf{r}) - \overset{*}{\epsilon}_{\text{I}})$ is the same for all sites on the heterogeneous and the homogeneous surfaces." In other words, he assumes that the potential function that had been found to be characteristic of the cleavage-plane surface of graphite applies equally well to *every* site on the surface of the heterogeneous carbon black B.P.

If, for the time being, we accept these assumptions, equations (Ad-3) and (Ad-4) can be combined to give

$$\ln \frac{Z_{\text{B.P.}}}{Z_{\text{P-33}}} = \ln \frac{V_{\beta\ \text{B.P.}}}{V_{\beta\ \text{P-33}}} + \frac{1}{4\gamma\ (RT)^2} \tag{Ad-5}$$

Equation (Ad-5) is identical with the following equations of Steele, numbered 4.4 and 4.5 in his paper:

$$L = \ln \left\{ \frac{[V_{\text{app}}(0) - V_{\text{He}} + 2As_0]_{\text{B.P.}}}{[V_{\text{app}}(0) - V_{\text{He}} + 2As_0]_{\text{P-33}}} \right\} \tag{4.4}$$

$$= \ln \{ [N_s]_{\text{B.P.}}/[N_s]_{\text{P-33}} \} + (u/2\,kT)^2 \tag{4.5}$$

By handling the experimental data according to these equations, Steele is able to evaluate the heterogeneity parameter u and the monolayer capacity N_s for the B.P. carbon black, with neon and argon as adsorbents. Since our equation (Ad-5) is actually the same as Steele's equations 4.4 and 4.5, we, too, can handle the data in exactly the same way to evaluate the temperature-independent parameters γ and V_β for neon and argon on B.P. carbon black, which we report in Table Ad-II. This is really no more than translating the adsorption parameters given by Steele into our own symbols. Considering the proliferation of theories of adsorption now current, every such translation is a benefit to investigators.

TABLE Ad-II
Adsorption Parameters for B.P. Carbon Black

	V_β, cc. (STP)/g.	γ, kcal^{-2}/mole^{-2}	U_0', kcal/mole
Neon	202.5	3.5	0.681
Argon	133.3	0.45	2.066

We do not offer these parameters with any too firm a conviction that they accurately describe the adsorption of neon and argon by this carbon black; they give, however, the same Gaussian distribution functions that were obtained by Steele. We conclude, therefore, that whatever useful future applications are yet to be obtained from Dr. Steele's exercise in statistical mechanics its present suggested mode of handling data does not introduce anything new beyond the stage reached by us three years ago.

References

1. W. A. Steele, *J. Phys. Chem.*, **67,** 2016 (1963).
2. S. Ross and J. P. Olivier, *ibid.*, **65,** 608 (1961).
3. S. Ross and J. P. Olivier, *Advances in Chemistry Series*, **33,** 301 (1961).
4. J. P. Olivier and S. Ross, *Proc. Roy. Soc.* (*London*), **265** A, 447 (1962).
5. S. Ross in Kirk-Othmer's *Encyclopedia of Chemical Technology*, second edition, Vol. **1,** John Wiley and Sons (Interscience), New York, N. Y., 1963, pp. 421–459.
6. W. A. Steele and G. D. Halsey, Jr., *J. Chem. Phys.*, **22,** 979 (1954).
7. M. P. Freeman and G. D. Halsey, Jr., *J. Phys. Chem.*, **59,** 181 (1955).
8. J. R. Sams, Jr., G. Constabaris, and G. D. Halsey, Jr., *ibid.*, **64,** 1689 (1960).

Symbols

A	Total area of adsorbed phase, equation II-26.
A	Constant in equation I-10.
A	Constant in Sips' equation in Section IV-3.
A^0	Constant in equations I-32, I-33.
B	Virial coefficient, equation I-14.
B^{rot}	Energy barrier to free rotation of a molecule, equation VII-7.
C	Number of components in system, equation I-1.
C	Constant in B.E.T. theory, equation I-12.
C	Virial coefficient, equation I-14.
C	Attraction constant of London forces, Section I-7.
C	Chromatographic parameter in equation II-46.
C	Constant in equations IV-5, V-60.
C	Constant in equation VIII-36.
C_1, C_2, C_3	Coefficients of terms in potential energy function, equation VIII-1, defined respectively by equations VIII-3, 4, and 5.
${}_aC_v$	Heat capacity at constant volume of the adsorbed phase, equation III-14.
${}_gC_v$	Heat capacity at constant volume of the gas phase, equation III-13.
${}_sC_v$	Heat capacity at constant volume of the solid phase (i.e., container plus adsorbent), equation III-12.
E	"Characteristic energy" in equation VI-6.
${}_aE^{\text{kin}}$	Kinetic energy of translation and rotation of an adsorbed molecule.
${}_aE^{\text{rot}}$	Kinetic energy of rotation of a molecule in the adsorbed film.
${}_aE^{\text{vib}}$	Average vibrational energy of an adsorbed molecule at temperature T, equation I-29.
${}_aE_0{}^{\text{vib}}$	Average vibrational energy of an adsorbed molecule at 0°K, equation I-31.
${}^0_aE$	Energy per mole, kinetic and potential, of a molecule in the surface film at infinite dilution.
${}_gE^{\text{kin}}$	Total kinetic energy per mole of a molecule in the gas phase: for an ideal monatomic gas ${}_gE^{\text{kin}} = (3/2)RT$.

${}_gE^{\text{rot}}$	Kinetic energy of rotation of a molecule in the gas phase.
${}_gE^{\text{tr}}$	Kinetic energy of translation of a molecule in the gas phase.
${}^0_gE$	Total energy per mole, kinetic and potential, of a molecule in the gas phase, at infinite dilution.
${}_sE$	Energy of the solid phase, equation III-11.
${}_a\mathbf{E}^{\text{vib}}$	Effective value of ${}_aE^{\text{vib}}$ for the adsorbate on a heterogeneous surface, equation IV-17.
${}_a\mathbf{E}_0{}^{\text{vib}}$	Effective value of ${}_aE_0{}^{\text{vib}}$ for the adsorbate on a heterogeneous surface, equation IV-18.
ΔE^{kin} =	${}_aE^{\text{kin}} - {}_gE^{\text{kin}}$.
ΔE^{rot} =	${}_aE^{\text{rot}} - {}_gE^{\text{rot}}$, equation I-29.
ΔE^{tr} =	${}_aE^{\text{tr}} - {}_gE^{\text{tr}}$, equation I-29.
F	Number of degrees of freedom in system, equation I-1.
F	Surface electric field, Section VI-2-B.
F_c	Same as F_r, except as measured at the column temperature, equation II-54.
F_{dose}	Constant in equation II-5, dosing factor.
F_{eq}	Constant in equation II-6, dead space factor.
F_r	Flow rate of carrier gas through a chromatographic column as measured at room temperature, equation II-54.
${}_aF$	Molar integral Gibbs free energy of adsorbed phase, equation II-26.
${}_aF^{\text{vib}}$	Additional Gibbs free energy of the adsorbed phase due to molecular vibrations with respect to the surface, equation V-26.
${}_gF$	Gibbs free energy of gas phase, equation II-25.
ΔF^{ads}	Molar integral change in Gibbs free energy on adsorption.
$\Delta F_s{}^{\text{ads}}$	Standard integral Gibbs free energy change on adsorption, equation V-41.
$\Delta \dot{F}^{\text{ads}}$	Differential change in Gibbs free energy on adsorption, equation III-76.
$\Delta \dot{F}_s{}^{\text{ads}}$	Standard differential change in Gibbs free energy on adsorption.
${}_aH$	Molar integral enthalpy of the adsorbed phase.
${}_gH$	Molar integral enthalpy of the gas phase.
ΔH^{ads}	Molar integral change of enthalpy of the system on adsorption.
$\Delta \dot{H}^{\text{ads}}$	Molar differential change of enthalpy on adsorption.
ΔH^{vap}	Molar heat of vaporization.
I_a, I_b, I_g	Moments of inertia about axes A, B, . . ., G, equation V-24.

K	Constant in adsorption isotherm equation, equation I-27.
K_i	Particular value of K for the ith patch of a heterogeneous surface, Section IV-3.
L	Length of chromatographic column, equation II-46.
M	Molecular weight, g/mole.
N	Avogadro's number (6.02×10^{23} molecules/mole).
P	Number of phases in system, equation I-1.
P_{ij}	Potential energy lost by two particles i and j on approaching each other from an infinite separation to a separation s, equation VIII-1.
${}_aP^{ia}$	Potential energy of interaction of the adsorbate, equation I-39.
${}_gP$	Adsorptive potential between a gas molecule and the surface of an adsorbent, Section I-4.
${}_gP^{ads}$	Adsorptive potential between a gas molecule and the surface of an adsorbent at equilibrium separation, equation I-29.
${}_gP^{disp}$	That part of ${}_gP^{ads}$ due to dispersion forces.
${}_gP^{elec}$	That part of ${}_gP^{ads}$ due to interaction with surface electric field.
${}_gP^{quad}$	That part of ${}_gP^{ads}$ due to interaction of the gradient of the surface electric field with the quadrupole moment of an adsorbate.
${}_a\mathbf{P}^{ia}$	Over-all adsorbate-adsorbate interaction potential on a heterogeneous surface, equation IV-20.
Q	Permanent quadrupole moment of a molecule, equation VIII-35.
R	Gas constant; 8.314×10^7 ergs/mole/degree or 1.987 calories/mole/degree.
$\mathcal{R}$	Repulsion coefficient in equation VIII-1.
S^{rot}	General rotational entropy, equation V-25.
${}_aS$	Molar integral entropy of adsorbed phase, equation II-27.
${}_aS^{fig}$	Molar configurational entropy, equation V-44.
${}_aS^{gg}$	Molar entropy of congregation, equation V-35.
${}_aS^{vib}$	Additional entropy of the adsorbed phase due to vibrations with respect to the surface, equation V-15.
${}_aS_s^{tr}$	Translational entropy per mole of the adsorbed phase in its standard state, equation V-16.
${}_gS$	Molar integral entropy of gas phase, equation II-25.
${}_gS_s^{tr}$	Translational entropy per mole of a gas in its standard state, equation V-16.
ΔS^{ads}	Integral change in entropy on adsorption, equation III-77.

ΔS_s^{ads}	Total standard integral entropy change on adsorption, equation V-37.
$\Delta \dot{S}^{\text{ads}}$	Differential change in entropy on adsorption, equation III-78.
$\Delta \dot{S}_s^{\text{ads}}$	Differential standard change in entropy on adsorption.
ΔS^{rot}	Change in rotational entropy on adsorption, equation V-15.
ΔS^{tr}	Change in translational entropy on adsorption, equation V-15.
ΔS_s^{tr}	${}_aS_s^{\text{tr}} - {}_gS_s^{\text{tr}}$.
T	Temperature in degrees Kelvin.
T_c	Temperature of the chromatographic column, equation II-48.
T_r	Room temperature.
${}_aT_c$	Observed critical temperature of the adsorbed phase.
${}_aT_c^{id}$	Ideal critical temperature of the adsorbed phase: ${}_aT_c^{id} = 8\alpha^{id}/27R\beta^{id}$.
${}_gT_c$	Critical temperature of the gas.
U_0	Adsorptive potential of a homotattic surface, equation I-30.
U_{0i}	Adsorptive potential of the ith patch of a heterogeneous surface, Section IV-2.
U_0'	Adsorptive potential of the average patch of a heterogeneous surface, Section IV-3, and equation IV-7.
$\mathbf{U}_0^{\text{diff}}$	Differential adsorptive potential, Section IV-5.
$\mathbf{U}_0^{\text{int}}$	Integral adsorptive potential, Section IV-5 and equation IV-12.
V	Amount adsorbed per gram of adsorbent (cm^3 STP/g), equation I-3.
V_1, V_2, V_c, V_d	Volumes of particular parts of an adsorption system, Section II-B-1.
V_c	Volume of an empty chromatographic column, equation II-42.
V_g	Interstitial volume of a chromatographic column, equation II-42.
V_{geo}	Geometrical volume of sample bulb, equation II-2.
V_m	Amount adsorbed per gram of adsorbent when surface is saturated, equation I-12.
V_s	Apparent volume of sample bulb, equation II-1.
V_{unads}	Amount of adsorbate in gas phase at equilibrium, equation II-6.
V_β	Amount adsorbed per gram of adsorbent when area occupied by each molecule equals β.
${}_aV$	Volume of adsorbed phase, equation II-27.
${}_gV$	Volume of gas phase.

ΔV_{dose}	Increment of adsorbate introduced, equation II-5.
W	Parameter in equation V-1.
W_i	Probability of finding an adsorbate molecule on the ith surface patch, Section IV-5.
Y	Parameter in equation V-5.
Z	Initial slope of an adsorption isotherm, equation V-61.
a	Van der Waals constant of a nonideal gas.
a	Interstitial cross sectional area of a packed column, equation II-55.
a	Constant in equation VIII-7.
a_μ	Contribution to the van der Waals a caused by Keesom alignment effect, equation VI-14.
b	Van der Waals constant of a nonideal gas.
b	Constant in equation VIII-7.
c	Constant in equations I-20a and I-20b.
c	Concentration of bulk phase, equation I-6.
c	Constant in equation II-9.
c	Constant in Sips' equation in Section IV-3.
c	Velocity of light (3×10^{10} cm/sec), equation VIII-3.
c	Parameter in equation I-20 (see equation I-43).
d	Effective diameter of a molecule, p. 37.
d	Interlamellar spacing of a layer-lattice structure, Section VIII-2-B-1.
$\mathbf{d}$	Diameter of connecting tube (pp 35–8).
e	Base of Napierian logarithms (2.7182...).
f	Fractional void space in a chromatographic column, equation II-42.
f_1	Fraction of molecules striking a surface that remain long enough to change their kinetic energy, equation I-7.
f_i	Frequency of particular surface patches per unit energy interval, equation IV-1.
f^{rot}	General rotational partition function, equation V-24.
f^{tr}	General translational partition function, equation V-18.
f^{vib}	Vibrational partition function, equation V-29.
${}_af^{\text{tr}}$	Translational partition function of an adsorbed phase, equation V-20.
${}^0_af$	Free energy of clean surface in equilibrium with its own vapor (ergs/cm^2), equation I-4.

${}^{\theta}_{a}f$	Free energy of interface of which a fraction θ is covered with adsorbate (ergs/cm^2), equation I-4.
${}_{g}f^{\text{tr}}$	Translational partition function of a gas, equation V-19.
h	Planck's constant (6.624×10^{-27} ergs/seconds).
h_f	Residual heat of immersion, p. 89.
h_s	Total heat of immersion, p. 89.
i	Number of interfaces to be considered in a system, equation I-1.
k	Boltzmann constant R/N, ($= 1.38 \times 10^{-16}$ ergs/deg).
k	Constant in Freundlich equation, equation I-6.
k'	Constant in Freundlich equation on p. 127.
k_0	Constant in equation I-8.
l	Number of dimensions in n-dimensional space, equation V-18.
m	Mass of adsorbent, equation I-6.
m	Mass of a molecule, equation V-18.
m	Mass of an electron, equation VIII-3.
n	Constant in Freundlich equation, equation I-6.
n	Normalizing factor of the Gaussian distribution, equation IV-8.
n	Number of independent ways in which a molecule is free to rotate, equations V-24 and V-25.
${}_{a}n$	Number of moles adsorbed.
${}_{a}n_m$	Number of moles adsorbed at completion of a monolayer, equation II-31.
${}_{a}n_{\beta}$	Number of moles adsorbed at a surface concentration corresponding to $\sigma = \beta$; also taken as completion of a monolayer; hence, ${}_{a}n_{\beta} = {}_{a}n_m$.
${}_{g}n$	Number of moles of adsorbate in gas phase.
p	Equilibrium pressure in a system.
p_0	Saturation vapor pressure of adsorbate.
p_s	Pressure defining the standard state of the gas ($p_s = 760$ mm Hg).
$\bar{p}$	Average equilibrium pressure in chromatographic column, equation II-48.
Δp	Pressure difference due to thermal transpiration, equation II-10.
Δp_{dose}	Effective dosing pressure, equation II-5.
q^{ab}	Adiabatic heat of adsorption, equation III-16.
q^{diff}	Differential heat (liberated) of adsorption, equations I-29 and III-3.

q^{int}	Integral heat of adsorption, Section III-1.
q^{st}	Isosteric heat of adsorption, equations I-34 and III-24.
q^{th}	Isothermal heat of adsorption, equation III-5.
q_s^{diff}	Differential heat of adsorption between standard states, equation V-39.
r	Degree of heterogeneity of a substrate ($r = 0.4769/\sqrt{\gamma}$), p. 131.
r_e	Radius of an adsorbate molecule, derived either from the crystal structure or from the second virial coefficient, p. 259.
s	Number of modes of vibration of a molecule, equation V-29.
s	Separation between two particles.
s_{ij}	Separation between two particles i and j.
t	Retention time of pulse in chromatographic column, equation II-46.
u	Rate of adsorption (molecules/cm²/sec), equation I-7.
u	Interstitial gas velocity, equation II-46.
v	Rate of desorption (molecules/cm²/sec), equation I-8.
w	Grams of packed adsorbent per cm³ of column space, equation II-43.
w	Work, Section III-1.
x	Amount adsorbed in grams, equation I-6.
x	Parameter in equation VIII-9.
x_e	Parameter in equation VIII-14: $x_e = z_e/d$.
y	Parameter in equation II-12.
z	Number of nearest neighbors sites to a given site in an arbitrary lattice.
z	Distance between center of an adsorbate particle and the mathematical plane in which lie the centers of the surface atoms of the adsorbent, Section VIII-2-A.
z_e	Equilibrium value of $z (z_e = r_e + d/2)$.
Γ	Equilibrium surface concentration of adsorbate in moles/cm², equation II-43.
Π	Two-dimensional spreading pressure in dynes/cm, equation I-2.
Π_s	Two-dimensional spreading pressure defining the standard state of the adsorbed phase ($\Pi_s = 0.338$ dynes/cm).
Σ	Specific surface area of adsorbent, cm²/g, equation I-3.
$\Psi(x)$	Psi function of x: $\Psi(x) = (d/dx) \ln x!$
$\Psi^k(x)$	k^{th} derivative of $\Psi(x)$.
α	Two-dimensional van der Waals constant, corresponding to

a. The same symbol is used for both molar and molecular quantities.

α Constant in equation II-12.

α^{id} Ideal value of α, equation VI-4.

β Two-dimensional van der Waals constant, corresponding to b. The same symbol is used for both molar and molecular quantities.

β Constant in equation II-12.

β^{id} Ideal value of β, equation VI-5.

γ Heterogeneity parameter in equation IV-7.

δ_i $d\delta_i$ is the fraction of surface having energies between U_{0i} and $U_{0i} + dU_0$.

θ Fraction of the whole surface of adsorbent covered by an adsorbed layer of which each molecule has an area β, hence, $\theta = \beta/\sigma$, equation II-32.

$\bar{\theta}$ Average fraction of adsorbent surface covered at pressure $\bar{p}$ in chromatographic column, equation II-49.

θ_i Fraction of the ith patch of the adsorbent surface covered by the adsorbate, equation IV-3.

θ_s Fraction of surface covered at the standard state of the adsorbed phase, equation V-34.

κ Vibrational force constant, equation VIII-17.

λ Polarization factor, equation VI-15.

λ Mean free path of a molecule, p. 37.

μ Constant in equation II-12.

μ Permanent dipole moment.

$\bar{\mu}$ Component of the permanent dipole moment μ in the direction of the electric field, equation VI-19.

μ^{ind} Dipole moment induced by the surface electric field, equation VI-17.

${}_a\mu$ Chemical potential of the adsorbed phase, Section III-4.

${}_a\mu$ Total effective dipole moment of the adsorbed molecule normal to the surface, Section VI-2-B.

${}_g\mu$ Chemical potential of the gas phase, Section III-4.

ν Vibrational frequency with respect to the surface acquired by a molecule on adsorption, equation I-31.

$\nu^{\perp}$ Vibrational frequency of an adsorbate molecule normal to the surface, equation V-12.

$\nu^{\parallel}$	Vibrational frequency of an adsorbate molecule parallel to the surface, equation V-12.
ξ	Electron polarizability of a molecule.
ξ_1, ξ_2, ξ_3	The three principal polarizabilities of a molecule.
π	3.1415....
ρ	Number of atoms per cm^2 in a crystallographic plane parallel to the surface, equation VIII-8.
ρ_v	Number of atoms per cm^3 of adsorbent, equation VIII-27.
σ	Area per molecule, equation I-2.
σ	Symmetry factor, equation V-24.
σ_0	Area per molecule at infinite compression (see also β), p. 18.
σ_s	Area per molecule in the standard state of the adsorbed phase, equation V-33.
τ'	Time of residence of a molecule on the average surface patch, equation IV-14.
τ_i	Time of residence of a molecule on the ith surface patch, equation IV-13.
χ	Energy barrier to lateral translation of an adsorbed molecule, Section I-4.
χ	Diamagnetic susceptibility of a molecule, equation VIII-3.
ω	Interaction energy per pair of nearest neighboring atoms, equation I-20.
ω	Orientation factor defined by equation VI-11.

Computed Tables of Model Isotherms for $2\alpha/RT\beta = 0.10$ to 10.00 at Intervals of One-tenth

$2\alpha/RT\beta = 0.10$

p/K'	γ-Values 1.00	1.50	2.00	3.00	4.00	5.00	7.00	8.00	10.0	20.0	40.0	70.0	100	200	400	1000
00.001	0621	0380	0255	0138	0086	0061	0037	0030	0023	0011	0008	0006	0005	0005	0005	0004
00.002	0796	0526	0376	0223	0151	0113	0074	0063	0049	0028	0021	0018	0017	0016	0015	0015
00.004	1007	0712	0539	0352	0254	0198	0138	0121	0098	0062	0047	0041	0039	0037	0036	0035
00.006	1146	0841	0657	0448	0337	0270	0196	0173	0144	0093	0072	0064	0062	0057	0056	0055
00.008	1253	0943	0751	0529	0408	0333	0248	0222	0187	0125	0098	0088	0084	0079	0076	0075
00.010	1340	1027	0831	0601	0472	0390	0296	0267	0228	0155	0122	0110	0105	0100	0098	0095
00.020	1635	1321	1118	0867	0719	0620	0500	0462	0406	0294	0240	0220	0210	0202	0199	0195
00.030	1824	1518	1317	1060	0903	0796	0664	0619	0555	0420	0351	0321	0310	0296	0290	0287
00.040	1965	1668	1467	1212	1053	0946	0803	0755	0686	0533	0453	0419	0405	0388	0381	0375
00.050	2080	1789	1592	1339	1180	1070	0925	0876	0802	0639	0550	0510	0495	0476	0467	0462
00.060	2174	1891	1701	1452	1292	1181	1034	0985	0908	0737	0642	0599	0582	0561	0550	0544
00.070	2257	1981	1795	1548	1391	1281	1133	1082	1006	0831	0729	0683	0666	0643	0630	0623
00.080	2330	2061	1879	1636	1481	1372	1225	1175	1097	0917	0814	0765	0744	0722	0710	0702
00.090	2395	2134	1953	1715	1563	1453	1309	1259	1180	1000	0893	0844	0823	0799	0783	0777
00.100	2456	2197	2022	1790	1639	1532	1388	1337	1259	1077	0970	0919	0899	0872	0859	0850
00.130	2603	2365	2200	1977	1834	1732	1594	1545	1470	1292	1182	1129	1107	1082	1068	1061
00.170	2760	2542	2388	2184	2050	1957	1825	1780	1707	1538	1430	1379	1357	1332	1318	1310
00.200	2860	2649	2502	2312	2186	2094	1974	1931	1863	1699	1597	1547	1525	1500	1488	1479
00.250	2995	2802	2671	2496	2381	2300	2187	2147	2084	1936	1843	1797	1778	1754	1740	1732
00.300	3105	2929	2809	2649	2544	2469	2369	2332	2276	2143	2058	2018	2001	1979	1969	1962
00.400	3288	3135	3033	2899	2812	2750	2668	2637	2594	2488	2424	2392	2377	2363	2355	2349
00.500	3430	3297	3211	3095	3025	2974	2908	2885	2850	2769	2721	2698	2688	2677	2672	2667
00.600	3546	3429	3354	3260	3201	3159	3109	3093	3065	3004	2971	2956	2948	2943	2940	2939
00.700	3644	3547	3481	3403	3356	3324	3284	3268	3248	3206	3186	3178	3175	3172	3171	3171
00.800	3732	3646	3592	3525	3488	3464	3434	3423	3408	3385	3375	3372	3372	3372	3372	3372
00.900	3812	3733	3688	3635	3605	3587	3568	3560	3552	3543	3542	3543	3543	3547	3550	3549
01.000	3879	3812	3776	3734	3712	3700	3688	3682	3680	3682	3691	3697	3701	3704	3705	3708
01.200	3999	3950	3925	3906	3898	3893	3895	3896	3904	3925	3946	3960	3968	3975	3980	3982
01.400	4099	4069	4057	4047	4053	4058	4071	4078	4091	4127	4163	4178	4188	4197	4203	4208
01.600	4187	4170	4168	4176	4187	4201	4223	4233	4253	4304	4344	4365	4373	4388	4396	4398
01.800	4265	4261	4266	4284	4307	4323	4357	4369	4394	4458	4504	4529	4540	4553	4559	4562
02.000	4338	4340	4354	4384	4411	4437	4473	4492	4516	4586	4642	4670	4680	4695	4704	4708
04.000	4795	4863	4925	5019	5091	5145	5226	5258	5303	5420	5491	5526	5541	5559	5566	5571
06.000	5057	5165	5250	5380	5470	5540	5636	5669	5722	5847	5925	5958	5973	5991	6000	6007
08.000	5246	5373	5479	5628	5728	5800	5908	5944	6001	6129	6202	6237	6249	6265	6273	6277
10.000	5385	5537	5650	5810	5921	6000	6107	6145	6204	6332	6402	6432	6447	6459	6467	6472
20.000	5818	6016	6156	6350	6471	6558	6669	6706	6759	6877	6938	6961	6972	6985	6993	6998
30.000	6062	6277	6435	6638	6762	6847	6955	6988	7039	7146	7197	7218	7228	7239	7242	7243
40.000	6229	6461	6619	6826	6950	7035	7138	7169	7218	7312	7361	7380	7388	7397	7402	7406

$2\alpha/RT\beta = 0.20$

p/K'	γ-Values															
	1.00	1.50	2.00	3.00	4.00	5.00	7.00	8.00	10.0	20.0	40.0	70.0	100	200	400	1000
00.001	0627	0385	0258	0139	0087	0061	0037	0030	0023	0011	0008	0006	0005	0005	0005	0004
00.002	0804	0531	0380	0225	0152	0114	0074	0063	0049	0028	0021	0018	0017	0016	0015	0015
00.004	1017	0720	0544	0355	0256	0199	0139	0121	0098	0062	0047	0041	0039	0037	0036	0035
00.006	1157	0850	0663	0452	0340	0273	0198	0174	0145	0093	0072	0064	0062	0057	0056	0055
00.008	1265	0953	0760	0535	0412	0336	0250	0223	0188	0125	0098	0088	0084	0079	0076	0075
00.010	1353	1037	0840	0607	0477	0394	0299	0269	0230	0155	0122	0110	0105	0100	0098	0095
00.020	1650	1335	1130	0877	0729	0627	0504	0466	0409	0296	0241	0221	0211	0202	0199	0196
00.030	1839	1532	1330	1072	0914	0806	0670	0625	0560	0422	0352	0322	0311	0298	0292	0288
00.040	1982	1683	1483	1225	1066	0955	0812	0763	0692	0537	0455	0421	0406	0390	0383	0376
00.050	2098	1806	1610	1354	1194	1082	0936	0885	0810	0645	0554	0512	0497	0477	0470	0464
00.060	2194	1911	1716	1467	1306	1194	1046	0995	0918	0745	0647	0602	0584	0563	0553	0546
00.070	2277	2001	1811	1566	1406	1296	1145	1093	1016	0838	0735	0688	0669	0645	0634	0626
00.080	2349	2081	1897	1655	1498	1387	1239	1187	1110	0927	0819	0772	0751	0726	0714	0706
00.090	2416	2154	1973	1734	1580	1469	1324	1272	1193	1011	0902	0850	0828	0804	0790	0783
00.100	2476	2217	2043	1810	1658	1550	1403	1352	1273	1089	0978	0928	0906	0879	0866	0857
00.130	2625	2386	2221	1999	1856	1752	1614	1565	1489	1307	1194	1141	1115	1091	1078	1071
00.170	2783	2564	2409	2207	2075	1979	1848	1802	1730	1555	1446	1395	1373	1345	1331	1323
00.200	2882	2671	2528	2337	2211	2122	2000	1954	1884	1720	1615	1566	1545	1519	1503	1495
00.250	3015	2827	2696	2523	2408	2326	2215	2173	2110	1962	1867	1819	1800	1774	1764	1754
00.300	3130	2954	2834	2676	2572	2498	2397	2361	2307	2170	2085	2043	2027	2005	1992	1988
00.400	3312	3161	3059	2929	2843	2779	2699	2670	2628	2522	2456	2424	2412	2395	2386	2382
00.500	3452	3325	3240	3128	3058	3010	2944	2920	2886	2806	2758	2736	2726	2714	2710	2706
00.600	3569	3460	3385	3294	3238	3197	3146	3130	3102	3046	3014	3000	2994	2987	2985	2983
00.700	3671	3576	3512	3436	3391	3361	3320	3308	3289	3251	3232	3225	3223	3221	3217	3219
00.800	3760	3675	3624	3558	3523	3500	3473	3464	3450	3430	3425	3421	3421	3421	3423	3423
00.900	3837	3765	3720	3670	3642	3627	3609	3601	3597	3586	3592	3594	3599	3599	3602	3603
01.000	3905	3843	3809	3770	3750	3738	3728	3727	3726	3732	3741	3750	3755	3758	3763	3764
01.200	4026	3985	3960	3942	3935	3931	3939	3942	3949	3977	4001	4015	4022	4030	4038	4040
01.400	4126	4100	4089	4083	4091	4100	4116	4125	4137	4182	4218	4237	4245	4258	4264	4268
01.600	4215	4201	4202	4212	4228	4243	4270	4280	4301	4356	4403	4424	4438	4448	4458	4460
01.800	4293	4293	4301	4323	4344	4367	4403	4417	4441	4512	4559	4588	4600	4614	4622	4626
02.000	4366	4371	4388	4421	4454	4480	4523	4538	4569	4646	4702	4731	4743	4758	4767	4772
04.000	4825	4896	4961	5062	5136	5195	5276	5307	5354	5473	5553	5586	5601	5621	5630	5634
06.000	5088	5199	5289	5419	5515	5585	5686	5721	5777	5906	5983	6019	6032	6051	6058	6066
08.000	5275	5406	5515	5666	5774	5850	5954	5993	6051	6182	6257	6294	6307	6320	6328	6333
10.000	5414	5571	5685	5853	5964	6045	6154	6195	6254	6379	6455	6488	6499	6515	6522	6525
20.000	5844	6047	6191	6388	6514	6601	6713	6749	6806	6921	6982	7008	7018	7031	7038	7042
30.000	6091	6310	6468	6675	6802	6888	6997	7031	7080	7184	7238	7260	7270	7278	7282	7284
40.000	6258	6491	6656	6861	6990	7074	7176	7210	7256	7355	7401	7418	7428	7436	7442	7445

$2\alpha/RT\beta = 0.30$

p/K′	1.00	1.50	2.00	3.00	4.00	5.00	7.00	8.00	10.0	20.0	40.0	70.0	100	200	400	1000
	γ-Values															
00.001	0633	0388	0261	0141	0088	0061	0038	0030	0024	0011	0008	0006	0005	0005	0005	0004
00.002	0812	0537	0383	0227	0154	0114	0074	0063	0049	0028	0021	0018	0017	0016	0015	0015
00.004	1027	0727	0550	0359	0258	0201	0140	0122	0098	0062	0047	0041	0039	0037	0036	0035
00.006	1167	0858	0670	0457	0343	0275	0200	0175	0146	0093	0072	0064	0062	0057	0056	0055
00.008	1276	0962	0768	0541	0416	0339	0252	0225	0190	0125	0098	0088	0084	0079	0076	0075
00.010	1365	1049	0850	0613	0482	0397	0301	0271	0231	0155	0123	0110	0105	0100	0098	0095
00.020	1665	1348	1142	0887	0736	0634	0510	0469	0412	0296	0242	0221	0212	0202	0199	0196
00.030	1855	1547	1345	1083	0924	0815	0677	0631	0565	0423	0353	0323	0312	0298	0292	0288
00.040	2000	1700	1497	1239	1077	0966	0821	0770	0700	0540	0458	0423	0408	0391	0383	0377
00.050	2114	1824	1625	1369	1207	1093	0946	0896	0818	0649	0556	0516	0500	0479	0472	0466
00.060	2211	1929	1736	1484	1321	1207	1057	1006	0928	0751	0653	0605	0589	0565	0555	0550
00.070	2294	2020	1831	1584	1423	1311	1160	1106	1027	0844	0741	0693	0674	0650	0637	0630
00.080	2368	2101	1916	1673	1516	1404	1253	1200	1123	0937	0827	0778	0756	0730	0719	0711
00.090	2435	2174	1993	1755	1600	1487	1341	1289	1206	1021	0909	0858	0836	0808	0794	0788
00.100	2496	2238	2063	1828	1677	1567	1420	1369	1288	1100	0988	0935	0913	0886	0872	0863
00.130	2645	2407	2242	2021	1877	1773	1634	1585	1509	1321	1207	1152	1128	1101	1088	1081
00.170	2804	2586	2435	2232	2097	2003	1870	1824	1751	1575	1464	1409	1387	1358	1345	1338
00.200	2904	2696	2553	2361	2237	2147	2023	1977	1907	1742	1636	1583	1563	1535	1520	1511
00.250	3040	2852	2721	2549	2436	2353	2241	2202	2139	1988	1891	1844	1823	1798	1785	1776
00.300	3155	2981	2865	2705	2601	2525	2428	2390	2334	2200	2115	2072	2053	2030	2020	2015
00.400	3337	3187	3089	2961	2875	2813	2733	2704	2660	2555	2493	2460	2445	2430	2422	2417
00.500	3479	3352	3270	3160	3092	3043	2980	2957	2923	2844	2796	2777	2768	2755	2750	2746
00.600	3597	3489	3417	3328	3273	3235	3185	3169	3142	3086	3056	3045	3038	3033	3029	3027
00.700	3697	3604	3543	3470	3427	3398	3361	3349	3330	3296	3278	3273	3270	3269	3269	3267
00.800	3786	3703	3655	3595	3562	3538	3515	3506	3494	3477	3472	3473	3474	3473	3474	3476
00.900	3862	3795	3751	3705	3683	3666	3651	3646	3639	3638	3643	3647	3651	3653	3657	3660
01.000	3929	3873	3841	3808	3788	3780	3773	3773	3771	3782	3798	3806	3809	3816	3820	3822
01.200	4050	4015	3991	3978	3977	3976	3980	3990	3998	4029	4057	4074	4083	4093	4098	4103
01.400	4156	4131	4125	4123	4133	4143	4163	4169	4188	4236	4275	4297	4306	4321	4325	4330
01.600	4245	4235	4235	4252	4269	4286	4316	4331	4348	4414	4459	4486	4497	4513	4521	4524
01.800	4323	4324	4332	4360	4388	4414	4450	4468	4493	4564	4621	4648	4663	4678	4685	4690
02.000	4390	4402	4422	4461	4495	4526	4568	4589	4621	4703	4763	4791	4805	4824	4832	4836
04.000	4853	4932	4999	5103	5181	5239	5326	5358	5411	5534	5611	5648	5665	5684	5693	5698
06.000	5118	5234	5325	5463	5559	5634	5734	5771	5826	5963	6041	6075	6091	6108	6119	6125
08.000	5304	5446	5554	5708	5816	5897	6004	6040	6100	6236	6313	6347	6359	6376	6385	6389
10.000	5445	5604	5725	5893	6008	6091	6205	6244	6302	6437	6507	6538	6552	6568	6575	6578
20.000	5877	6081	6230	6425	6557	6644	6757	6797	6851	6966	7027	7054	7065	7077	7082	7087
30.000	6121	6348	6502	6713	6839	6927	7035	7072	7124	7228	7280	7301	7308	7320	7324	7328
40.000	6282	6524	6690	6900	7029	7112	7214	7250	7297	7391	7439	7458	7466	7474	7480	7482

$2\alpha/RT\beta = 0.40$

p/K'	γ-Values 1.00	1.50	2.00	3.00	4.00	5.00	7.00	8.00	10.0	20.0	40.0	70.0	100	200	400	1000
00.001	0640	0393	0264	0142	0088	0062	0038	0031	0024	0011	0008	0006	0005	0005	0005	0004
00.002	0820	0543	0388	0229	0155	0115	0075	0064	0049	0028	0021	0018	0017	0016	0015	0015
00.004	1036	0734	0555	0362	0261	0202	0140	0123	0099	0062	0047	0041	0039	0037	0036	0035
00.006	1179	0867	0678	0462	0347	0277	0201	0176	0147	0093	0072	0064	0062	0057	0056	0055
00.008	1287	0972	0777	0547	0420	0342	0254	0226	0191	0125	0098	0088	0084	0079	0076	0075
00.010	1377	1059	0859	0620	0487	0402	0303	0273	0232	0156	0123	0110	0105	0100	0098	0095
00.020	1680	1362	1154	0897	0743	0641	0514	0474	0415	0298	0242	0222	0212	0203	0200	0196
00.030	1871	1564	1358	1096	0935	0824	0685	0638	0572	0427	0355	0325	0313	0299	0293	0289
00.040	2017	1718	1513	1253	1091	0977	0831	0779	0706	0546	0460	0425	0410	0393	0385	0379
00.050	2131	1842	1643	1385	1221	1106	0957	0906	0827	0655	0560	0519	0502	0482	0474	0467
00.060	2230	1947	1754	1500	1337	1222	1070	1018	0939	0758	0657	0610	0592	0570	0559	0552
00.070	2314	2038	1850	1602	1440	1327	1173	1119	1041	0856	0748	0697	0678	0654	0642	0634
00.080	2390	2121	1936	1691	1533	1420	1269	1216	1134	0946	0834	0783	0760	0737	0724	0715
00.090	2455	2194	2014	1775	1618	1505	1356	1304	1222	1032	0919	0864	0842	0816	0800	0793
00.100	2515	2259	2086	1848	1697	1587	1438	1386	1304	1113	0998	0943	0922	0892	0879	0869
00.130	2668	2430	2266	2043	1898	1795	1653	1601	1527	1337	1220	1163	1139	1111	1096	1089
00.170	2829	2611	2456	2257	2121	2026	1893	1846	1773	1594	1481	1426	1402	1374	1360	1349
00.200	2928	2720	2577	2389	2262	2172	2049	2003	1933	1764	1657	1603	1580	1553	1538	1530
00.250	3062	2877	2749	2577	2463	2380	2270	2228	2165	2014	1915	1866	1847	1820	1808	1797
00.300	3178	3008	2888	2732	2632	2558	2458	2421	2365	2230	2144	2100	2083	2058	2049	2042
00.400	3360	3214	3118	2989	2906	2848	2766	2738	2696	2591	2528	2496	2480	2465	2458	2452
00.500	3504	3378	3300	3193	3125	3078	3017	2994	2961	2885	2839	2818	2808	2797	2793	2789
00.600	3624	3517	3449	3363	3309	3271	3222	3207	3183	3129	3101	3090	3084	3079	3075	3077
00.700	3724	3634	3576	3505	3462	3435	3402	3391	3372	3340	3326	3321	3320	3319	3319	3319
00.800	3814	3734	3688	3632	3600	3580	3556	3548	3539	3526	3522	3524	3526	3529	3528	3529
00.900	3889	3823	3786	3742	3719	3707	3693	3690	3686	3686	3695	3702	3707	3711	3714	3715
01.000	3959	3904	3875	3844	3828	3820	3817	3815	3819	3831	3849	3861	3867	3873	3880	3882
01.200	4080	4045	4028	4017	4015	4020	4030	4036	4045	4080	4112	4132	4139	4154	4158	4163
01.400	4183	4163	4158	4163	4174	4186	4209	4220	4236	4290	4335	4358	4368	4383	4391	4393
01.600	4272	4265	4271	4290	4311	4330	4364	4378	4400	4469	4521	4547	4560	4577	4584	4587
01.800	4349	4356	4370	4401	4432	4458	4501	4515	4544	4624	4682	4712	4725	4744	4751	4757
02.000	4419	4438	4458	4499	4540	4572	4619	4638	4669	4760	4822	4858	4871	4889	4896	4905
04.000	4879	4966	5037	5146	5225	5287	5372	5412	5461	5593	5674	5712	5728	5745	5755	5761
06.000	5146	5267	5365	5504	5605	5681	5784	5822	5879	6015	6099	6136	6151	6168	6177	6182
08.000	5332	5479	5589	5752	5860	5942	6052	6095	6155	6294	6367	6403	6416	6431	6440	6445
10.000	5474	5638	5760	5934	6053	6138	6252	6291	6352	6487	6562	6590	6606	6621	6629	6633
20.000	5907	6116	6267	6470	6596	6686	6800	6838	6895	7016	7072	7098	7107	7119	7127	7129
30.000	6151	6381	6539	6748	6881	6968	7079	7113	7167	7269	7321	7342	7350	7360	7366	7369
40.000	6314	6559	6724	6937	7067	7151	7255	7289	7337	7432	7477	7498	7503	7513	7516	7520

$2\alpha/RT\beta = 0.50$

p/K'	γ-Values 1.00	1.50	2.00	3.00	4.00	5.00	7.00	8.00	10.0	20.0	40.0	70.0	100	200	400	1000
00.001	0647	0397	0266	0143	0089	0062	0038	0031	0024	0011	0008	0006	0005	0005	0005	0004
00.002	0828	0549	0392	0232	0156	0116	0075	0064	0049	0028	0021	0018	0017	0016	0015	0015
00.004	1046	0742	0562	0366	0263	0204	0141	0124	0099	0062	0047	0041	0039	0037	0036	0035
00.006	1190	0877	0686	0467	0351	0280	0202	0178	0147	0093	0072	0064	0062	0057	0056	0055
00.008	1299	0982	0785	0554	0425	0345	0256	0228	0192	0126	0098	0088	0084	0079	0076	0075
00.010	1390	1070	0868	0627	0491	0406	0305	0275	0234	0156	0123	0110	0105	0100	0098	0095
00.020	1694	1377	1167	0907	0753	0647	0520	0478	0418	0300	0243	0222	0213	0203	0200	0196
00.030	1889	1578	1373	1108	0945	0834	0692	0645	0577	0429	0356	0326	0314	0300	0294	0290
00.040	2035	1735	1530	1268	1102	0987	0839	0787	0714	0551	0463	0427	0412	0394	0387	0380
00.050	2151	1860	1660	1401	1234	1119	0967	0915	0836	0662	0564	0522	0504	0484	0476	0470
00.060	2249	1965	1773	1517	1352	1235	1083	1029	0950	0766	0661	0614	0596	0572	0561	0555
00.070	2333	2059	1870	1618	1457	1342	1188	1132	1050	0865	0754	0702	0683	0659	0645	0637
00.080	2410	2142	1956	1711	1551	1437	1284	1229	1147	0955	0843	0790	0767	0740	0727	0720
00.090	2476	2216	2034	1795	1637	1523	1373	1318	1235	1042	0927	0872	0849	0821	0805	0799
00.100	2536	2280	2106	1870	1716	1606	1456	1403	1320	1126	1007	0952	0929	0899	0887	0877
00.130	2690	2452	2288	2067	1922	1816	1674	1625	1545	1352	1234	1175	1151	1122	1107	1098
00.170	2851	2634	2483	2281	2146	2051	1917	1870	1796	1615	1499	1442	1419	1389	1374	1365
00.200	2949	2745	2604	2415	2289	2199	2074	2028	1958	1785	1679	1621	1600	1571	1556	1547
00.250	3088	2902	2774	2603	2493	2410	2298	2257	2195	2043	1940	1893	1871	1845	1831	1821
00.300	3203	3034	2918	2764	2661	2587	2488	2453	2397	2261	2173	2130	2110	2088	2076	2070
00.400	3385	3244	3148	3021	2937	2881	2800	2772	2729	2628	2563	2531	2518	2502	2495	2487
00.500	3529	3408	3330	3226	3159	3113	3054	3032	3000	2923	2882	2859	2852	2841	2837	2834
00.600	3649	3547	3478	3395	3344	3309	3260	3247	3224	3172	3146	3137	3133	3127	3125	3125
00.700	3751	3664	3608	3540	3501	3474	3442	3429	3416	3389	3377	3374	3372	3371	3371	3373
00.800	3840	3764	3718	3667	3638	3619	3596	3592	3582	3573	3576	3578	3581	3584	3586	3586
00.900	3914	3854	3818	3780	3759	3747	3738	3735	3732	3738	3750	3759	3763	3769	3774	3775
01.000	3986	3936	3909	3877	3867	3861	3861	3863	3866	3885	3906	3917	3926	3933	3939	3943
01.200	4109	4076	4061	4053	4058	4061	4076	4083	4094	4137	4171	4193	4202	4216	4223	4226
01.400	4211	4195	4193	4200	4213	4230	4256	4265	4286	4346	4395	4420	4431	4447	4454	4459
01.600	4300	4295	4306	4329	4353	4376	4412	4427	4454	4524	4581	4612	4625	4641	4651	4654
01.800	4376	4390	4405	4441	4474	4501	4547	4566	4596	4681	4742	4775	4790	4808	4819	4824
02.000	4449	4470	4495	4543	4581	4615	4667	4690	4723	4819	4888	4918	4934	4954	4963	4972
04.000	4911	5000	5072	5185	5267	5334	5427	5458	5516	5648	5731	5771	5788	5808	5818	5826
06.000	5178	5299	5403	5547	5649	5726	5836	5872	5933	6075	6155	6191	6207	6226	6237	6240
08.000	5362	5512	5626	5791	5906	5990	6102	6143	6205	6344	6422	6458	6472	6489	6496	6501
10.000	5506	5671	5798	5977	6097	6185	6297	6343	6404	6537	6614	6647	6660	6672	6682	6686
20.000	5937	6149	6303	6507	6638	6730	6845	6885	6939	7056	7119	7144	7152	7166	7171	7174
30.000	6179	6410	6575	6789	6919	7011	7119	7154	7205	7309	7362	7384	7391	7402	7405	7409
40.000	6342	6592	6759	6975	7103	7187	7294	7328	7374	7470	7515	7534	7542	7549	7554	7556

$2\alpha/RT\beta = 0.60$

p/K′	1.00	1.50	2.00	3.00	4.00	5.00	7.00	8.00	10.0	20.0	40.0	70.0	100	200	400	1000
	γ-Values															
00.001	0653	0401	0269	0144	0089	0062	0038	0031	0024	0011	0008	0006	0005	0005	0005	0004
00.002	0837	0554	0396	0234	0157	0117	0075	0064	0049	0028	0021	0018	0017	0016	0015	0015
00.004	1056	0751	0568	0370	0266	0205	0142	0124	0099	0062	0047	0041	0039	0037	0036	0035
00.006	1201	0886	0693	0471	0354	0282	0204	0178	0148	0093	0072	0064	0062	0057	0056	0055
00.008	1312	0993	0794	0560	0428	0349	0258	0229	0193	0126	0098	0088	0084	0079	0076	0075
00.010	1403	1082	0878	0633	0498	0410	0308	0277	0235	0157	0123	0110	0106	0100	0098	0095
00.020	1710	1390	1180	0918	0761	0655	0524	0483	0422	0302	0244	0223	0213	0204	0200	0196
00.030	1904	1595	1388	1122	0958	0843	0700	0651	0582	0433	0358	0327	0315	0301	0294	0291
00.040	2052	1753	1547	1281	1115	0999	0849	0796	0721	0555	0466	0429	0413	0396	0387	0381
00.050	2169	1876	1679	1416	1250	1132	0978	0925	0846	0667	0568	0524	0508	0486	0478	0471
00.060	2268	1984	1791	1536	1369	1251	1096	1042	0960	0773	0667	0618	0599	0576	0564	0558
00.070	2354	2078	1889	1638	1475	1358	1202	1147	1064	0871	0758	0708	0688	0661	0649	0641
00.080	2430	2163	1976	1730	1569	1455	1300	1244	1162	0966	0849	0795	0772	0747	0733	0724
00.090	2494	2236	2057	1815	1656	1541	1389	1335	1251	1053	0936	0879	0856	0828	0812	0805
00.100	2557	2302	2126	1891	1737	1626	1474	1419	1337	1139	1016	0960	0938	0907	0891	0883
00.130	2711	2475	2312	2089	1945	1839	1696	1645	1564	1368	1246	1189	1163	1133	1119	1109
00.170	2873	2659	2510	2307	2170	2076	1942	1895	1820	1636	1518	1460	1435	1405	1390	1381
00.200	2973	2770	2631	2442	2315	2226	2101	2056	1984	1811	1701	1644	1619	1590	1576	1564
00.250	3112	2927	2802	2633	2519	2439	2329	2286	2224	2068	1967	1918	1896	1870	1855	1845
00.300	3227	3061	2946	2794	2692	2618	2519	2483	2430	2292	2204	2161	2141	2118	2106	2100
00.400	3410	3269	3176	3054	2971	2914	2836	2809	2767	2663	2602	2571	2555	2540	2531	2525
00.500	3557	3438	3361	3259	3195	3149	3091	3069	3038	2967	2926	2906	2897	2886	2882	2879
00.600	3675	3578	3510	3430	3379	3345	3302	3287	3267	3219	3194	3187	3181	3178	3175	3175
00.700	3777	3691	3641	3574	3537	3513	3484	3474	3461	3436	3426	3426	3425	3424	3426	3427
00.800	3865	3794	3751	3703	3677	3661	3641	3635	3630	3624	3629	3633	3636	3640	3643	3643
00.900	3943	3886	3853	3816	3796	3789	3781	3779	3778	3788	3804	3817	3821	3828	3833	3835
01.000	4014	3966	3943	3916	3908	3905	3906	3906	3911	3937	3963	3978	3985	3993	4001	4003
01.200	4136	4109	4096	4092	4099	4107	4123	4129	4145	4192	4231	4255	4263	4278	4284	4293
01.400	4240	4226	4228	4240	4255	4273	4303	4315	4336	4405	4453	4482	4494	4511	4521	4523
01.600	4328	4328	4343	4370	4395	4420	4461	4475	4505	4584	4644	4675	4690	4707	4715	4722
01.800	4407	4421	4442	4481	4518	4547	4597	4616	4648	4740	4805	4841	4853	4875	4886	4893
02.000	4478	4501	4530	4584	4623	4661	4714	4742	4778	4878	4947	4985	5000	5021	5030	5039
04.000	4940	5034	5111	5226	5315	5379	5479	5514	5569	5709	5796	5836	5851	5871	5882	5890
06.000	5208	5335	5439	5587	5695	5772	5884	5925	5985	6131	6216	6251	6267	6285	6293	6298
08.000	5393	5548	5666	5832	5951	6036	6153	6193	6257	6400	6477	6512	6529	6543	6550	6556
10.000	5538	5707	5837	6015	6141	6230	6348	6389	6454	6592	6667	6699	6710	6727	6735	6738
20.000	5968	6183	6338	6547	6680	6773	6889	6928	6984	7104	7162	7186	7198	7209	7213	7215
30.000	6207	6445	6613	6825	6958	7047	7161	7196	7246	7352	7402	7422	7428	7440	7443	7448
40.000	6374	6621	6792	7009	7143	7228	7333	7366	7413	7507	7552	7571	7577	7584	7587	7592

$2\alpha/RT\beta = 0.70$

p/K'	γ-Values 1.00	1.50	2.00	3.00	4.00	5.00	7.00	8.00	10.0	20.0	40.0	70.0	100	200	400	1000
00.001	0659	0406	0272	0146	0090	0063	0038	0031	0024	0011	0008	0006	0005	0005	0005	0004
00.002	0844	0560	0401	0237	0159	0117	0076	0064	0050	0028	0021	0018	0017	0016	0015	0015
00.004	1067	0758	0574	0374	0269	0207	0143	0125	0100	0062	0047	0041	0039	0037	0036	0035
00.006	1213	0894	0701	0477	0358	0285	0205	0180	0149	0094	0072	0064	0062	0057	0056	0055
00.008	1325	1003	0802	0566	0433	0352	0260	0231	0194	0126	0098	0088	0084	0079	0076	0075
00.010	1417	1093	0887	0641	0503	0414	0310	0280	0237	0158	0123	0110	0106	0100	0098	0095
00.020	1725	1405	1193	0928	0770	0661	0530	0487	0426	0303	0245	0223	0213	0204	0201	0196
00.030	1921	1612	1402	1135	0968	0853	0707	0658	0587	0436	0360	0329	0316	0302	0296	0292
00.040	2069	1769	1563	1296	1129	1013	0859	0806	0728	0558	0469	0432	0415	0397	0389	0383
00.050	2187	1895	1697	1434	1264	1146	0990	0937	0856	0673	0572	0528	0510	0488	0480	0474
00.060	2288	2005	1810	1552	1385	1267	1108	1054	0970	0781	0672	0622	0603	0578	0568	0560
00.070	2373	2098	1908	1656	1491	1374	1217	1161	1077	0881	0766	0714	0692	0666	0653	0644
00.080	2448	2183	1996	1751	1588	1472	1316	1260	1176	0975	0857	0803	0777	0751	0738	0729
00.090	2516	2256	2078	1836	1677	1561	1407	1351	1266	1066	0944	0887	0864	0834	0819	0810
00.100	2577	2323	2149	1913	1758	1646	1493	1438	1352	1152	1028	0969	0946	0914	0901	0890
00.130	2733	2498	2335	2114	1967	1862	1718	1667	1584	1387	1261	1201	1173	1145	1130	1119
00.170	2895	2682	2532	2331	2197	2101	1967	1919	1843	1657	1535	1477	1453	1421	1406	1394
00.200	2995	2795	2655	2469	2343	2254	2127	2081	2011	1835	1723	1664	1641	1610	1594	1583
00.250	3137	2954	2829	2660	2550	2469	2358	2317	2253	2097	1996	1946	1922	1894	1880	1870
00.300	3252	3088	2974	2823	2723	2650	2554	2516	2462	2324	2236	2191	2172	2147	2137	2130
00.400	3437	3298	3206	3083	3006	2948	2871	2846	2804	2703	2639	2609	2595	2579	2571	2565
00.500	3582	3466	3392	3292	3231	3186	3130	3110	3077	3009	2968	2950	2941	2931	2927	2925
00.600	3702	3605	3541	3466	3417	3384	3344	3330	3309	3263	3243	3235	3231	3228	3228	3226
00.700	3802	3722	3673	3611	3577	3553	3526	3516	3505	3483	3477	3479	3478	3481	3480	3481
00.800	3894	3825	3784	3738	3716	3701	3686	3681	3674	3674	3682	3689	3694	3698	3702	3703
00.900	3971	3916	3886	3852	3838	3831	3826	3827	3826	3841	3862	3875	3880	3889	3893	3898
01.000	4042	3997	3977	3956	3948	3948	3949	3954	3961	3991	4019	4038	4049	4057	4065	4069
01.200	4163	4141	4132	4132	4140	4148	4170	4177	4193	4246	4292	4318	4330	4344	4350	4356
01.400	4269	4257	4261	4279	4301	4321	4350	4368	4391	4461	4517	4547	4561	4578	4588	4593
01.600	4356	4363	4378	4408	4439	4466	4509	4527	4557	4641	4709	4741	4754	4774	4784	4791
01.800	4435	4454	4477	4523	4562	4594	4645	4668	4701	4798	4869	4905	4923	4942	4953	4959
02.000	4510	4536	4567	4623	4669	4708	4766	4791	4831	4937	5011	5050	5067	5086	5100	5108
04.000	4970	5068	5147	5271	5360	5427	5529	5565	5626	5768	5855	5897	5914	5934	5945	5954
06.000	5237	5371	5476	5629	5741	5821	5935	5977	6038	6185	6271	6310	6325	6343	6353	6357
08.000	5425	5583	5700	5876	5995	6083	6202	6245	6306	6453	6533	6568	6583	6599	6605	6610
10.000	5566	5740	5872	6058	6182	6277	6397	6441	6502	6643	6716	6751	6762	6778	6786	6791
20.000	5997	6215	6372	6588	6722	6813	6935	6972	7031	7147	7205	7230	7241	7251	7257	7258
30.000	6235	6480	6647	6864	7000	7091	7198	7235	7288	7391	7442	7459	7469	7477	7484	7487
40.000	6405	6654	6826	7046	7178	7266	7369	7403	7452	7547	7588	7606	7614	7622	7625	7626

$2\alpha/RT\beta = 0.80$

p/K'	γ-Values 1.00	1.50	2.00	3.00	4.00	5.00	7.00	8.00	10.0	20.0	40.0	70.0	100	200	400	1000
00.001	0667	0410	0275	0147	0091	0063	0038	0031	0024	0011	0008	0006	0005	0005	0005	0004
00.002	0852	0567	0405	0239	0160	0118	0076	0064	0050	0028	0021	0018	0017	0016	0015	0015
00.004	1076	0767	0580	0378	0272	0209	0144	0126	0100	0062	0047	0041	0039	0037	0036	0035
00.006	1224	0905	0708	0483	0361	0287	0206	0181	0149	0094	0072	0064	0062	0057	0056	0055
00.008	1337	1014	0812	0571	0438	0356	0262	0233	0195	0127	0098	0088	0084	0079	0076	0075
00.010	1430	1104	0897	0648	0508	0418	0314	0282	0239	0158	0123	0110	0106	0100	0098	0095
00.020	1741	1421	1206	0940	0778	0669	0535	0492	0430	0304	0245	0224	0214	0204	0201	0196
00.030	1939	1628	1419	1148	0979	0862	0715	0665	0593	0439	0362	0329	0317	0303	0297	0293
00.040	2089	1787	1580	1313	1142	1023	0868	0814	0736	0563	0472	0433	0417	0399	0391	0384
00.050	2207	1914	1714	1451	1280	1160	1001	0948	0864	0678	0576	0531	0513	0491	0482	0476
00.060	2307	2024	1828	1571	1403	1281	1122	1067	0982	0788	0678	0626	0607	0582	0570	0564
00.070	2393	2118	1930	1677	1510	1393	1232	1173	1089	0890	0773	0718	0698	0670	0657	0649
00.080	2469	2205	2018	1771	1609	1490	1333	1275	1191	0986	0866	0809	0785	0758	0743	0734
00.090	2538	2278	2098	1857	1697	1579	1426	1368	1281	1079	0955	0895	0870	0841	0824	0817
00.100	2600	2346	2172	1934	1777	1668	1511	1457	1369	1166	1039	0979	0954	0923	0908	0898
00.130	2756	2523	2359	2138	1991	1884	1739	1688	1605	1404	1277	1214	1187	1156	1140	1131
00.170	2919	2707	2559	2358	2224	2127	1994	1945	1867	1679	1556	1495	1469	1438	1420	1410
00.200	3019	2821	2682	2496	2370	2280	2154	2109	2037	1861	1746	1686	1661	1631	1613	1603
00.250	3159	2980	2857	2691	2581	2499	2389	2346	2283	2129	2025	1971	1949	1921	1906	1896
00.300	3277	3115	3002	2853	2755	2682	2586	2549	2495	2358	2268	2223	2204	2181	2168	2160
00.400	3461	3327	3238	3118	3040	2984	2908	2882	2841	2742	2681	2650	2635	2619	2611	2606
00.500	3608	3495	3422	3325	3268	3225	3169	3149	3121	3054	3013	2995	2990	2980	2976	2974
00.600	3728	3637	3573	3500	3456	3423	3385	3371	3353	3312	3294	3285	3284	3282	3281	3282
00.700	3829	3753	3704	3648	3616	3594	3568	3561	3550	3535	3533	3534	3536	3538	3537	3541
00.800	3922	3857	3820	3777	3756	3742	3730	3725	3724	3725	3739	3746	3753	3759	3764	3765
00.900	4001	3947	3919	3890	3880	3873	3871	3872	3877	3896	3919	3935	3944	3953	3956	3962
01.000	4070	4030	4010	3993	3990	3990	3997	4004	4011	4046	4079	4100	4106	4123	4130	4134
01.200	4193	4171	4166	4171	4179	4195	4216	4227	4246	4304	4353	4380	4394	4409	4418	4425
01.400	4294	4291	4300	4321	4340	4363	4401	4416	4444	4519	4579	4612	4627	4644	4656	4663
01.600	4385	4395	4414	4447	4484	4512	4560	4579	4609	4702	4768	4805	4824	4843	4853	4860
01.800	4464	4487	4513	4564	4602	4640	4695	4718	4757	4858	4931	4973	4990	5010	5021	5030
02.000	4537	4572	4602	4663	4713	4755	4819	4844	4885	4997	5077	5118	5136	5155	5169	5176
04.000	5002	5104	5185	5312	5404	5478	5581	5619	5680	5827	5918	5959	5976	5997	6009	6018
06.000	5266	5404	5513	5673	5785	5868	5984	6027	6092	6242	6331	6367	6385	6401	6410	6415
08.000	5453	5618	5742	5918	6039	6129	6248	6292	6358	6503	6587	6622	6636	6652	6660	6664
10.000	5597	5776	5910	6101	6228	6318	6446	6488	6550	6693	6768	6803	6813	6832	6839	6843
20.000	6028	6252	6407	6625	6764	6858	6975	7015	7072	7188	7250	7274	7283	7293	7300	7302
30.000	6267	6512	6681	6902	7039	7128	7240	7276	7328	7430	7480	7500	7507	7516	7523	7524
40.000	6433	6687	6862	7085	7217	7306	7405	7443	7488	7582	7626	7642	7651	7657	7662	7662

$2\alpha/RT\beta = 0.90$

p/K'	1.00	1.50	2.00	3.00	4.00	5.00	7.00	8.00	10.0	20.0	40.0	70.0	100	200	400	1000
								γ-Values								
00.001	0673	0414	0278	0148	0091	0063	0038	0031	0024	0011.	0008	0006	0005	0005	0005	0004
00.002	0862	0572	0410	0241	0162	0119	0077	0065	0050	0028	0021	0018	0017	0016	0015	0015
00.004	1087	0775	0588	0381	0274	0211	0145	0126	0101	0062	0047	0041	0039	0037	0036	0035
00.006	1236	0914	0717	0488	0365	0290	0208	0182	0150	0094	0072	0064	0062	0057	0056	0055
00.008	1349	1025	0821	0579	0443	0360	0264	0235	0196	0127	0098	0088	0084	0079	0076	0075
00.010	1444	1117	0907	0656	0514	0423	0316	0284	0240	0158	0124	0110	0106	0100	0098	0095
00.020	1757	1435	1220	0951	0787	0676	0541	0497	0434	0306	0246	0224	0214	0204	0201	0196
00.030	1958	1644	1433	1161	0990	0874	0723	0672	0599	0442	0363	0330	0318	0304	0298	0294
00.040	2106	1805	1597	1328	1157	1035	0878	0824	0744	0568	0474	0435	0419	0401	0392	0385
00.050	2226	1933	1734	1467	1297	1176	1015	0959	0874	0685	0579	0533	0517	0494	0484	0478
00.060	2327	2044	1850	1589	1420	1297	1136	1081	0995	0796	0682	0630	0611	0585	0573	0566
00.070	2412	2140	1950	1695	1529	1410	1247	1189	1102	0899	0781	0724	0704	0676	0661	0652
00.080	2490	2226	2038	1790	1627	1510	1349	1293	1206	0998	0874	0816	0790	0762	0748	0739
00.090	2558	2300	2120	1879	1718	1601	1443	1386	1299	1091	0965	0902	0879	0847	0831	0823
00.100	2620	2369	2194	1958	1800	1686	1529	1476	1387	1180	1048	0989	0962	0932	0915	0905
00.130	2778	2546	2384	2163	2016	1909	1762	1708	1628	1422	1290	1227	1199	1167	1151	1142
00.170	2941	2731	2584	2384	2252	2153	2020	1971	1893	1703	1577	1514	1488	1454	1438	1426
00.200	3044	2846	2710	2524	2399	2311	2184	2138	2065	1886	1771	1709	1684	1652	1635	1623
00.250	3182	3007	2885	2720	2612	2529	2419	2380	2314	2159	2051	2001	1979	1948	1933	1923
00.300	3300	3142	3030	2884	2786	2714	2620	2584	2528	2392	2302	2258	2239	2214	2199	2193
00.400	3489	3355	3268	3150	3075	3019	2947	2918	2880	2785	2722	2691	2679	2661	2652	2648
00.500	3634	3527	3454	3361	3300	3262	3207	3190	3162	3099	3062	3045	3037	3027	3025	3024
00.600	3756	3664	3605	3537	3494	3464	3426	3416	3397	3362	3346	3341	3339	3337	3333	3338
00.700	3857	3783	3739	3685	3655	3635	3611	3605	3596	3585	3587	3588	3593	3596	3599	3601
00.800	3946	3887	3853	3814	3795	3785	3775	3773	3771	3781	3796	3807	3814	3819	3822	3828
00.900	4027	3980	3953	3929	3921	3917	3917	3919	3924	3951	3976	3995	4006	4018	4021	4027
01.000	4098	4063	4043	4031	4030	4034	4046	4051	4062	4102	4142	4164	4173	4186	4196	4202
01.200	4221	4204	4201	4210	4224	4239	4265	4278	4297	4361	4417	4445	4461	4478	4488	4493
01.400	4323	4323	4335	4360	4387	4411	4451	4468	4496	4579	4642	4677	4696	4714	4725	4731
01.600	4415	4428	4448	4490	4529	4557	4608	4631	4662	4761	4835	4874	4893	4913	4924	4932
01.800	4496	4520	4548	4604	4648	4688	4747	4771	4811	4921	4998	5039	5058	5081	5091	5101
02.000	4565	4604	4638	4704	4758	4803	4871	4896	4939	5060	5142	5184	5202	5224	5238	5245
04.000	5032	5138	5224	5355	5449	5525	5631	5671	5733	5884	5978	6024	6039	6061	6073	6081
06.000	5294	5438	5550	5714	5829	5918	6035	6079	6145	6299	6387	6424	6441	6458	6468	6474
08.000	5485	5651	5779	5959	6086	6178	6300	6344	6411	6561	6640	6677	6690	6706	6713	6718
10.000	5628	5809	5947	6140	6272	6365	6490	6537	6601	6743	6819	6853	6866	6881	6891	6895
20.000	6057	6285	6447	6663	6802	6897	7017	7057	7116	7237	7293	7315	7326	7335	7342	7345
30.000	6298	6546	6716	6941	7075	7165	7281	7316	7368	7469	7518	7539	7544	7554	7557	7562
40.000	6463	6718	6895	7120	7253	7341	7446	7479	7525	7618	7661	7677	7685	7693	7694	7696

$2\alpha/RT\beta = 1.00$

p/K'	γ-Values 1.00	1.50	2.00	3.00	4.00	5.00	7.00	8.00	10.0	20.0	40.0	70.0	100	200	400	1000
00.001	0679	0419	0281	0150	0092	0064	0038	0031	0024	0011	0008	0006	0005	0005	0005	0004
00.002	0870	0579	0414	0244	0163	0120	0077	0065	0050	0028	0021	0018	0017	0016	0015	0015
00.004	1097	0783	0594	0386	0277	0213	0146	0127	0101	0062	0047	0041	0039	0037	0036	0035
00.006	1247	0923	0725	0493	0369	0293	0209	0184	0151	0094	0072	0064	0062	0057	0056	0055
00.008	1362	1037	0830	0586	0448	0363	0266	0237	0197	0127	0099	0088	0084	0079	0076	0075
00.010	1457	1128	0918	0664	0520	0427	0319	0286	0242	0159	0124	0110	0106	0100	0098	0095
00.020	1774	1450	1233	0961	0797	0684	0548	0502	0438	0308	0247	0225	0214	0204	0201	0196
00.030	1973	1661	1451	1176	1002	0884	0730	0680	0605	0445	0365	0332	0319	0304	0298	0295
00.040	2125	1822	1616	1343	1171	1049	0888	0834	0753	0574	0477	0438	0420	0402	0394	0387
00.050	2245	1952	1751	1485	1312	1189	1027	0971	0884	0691	0584	0538	0519	0495	0487	0480
00.060	2347	2066	1870	1608	1436	1314	1151	1093	1007	0804	0688	0636	0614	0588	0576	0569
00.070	2433	2160	1972	1715	1547	1428	1263	1204	1115	0910	0786	0730	0707	0679	0666	0657
00.080	2512	2246	2062	1811	1648	1528	1367	1309	1221	1009	0882	0823	0797	0768	0753	0744
00.090	2581	2324	2142	1901	1739	1620	1463	1404	1315	1104	0975	0912	0884	0855	0838	0828
00.100	2643	2390	2217	1981	1822	1709	1550	1495	1407	1194	1060	0998	0971	0938	0923	0911
00.130	2800	2571	2409	2186	2040	1932	1785	1733	1650	1441	1307	1241	1213	1179	1163	1154
00.170	2966	2758	2611	2411	2279	2181	2045	1998	1920	1727	1598	1534	1507	1471	1454	1443
00.200	3069	2871	2737	2553	2428	2336	2212	2166	2094	1914	1793	1732	1706	1673	1655	1643
00.250	3209	3035	2913	2751	2643	2560	2452	2411	2346	2192	2084	2030	2007	1978	1959	1950
00.300	3325	3169	3060	2916	2820	2749	2655	2618	2564	2430	2339	2293	2273	2247	2234	2227
00.400	3515	3384	3301	3185	3111	3058	2986	2960	2920	2825	2766	2737	2722	2705	2698	2692
00.500	3660	3556	3486	3395	3340	3301	3251	3230	3202	3146	3110	3093	3089	3081	3077	3075
00.600	3782	3695	3640	3573	3533	3502	3470	3459	3443	3411	3397	3395	3394	3393	3393	3393
00.700	3885	3815	3774	3719	3694	3676	3658	3650	3645	3637	3643	3649	3652	3657	3661	3662
00.800	3977	3919	3886	3852	3836	3826	3819	3821	3821	3835	3854	3869	3875	3883	3890	3894
00.900	4056	4012	3987	3968	3962	3961	3964	3964	3975	4007	4037	4057	4069	4080	4090	4095
01.000	4126	4095	4079	4073	4073	4078	4095	4101	4115	4160	4203	4228	4238	4256	4265	4270
01.200	4248	4236	4239	4250	4268	4285	4315	4325	4352	4422	4482	4512	4528	4546	4558	4564
01.400	4353	4356	4368	4399	4430	4456	4502	4519	4550	4640	4709	4746	4765	4786	4798	4804
01.600	4443	4460	4484	4531	4568	4605	4659	4681	4721	4824	4901	4941	4959	4983	4996	5003
01.800	4524	4555	4584	4644	4694	4735	4800	4824	4863	4980	5066	5108	5127	5151	5161	5171
02.000	4595	4638	4677	4747	4804	4851	4921	4950	4996	5120	5207	5252	5272	5295	5308	5315
04.000	5061	5172	5263	5398	5496	5575	5684	5727	5787	5945	6039	6085	6102	6126	6134	6143
06.000	5326	5471	5587	5757	5875	5964	6088	6130	6200	6351	6443	6482	6497	6516	6525	6530
08.000	5516	5687	5814	6002	6129	6222	6349	6394	6462	6611	6693	6731	6744	6761	6768	6771
10.000	5657	5843	5985	6183	6318	6412	6539	6584	6651	6795	6870	6903	6917	6932	6941	6948
20.000	6087	6316	6484	6704	6845	6940	7062	7104	7160	7277	7335	7358	7367	7379	7382	7386
30.000	6326	6577	6748	6975	7115	7208	7320	7356	7407	7507	7556	7576	7582	7592	7595	7598
40.000	6490	6749	6932	7156	7291	7376	7483	7516	7563	7653	7696	7713	7721	7729	7730	7732

$2\alpha/RT\beta = 1.10$

p/K'	1.00	1.50	2.00	3.00	4.00	5.00	7.00	8.00	10.0	20.0	40.0	70.0	100	200	400	1000
									γ-Values							
00.001	0687	0423	0284	0151	0093	0064	0038	0031	0024	0011	0008	0006	0005	0005	0005	0004
00.002	0879	0585	0419	0247	0165	0121	0077	0065	0050	0028	0021	0018	0017	0016	0015	0015
00.004	1108	0792	0601	0391	0280	0215	0147	0127	0101	0063	0047	0041	0039	0037	0036	0035
00.006	1260	0934	0734	0499	0373	0296	0212	0185	0152	0094	0072	0064	0062	0057	0056	0055
00.008	1375	1047	0839	0592	0453	0367	0269	0238	0199	0128	0099	0088	0084	0079	0076	0075
00.010	1471	1139	0928	0672	0526	0432	0322	0288	0244	0159	0124	0110	0106	0100	0098	0095
00.020	1790	1464	1248	0973	0807	0692	0554	0508	0441	0310	0248	0225	0214	0205	0201	0196
00.030	1991	1679	1465	1189	1015	0895	0739	0688	0612	0448	0366	0333	0321	0305	0299	0296
00.040	2144	1841	1634	1360	1186	1061	0900	0843	0761	0578	0481	0440	0423	0404	0395	0388
00.050	2264	1972	1771	1503	1328	1204	1040	0983	0896	0699	0588	0541	0522	0499	0489	0483
00.060	2366	2086	1891	1627	1454	1331	1164	1107	1018	0813	0694	0640	0619	0592	0580	0573
00.070	2454	2184	1992	1735	1567	1445	1280	1222	1130	0919	0795	0736	0714	0685	0670	0661
00.080	2533	2269	2084	1834	1669	1548	1384	1325	1237	1021	0892	0830	0804	0774	0759	0749
00.090	2602	2345	2164	1925	1762	1642	1481	1424	1332	1116	0986	0920	0894	0862	0844	0836
00.100	2665	2413	2242	2002	1845	1730	1572	1514	1425	1208	1073	1008	0981	0948	0931	0921
00.130	2822	2596	2434	2211	2065	1958	1810	1755	1672	1459	1324	1256	1226	1192	1176	1164
00.170	2990	2781	2636	2440	2305	2209	2075	2024	1946	1751	1620	1554	1527	1490	1473	1461
00.200	3092	2895	2764	2580	2458	2368	2241	2196	2123	1942	1820	1757	1730	1697	1678	1664
00.250	3234	3063	2942	2781	2673	2591	2485	2442	2379	2220	2115	2061	2036	2005	1990	1980
00.300	3351	3198	3089	2948	2853	2783	2689	2655	2600	2462	2374	2328	2310	2283	2270	2263
00.400	3541	3412	3330	3218	3146	3093	3025	3001	2961	2867	2809	2780	2767	2751	2744	2737
00.500	3688	3586	3518	3431	3379	3341	3291	3274	3248	3193	3161	3147	3141	3135	3132	3131
00.600	3808	3728	3674	3612	3570	3545	3513	3503	3489	3464	3452	3450	3450	3450	3453	3454
00.700	3913	3846	3808	3760	3735	3719	3703	3700	3693	3691	3697	3709	3714	3721	3724	3727
00.800	4005	3952	3921	3888	3877	3871	3868	3868	3869	3890	3914	3931	3939	3951	3955	3962
00.900	4083	4044	4022	4008	4003	4004	4013	4017	4028	4064	4101	4124	4135	4151	4157	4165
01.000	4155	4127	4112	4112	4114	4123	4143	4151	4166	4219	4266	4295	4308	4324	4333	4341
01.200	4277	4270	4275	4290	4312	4329	4365	4378	4406	4481	4545	4580	4595	4619	4629	4635
01.400	4381	4389	4405	4440	4475	4504	4551	4571	4605	4700	4775	4813	4833	4854	4870	4877
01.600	4475	4498	4523	4573	4616	4651	4709	4733	4775	4885	4968	5009	5030	5054	5066	5075
01.800	4554	4590	4620	4687	4740	4784	4851	4879	4922	5044	5131	5177	5197	5221	5235	5243
02.000	4624	4670	4717	4790	4850	4898	4974	5002	5052	5182	5272	5322	5343	5366	5378	5386
04.000	5091	5204	5301	5442	5543	5622	5736	5780	5845	6008	6102	6147	6164	6188	6198	6205
06.000	5356	5512	5627	5802	5921	6012	6137	6184	6250	6410	6498	6540	6556	6575	6582	6588
08.000	5545	5720	5852	6044	6172	6268	6399	6444	6512	6664	6748	6783	6797	6813	6822	6830
10.000	5687	5878	6020	6226	6361	6458	6589	6634	6700	6847	6920	6954	6967	6982	6991	6997
20.000	6115	6349	6521	6744	6883	6983	7105	7145	7204	7318	7377	7399	7408	7420	7424	7428
30.000	6358	6610	6783	7013	7152	7244	7357	7395	7444	7546	7593	7613	7621	7629	7632	7634
40.000	6518	6781	6966	7192	7325	7415	7519	7553	7599	7689	7731	7748	7756	7762	7766	7769

$2\alpha/RT\beta = 1.20$

p/K'	γ-Values 1.00	1.50	2.00	3.00	4.00	5.00	7.00	8.00	10.0	20.0	40.0	70.0	100	200	400	1000
00.001	0694	0428	0288	0153	0094	0065	0039	0031	0024	0011	0008	0006	0005	0005	0005	0004
00.002	0889	0592	0424	0249	0166	0122	0078	0065	0050	0028	0021	0018	0017	0016	0015	0015
00.004	1120	0801	0607	0395	0282	0217	0148	0129	0102	0063	0047	0041	0039	0037	0036	0035
00.006	1272	0944	0742	0505	0377	0299	0213	0186	0153	0094	0072	0064	0062	0057	0056	0055
00.008	1389	1059	0850	0599	0458	0371	0271	0240	0200	0128	0099	0088	0084	0079	0076	0075
00.010	1485	1152	0939	0679	0532	0437	0325	0291	0245	0160	0124	0110	0106	0100	0098	0095
00.020	1806	1479	1262	0985	0817	0701	0559	0512	0446	0312	0249	0226	0215	0205	0201	0196
00.030	2009	1696	1482	1203	1027	0905	0749	0696	0619	0452	0368	0334	0322	0306	0300	0297
00.040	2162	1860	1651	1376	1201	1076	0911	0854	0771	0583	0483	0442	0425	0406	0397	0390
00.050	2285	1992	1790	1521	1344	1220	1053	0996	0907	0705	0593	0544	0524	0501	0492	0485
00.060	2386	2107	1911	1647	1473	1348	1179	1121	1031	0821	0700	0645	0622	0596	0583	0576
00.070	2475	2203	2013	1756	1586	1465	1296	1237	1144	0929	0801	0742	0718	0690	0675	0664
00.080	2555	2290	2106	1855	1688	1569	1402	1343	1253	1034	0899	0837	0810	0779	0763	0754
00.090	2624	2367	2188	1947	1784	1664	1502	1444	1351	1132	0996	0927	0900	0867	0851	0842
00.100	2686	2438	2265	2027	1868	1753	1591	1534	1442	1224	1085	1017	0991	0956	0937	0927
00.130	2846	2619	2456	2237	2092	1984	1835	1782	1693	1481	1342	1271	1240	1206	1188	1177
00.170	3013	2807	2664	2466	2334	2238	2101	2052	1973	1777	1642	1575	1547	1509	1489	1477
00.200	3116	2925	2793	2610	2490	2397	2273	2226	2152	1970	1847	1781	1756	1720	1699	1688
00.250	3257	3090	2971	2814	2704	2627	2518	2477	2415	2257	2147	2093	2068	2037	2020	2010
00.300	3376	3225	3120	2979	2887	2819	2726	2693	2637	2503	2414	2365	2347	2320	2306	2299
00.400	3567	3444	3360	3253	3183	3132	3063	3041	3003	2912	2855	2827	2814	2799	2792	2786
00.500	3714	3617	3549	3468	3417	3379	3335	3320	3294	3240	3211	3199	3195	3189	3186	3186
00.600	3838	3757	3708	3647	3609	3586	3555	3549	3536	3513	3508	3508	3510	3513	3514	3516
00.700	3941	3877	3842	3799	3774	3760	3748	3747	3742	3748	3760	3771	3774	3784	3789	3792
00.800	4032	3983	3954	3927	3920	3914	3915	3915	3919	3948	3976	3995	4004	4018	4023	4030
00.900	4112	4077	4056	4047	4045	4049	4060	4068	4079	4123	4163	4191	4203	4219	4227	4234
01.000	4182	4157	4150	4150	4160	4170	4191	4202	4220	4280	4329	4362	4376	4395	4405	4411
01.200	4306	4302	4310	4330	4356	4376	4413	4433	4458	4545	4612	4652	4669	4692	4702	4708
01.400	4411	4424	4442	4481	4517	4552	4604	4624	4660	4763	4843	4885	4904	4929	4941	4951
01.600	4504	4531	4557	4613	4660	4701	4763	4789	4829	4946	5036	5081	5103	5126	5140	5149
01.800	4583	4622	4660	4729	4786	4832	4903	4931	4978	5107	5200	5247	5269	5295	5307	5316
02.000	4655	4706	4755	4829	4895	4948	5027	5056	5108	5244	5342	5389	5411	5437	5451	5458
04.000	5120	5241	5339	5483	5590	5671	5788	5833	5899	6063	6165	6210	6229	6250	6259	6267
06.000	5389	5547	5670	5844	5968	6061	6188	6234	6303	6466	6558	6597	6612	6630	6640	6644
08.000	5575	5756	5891	6085	6219	6315	6446	6494	6564	6719	6801	6834	6851	6866	6874	6883
10.000	5716	5910	6061	6266	6406	6505	6636	6683	6748	6894	6970	7002	7016	7031	7040	7046
20.000	6145	6387	6557	6784	6928	7025	7144	7187	7245	7362	7417	7440	7449	7459	7465	7469
30.000	6388	6644	6821	7053	7192	7284	7395	7433	7484	7583	7629	7648	7656	7664	7666	7669
40.000	6546	6816	6999	7228	7362	7451	7553	7588	7633	7723	7764	7782	7788	7796	7800	7804

$2\alpha/RT\beta = 1.30$

p/K'	γ-Values 1.00	1.50	2.00	3.00	4.00	5.00	7.00	8.00	10.0	20.0	40.0	70.0	100	200	400	1000
00.001	0701	0433	0291	0154	0095	0065	0039	0031	0024	0011	0008	0006	0005	0005	0005	0004
00.002	0898	0599	0429	0252	0168	0123	0078	0066	0050	0028	0021	0018	0017	0016	0015	0015
00.004	1130	0810	0614	0399	0285	0219	0149	0129	0103	0063	0047	0041	0039	0037	0036	0035
00.006	1284	0954	0751	0511	0381	0302	0215	0187	0153	0094	0072	0064	0062	0057	0056	0055
00.008	1402	1071	0859	0607	0463	0375	0273	0242	0201	0128	0099	0088	0084	0079	0076	0075
00.010	1499	1165	0950	0688	0538	0442	0328	0294	0246	0160	0124	0110	0106	0100	0098	0095
00.020	1822	1496	1276	0997	0827	0710	0566	0518	0449	0313	0249	0226	0215	0205	0201	0196
00.030	2026	1713	1499	1217	1040	0918	0758	0704	0625	0455	0370	0335	0323	0307	0301	0298
00.040	2182	1879	1669	1394	1215	1090	0923	0864	0779	0589	0487	0444	0426	0407	0399	0391
00.050	2305	2012	1809	1540	1362	1236	1067	1008	0917	0713	0597	0548	0529	0504	0494	0487
00.060	2408	2127	1933	1667	1493	1366	1197	1135	1045	0830	0706	0649	0627	0599	0587	0579
00.070	2496	2224	2035	1777	1607	1484	1315	1254	1160	0940	0809	0749	0725	0694	0678	0669
00.080	2576	2312	2128	1878	1711	1589	1423	1361	1268	1047	0908	0845	0817	0785	0771	0760
00.090	2645	2389	2213	1970	1806	1686	1522	1463	1370	1145	1006	0937	0910	0875	0857	0848
00.100	2709	2462	2289	2052	1892	1776	1614	1555	1464	1239	1098	1029	1002	0963	0947	0936
00.130	2869	2642	2483	2266	2119	2010	1859	1805	1719	1500	1358	1286	1255	1219	1200	1188
00.170	3037	2833	2692	2496	2361	2266	2130	2081	2002	1804	1667	1597	1566	1529	1508	1498
00.200	3142	2953	2819	2638	2520	2429	2304	2258	2185	2000	1873	1808	1779	1744	1724	1710
00.250	3283	3116	3000	2844	2738	2660	2551	2509	2448	2290	2182	2126	2101	2068	2051	2042
00.300	3402	3253	3152	3012	2922	2854	2761	2729	2675	2541	2451	2404	2384	2358	2345	2337
00.400	3594	3474	3392	3288	3217	3168	3104	3082	3045	2956	2903	2874	2863	2848	2842	2836
00.500	3741	3647	3582	3504	3455	3422	3377	3363	3341	3291	3265	3256	3250	3246	3245	3245
00.600	3863	3787	3743	3685	3649	3628	3603	3595	3585	3566	3564	3569	3570	3575	3578	3579
00.700	3971	3910	3875	3835	3819	3804	3796	3795	3793	3801	3823	3833	3841	3850	3856	3861
00.800	4061	4014	3990	3969	3962	3960	3962	3965	3973	4006	4039	4060	4073	4087	4096	4102
00.900	4140	4110	4092	4087	4087	4096	4109	4119	4130	4181	4229	4258	4271	4291	4299	4307
01.000	4210	4192	4189	4192	4201	4216	4240	4252	4272	4340	4397	4432	4444	4465	4480	4486
01.200	4336	4335	4347	4370	4401	4425	4467	4485	4515	4605	4679	4720	4741	4764	4776	4786
01.400	4440	4459	4480	4524	4564	4596	4655	4677	4716	4827	4908	4957	4978	5003	5017	5027
01.600	4533	4564	4593	4658	4708	4747	4816	4843	4887	5010	5102	5150	5172	5198	5213	5221
01.800	4612	4655	4700	4770	4831	4881	4957	4985	5035	5171	5266	5317	5339	5365	5379	5387
02.000	4683	4741	4791	4872	4938	4996	5078	5113	5163	5309	5409	5461	5481	5508	5523	5529
04.000	5151	5273	5376	5528	5638	5722	5839	5888	5956	6124	6224	6272	6291	6313	6324	6329
06.000	5421	5581	5707	5887	6016	6107	6239	6286	6356	6522	6611	6653	6669	6687	6696	6701
08.000	5606	5790	5928	6126	6264	6366	6494	6544	6615	6769	6853	6887	6901	6918	6926	6935
10.000	5747	5950	6100	6309	6448	6548	6683	6731	6797	6945	7018	7051	7063	7081	7088	7091
20.000	6175	6420	6593	6820	6967	7064	7187	7227	7285	7401	7457	7480	7488	7497	7503	7508
30.000	6416	6674	6855	7090	7231	7323	7437	7472	7524	7621	7666	7684	7693	7701	7704	7704
40.000	6575	6849	7031	7264	7396	7487	7593	7626	7669	7757	7799	7815	7819	7829	7834	7839

$2\alpha/RT\beta = 1.40$

p/K′	γ-Values															
	1.00	1.50	2.00	3.00	4.00	5.00	7.00	8.00	10.0	20.0	40.0	70.0	100	200	400	1000
00.001	0709	0438	0294	0156	0095	0065	0039	0031	0024	0011	0008	0006	0005	0005	0005	0004
00.002	0907	0606	0433	0255	0170	0124	0078	0066	0051	0028	0021	0018	0017	0016	0015	0015
00.004	1142	0819	0622	0404	0288	0221	0150	0130	0103	0063	0047	0041	0039	0037	0036	0035
00.006	1298	0964	0759	0517	0386	0306	0216	0189	0154	0095	0072	0064	0062	0057	0056	0055
00.008	1416	1083	0869	0614	0469	0379	0275	0244	0202	0129	0099	0088	0084	0079	0076	0075
00.010	1513	1178	0961	0696	0544	0447	0331	0296	0248	0161	0124	0110	0106	0100	0098	0095
00.020	1839	1511	1290	1010	0837	0719	0571	0523	0455	0315	0250	0226	0216	0205	0201	0196
00.030	2045	1732	1516	1234	1055	0930	0767	0712	0632	0458	0372	0337	0323	0308	0301	0298
00.040	2201	1898	1688	1410	1232	1104	0935	0875	0789	0594	0490	0447	0428	0409	0400	0393
00.050	2323	2032	1831	1559	1380	1251	1081	1021	0927	0720	0601	0551	0531	0506	0497	0489
00.060	2428	2150	1952	1687	1511	1384	1212	1151	1058	0839	0713	0654	0631	0603	0589	0582
00.070	2517	2247	2059	1800	1626	1502	1331	1270	1175	0952	0817	0754	0730	0698	0683	0673
00.080	2597	2335	2149	1902	1732	1610	1441	1380	1287	1059	0919	0852	0825	0792	0775	0765
00.090	2666	2414	2238	1993	1828	1708	1542	1483	1387	1159	1017	0948	0919	0883	0864	0855
00.100	2732	2486	2314	2075	1916	1800	1637	1576	1484	1257	1111	1040	1010	0972	0956	0944
00.130	2892	2668	2509	2291	2145	2036	1884	1830	1744	1521	1377	1303	1270	1234	1212	1201
00.170	3063	2860	2719	2523	2391	2297	2160	2110	2030	1829	1692	1620	1588	1549	1528	1516
00.200	3166	2980	2849	2670	2550	2462	2336	2290	2216	2029	1904	1835	1805	1770	1748	1734
00.250	3309	3146	3031	2876	2771	2692	2586	2546	2484	2326	2216	2159	2133	2101	2084	2072
00.300	3430	3282	3182	3046	2956	2888	2800	2767	2712	2580	2492	2445	2424	2399	2384	2375
00.400	3620	3505	3426	3323	3256	3210	3145	3122	3087	3002	2950	2924	2913	2898	2893	2888
00.500	3771	3678	3617	3540	3495	3462	3422	3407	3386	3341	3319	3311	3307	3306	3304	3302
00.600	3892	3820	3776	3723	3690	3671	3649	3641	3634	3621	3625	3630	3634	3637	3643	3643
00.700	3998	3941	3910	3873	3860	3850	3844	3841	3844	3858	3883	3898	3907	3920	3926	3931
00.800	4089	4049	4026	4008	4005	4004	4012	4017	4027	4063	4104	4128	4141	4155	4167	4174
00.900	4167	4139	4131	4127	4131	4141	4160	4169	4185	4242	4293	4327	4342	4361	4372	4381
01.000	4238	4224	4225	4233	4247	4263	4288	4304	4327	4399	4463	4500	4520	4543	4554	4561
01.200	4365	4368	4380	4413	4445	4473	4518	4538	4569	4669	4748	4793	4812	4838	4851	4861
01.400	4471	4492	4515	4567	4609	4645	4707	4731	4771	4891	4981	5027	5050	5077	5090	5101
01.600	4562	4599	4633	4699	4753	4797	4868	4897	4944	5077	5172	5223	5246	5272	5289	5296
01.800	4640	4692	4738	4812	4877	4930	5010	5041	5091	5233	5335	5387	5411	5438	5454	5461
02.000	4712	4772	4828	4917	4987	5046	5129	5167	5224	5373	5476	5528	5552	5580	5594	5602
04.000	5180	5310	5413	5571	5684	5768	5894	5937	6009	6185	6287	6333	6353	6374	6386	6392
06.000	5451	5617	5745	5929	6058	6155	6291	6338	6410	6576	6670	6708	6723	6743	6750	6756
08.000	5635	5823	5967	6170	6311	6411	6547	6593	6663	6821	6903	6938	6953	6969	6978	6984
10.000	5781	5984	6139	6351	6491	6594	6730	6777	6845	6994	7070	7099	7112	7127	7133	7138
20.000	6206	6454	6628	6860	7007	7105	7232	7271	7328	7443	7497	7519	7529	7537	7542	7546
30.000	6444	6706	6893	7124	7269	7359	7474	7509	7560	7657	7702	7721	7729	7735	7740	7741
40.000	6605	6881	7063	7299	7437	7523	7630	7659	7705	7791	7831	7847	7853	7862	7867	7872

$2\alpha/RT\beta = 1.50$

p/K'	1.00	1.50	2.00	3.00	4.00	5.00	7.00	8.00	10.0	20.0	40.0	70.0	100	200	400	1000
									γ-Values							
00.001	0716	0443	0297	0158	0096	0066	0039	0031	0024	0011	0008	0006	0005	0005	0005	0004
00.002	0916	0613	0439	0258	0171	0125	0079	0066	0051	0028	0021	0018	0017	0016	0015	0015
00.004	1154	0828	0630	0409	0292	0223	0151	0131	0103	0063	0047	0041	0039	0037	0036	0035
00.006	1310	0976	0768	0523	0390	0308	0218	0190	0155	0095	0072	0064	0062	0057	0056	0055
00.008	1429	1094	0879	0622	0474	0383	0278	0246	0204	0129	0099	0088	0084	0079	0076	0075
00.010	1528	1191	0973	0704	0551	0452	0335	0299	0251	0161	0124	0110	0106	0100	0098	0095
00.020	1856	1528	1305	1023	0848	0728	0578	0530	0458	0317	0251	0227	0216	0205	0201	0196
00.030	2064	1749	1535	1250	1068	0941	0777	0720	0639	0463	0374	0338	0324	0309	0302	0298
00.040	2220	1918	1706	1428	1246	1119	0945	0885	0798	0599	0493	0449	0431	0410	0402	0395
00.050	2345	2054	1850	1578	1397	1269	1096	1035	0939	0727	0607	0554	0534	0509	0499	0491
00.060	2451	2168	1974	1706	1530	1403	1228	1167	1073	0849	0718	0659	0635	0607	0593	0585
00.070	2539	2270	2079	1822	1648	1524	1350	1286	1191	0963	0825	0760	0735	0704	0687	0678
00.080	2620	2358	2175	1924	1755	1631	1460	1399	1306	1072	0929	0861	0831	0799	0782	0771
00.090	2689	2438	2260	2017	1852	1730	1566	1503	1407	1176	1028	0956	0926	0891	0872	0862
00.100	2754	2509	2339	2099	1940	1824	1657	1599	1506	1274	1125	1051	1022	0982	0965	0952
00.130	2915	2692	2535	2319	2173	2063	1911	1854	1769	1544	1396	1318	1286	1247	1227	1215
00.170	3086	2888	2749	2553	2421	2325	2189	2137	2060	1856	1715	1642	1611	1569	1549	1535
00.200	3190	3007	2875	2702	2580	2492	2368	2323	2248	2062	1933	1863	1834	1795	1775	1760
00.250	3335	3173	3061	2906	2805	2728	2623	2583	2521	2364	2254	2195	2168	2136	2119	2107
00.300	3456	3312	3215	3080	2993	2926	2838	2804	2752	2623	2534	2485	2467	2440	2425	2418
00.400	3646	3534	3458	3356	3294	3249	3187	3166	3131	3052	3002	2975	2963	2953	2946	2943
00.500	3798	3709	3651	3579	3533	3505	3465	3452	3435	3395	3376	3369	3369	3364	3367	3366
00.600	3921	3852	3809	3761	3732	3714	3696	3692	3685	3679	3686	3692	3699	3708	3711	3714
00.700	4027	3975	3944	3915	3902	3893	3889	3892	3896	3917	3946	3966	3975	3989	3997	4005
00.800	4117	4080	4061	4050	4047	4051	4058	4069	4077	4123	4169	4195	4211	4231	4242	4248
00.900	4196	4173	4165	4169	4176	4187	4211	4219	4239	4306	4358	4398	4414	4437	4448	4456
01.000	4271	4256	4261	4271	4292	4309	4342	4357	4384	4462	4533	4573	4592	4616	4631	4638
01.200	4394	4403	4418	4454	4486	4521	4570	4588	4626	4732	4820	4866	4886	4913	4929	4939
01.400	4499	4526	4552	4610	4656	4695	4759	4786	4831	4953	5048	5100	5125	5151	5169	5178
01.600	4589	4630	4672	4740	4800	4846	4921	4952	5002	5140	5241	5295	5318	5347	5363	5370
01.800	4669	4724	4776	4857	4923	4979	5064	5098	5150	5301	5406	5461	5484	5511	5526	5534
02.000	4744	4807	4863	4959	5033	5096	5186	5224	5282	5435	5546	5602	5623	5653	5668	5675
04.000	5212	5345	5452	5614	5727	5821	5948	5993	6070	6243	6347	6395	6414	6435	6447	6453
06.000	5481	5648	5782	5974	6102	6203	6341	6390	6464	6631	6721	6764	6779	6798	6806	6811
08.000	5665	5859	6002	6211	6356	6458	6596	6644	6716	6872	6954	6989	7003	7020	7027	7034
10.000	5811	6021	6176	6392	6536	6641	6779	6823	6896	7040	7115	7147	7159	7173	7181	7184
20.000	6237	6487	6662	6898	7047	7148	7273	7313	7370	7483	7537	7559	7568	7575	7582	7584
30.000	6472	6741	6925	7161	7305	7398	7511	7547	7596	7692	7739	7756	7763	7770	7775	7778
40.000	6636	6913	7098	7335	7471	7557	7660	7695	7739	7824	7863	7878	7884	7894	7899	7903

$2\alpha/RT\beta = 1.60$

p/K'	γ-Values 1.00	1.50	2.00	3.00	4.00	5.00	7.00	8.00	10.0	20.0	40.0	70.0	100	200	400	1000
00.001	0723	0448	0301	0159	0097	0066	0039	0031	0024	0011	0008	0006	0005	0005	0005	0004
00.002	0925	0619	0444	0261	0173	0126	0079	0066	0051	0028	0021	0018	0017	0016	0015	0015
00.004	1165	0838	0637	0414	0295	0226	0152	0131	0104	0063	0047	0041	0039	0037	0036	0035
00.006	1323	0988	0778	0529	0395	0312	0220	0191	0156	0095	0072	0064	0062	0057	0056	0055
00.008	1442	1107	0890	0629	0480	0388	0281	0248	0205	0129	0099	0088	0084	0079	0076	0075
00.010	1541	1204	0985	0713	0558	0457	0338	0301	0252	0162	0124	0110	0106	0100	0098	0095
00.020	1873	1544	1321	1036	0858	0737	0586	0536	0463	0319	0252	0227	0216	0205	0201	0196
00.030	2083	1769	1552	1265	1082	0954	0786	0730	0647	0465	0375	0339	0325	0309	0302	0298
00.040	2240	1938	1725	1444	1264	1132	0959	0898	0808	0605	0496	0452	0433	0412	0404	0397
00.050	2365	2075	1872	1598	1416	1286	1109	1050	0953	0734	0612	0558	0537	0512	0501	0494
00.060	2471	2192	1995	1729	1551	1421	1244	1182	1087	0858	0726	0665	0640	0611	0597	0589
00.070	2561	2292	2102	1844	1670	1545	1368	1305	1207	0975	0833	0767	0741	0709	0692	0682
00.080	2641	2381	2199	1948	1778	1654	1480	1419	1322	1087	0938	0869	0839	0805	0788	0777
00.090	2712	2462	2285	2042	1876	1754	1586	1525	1426	1191	1041	0968	0936	0899	0879	0870
00.100	2777	2533	2364	2126	1966	1848	1682	1623	1528	1291	1139	1065	1033	0993	0974	0963
00.130	2941	2721	2561	2347	2200	2091	1939	1883	1794	1566	1414	1336	1302	1262	1240	1228
00.170	3111	2914	2776	2582	2453	2358	2221	2172	2092	1886	1743	1667	1633	1592	1570	1556
00.200	3216	3033	2908	2733	2613	2525	2399	2355	2283	2095	1962	1893	1863	1822	1802	1786
00.250	3361	3203	3091	2941	2839	2764	2660	2619	2556	2401	2290	2233	2207	2172	2153	2142
00.300	3482	3343	3246	3115	3028	2964	2877	2845	2793	2664	2576	2530	2510	2482	2469	2461
00.400	3673	3562	3490	3396	3334	3288	3231	3209	3176	3101	3053	3030	3020	3007	3003	2999
00.500	3826	3740	3685	3618	3574	3546	3510	3500	3482	3449	3432	3432	3428	3428	3430	3431
00.600	3949	3885	3842	3800	3775	3758	3744	3739	3735	3735	3747	3759	3764	3776	3782	3784
00.700	4053	4007	3977	3952	3943	3941	3939	3941	3948	3977	4011	4036	4045	4062	4071	4078
00.800	4144	4113	4096	4089	4091	4098	4112	4119	4133	4185	4235	4269	4283	4305	4317	4323
00.900	4226	4207	4201	4208	4220	4234	4260	4273	4295	4366	4431	4469	4489	4514	4527	4533
01.000	4299	4289	4296	4313	4335	4357	4395	4410	4440	4528	4601	4647	4665	4693	4706	4716
01.200	4424	4439	4455	4497	4536	4566	4622	4646	4686	4800	4886	4937	4961	4991	5008	5017
01.400	4530	4560	4592	4652	4703	4745	4813	4841	4887	5018	5121	5176	5197	5227	5243	5255
01.600	4621	4667	4710	4784	4844	4896	4976	5009	5060	5204	5313	5368	5394	5422	5439	5447
01.800	4702	4760	4812	4899	4971	5027	5117	5149	5208	5367	5476	5532	5556	5585	5601	5609
02.000	4775	4844	4904	5005	5082	5146	5243	5278	5337	5502	5614	5671	5693	5724	5739	5748
04.000	5244	5382	5488	5657	5777	5870	6001	6047	6123	6301	6409	6454	6474	6497	6508	6514
06.000	5509	5686	5819	6015	6150	6251	6388	6442	6513	6688	6780	6818	6834	6853	6862	6869
08.000	5698	5895	6045	6255	6399	6502	6641	6693	6767	6925	7005	7038	7055	7070	7077	7084
10.000	5839	6054	6213	6433	6581	6687	6826	6873	6941	7090	7166	7194	7205	7219	7227	7229
20.000	6267	6520	6701	6936	7085	7188	7313	7353	7409	7524	7575	7596	7605	7615	7617	7620
30.000	6501	6774	6962	7201	7340	7435	7547	7585	7634	7729	7770	7789	7797	7806	7810	7814
40.000	6666	6944	7132	7368	7507	7594	7695	7728	7773	7856	7894	7911	7917	7925	7930	7934

$2\alpha/RT\beta = 1.70$

	γ-Values														
p/K′	1.00	1.50	2.00	3.00	4.00	5.00	7.00	8.00	10.0	20.0	40.0	70.0	100	200	400	1000
00.001	0731	0454	0305	0161	0098	0067	0039	0032	0024	0011	0008	0006	0005	0005	0005	0004
00.002	0934	0627	0449	0264	0175	0127	0080	0067	0051	0028	0021	0018	0017	0016	0015	0015
00.004	1176	0848	0645	0418	0298	0228	0153	0132	0104	0063	0047	0041	0039	0037	0036	0035
00.006	1336	0998	0787	0536	0399	0315	0222	0193	0157	0095	0072	0064	0062	0057	0056	0055
00.008	1457	1119	0901	0637	0486	0392	0284	0250	0206	0129	0099	0088	0084	0079	0076	0075
00.010	1557	1218	0996	0723	0565	0463	0341	0304	0254	0162	0125	0110	0106	0100	0098	0095
00.020	1890	1560	1336	1049	0870	0747	0592	0541	0469	0321	0252	0228	0216	0205	0201	0196
00.030	2101	1787	1570	1280	1096	0965	0797	0739	0654	0470	0378	0340	0326	0310	0302	0298
00.040	2260	1956	1745	1463	1279	1149	0971	0910	0818	0611	0500	0454	0434	0414	0405	0398
00.050	2386	2095	1892	1617	1434	1303	1125	1063	0965	0743	0617	0563	0541	0514	0504	0497
00.060	2493	2213	2018	1751	1571	1441	1264	1200	1102	0868	0732	0670	0644	0615	0601	0592
00.070	2584	2316	2126	1866	1692	1565	1387	1323	1225	0986	0843	0773	0747	0714	0697	0687
00.080	2664	2405	2222	1972	1802	1678	1502	1440	1341	1100	0949	0879	0847	0812	0794	0783
00.090	2735	2486	2310	2066	1899	1777	1609	1546	1449	1207	1053	0978	0946	0907	0887	0877
00.100	2801	2558	2387	2151	1992	1873	1708	1646	1550	1310	1153	1076	1044	1002	0983	0971
00.130	2964	2746	2589	2374	2229	2118	1966	1911	1821	1590	1435	1355	1318	1276	1255	1242
00.170	3134	2942	2804	2614	2484	2390	2252	2202	2122	1915	1769	1691	1658	1615	1591	1578
00.200	3241	3062	2936	2763	2646	2559	2436	2388	2316	2128	1995	1924	1892	1850	1828	1814
00.250	3387	3231	3123	2974	2874	2800	2695	2657	2595	2440	2329	2269	2244	2208	2190	2179
00.300	3509	3372	3277	3150	3065	3001	2917	2884	2834	2707	2622	2573	2554	2526	2514	2506
00.400	3701	3596	3525	3432	3373	3330	3275	3254	3224	3151	3108	3085	3076	3064	3060	3056
00.500	3853	3771	3720	3653	3616	3589	3558	3546	3531	3503	3490	3493	3493	3495	3495	3497
00.600	3976	3916	3878	3838	3816	3804	3792	3789	3786	3793	3813	3822	3836	3846	3854	3857
00.700	4083	4038	4013	3993	3985	3986	3991	3994	4002	4038	4079	4104	4119	4136	4147	4155
00.800	4174	4146	4134	4128	4136	4143	4162	4170	4189	4249	4306	4340	4356	4383	4395	4402
00.900	4256	4240	4238	4249	4265	4282	4309	4326	4348	4431	4501	4544	4563	4591	4605	4613
01.000	4327	4325	4333	4355	4379	4405	4447	4465	4496	4592	4672	4722	4743	4770	4784	4795
01.200	4454	4472	4492	4540	4582	4616	4677	4699	4740	4864	4961	5013	5036	5068	5086	5096
01.400	4559	4593	4629	4694	4749	4796	4868	4895	4947	5087	5187	5249	5275	5304	5322	5330
01.600	4651	4697	4748	4826	4891	4948	5029	5063	5121	5272	5382	5441	5467	5497	5514	5523
01.800	4732	4794	4851	4942	5020	5078	5172	5208	5269	5429	5544	5605	5630	5659	5676	5685
02.000	4807	4877	4942	5048	5130	5197	5295	5335	5395	5570	5685	5744	5767	5796	5812	5823
04.000	5274	5416	5530	5700	5824	5920	6052	6103	6180	6362	6468	6517	6536	6558	6568	6573
06.000	5540	5718	5859	6057	6197	6300	6441	6491	6570	6740	6831	6873	6887	6907	6916	6924
08.000	5729	5930	6082	6298	6443	6550	6692	6740	6815	6974	7056	7090	7103	7120	7125	7130
10.000	5870	6086	6248	6475	6626	6732	6872	6921	6992	7136	7210	7239	7252	7266	7272	7274
20.000	6296	6554	6736	6977	7127	7230	7349	7391	7450	7562	7615	7635	7644	7651	7656	7656
30.000	6530	6807	6993	7234	7379	7473	7585	7621	7669	7762	7807	7822	7831	7839	7844	7849
40.000	6693	6974	7165	7402	7539	7629	7731	7765	7805	7889	7926	7942	7948	7957	7961	7964

$2\alpha/RT\beta = 1.80$

p/K′	γ-Values 1.00	1.50	2.00	3.00	4.00	5.00	7.00	8.00	10.0	20.0	40.0	70.0	100	200	400	1000
00.001	0739	0459	0308	0163	0099	0067	0039	0032	0024	0011	0008	0006	0005	0005	0005	0004
00.002	0945	0634	0455	0267	0177	0128	0080	0067	0051	0028	0021	0018	0017	0016	0015	0015
00.004	1189	0857	0653	0424	0302	0230	0155	0133	0105	0063	0047	0041	0039	0037	0036	0035
00.006	1349	1010	0796	0543	0404	0319	0224	0194	0158	0095	0072	0064	0062	0057	0056	0055
00.008	1472	1132	0911	0645	0492	0396	0286	0253	0208	0130	0099	0088	0084	0079	0076	0075
00.010	1572	1230	1008	0732	0572	0468	0345	0307	0256	0163	0125	0111	0106	0100	0098	0095
00.020	1908	1577	1351	1062	0881	0756	0599	0548	0473	0323	0253	0228	0216	0205	0201	0196
00.030	2121	1806	1587	1296	1110	0980	0807	0748	0662	0473	0380	0341	0327	0310	0302	0298
00.040	2280	1979	1766	1482	1297	1164	0984	0921	0830	0617	0502	0456	0436	0415	0406	0398
00.050	2407	2117	1914	1637	1453	1321	1141	1079	0977	0751	0622	0566	0544	0517	0507	0500
00.060	2513	2236	2039	1772	1592	1461	1280	1216	1116	0879	0739	0675	0649	0619	0604	0596
00.070	2604	2338	2149	1890	1714	1586	1407	1341	1243	1000	0850	0780	0754	0719	0702	0691
00.080	2687	2430	2247	1994	1825	1698	1524	1459	1361	1115	0960	0887	0855	0819	0800	0789
00.090	2759	2510	2335	2092	1925	1802	1632	1569	1470	1223	1066	0989	0956	0916	0894	0884
00.100	2825	2584	2414	2178	2017	1899	1731	1670	1572	1327	1167	1088	1054	1011	0993	0979
00.130	2989	2771	2617	2401	2258	2148	1993	1937	1848	1613	1455	1373	1335	1294	1269	1255
00.170	3161	2969	2832	2645	2514	2420	2286	2235	2155	1946	1797	1719	1682	1638	1614	1598
00.200	3267	3090	2964	2793	2679	2593	2470	2423	2350	2163	2027	1954	1921	1881	1857	1842
00.250	3413	3260	3154	3008	2910	2834	2735	2697	2636	2480	2369	2310	2283	2247	2230	2216
00.300	3534	3402	3311	3186	3103	3041	2957	2926	2875	2752	2668	2623	2601	2575	2561	2551
00.400	3729	3626	3558	3467	3410	3373	3318	3298	3267	3203	3162	3142	3131	3124	3120	3118
00.500	3881	3804	3755	3693	3658	3633	3605	3595	3580	3559	3555	3556	3559	3564	3567	3566
00.600	4004	3950	3915	3879	3858	3851	3839	3840	3839	3852	3877	3895	3903	3919	3927	3933
00.700	4110	4072	4051	4033	4030	4032	4042	4047	4058	4100	4144	4176	4191	4212	4225	4235
00.800	4204	4177	4171	4171	4181	4192	4213	4226	4244	4311	4373	4415	4434	4459	4474	4482
00.900	4286	4276	4275	4291	4309	4329	4365	4382	4409	4496	4569	4618	4638	4668	4686	4694
01.000	4357	4357	4371	4396	4423	4454	4499	4519	4554	4658	4745	4797	4820	4849	4867	4876
01.200	4484	4506	4531	4584	4630	4667	4731	4755	4801	4931	5032	5090	5116	5147	5165	5175
01.400	4590	4627	4669	4737	4793	4844	4922	4954	5004	5151	5263	5324	5349	5381	5401	5408
01.600	4683	4736	4785	4870	4939	4995	5086	5120	5176	5339	5457	5514	5540	5573	5590	5599
01.800	4764	4829	4889	4988	5065	5128	5229	5263	5327	5496	5617	5678	5703	5733	5749	5760
02.000	4834	4913	4980	5092	5176	5246	5351	5391	5457	5633	5754	5815	5842	5868	5886	5896
04.000	5306	5453	5569	5746	5873	5972	6102	6156	6233	6423	6532	6577	6595	6617	6628	6634
06.000	5571	5755	5897	6101	6244	6349	6492	6544	6621	6794	6886	6926	6941	6959	6970	6976
08.000	5761	5966	6121	6340	6491	6597	6740	6790	6866	7023	7104	7138	7150	7167	7174	7176
10.000	5902	6121	6287	6515	6670	6775	6916	6965	7036	7186	7258	7286	7298	7312	7316	7321
20.000	6326	6586	6770	7015	7168	7267	7394	7434	7493	7599	7650	7673	7681	7688	7692	7693
30.000	6562	6839	7030	7272	7417	7509	7623	7657	7704	7797	7837	7855	7863	7872	7877	7882
40.000	6722	7006	7200	7437	7574	7664	7767	7796	7840	7919	7957	7972	7978	7986	7990	7993

$2\alpha/RT\beta = 1.90$

p/K′	1.00	1.50	2.00	3.00	4.00	5.00	7.00	8.00	10.0	20.0	40.0	70.0	100	200	400	1000
								γ-Values								
00.001	0747	0464	0312	0165	0100	0068	0040	0032	0024	0011	0008	0006	0005	0005	0005	0004
00.002	0955	0642	0461	0270	0179	0129	0080	0067	0051	0028	0021	0018	0017	0016	0015	0015
00.004	1202	0867	0661	0430	0305	0232	0156	0134	0105	0063	0047	0041	0039	0037	0036	0035
00.006	1363	1021	0806	0550	0410	0322	0226	0196	0159	0095	0072	0064	0062	0057	0056	0055
00.008	1486	1145	0923	0654	0498	0402	0289	0254	0209	0130	0099	0088	0084	0079	0076	0075
00.010	1587	1245	1020	0741	0579	0473	0349	0310	0258	0163	0125	0111	0106	0100	0098	0095
00.020	1926	1595	1368	1077	0894	0766	0607	0554	0478	0324	0254	0228	0216	0205	0201	0196
00.030	2139	1825	1606	1314	1126	0992	0818	0758	0669	0478	0382	0342	0328	0310	0302	0298
00.040	2301	1998	1785	1501	1314	1181	0999	0934	0840	0623	0505	0458	0438	0416	0406	0398
00.050	2429	2139	1936	1659	1473	1339	1157	1092	0991	0759	0627	0568	0546	0519	0509	0501
00.060	2535	2260	2062	1794	1613	1482	1298	1232	1132	0889	0746	0680	0653	0623	0608	0600
00.070	2626	2362	2173	1915	1738	1608	1427	1361	1261	1012	0859	0788	0760	0725	0707	0696
00.080	2709	2453	2272	2020	1850	1722	1545	1480	1381	1131	0971	0895	0863	0826	0807	0795
00.090	2782	2536	2359	2117	1951	1827	1655	1592	1491	1242	1078	0998	0965	0925	0903	0893
00.100	2846	2607	2438	2204	2045	1924	1756	1695	1596	1346	1182	1100	1066	1023	1003	0990
00.130	3011	2798	2644	2432	2286	2177	2024	1967	1876	1640	1476	1391	1353	1308	1286	1271
00.170	3187	2995	2862	2675	2547	2454	2318	2269	2186	1979	1826	1746	1710	1665	1637	1623
00.200	3293	3117	2995	2828	2713	2627	2505	2459	2388	2199	2062	1988	1956	1911	1887	1873
00.250	3438	3290	3185	3041	2945	2875	2772	2734	2676	2522	2411	2351	2324	2286	2270	2255
00.300	3561	3430	3342	3221	3139	3079	2997	2968	2919	2798	2716	2671	2650	2623	2610	2600
00.400	3758	3657	3592	3507	3452	3413	3363	3345	3319	3255	3219	3200	3195	3186	3182	3181
00.500	3909	3837	3791	3730	3699	3675	3652	3646	3632	3618	3617	3623	3628	3633	3638	3640
00.600	4032	3980	3948	3918	3901	3895	3891	3891	3892	3915	3944	3967	3978	3995	4003	4012
00.700	4142	4105	4088	4074	4077	4078	4091	4096	4112	4163	4217	4251	4267	4293	4305	4314
00.800	4235	4212	4206	4211	4225	4241	4265	4277	4299	4377	4448	4488	4509	4538	4554	4562
00.900	4312	4310	4314	4333	4355	4379	4418	4437	4465	4560	4643	4695	4720	4748	4766	4776
01.000	4387	4394	4407	4441	4471	4503	4552	4576	4611	4724	4816	4871	4896	4930	4949	4960
01.200	4512	4539	4569	4626	4675	4716	4784	4814	4861	4998	5106	5167	5192	5226	5244	5257
01.400	4618	4662	4705	4782	4841	4896	4975	5010	5064	5221	5336	5400	5424	5461	5478	5487
01.600	4713	4771	4825	4914	4989	5046	5141	5177	5239	5404	5526	5592	5619	5647	5666	5677
01.800	4795	4865	4926	5031	5115	5179	5282	5323	5387	5564	5688	5749	5774	5807	5825	5835
02.000	4866	4949	5020	5135	5224	5298	5403	5449	5519	5700	5826	5884	5913	5941	5957	5970
04.000	5336	5489	5608	5788	5921	6020	6162	6209	6291	6480	6587	6636	6656	6676	6688	6695
06.000	5602	5794	5934	6144	6293	6396	6543	6595	6674	6847	6938	6979	6994	7011	7022	7026
08.000	5791	6001	6157	6382	6533	6644	6792	6839	6913	7074	7155	7185	7198	7212	7222	7223
10.000	5932	6156	6325	6557	6711	6821	6966	7014	7084	7230	7301	7331	7342	7355	7362	7366
20.000	6356	6617	6806	7054	7206	7305	7435	7473	7530	7639	7688	7708	7716	7725	7727	7730
30.000	6591	6871	7062	7307	7453	7544	7657	7692	7739	7830	7872	7887	7895	7904	7910	7913
40.000	6750	7037	7232	7471	7609	7697	7799	7828	7872	7950	7986	8000	8007	8012	8019	8022

$2\alpha/RT\beta = 2.00$

p/K′	γ-Values 1.00	1.50	2.00	3.00	4.00	5.00	7.00	8.00	10.0	20.0	40.0	70.0	100	200	400	1000
00.001	0752	0471	0319	0170	0106	0074	0046	0039	0032	0021	0019	0019	0019	0019	0019	0019
00.002	0961	0650	0469	0278	0186	0137	0088	0077	0060	0040	0033	0031	0030	0029	0029	0029
00.004	1207	0876	0672	0437	0313	0240	0164	0142	0115	0073	0058	0053	0051	0049	0048	0048
00.006	1369	1032	0817	0561	0419	0331	0234	0205	0167	0105	0082	0075	0072	0069	0068	0068
00.008	1493	1154	0933	0664	0509	0411	0298	0264	0216	0138	0108	0097	0092	0088	0087	0086
00.010	1594	1255	1030	0753	0590	0485	0357	0319	0266	0169	0131	0117	0112	0106	0102	0100
00.020	1932	1606	1381	1088	0905	0781	0618	0563	0487	0330	0258	0230	0219	0207	0202	0199
00.030	2147	1836	1617	1328	1138	1006	0830	0769	0678	0483	0386	0346	0331	0313	0304	0301
00.040	2307	2011	1800	1517	1329	1194	1011	0947	0849	0629	0511	0460	0441	0420	0408	0402
00.050	2434	2151	1949	1673	1489	1356	1171	1104	1003	0767	0633	0573	0551	0525	0511	0503
00.060	2541	2270	2076	1809	1628	1498	1315	1248	1145	0899	0751	0685	0659	0627	0612	0603
00.070	2634	2373	2185	1927	1753	1623	1443	1379	1276	1023	0868	0795	0765	0731	0715	0703
00.080	2715	2464	2284	2035	1866	1739	1562	1497	1396	1142	0981	0902	0871	0832	0815	0803
00.090	2786	2545	2373	2133	1965	1845	1672	1609	1508	1255	1089	1009	0974	0933	0913	0900
00.100	2852	2619	2453	2220	2061	1942	1774	1710	1613	1362	1195	1111	1075	1032	1010	0996
00.130	3018	2808	2657	2449	2304	2197	2042	1986	1894	1658	1493	1405	1368	1319	1294	1279
00.170	3190	3006	2874	2692	2566	2472	2339	2287	2207	1999	1847	1766	1728	1679	1652	1639
00.200	3296	3128	3009	2842	2732	2647	2526	2482	2412	2222	2086	2010	1976	1931	1907	1892
00.250	3443	3298	3196	3058	2961	2893	2793	2758	2700	2548	2439	2380	2351	2315	2295	2283
00.300	3563	3438	3352	3234	3156	3098	3021	2991	2945	2828	2747	2703	2682	2656	2640	2633
00.400	3758	3664	3602	3521	3468	3433	3388	3369	3343	3287	3253	3238	3232	3226	3222	3220
00.500	3910	3841	3798	3744	3715	3696	3674	3667	3659	3649	3654	3662	3668	3677	3683	3684
00.600	4033	3985	3959	3930	3918	3913	3912	3915	3920	3946	3981	4006	4019	4037	4049	4056
00.700	4139	4108	4094	4087	4088	4094	4112	4121	4136	4194	4252	4290	4309	4334	4348	4358
00.800	4231	4216	4213	4223	4237	4255	4285	4301	4323	4407	4482	4529	4550	4581	4597	4608
00.900	4310	4309	4318	4342	4369	4393	4438	4456	4489	4590	4678	4732	4758	4789	4806	4817
01.000	4384	4395	4411	4448	4485	4518	4571	4595	4635	4751	4852	4908	4935	4968	4986	4998
01.200	4509	4540	4571	4634	4685	4731	4802	4831	4879	5023	5136	5197	5225	5260	5278	5290
01.400	4614	4662	4707	4787	4852	4907	4992	5026	5083	5242	5362	5427	5455	5489	5505	5517
01.600	4706	4768	4824	4921	4996	5056	5154	5192	5255	5427	5550	5613	5641	5674	5691	5702
01.800	4788	4863	4927	5035	5120	5189	5293	5335	5402	5580	5707	5770	5796	5830	5845	5855
02.000	4858	4945	5019	5138	5230	5305	5416	5460	5530	5717	5842	5904	5930	5959	5974	5985
04.000	5323	5478	5601	5785	5917	6017	6159	6211	6290	6481	6588	6633	6653	6674	6684	6690
06.000	5589	5779	5925	6134	6281	6389	6536	6588	6666	6838	6928	6966	6978	6995	7004	7011
08.000	5774	5986	6144	6369	6521	6630	6777	6826	6900	7055	7136	7167	7179	7194	7200	7205
10.000	5913	6140	6309	6541	6697	6805	6947	6995	7065	7209	7278	7307	7318	7330	7336	7340
20.000	6332	6597	6785	7028	7179	7280	7404	7445	7498	7603	7651	7672	7678	7687	7692	7694
30.000	6563	6842	7035	7277	7419	7513	7622	7656	7702	7790	7828	7844	7850	7858	7861	7864
40.000	6722	7009	7201	7438	7574	7660	7758	7789	7831	7905	7940	7953	7960	7965	7968	7970

$2\alpha/RT\beta = 2.10$

p/K'	1.00	1.50	2.00	3.00	4.00	5.00	7.00	8.00	10.0	20.0	40.0	70.0	100	200	400	1000
	γ-Values															
00.001	0766	0479	0324	0174	0109	0076	0047	0040	0033	0022	0019	0019	0019	0019	0019	0019
00.002	0979	0659	0476	0283	0189	0140	0090	0078	0061	0041	0033	0031	0030	0029	0029	0029
00.004	1230	0888	0682	0445	0319	0246	0168	0146	0118	0075	0060	0054	0052	0049	0048	0048
00.006	1392	1045	0829	0571	0427	0337	0239	0210	0170	0108	0085	0077	0074	0071	0069	0068
00.008	1520	1169	0947	0675	0518	0419	0305	0270	0222	0142	0112	0101	0097	0092	0090	0088
00.010	1622	1271	1044	0765	0601	0494	0365	0326	0272	0175	0136	0123	0119	0113	0109	0108
00.020	1965	1625	1399	1105	0920	0794	0630	0575	0497	0338	0267	0239	0229	0217	0212	0209
00.030	2181	1857	1639	1349	1157	1023	0845	0783	0692	0493	0395	0355	0340	0323	0314	0311
00.040	2346	2033	1823	1539	1350	1214	1029	0964	0866	0641	0521	0469	0450	0429	0417	0412
00.050	2473	2174	1973	1698	1512	1378	1192	1124	1022	0780	0643	0583	0559	0533	0520	0512
00.060	2582	2294	2101	1834	1653	1522	1337	1269	1165	0914	0762	0695	0668	0635	0619	0610
00.070	2675	2398	2212	1954	1779	1649	1468	1402	1298	1040	0881	0806	0774	0737	0720	0706
00.080	2760	2490	2312	2063	1893	1766	1589	1521	1420	1162	0996	0913	0880	0840	0820	0806
00.090	2829	2571	2400	2161	1995	1873	1701	1635	1533	1276	1105	1020	0986	0943	0920	0906
00.100	2898	2646	2480	2250	2090	1972	1803	1740	1640	1386	1213	1126	1088	1043	1020	1006
00.130	3064	2836	2689	2479	2336	2229	2074	2018	1926	1686	1516	1427	1385	1335	1311	1295
00.170	3241	3035	2906	2725	2600	2508	2376	2325	2243	2033	1878	1794	1755	1705	1678	1662
00.200	3348	3159	3041	2878	2768	2685	2565	2521	2452	2262	2123	2047	2011	1965	1940	1923
00.250	3495	3329	3230	3094	3001	2931	2836	2800	2743	2591	2483	2423.	2395	2359	2339	2325
00.300	3618	3470	3386	3273	3195	3141	3065	3035	2991	2877	2798	2752	2734	2709	2694	2686
00.400	3816	3697	3637	3561	3512	3477	3433	3417	3394	3341	3313	3300	3296	3291	3290	3289
00.500	3966	3875	3834	3787	3760	3742	3724	3718	3712	3709	3720	3731	3739	3750	3758	3761
00.600	4094	4021	3995	3973	3964	3961	3963	3966	3974	4009	4051	4080	4095	4117	4130	4138
00.700	4203	4143	4134	4129	4135	4144	4165	4175	4192	4260	4324	4367	4387	4416	4431	4441
00.800	4294	4249	4252	4265	4284	4304	4339	4354	4382	4472	4555	4606	4630	4662	4681	4692
00.900	4374	4345	4356	4386	4416	4444	4489	4512	4548	4658	4753	4809	4837	4872	4891	4902
01.000	4446	4431	4449	4492	4532	4567	4626	4650	4693	4819	4925	4986	5013	5050	5070	5080
01.200	4574	4576	4612	4677	4732	4780	4857	4889	4940	5091	5210	5274	5304	5340	5359	5371
01.400	4684	4698	4747	4831	4901	4959	5048	5084	5144	5311	5436	5504	5532	5566	5585	5596
01.600	4776	4804	4865	4964	5043	5109	5209	5251	5314	5494	5623	5688	5717	5750	5767	5778
01.800	4856	4898	4966	5079	5168	5241	5350	5393	5461	5649	5777	5844	5869	5902	5918	5928
02.000	4927	4980	5058	5183	5279	5356	5473	5518	5589	5781	5911	5975	6000	6032	6046	6056
04.000	5400	5515	5642	5828	5965	6066	6212	6264	6345	6539	6647	6692	6709	6732	6741	6748
06.000	5668	5814	5963	6178	6328	6437	6585	6637	6715	6889	6979	7017	7031	7046	7055	7060
08.000	5852	6022	6181	6411	6566	6676	6823	6873	6948	7105	7181	7213	7226	7239	7246	7249
10.000	5993	6176	6346	6583	6741	6850	6992	7040	7111	7253	7322	7349	7361	7372	7378	7383
20.000	6414	6629	6819	7065	7217	7318	7443	7482	7536	7640	7686	7706	7713	7722	7726	7728
30.000	6650	6874	7068	7312	7455	7549	7656	7691	7736	7823	7860	7876	7881	7888	7891	7894
40.000	6807	7039	7231	7470	7608	7693	7791	7821	7861	7936	7968	7982	7988	7994	7996	7998

$2\alpha/RT\beta = 2.20$

p/K′	1.00	1.50	2.00	3.00	4.00	5.00	7.00	8.00	10.0	20.0	40.0	70.0	100	200	400	1000
	γ-Values															
00.001	0769	0484	0328	0176	0110	0077	0047	0040	0033	0016	0019	0019	0019	0019	0019	0019
00.002	0982	0667	0482	0287	0191	0141	0090	0079	0061	0036	0033	0031	0030	0029	0029	0029
00.004	1233	0898	0691	0450	0323	0248	0169	0146	0118	0072	0060	0054	0052	0049	0048	0048
00.006	1398	1057	0839	0578	0432	0341	0241	0211	0172	0105	0085	0077	0074	0071	0069	0068
00.008	1524	1182	0959	0684	0525	0424	0308	0272	0223	0140	0112	0101	0097	0092	0090	0088
00.010	1626	1285	1057	0775	0608	0501	0369	0330	0275	0174	0136	0123	0119	0113	0109	0108
00.020	1970	1643	1416	1120	0933	0805	0638	0582	0503	0339	0267	0240	0229	0217	0212	0209
00.030	2188	1877	1658	1366	1173	1038	0856	0793	0700	0497	0397	0356	0341	0323	0314	0311
00.040	2349	2053	1845	1558	1370	1232	1044	0978	0877	0648	0524	0471	0452	0430	0418	0412
00.050	2479	2197	1996	1719	1533	1398	1209	1141	1037	0792	0649	0586	0562	0535	0521	0513
00.060	2588	2319	2125	1857	1676	1543	1357	1287	1183	0930	0771	0700	0673	0639	0623	0613
00.070	2681	2423	2237	1979	1803	1671	1490	1423	1318	1059	0891	0814	0781	0744	0727	0714
00.080	2762	2515	2337	2088	1919	1792	1613	1545	1442	1182	1008	0924	0890	0849	0829	0816
00.090	2835	2598	2427	2189	2022	1900	1726	1660	1558	1301	1121	1034	0997	0953	0931	0917
00.100	2901	2673	2508	2277	2118	1999	1831	1767	1666	1412	1231	1141	1102	1056	1032	1017
00.130	3068	2863	2717	2510	2368	2260	2106	2049	1957	1723	1542	1449	1407	1355	1328	1311
00.170	3242	3064	2936	2758	2635	2542	2411	2359	2281	2079	1912	1825	1785	1733	1705	1688
00.200	3350	3188	3072	2912	2803	2721	2603	2558	2489	2313	2163	2084	2047	2000	1974	1957
00.250	3498	3360	3262	3129	3039	2972	2875	2841	2786	2655	2531	2471	2442	2406	2384	2371
00.300	3619	3501	3420	3309	3235	3181	3109	3080	3036	2947	2849	2808	2788	2765	2750	2742
00.400	3815	3729	3674	3599	3552	3522	3481	3466	3443	3424	3374	3366	3364	3361	3360	3360
00.500	3968	3907	3870	3826	3803	3789	3773	3771	3766	3799	3788	3805	3813	3829	3836	3841
00.600	4093	4053	4032	4013	4008	4008	4016	4020	4031	4104	4122	4156	4175	4198	4213	4223
00.700	4198	4176	4170	4171	4180	4192	4215	4230	4251	4361	4398	4445	4467	4498	4516	4528
00.800	4292	4286	4289	4307	4330	4352	4393	4411	4441	4578	4632	4685	4712	4747	4766	4779
00.900	4371	4379	4395	4428	4462	4493	4545	4568	4606	4766	4828	4890	4918	4953	4974	4987
01.000	4445	4464	4488	4536	4580	4618	4681	4707	4752	4930	5001	5066	5096	5133	5152	5165
01.200	4570	4610	4650	4721	4781	4831	4912	4946	5002	5205	5285	5353	5383	5421	5440	5452
01.400	4676	4734	4785	4875	4948	5009	5104	5141	5205	5426	5510	5579	5608	5644	5663	5675
01.600	4768	4840	4905	5009	5092	5159	5265	5307	5374	5609	5694	5762	5791	5826	5843	5854
01.800	4848	4933	5006	5125	5217	5292	5405	5449	5523	5766	5849	5915	5942	5975	5992	6002
02.000	4919	5015	5097	5228	5326	5407	5527	5575	5650	5900	5982	6044	6070	6101	6118	6127
04.000	5387	5552	5679	5872	6011	6116	6264	6318	6400	6654	6703	6749	6766	6788	6798	6804
06.000	5651	5850	6002	6221	6373	6483	6633	6688	6767	7004	7030	7066	7079	7096	7104	7109
08.000	5835	6055	6220	6454	6609	6721	6869	6921	6994	7218	7227	7259	7269	7283	7291	7295
10.000	5975	6210	6385	6624	6782	6893	7038	7086	7155	7364	7365	7392	7401	7414	7421	7424
20.000	6392	6663	6854	7103	7255	7357	7480	7519	7575	7747	7721	7738	7748	7756	7759	7761
30.000	6621	6906	7102	7345	7492	7583	7691	7724	7770	7926	7889	7904	7912	7919	7921	7923
40.000	6779	7071	7263	7504	7640	7725	7823	7852	7891	8041	7996	8008	8015	8021	8023	8025

$2\alpha/RT\beta = 2.30$

p/K′	1.00	1.50	2.00	3.00	4.00	5.00	7.00	8.00	10.0	20.0	40.0	70.0	100	200	400	1000
	γ-Values															
00.001	0778	0490	0333	0178	0111	0077	0047	0040	0033	0022	0019	0019	0019	0019	0019	0019
00.002	0993	0675	0488	0290	0193	0142	0091	0079	0061	0041	0033	0031	0030	0029	0029	0029
00.004	1247	0910	0699	0456	0327	0251	0170	0147	0119	0075	0060	0054	0052	0049	0048	0048
00.006	1412	1069	0850	0586	0438	0345	0243	0213	0172	0109	0085	0077	0074	0071	0069	0068
00.008	1540	1197	0970	0693	0532	0430	0311	0275	0225	0142	0112	0101	0097	0092	0090	0088
00.010	1642	1300	1070	0786	0616	0507	0373	0333	0277	0175	0136	0123	0119	0113	0109	0108
00.020	1988	1661	1433	1135	0946	0816	0647	0589	0508	0342	0268	0240	0229	0217	0212	0209
00.030	2207	1897	1677	1384	1189	1052	0868	0804	0710	0502	0399	0357	0341	0323	0314	0311
00.040	2371	2076	1866	1579	1388	1249	1059	0992	0890	0655	0527	0473	0453	0430	0418	0412
00.050	2500	2219	2019	1742	1554	1418	1226	1158	1051	0800	0654	0590	0565	0536	0522	0513
00.060	2610	2342	2149	1882	1699	1565	1377	1307	1200	0939	0778	0706	0677	0642	0625	0614
00.070	2704	2447	2262	2004	1828	1696	1511	1444	1338	1070	0902	0821	0787	0749	0731	0717
00.080	2785	2539	2363	2114	1945	1817	1636	1569	1464	1195	1021	0934	0898	0856	0836	0823
00.090	2859	2623	2454	2215	2049	1926	1753	1686	1581	1316	1136	1046	1008	0963	0940	0926
00.100	2925	2699	2535	2307	2147	2028	1858	1794	1691	1430	1248	1156	1116	1068	1043	1028
00.130	3094	2891	2746	2542	2398	2292	2136	2080	1987	1743	1567	1471	1427	1374	1347	1329
00.170	3268	3093	2966	2790	2669	2578	2447	2396	2317	2103	1946	1858	1816	1763	1734	1716
00.200	3377	3218	3104	2944	2838	2759	2640	2597	2529	2341	2203	2122	2085	2037	2011	1991
00.250	3525	3390	3294	3166	3077	3010	2918	2884	2830	2683	2578	2519	2491	2454	2432	2419
00.300	3646	3532	3453	3346	3274	3222	3154	3125	3083	2979	2905	2865	2846	2824	2809	2801
00.400	3843	3761	3707	3639	3595	3566	3529	3516	3495	3457	3438	3433	3432	3433	3435	3434
00.500	3996	3940	3906	3867	3847	3835	3824	3822	3821	3835	3858	3878	3889	3907	3918	3925
00.600	4121	4086	4068	4055	4052	4056	4066	4074	4088	4140	4196	4234	4254	4282	4299	4309
00.700	4228	4211	4207	4213	4226	4242	4271	4285	4309	4394	4473	4524	4550	4583	4603	4616
00.800	4320	4318	4326	4349	4376	4403	4446	4466	4499	4610	4707	4766	4794	4832	4852	4867
00.900	4402	4414	4433	4471	4510	4543	4600	4625	4666	4797	4905	4970	4999	5039	5061	5074
01.000	4474	4498	4526	4579	4627	4668	4737	4764	4812	4958	5078	5145	5176	5216	5236	5249
01.200	4600	4644	4689	4766	4828	4883	4968	5004	5061	5230	5360	5431	5462	5501	5521	5532
01.400	4706	4769	4826	4920	4997	5060	5161	5200	5265	5449	5584	5655	5685	5723	5741	5752
01.600	4798	4875	4942	5054	5141	5211	5322	5365	5436	5629	5767	5836	5866	5900	5918	5929
01.800	4880	4970	5045	5169	5265	5343	5462	5507	5583	5783	5920	5987	6014	6048	6063	6075
02.000	4949	5050	5137	5271	5375	5458	5583	5631	5710	5913	6048	6114	6141	6173	6188	6198
04.000	5416	5585	5719	5915	6058	6165	6316	6372	6456	6652	6760	6805	6823	6844	6854	6858
06.000	5683	5886	6040	6263	6418	6530	6683	6737	6818	6992	7079	7114	7129	7144	7152	7157
08.000	5866	6091	6258	6494	6652	6768	6916	6967	7042	7197	7273	7303	7314	7327	7334	7337
10.000	6006	6245	6422	6663	6823	6935	7082	7129	7201	7342	7408	7433	7444	7455	7462	7465
20.000	6421	6695	6889	7140	7293	7394	7517	7555	7610	7710	7754	7772	7781	7788	7792	7793
30.000	6651	6939	7134	7380	7526	7617	7725	7758	7804	7885	7921	7934	7941	7945	7950	7953
40.000	6807	7101	7295	7536	7672	7757	7853	7883	7922	7991	8024	8036	8042	8046	8049	8052

$2\alpha/RT\beta = 2.40$

p/K'	γ-Values 1.00	1.50	2.00	3.00	4.00	5.00	7.00	8.00	10.0	20.0	40.0	70.0	100	200	400	1000
00.001	0787	0497	0337	0180	0112	0078	0048	0040	0033	0022	0019	0019	0019	0019	0019	0019
00.002	1003	0684	0494	0294	0196	0144	0091	0079	0062	0041	0033	0031	0030	0029	0029	0029
00.004	1259	0920	0708	0462	0331	0254	0171	0148	0119	0075	0060	0054	0052	0049	0048	0048
00.006	1427	1083	0860	0593	0443	0350	0246	0215	0174	0109	0085	0077	0074	0071	0069	0068
00.008	1555	1210	0983	0702	0539	0435	0314	0277	0227	0143	0112	0101	0097	0092	0090	0088
00.010	1658	1314	1084	0796	0625	0513	0377	0336	0279	0176	0136	0123	0119	0113	0109	0108
00.020	2006	1678	1451	1150	0958	0828	0656	0596	0514	0344	0268	0240	0229	0217	0212	0209
00.030	2228	1917	1697	1403	1204	1067	0881	0815	0719	0506	0401	0358	0342	0323	0314	0311
00.040	2392	2098	1887	1600	1407	1267	1073	1006	0902	0662	0531	0476	0455	0431	0418	0412
00.050	2523	2243	2042	1764	1577	1438	1245	1175	1067	0809	0660	0594	0569	0539	0524	0515
00.060	2632	2365	2173	1905	1722	1589	1398	1327	1219	0951	0787	0712	0682	0647	0630	0620
00.070	2727	2472	2286	2029	1854	1720	1535	1466	1359	1086	0912	0829	0795	0755	0737	0724
00.080	2809	2565	2389	2141	1972	1844	1662	1593	1488	1214	1034	0944	0908	0864	0843	0829
00.090	2883	2649	2481	2244	2077	1954	1779	1713	1608	1336	1151	1059	1019	0972	0949	0934
00.100	2950	2726	2563	2335	2176	2056	1886	1822	1719	1453	1267	1172	1130	1080	1053	1037
00.130	3119	2919	2774	2572	2432	2323	2170	2113	2020	1772	1592	1494	1450	1394	1365	1347
00.170	3294	3121	2996	2825	2704	2614	2484	2433	2355	2141	1981	1891	1848	1794	1763	1745
00.200	3402	3247	3135	2981	2874	2796	2681	2638	2570	2384	2244	2163	2124	2075	2048	2029
00.250	3552	3420	3327	3201	3115	3051	2960	2927	2874	2732	2627	2570	2541	2504	2483	2469
00.300	3674	3563	3487	3385	3315	3265	3197	3171	3133	3032	2961	2922	2906	2884	2870	2863
00.400	3871	3793	3743	3677	3638	3610	3579	3565	3547	3516	3505	3502	3504	3508	3510	3512
00.500	4025	3973	3942	3909	3889	3882	3876	3875	3877	6898	3931	3955	3969	3990	4003	4011
00.600	4150	4119	4106	4097	4099	4105	4120	4128	4145	4207	4269	4313	4337	4366	4386	4398
00.700	4256	4244	4243	4257	4272	4290	4325	4340	4369	4464	4550	4605	4633	4671	4692	4705
00.800	4349	4352	4364	4393	4422	4452	4502	4522	4560	4678	4783	4847	4879	4918	4941	4954
00.900	4432	4448	4470	4514	4557	4593	4655	4683	4727	4866	4984	5052	5084	5124	5147	5161
01.000	4504	4533	4564	4622	4676	4720	4792	4822	4875	5029	5154	5226	5258	5301	5322	5336
01.200	4632	4682	4728	4810	4876	4932	5024	5062	5123	5299	5436	5509	5541	5582	5604	5615
01.400	4737	4803	4863	4964	5046	5113	5216	5258	5327	5519	5658	5731	5762	5800	5820	5831
01.600	4827	4909	4981	5098	5188	5263	5379	5424	5496	5699	5839	5909	5940	5976	5994	6004
01.800	4910	5004	5084	5213	5315	5396	5518	5566	5642	5851	5991	6059	6088	6119	6137	6147
02.000	4981	5087	5177	5317	5425	5511	5640	5690	5769	5979	6117	6184	6211	6241	6257	6267
04.000	5448	5621	5757	5960	6105	6215	6368	6425	6510	6708	6816	6861	6878	6898	6907	6915
06.000	5713	5921	6079	6306	6463	6578	6732	6787	6867	7043	7129	7164	7176	7192	7200	7204
08.000	5896	6124	6297	6536	6698	6811	6962	7014	7089	7244	7317	7348	7358	7370	7377	7382
10.000	6036	6279	6459	6705	6868	6981	7125	7174	7243	7384	7448	7474	7485	7495	7502	7504
20.000	6451	6729	6922	7177	7329	7432	7553	7593	7646	7744	7789	7807	7812	7821	7824	7825
30.000	6679	6970	7167	7415	7560	7653	7757	7790	7834	7915	7950	7964	7969	7975	7979	7981
40.000	6836	7132	7327	7569	7704	7790	7885	7914	7953	8021	8051	8063	8068	8073	8076	8078

$2\alpha/RT\beta = 2.50$

p/K'	1.00	1.50	2.00	3.00	4.00	5.00	7.00	8.00	10.0	20.0	40.0	70.0	100	200	400	1000
									γ-Values							
00.001	0801	0504	0341	0181	0110	0075	0045	0037	0029	0016	0013	0011	0010	0010	0010	0009
00.002	1021	0694	0501	0296	0197	0143	0090	0075	0059	0036	0029	0027	0026	0025	0025	0025
00.004	1281	0935	0718	0470	0334	0255	0170	0147	0116	0072	0056	0051	0049	0047	0046	0045
00.006	1451	1099	0873	0601	0448	0353	0247	0214	0173	0106	0082	0074	0072	0067	0066	0065
00.008	1580	1230	1000	0713	0545	0439	0316	0277	0228	0141	0109	0098	0094	0089	0086	0085
00.010	1686	1336	1104	0809	0635	0519	0381	0338	0280	0175	0135	0121	0116	0110	0108	0105
00.020	2040	1707	1474	1169	0975	0839	0663	0605	0519	0346	0266	0239	0227	0215	0211	0206
00.030	2263	1949	1729	1427	1228	1086	0894	0829	0731	0510	0400	0356	0340	0321	0313	0308
00.040	2430	2132	1919	1628	1435	1290	1093	1023	0918	0670	0534	0478	0455	0430	0419	0409
00.050	2562	2278	2076	1795	1606	1463	1268	1198	1086	0822	0664	0596	0571	0539	0526	0517
00.060	2674	2406	2212	1943	1757	1619	1424	1354	1242	0968	0798	0718	0687	0650	0632	0622
00.070	2769	2510	2328	2069	1890	1756	1565	1495	1383	1106	0925	0837	0804	0761	0740	0726
00.080	2854	2608	2431	2181	2011	1880	1696	1627	1518	1237	1050	0958	0917	0873	0850	0836
00.090	2928	2691	2522	2285	2119	1993	1816	1749	1639	1363	1173	1075	1034	0984	0957	0944
00.100	2995	2770	2608	2378	2218	2098	1925	1860	1757	1483	1291	1191	1148	1094	1068	1053
00.130	3167	2966	2821	2615	2476	2371	2216	2160	2066	1812	1627	1526	1478	1421	1391	1372
00.170	3344	3171	3049	2875	2756	2667	2535	2488	2405	2193	2029	1936	1894	1836	1804	1784
00.200	3452	3300	3188	3036	2929	2853	2738	2695	2626	2439	2299	2219	2180	2129	2101	2082
00.250	3606	3475	3385	3260	3176	3112	3026	2989	2939	2801	2698	2640	2612	2574	2555	2540
00.300	3730	3620	3546	3447	3380	3330	3268	3242	3202	3106	3042	3006	2990	2968	2958	2951
00.400	3929	3854	3805	3746	3710	3683	3654	3643	3629	3605	3601	3604	3607	3615	3619	3624
00.500	4085	4032	4006	3977	3965	3958	3955	3959	3963	3992	4032	4066	4084	4108	4125	4134
00.600	4213	4184	4173	4169	4176	4184	4205	4214	4233	4306	4380	4432	4459	4495	4514	4527
00.700	4321	4310	4313	4331	4353	4372	4412	4433	4463	4569	4667	4726	4759	4799	4823	4839
00.800	4413	4419	4436	4470	4504	4537	4593	4614	4658	4788	4904	4975	5006	5048	5074	5090
00.900	4493	4517	4541	4593	4640	4682	4750	4777	4828	4978	5105	5179	5213	5257	5281	5296
01.000	4569	4601	4638	4704	4759	4806	4889	4920	4975	5143	5277	5356	5388	5432	5456	5470
01.200	4697	4751	4800	4892	4964	5025	5122	5163	5226	5417	5562	5640	5673	5713	5735	5749
01.400	4805	4878	4941	5045	5135	5205	5319	5357	5432	5637	5783	5858	5891	5932	5953	5966
01.600	4899	4984	5059	5185	5282	5359	5482	5528	5605	5815	5964	6040	6068	6104	6123	6134
01.800	4977	5079	5164	5302	5407	5492	5619	5670	5752	5970	6116	6186	6216	6247	6265	6276
02.000	5052	5164	5258	5405	5518	5608	5745	5798	5881	6101	6244	6310	6337	6369	6384	6394
04.000	5524	5705	5844	6055	6205	6320	6477	6537	6622	6825	6934	6980	6998	7019	7029	7036
06.000	5793	6003	6169	6405	6565	6683	6840	6893	6979	7157	7245	7278	7292	7308	7313	7319
08.000	5979	6213	6387	6635	6801	6919	7071	7121	7200	7355	7429	7459	7470	7484	7488	7493
10.000	6122	6367	6552	6805	6969	7086	7234	7284	7354	7496	7560	7585	7596	7605	7611	7613
20.000	6543	6821	7021	7278	7437	7539	7660	7698	7753	7850	7894	7912	7922	7930	7935	7938
30.000	6774	7065	7263	7516	7663	7757	7863	7896	7939	8020	8055	8067	8075	8081	8083	8088
40.000	6931	7227	7427	7672	7809	7894	7988	8017	8054	8124	8154	8165	8170	8174	8174	8174

$2\alpha/RT\beta = 2.60$

γ-Values

p/K′	1.00	1.50	2.00	3.00	4.00	5.00	7.00	8.00	10.0	20.0	40.0	70.0	100	200	400	1000
00.001	0804	0509	0346	0184	0114	0079	0048	0040	0033	0022	0019	0019	0019	0019	0019	0019
00.002	1025	0701	0508	0302	0200	0147	0092	0080	0062	0041	0033	0031	0030	0029	0029	0029
00.004	1285	0942	0727	0475	0339	0260	0174	0151	0121	0076	0060	0054	0052	0049	0048	0048
00.006	1455	1108	0883	0610	0455	0359	0251	0219	0176	0109	0085	0077	0074	0071	0069	0068
00.008	1585	1238	1008	0722	0554	0446	0321	0283	0231	0144	0112	0101	0097	0092	0090	0088
00.010	1692	1346	1111	0818	0642	0528	0386	0344	0284	0177	0137	0123	0119	0113	0109	0108
00.020	2045	1717	1487	1182	0987	0851	0674	0612	0527	0350	0270	0241	0230	0217	0212	0209
00.030	2269	1959	1739	1441	1241	1099	0907	0840	0739	0516	0405	0360	0344	0325	0315	0311
00.040	2435	2142	1933	1644	1448.	1305	1108	1037	0929	0678	0540	0480	0460	0435	0422	0415
00.050	2567	2289	2089	1811	1620	1482	1285	1212	1101	0832	0673	0603	0576	0546	0531	0522
00.060	2679	2415	2223	1956	1772	1636	1442	1369	1258	0980	0806	0725	0694	0656	0638	0628
00.070	2775	2523	2339	2082	1905	1772	1584	1514	1403	1119	0936	0847	0810	0768	0748	0734
00.080	2857	2616	2443	2197	2027	1899	1715	1644	1535	1252	1063	0967	0928	0881	0858	0843
00.090	2932	2702	2535	2302	2135	2012	1834	1767	1660	1381	1186	1088	1044	0994	0968	0952
00.100	2999	2779	2619	2395	2237	2117	1946	1882	1776	1502	1308	1205	1159	1105	1077	1060
00.130	3170	2976	2835	2635	2498	2391	2238	2181	2088	1834	1650	1545	1497	1435	1403	1384
00.170	3347	3181	3059	2892	2775	2688	2560	2511	2432	2219	2058	1963	1919	1859	1826	1805
00.200	3457	3307	3200	3052	2950	2874	2761	2720	2655	2472	2333	2251	2211	2159	2129	2110
00.250	3607	3482	3394	3275	3194	3134	3048	3017	2967	2833	2734	2678	2651	2615	2592	2579
00.300	3730	3626	3557	3461	3399	3352	3290	3267	3233	3142	3081	3047	3033	3015	3002	2997
00.400	3928	3858	3814	3760	3724	3702	3678	3669	3658	3640	3642	3650	3657	3667	3674	3678
00.500	4083	4040	4015	3989	3980	3978	3980	3984	3991	4032	4078	4113	4132	4163	4180	4191
00.600	4210	4189	4180	4182	4191	4203	4228	4241	4263	4345	4426	4479	4507	4545	4568	4582
00.700	4316	4313	4319	4341	4366	4391	4434	4454	4490	4601	4706	4772	4805	4849	4874	4888
00.800	4410	4422	4440	4480	4518	4555	4614	4639	4683	4819	4941	5015	5050	5094	5119	5134
00.900	4492	4518	4547	4603	4654	4697	4769	4800	4850	5007	5141	5217	5252	5298	5321	5336
01.000	4565	4603	4641	4711	4771	4822	4906	4940	4999	5170	5310	5390	5425	5470	5493	5507
01.200	4692	4752	4806	4899	4974	5037	5138	5181	5248	5442	5590	5669	5702	5744	5765	5778
01.400	4798	4875	4942	5054	5142	5217	5330	5376	5450	5657	5808	5883	5917	5954	5975	5986
01.600	4890	4982	5060	5187	5287	5369	5493	5541	5620	5835	5984	6058	6089	6124	6142	6152
01.800	4972	5077	5162	5303	5411	5500	5631	5683	5766	5986	6132	6202	6231	6262	6278	6288
02.000	5042	5158	5255	5406	5522	5614	5752	5807	5890	6112	6253	6322	6349	6379	6395	6402
04.000	5509	5692	5837	6048	6199	6314	6474	6531	6620	6821	6925	6969	6985	7005	7015	7021
06.000	5774	5991	6155	6391	6555	6670	6829	6885	6966	7140	7224	7257	7269	7285	7292	7297
08.000	5958	6195	6372	6618	6785	6900	7054	7105	7181	7334	7403	7431	7442	7454	7461	7465
10.000	6095	6346	6531	6786	6950	7066	7212	7260	7329	7467	7529	7554	7563	7574	7579	7582
20.000	6509	6793	6992	7249	7405	7508	7627	7665	7717	7811	7853	7869	7877	7883	7886	7890
30.000	6737	7033	7232	7481	7627	7719	7822	7855	7897	7973	8006	8020	8026	8033	8034	8036
40.000	6893	7194	7391	7633	7767	7853	7944	7973	8008	8074	8104	8116	8119	8125	8128	8129

$2\alpha/RT\beta = 2.70$

p/K′	1.00	1.50	2.00	3.00	4.00	5.00	7.00	8.00	10.0	20.0	40.0	70.0	100	200	400	1000
									γ-Values							
00.001	0813	0515	0350	0186	0115	0080	0048	0040	0033	0022	0019	0019	0019	0019	0019	0019
00.002	1036	0709	0514	0306	0203	0148	0093	0081	0062	0041	0033	0031	0030	0029	0029	0029
00.004	1299	0954	0736	0481	0344	0263	0176	0152	0121	0076	0060	0054	0052	0049	0048	0048
00.006	1471	1121	0894	0618	0461	0363	0254	0221	0177	0109	0085	0077	0074	0071	0069	0068
00.008	1602	1253	1021	0732	0562	0453	0325	0286	0233	0144	0112	0101	0097	0092	0090	0088
00.010	1708	1361	1126	0830	0652	0534	0391	0348	0287	0178	0137	0123	0119	0113	0109	0108
00.020	2064	1736	1505	1199	1001	0865	0684	0622	0533	0352	0271	0241	0230	0217	0212	0209
00.030	2289	1980	1760	1461	1259	1115	0921	0853	0751	0522	0408	0362	0345	0325	0315	0311
00.040	2456	2164	1954	1665	1469	1324	1125	1054	0943	0686	0544	0484	0462	0437	0424	0418
00.050	2590	2313	2113	1836	1645	1504	1305	1231	1119	0842	0680	0609	0581	0549	0534	0525
00.060	2702	2440	2249	1982	1797	1660	1465	1391	1279	0993	0815	0732	0700	0661	0643	0632
00.070	2798	2548	2366	2109	1933	1799	1608	1537	1425	1136	0948	0856	0818	0774	0755	0740
00.080	2881	2643	2470	2225	2056	1927	1742	1670	1562	1274	1079	0978	0940	0889	0866	0850
00.090	2957	2730	2564	2331	2165	2041	1865	1796	1687	1404	1205	1103	1057	1004	0978	0961
00.100	3025	2808	2648	2424	2266	2149	1977	1912	1806	1528	1330	1222	1175	1119	1089	1072
00.130	3196	3004	2865	2668	2532	2425	2273	2215	2123	1868	1679	1572	1523	1457	1424	1404
00.170	3374	3210	3093	2926	2811	2725	2600	2551	2473	2259	2097	2002	1956	1894	1859	1838
00.200	3483	3337	3234	3088	2988	2915	2804	2764	2698	2517	2378	2297	2257	2202	2173	2151
00.250	3635	3514	3429	3313	3234	3175	3094	3063	3015	2886	2790	2736	2709	2673	2652	2638
00.300	3758	3658	3592	3501	3439	3397	3340	3317	3284	3201	3143	3113	3099	3085	3074	3069
00.400	3957	3891	3850	3799	3769	3750	3731	3722	3714	3707	3715	3727	3736	3750	3759	3766
00.500	4114	4075	4052	4032	4025	4026	4033	4038	4049	4098	4155	4196	4219	4252	4271	4286
00.600	4239	4222	4217	4223	4238	4253	4282	4297	4323	4414	4504	4563	4595	4636	4661	4677
00.700	4347	4348	4359	4385	4414	4442	4492	4514	4551	4675	4787	4858	4893	4939	4964	4982
00.800	4440	4457	4477	4524	4569	4605	4670	4697	4746	4893	5022	5099	5137	5184	5210	5225
00.900	4522	4553	4585	4647	4702	4748	4827	4858	4913	5080	5221	5301	5339	5384	5410	5426
01.000	4595	4638	4681	4755	4821	4875	4962	5000	5061	5244	5390	5472	5507	5555	5577	5592
01.200	4723	4788	4844	4944	5024	5091	5197	5239	5310	5512	5667	5747	5783	5825	5845	5859
01.400	4828	4910	4980	5100	5193	5271	5388	5435	5513	5728	5881	5960	5993	6031	6051	6062
01.600	4921	5018	5098	5233	5337	5422	5549	5600	5682	5904	6057	6130	6162	6197	6215	6226
01.800	5002	5112	5202	5348	5462	5551	5688	5741	5826	6051	6201	6271	6299	6331	6350	6357
02.000	5074	5195	5294	5453	5571	5666	5809	5863	5952	6177	6323	6389	6417	6446	6462	6470
04.000	5541	5729	5875	6093	6247	6362	6524	6584	6673	6873	6980	7022	7040	7058	7066	7073
06.000	5806	6026	6195	6434	6598	6718	6876	6933	7014	7187	7270	7303	7317	7331	7337	7342
08.000	5988	6230	6408	6660	6826	6945	7099	7151	7226	7378	7446	7473	7484	7495	7503	7506
10.000	6127	6382	6569	6825	6993	7108	7255	7303	7372	7509	7568	7592	7601	7612	7616	7620
20.000	6539	6826	7028	7284	7440	7543	7663	7699	7750	7843	7884	7900	7906	7913	7917	7921
30.000	6766	7064	7266	7515	7661	7751	7854	7886	7928	8001	8034	8047	8053	8059	8062	8063
40.000	6920	7223	7422	7665	7799	7882	7974	8001	8038	8100	8128	8141	8144	8150	8153	8154

$2\alpha/RT\beta = 2.80$

p/K'	γ-Values 1.00	1.50	2.00	3.00	4.00	5.00	7.00	8.00	10.0	20.0	40.0	70.0	100	200	400	1000
00.001	0822	0522	0355	0189	0117	0080	0048	0040	0033	0022	0019	0019	0019	0019	0019	0019
00.002	1048	0719	0521	0310	0205	0150	0093	0081	0062	0041	0033	0031	0030	0029	0029	0029
00.004	1313	0966	0746	0488	0349	0266	0178	0153	0122	0076	0060	0054	0052	0049	0048	0048
00.006	1486	1135	0906	0627	0468	0368	0256	0223	0179	0110	0085	0077	0074	0071	0069	0068
00.008	1618	1267	1034	0742	0570	0459	0329	0290	0234	0145	0112	0101	0097	0092	0090	0088
00.010	1725	1377	1141	0842	0661	0543	0397	0352	0290	0179	0137	0123	0119	0113	0109	0108
00.020	2085	1756	1524	1216	1017	0877	0693	0630	0540	0356	0272	0242	0230	0217	0212	0209
00.030	2311	2002	1781	1481	1278	1133	0935	0864	0762	0527	0411	0364	0347	0327	0316	0312
00.040	2479	2188	1979	1687	1491	1344	1142	1069	0958	0695	0549	0487	0465	0439	0426	0420
00.050	2613	2338	2139	1860	1668	1527	1325	1250	1136	0856	0688	0614	0585	0553	0537	0527
00.060	2726	2465	2275	2009	1823	1685	1489	1414	1300	1010	0824	0740	0706	0666	0647	0636
00.070	2822	2574	2393	2137	1961	1826	1635	1563	1450	1156	0962	0866	0827	0782	0762	0746
00.080	2906	2670	2498	2255	2084	1955	1771	1698	1588	1295	1095	0991	0949	0898	0874	0858
00.090	2981	2755	2593	2360	2195	2073	1895	1826	1717	1429	1224	1119	1073	1015	0988	0971
00.100	3049	2834	2678	2455	2299	2180	2008	1944	1837	1555	1350	1241	1192	1133	1103	1084
00.130	3222	3032	2896	2702	2565	2461	2309	2252	2159	1903	1710	1599	1548	1484	1446	1425
00.170	3400	3240	3124	2963	2849	2765	2640	2591	2515	2302	2137	2042	1995	1930	1895	1872
00.200	3511	3368	3266	3124	3026	2955	2848	2807	2742	2565	2428	2344	2304	2249	2219	2197
00.250	3662	3545	3463	3351	3276	3219	3140	3110	3064	2940	2848	2795	2770	2734	2713	2700
00.300	3787	3691	3628	3539	3481	3442	3388	3366	3335	3260	3209	3182	3170	3157	3149	3146
00.400	3987	3925	3888	3841	3813	3800	3781	3775	3770	3772	3788	3807	3818	3837	3849	3858
00.500	4143	4109	4091	4074	4073	4076	4090	4094	4108	4169	4235	4280	4305	4344	4367	4382
00.600	4268	4256	4255	4268	4284	4303	4339	4355	4385	4486	4585	4650	4683	4730	4757	4772
00.700	4376	4382	4396	4429	4462	4494	4547	4573	4614	4746	4868	4946	4981	5030	5058	5075
00.800	4471	4492	4518	4569	4616	4658	4727	4757	4808	4966	5104	5185	5224	5273	5299	5316
00.900	4553	4589	4626	4692	4751	4800	4883	4918	4976	5153	5301	5385	5424	5473	5499	5514
01.000	4626	4674	4720	4802	4870	4928	5021	5060	5126	5317	5469	5555	5593	5639	5663	5678
01.200	4753	4823	4884	4988	5073	5144	5255	5298	5374	5585	5744	5825	5861	5905	5925	5940
01.400	4859	4946	5019	5145	5243	5323	5448	5496	5576	5798	5956	6036	6069	6108	6127	6138
01.600	4952	5054	5140	5278	5387	5473	5606	5658	5744	5973	6128	6202	6234	6270	6285	6298
01.800	5033	5148	5243	5395	5510	5604	5744	5799	5888	6119	6269	6339	6370	6402	6418	6427
02.000	5105	5231	5334	5498	5620	5719	5865	5921	6013	6244	6391	6456	6483	6513	6526	6538
04.000	5573	5765	5916	6137	6294	6412	6576	6636	6726	6929	7033	7075	7091	7109	7118	7122
06.000	5836	6062	6234	6477	6641	6763	6925	6981	7063	7234	7316	7349	7361	7375	7382	7386
08.000	6019	6264	6446	6701	6870	6989	7144	7195	7271	7420	7488	7515	7524	7536	7543	7545
10.000	6158	6417	6605	6867	7034	7149	7298	7346	7414	7548	7607	7631	7639	7650	7654	7658
20.000	6568	6858	7061	7319	7476	7578	7697	7735	7785	7876	7915	7932	7936	7943	7946	7950
30.000	6794	7095	7297	7549	7693	7783	7886	7916	7959	8031	8062	8075	8079	8084	8087	8090
40.000	6947	7252	7454	7695	7831	7912	8002	8029	8064	8127	8155	8166	8169	8174	8176	8178

$2\alpha/RT\beta = 2.90$

p/K'	γ-Values 1.00	1.50	2.00	3.00	4.00	5.00	7.00	8.00	10.0	20.0	40.0	70.0	100	200	400	1000
00.001	0832	0529	0360	0191	0118	0081	0049	0041	0033	0022	0019	0019	0019	0019	0019	0019
00.002	1059	0728	0529	0314	0207	0151	0094	0081	0062	0041	0033	0031	0030	0029	0029	0029
00.004	1326	0977	0756	0495	0354	0269	0180	0154	0123	0076	0060	0054	0052	0049	0048	0048
00.006	1500	1148	0918	0636	0475	0373	0259	0225	0180	0110	0086	0077	0074	0071	0069	0068
00.008	1635	1283	1049	0753	0578	0466	0333	0293	0236	0145	0112	0101	0097	0092	0090	0088
00.010	1741	1393	1155	0854	0671	0550	0402	0357	0293	0179	0137	0123	0119	0113	0109	0108
00.020	2104	1775	1543	1232	1031	0892	0704	0639	0549	0358	0273	0242	0230	0217	0212	0209
00.030	2333	2025	1803	1502	1297	1150	0951	0879	0772	0532	0414	0366	0348	0328	0317	0312
00.040	2501	2211	2002	1711	1513	1365	1161	1088	0974	0704	0554	0490	0468	0442	0428	0422
00.050	2637	2363	2164	1885	1693	1551	1347	1271	1154	0868	0696	0619	0589	0556	0540	0531
00.060	2750	2490	2300	2035	1849	1713	1513	1438	1321	1025	0835	0746	0713	0672	0652	0641
00.070	2846	2600	2420	2166	1990	1853	1661	1590	1474	1174	0974	0877	0836	0789	0768	0752
00.080	2930	2696	2527	2284	2114	1985	1801	1727	1615	1319	1111	1004	0961	0908	0882	0867
00.090	3007	2784	2621	2391	2227	2103	1925	1856	1747	1454	1243	1135	1086	1028	0999	0981
00.100	3075	2862	2707	2487	2332	2214	2042	1977	1869	1583	1374	1261	1210	1149	1115	1096
00.130	3248	3062	2927	2736	2600	2497	2345	2289	2196	1939	1744	1629	1576	1507	1469	1446
00.170	3428	3271	3157	2999	2887	2802	2682	2634	2558	2347	2183	2086	2037	1971	1933	1910
00.200	3539	3400	3299	3162	3067	2996	2892	2851	2788	2614	2479	2394	2355	2300	2268	2245
00.250	3691	3578	3498	3391	3317	3263	3188	3160	3115	2995	2908	2859	2833	2800	2781	2767
00.300	3816	3725	3663	3579	3526	3489	3437	3418	3389	3320	3275	3253	3244	3235	3228	3226
00.400	4017	3959	3923	3884	3859	3846	3835	3832	3826	3838	3865	3889	3904	3927	3944	3953
00.500	4172	4142	4128	4119	4121	4125	4142	4152	4169	4239	4316	4367	4396	4438	4464	4480
00.600	4299	4291	4293	4311	4332	4355	4393	4414	4448	4558	4668	4739	4774	4825	4852	4870
00.700	4407	4418	4435	4474	4511	4545	4606	4630	4675	4821	4951	5032	5072	5124	5153	5171
00.800	4500	4527	4556	4613	4665	4710	4786	4817	4872	5041	5186	5272	5313	5364	5392	5409
00.900	4584	4625	4664	4737	4800	4855	4943	4979	5041	5228	5383	5470	5511	5562	5585	5602
01.000	4656	4709	4760	4845	4920	4981	5079	5120	5189	5390	5550	5637	5676	5724	5749	5764
01.200	4784	4858	4925	5033	5123	5196	5313	5360	5438	5657	5821	5905	5941	5985	6006	6019
01.400	4889	4981	5059	5191	5291	5376	5505	5555	5639	5869	6030	6111	6143	6185	6201	6214
01.600	4982	5089	5179	5324	5436	5528	5665	5719	5806	6042	6201	6275	6306	6342	6358	6368
01.800	5064	5183	5282	5440	5561	5658	5803	5859	5949	6186	6339	6410	6437	6470	6486	6496
02.000	5136	5267	5375	5543	5671	5771	5922	5980	6073	6309	6456	6523	6549	6578	6594	6602
04.000	5605	5802	5953	6179	6341	6461	6628	6688	6777	6981	7085	7125	7141	7160	7168	7174
06.000	5868	6097	6270	6518	6687	6809	6973	7029	7111	7281	7362	7392	7403	7419	7424	7428
08.000	6049	6299	6484	6741	6913	7032	7187	7240	7315	7462	7527	7554	7565	7575	7582	7585
10.000	6188	6450	6643	6905	7075	7192	7339	7388	7455	7587	7645	7667	7676	7686	7690	7694
20.000	6598	6891	7095	7355	7512	7613	7732	7768	7818	7907	7946	7960	7966	7973	7975	7979
30.000	6822	7125	7330	7582	7726	7817	7917	7947	7987	8059	8088	8100	8105	8110	8114	8115
40.000	6975	7282	7483	7725	7860	7941	8031	8057	8092	8152	8179	8189	8193	8198	8200	8202

$2\alpha/RT\beta = 3.00$

p/K'	γ-Values 1.00	1.50	2.00	3.00	4.00	5.00	7.00	8.00	10.0	20.0	40.0	70.0	100	200	400	1000
00.001	0841	0536	0365	0194	0119	0082	0049	0041	0033	0022	0019	0019	0019	0019	0019	0019
00.002	1070	0737	0536	0318	0210	0153	0095	0082	0063	0041	0033	0031	0030	0029	0029	0029
00.004	1341	0990	0766	0503	0359	0273	0181	0156	0124	0076	0060	0054	0052	0049	0048	0048
00.006	1516	1162	0930	0645	0482	0379	0262	0227	0181	0110	0086	0077	0074	0071	0069	0068
00.008	1651	1298	1062	0764	0587	0472	0338	0296	0239	0145	0112	0101	0097	0092	0090	0088
00.010	1760	1410	1170	0867	0680	0559	0407	0361	0296	0181	0137	0123	0119	0113	0109	0108
00.020	2124	1795	1563	1251	1047	0905	0715	0649	0556	0361	0274	0243	0231	0218	0212	0209
00.030	2353	2046	1824	1522	1316	1168	0966	0893	0785	0540	0417	0367	0349	0329	0318	0314
00.040	2523	2234	2026	1735	1537	1388	1180	1106	0990	0713	0559	0494	0471	0444	0430	0423
00.050	2660	2388	2190	1911	1718	1575	1369	1293	1175	0881	0703	0623	0594	0560	0543	0534
00.060	2773	2515	2329	2062	1878	1738	1537	1462	1344	1042	0847	0755	0720	0677	0656	0645
00.070	2871	2627	2447	2195	2018	1882	1690	1617	1500	1195	0989	0888	0846	0797	0774	0758
00.080	2956	2724	2555	2313	2145	2016	1830	1757	1644	1342	1130	1018	0973	0918	0892	0875
00.090	3033	2812	2651	2422	2259	2134	1957	1889	1778	1482	1264	1152	1102	1041	1011	0991
00.100	3102	2892	2736	2519	2365	2247	2075	2010	1902	1613	1400	1282	1229	1164	1130	1110
00.130	3275	3091	2959	2770	2636	2533	2384	2327	2235	1977	1778	1659	1605	1532	1493	1469
00.170	3456	3302	3189	3035	2925	2845	2724	2677	2601	2392	2227	2129	2080	2013	1973	1948
00.200	3566	3431	3334	3199	3108	3038	2937	2897	2837	2666	2532	2448	2408	2353	2321	2297
00.250	3719	3610	3534	3429	3359	3308	3236	3209	3167	3053	2970	2923	2900	2869	2850	2837
00.300	3844	3756	3699	3621	3570	3535	3488	3470	3444	3383	3345	3327	3320	3316	3309	3310
00.400	4045	3993	3962	3925	3904	3896	3889	3887	3888	3908	3943	3973	3993	4020	4039	4052
00.500	4203	4178	4166	4162	4167	4177	4200	4210	4230	4312	4399	4457	4488	4535	4563	4582
00.600	4329	4325	4333	4356	4380	4407	4453	4473	4511	4633	4750	4829	4867	4920	4951	4970
00.700	4436	4452	4474	4517	4560	4598	4664	4691	4740	4896	5034	5121	5164	5219	5247	5267
00.800	4531	4563	4596	4659	4715	4763	4844	4873	4936	5116	5269	5360	5402	5455	5482	5501
00.900	4614	4659	4705	4782	4850	4907	5001	5039	5103	5303	5463	5555	5597	5648	5675	5690
01.000	4687	4745	4799	4892	4970	5035	5139	5182	5253	5464	5629	5722	5761	5810	5833	5848
01.200	4816	4895	4964	5080	5173	5251	5372	5421	5500	5729	5897	5984	6021	6063	6084	6098
01.400	4921	5018	5102	5236	5343	5430	5563	5616	5702	5938	6104	6186	6219	6257	6277	6288
01.600	5015	5127	5219	5369	5488	5580	5724	5779	5868	6111	6270	6347	6376	6413	6429	6438
01.800	5096	5221	5321	5484	5610	5710	5860	5918	6010	6252	6408	6478	6506	6538	6553	6562
02.000	5169	5304	5415	5589	5720	5825	5979	6038	6133	6374	6523	6589	6613	6643	6658	6665
04.000	5635	5837	5994	6225	6388	6509	6679	6739	6832	7035	7136	7176	7190	7209	7217	7222
06.000	5899	6133	6308	6562	6732	6856	7019	7076	7157	7326	7405	7436	7447	7461	7467	7471
08.000	6080	6334	6520	6783	6955	7075	7233	7283	7358	7503	7567	7593	7603	7614	7620	7623
10.000	6219	6485	6679	6945	7115	7232	7381	7427	7496	7625	7681	7702	7712	7721	7726	7729
20.000	6627	6923	7130	7390	7548	7649	7765	7802	7850	7938	7976	7989	7994	8003	8005	8007
30.000	6852	7158	7362	7615	7757	7846	7947	7977	8015	8084	8114	8127	8131	8137	8139	8141
40.000	7002	7311	7514	7757	7891	7971	8059	8085	8118	8177	8203	8213	8216	8223	8224	8226

$2\alpha/RT\beta = 3.10$

p/K'	γ-Values 1.00	1.50	2.00	3.00	4.00	5.00	7.00	8.00	10.0	20.0	40.0	70.0	100	200	400	1000
00.001	0851	0543	0370	0197	0121	0082	0049	0041	0034	0022	0019	0019	0019	0019	0019	0019
00.002	1082	0746	0543	0323	0213	0155	0096	0082	0063	0041	0033	0031	0030	0029	0029	0029
00.004	1354	1002	0777	0510	0363	0277	0183	0157	0124	0076	0060	0054	0052	0049	0048	0048
00.006	1532	1177	0943	0655	0489	0384	0266	0230	0183	0110	0086	0077	0074	0071	0069	0068
00.008	1668	1314	1076	0775	0596	0478	0342	0299	0241	0146	0112	0101	0097	0092	0090	0088
00.010	1778	1427	1186	0879	0691	0567	0413	0365	0300	0181	0137	0123	0119	0113	0109	0108
00.020	2144	1815	1582	1268	1063	0919	0726	0659	0563	0365	0276	0243	0232	0218	0213	0209
00.030	2375	2068	1847	1544	1337	1187	0982	0908	0797	0547	0420	0369	0351	0330	0319	0315
00.040	2547	2259	2051	1760	1558	1410	1199	1125	1006	0724	0564	0497	0473	0446	0432	0425
00.050	2684	2413	2215	1937	1744	1598	1393	1315	1194	0895	0712	0629	0598	0564	0547	0536
00.060	2798	2543	2355	2090	1904	1766	1563	1487	1368	1060	0858	0764	0725	0683	0661	0648
00.070	2895	2653	2475	2224	2046	1911	1717	1645	1527	1216	1003	0899	0854	0804	0780	0763
00.080	2981	2751	2584	2343	2177	2046	1859	1786	1673	1366	1148	1033	0985	0929	0900	0884
00.090	3058	2840	2679	2454	2290	2169	1990	1921	1809	1508	1285	1169	1116	1055	1022	1002
00.100	3127	2920	2767	2551	2399	2280	2109	2042	1936	1644	1424	1302	1248	1181	1145	1122
00.130	3301	3121	2992	2804	2673	2572	2422	2366	2274	2015	1812	1693	1636	1558	1518	1493
00.170	3483	3333	3223	3072	2965	2885	2766	2720	2647	2441	2276	2175	2126	2056	2014	1991
00.200	3594	3462	3367	3236	3148	3082	2982	2944	2884	2717	2586	2505	2465	2408	2374	2351
00.250	3747	3642	3569	3469	3402	3352	3285	3259	3219	3113	3035	2991	2970	2941	2922	2912
00.300	3873	3789	3734	3661	3615	3580	3540	3524	3500	3447	3418	3405	3400	3400	3397	3400
00.400	4075	4026	3998	3968	3951	3945	3943	3945	3946	3979	4024	4061	4083	4115	4139	4153
00.500	4231	4210	4203	4204	4214	4228	4255	4270	4294	4384	4480	4547	4582	4635	4666	4685
00.600	4360	4361	4371	4400	4431	4461	4509	4534	4572	4706	4835	4920	4961	5018	5050	5070
00.700	4468	4488	4513	4564	4609	4651	4721	4753	4805	4971	5119	5211	5255	5313	5342	5363
00.800	4562	4598	4634	4703	4765	4817	4902	4939	5000	5189	5352	5447	5491	5546	5575	5592
00.900	4644	4695	4744	4828	4901	4960	5060	5100	5169	5377	5545	5641	5683	5735	5762	5778
01.000	4719	4782	4837	4936	5020	5089	5198	5242	5318	5537	5710	5804	5846	5893	5917	5933
01.200	4845	4930	5004	5126	5224	5304	5430	5481	5565	5800	5973	6062	6099	6142	6164	6176
01.400	4952	5054	5140	5284	5393	5484	5621	5676	5766	6008	6177	6260	6292	6331	6350	6360
01.600	5045	5162	5260	5416	5537	5633	5781	5838	5931	6178	6342	6417	6447	6481	6498	6509
01.800	5127	5256	5362	5530	5660	5765	5918	5976	6071	6318	6475	6547	6572	6605	6620	6628
02.000	5199	5339	5455	5633	5770	5878	6036	6097	6193	6438	6587	6653	6678	6707	6720	6729
04.000	5665	5871	6030	6267	6435	6559	6731	6791	6883	7085	7186	7225	7241	7257	7266	7271
06.000	5929	6167	6348	6603	6775	6901	7066	7123	7204	7372	7449	7479	7490	7503	7509	7512
08.000	6111	6369	6558	6822	6997	7119	7276	7326	7401	7544	7606	7631	7641	7652	7655	7660
10.000	6248	6519	6716	6984	7156	7273	7420	7470	7537	7664	7718	7740	7746	7756	7761	7764
20.000	6653	6952	7164	7425	7582	7683	7798	7836	7883	7968	8004	8018	8022	8031	8033	8035
30.000	6878	7186	7394	7647	7790	7878	7977	8005	8045	8111	8140	8151	8156	8162	8164	8166
40.000	7030	7341	7545	7787	7918	7999	8086	8111	8144	8201	8227	8236	8239	8245	8248	8248

$2\alpha/RT\beta = 3.20$

p/K′	γ-Values 1.00	1.50	2.00	3.00	4.00	5.00	7.00	8.00	10.0	20.0	40.0	70.0	100	200	400	1000
00.001	0860	0550	0375	0199	0122	0083	0049	0041	0034	0022	0019	0019	0019	0019	0019	0019
00.002	1094	0756	0550	0328	0216	0156	0096	0083	0063	0041	0033	0031	0030	0029	0029	0029
00.004	1369	1014	0788	0517	0369	0281	0185	0158	0125	0076	0060	0054	0052	0049	0048	0048
00.006	1548	1191	0955	0665	0496	0390	0269	0233	0185	0111	0086	0077	0074	0071	0069	0068
00.008	1685	1330	1090	0787	0605	0486	0347	0304	0244	0146	0113	0101	0097	0092	0090	0088
00.010	1795	1444	1201	0892	0702	0575	0419	0370	0303	0182	0138	0124	0119	0113	0109	0108
00.020	2165	1836	1601	1286	1079	0933	0738	0670	0572	0368	0277	0244	0232	0218	0213	0209
00.030	2398	2092	1869	1566	1356	1206	0998	0923	0810	0552	0422	0372	0353	0332	0321	0317
00.040	2570	2283	2075	1783	1582	1431	1220	1143	1024	0734	0570	0501	0476	0449	0434	0427
00.050	2707	2438	2241	1963	1771	1626	1416	1338	1215	0909	0720	0635	0604	0568	0549	0539
00.060	2822	2569	2382	2119	1933	1793	1589	1513	1392	1077	0870	0772	0733	0688	0666	0654
00.070	2921	2682	2505	2253	2076	1942	1746	1672	1553	1239	1020	0910	0865	0812	0788	0771
00.080	3007	2780	2614	2374	2208	2079	1890	1818	1703	1391	1167	1048	1000	0939	0909	0891
00.090	3084	2869	2710	2486	2324	2201	2025	1953	1841	1536	1311	1188	1133	1067	1034	1014
00.100	3153	2949	2798	2584	2432	2315	2144	2078	1971	1677	1451	1325	1268	1198	1160	1138
00.130	3328	3150	3024	2840	2709	2608	2462	2406	2314	2055	1849	1728	1669	1589	1546	1518
00.170	3510	3363	3257	3109	3005	2926	2812	2764	2694	2488	2324	2224	2174	2103	2060	2032
00.200	3622	3494	3404	3275	3189	3124	3029	2992	2935	2772	2643	2563	2523	2467	2433	2409
00.250	3775	3674	3604	3510	3445	3400	3334	3311	3272	3172	3101	3062	3043	3016	3001	2990
00.300	3903	3824	3772	3703	3660	3629	3592	3577	3558	3511	3492	3484	3485	3487	3488	3493
00.400	4104	4060	4037	4011	3998	3995	4000	4001	4009	4049	4105	4150	4176	4215	4241	4259
00.500	4262	4247	4243	4250	4263	4282	4313	4329	4356	4461	4566	4641	4678	4735	4767	4790
00.600	4391	4398	4410	4445	4479	4511	4568	4594	4638	4784	4922	5011	5055	5116	5150	5171
00.700	4498	4523	4553	4608	4659	4704	4782	4812	4868	5048	5204	5301	5347	5407	5440	5458
00.800	4593	4634	4673	4749	4814	4870	4962	5001	5065	5267	5436	5535	5580	5637	5666	5683
00.900	4675	4731	4782	4873	4951	5015	5119	5162	5235	5452	5628	5725	5769	5822	5849	5865
01.000	4750	4818	4879	4983	5071	5142	5257	5303	5384	5613	5792	5886	5927	5976	6000	6016
01.200	4878	4968	5044	5173	5273	5359	5489	5542	5629	5874	6051	6139	6176	6220	6241	6253
01.400	4983	5090	5182	5328	5444	5537	5679	5735	5828	6079	6251	6332	6366	6404	6422	6434
01.600	5078	5200	5300	5461	5588	5687	5838	5898	5993	6246	6411	6486	6516	6550	6566	6576
01.800	5159	5293	5403	5577	5710	5816	5974	6036	6134	6384	6543	6611	6639	6670	6683	6693
02.000	5230	5376	5495	5680	5820	5929	6093	6155	6253	6503	6652	6716	6740	6769	6783	6791
04.000	5697	5908	6072	6312	6481	6606	6781	6842	6936	7137	7235	7274	7289	7305	7312	7317
06.000	5960	6203	6385	6643	6819	6946	7111	7169	7251	7417	7491	7520	7531	7543	7550	7553
08.000	6142	6404	6597	6864	7038	7160	7317	7370	7444	7585	7646	7670	7678	7688	7693	7697
10.000	6278	6552	6752	7023	7198	7315	7460	7509	7575	7701	7753	7774	7782	7791	7794	7798
20.000	6684	6987	7196	7460	7616	7717	7831	7867	7914	7997	8033	8046	8050	8056	8060	8062
30.000	6906	7217	7425	7678	7820	7909	8005	8034	8072	8137	8165	8178	8180	8186	8188	8190
40.000	7057	7370	7574	7818	7949	8027	8113	8137	8170	8226	8250	8261	8263	8267	8271	8271

$2\alpha/RT\beta = 3.30$

p/K′	1.00	1.50	2.00	3.00	4.00	5.00	7.00	8.00	10.0	20.0	40.0	70.0	100	200	400	1000
								γ-Values								
00.001	0871	0558	0380	0202	0107	0084	0050	0041	0034	0022	0019	0019	0019	0019	0019	0019
00.002	1107	0766	0558	0333	0191	0158	0097	0083	0063	0041	0033	0031	0030	0029	0029	0029
00.004	1383	1027	0799	0525	0324	0285	0188	0160	0126	0076	0060	0054	0052	0049	0048	0048
00.006	1564	1206	0968	0675	0436	0395	0272	0236	0186	0111	0086	0077	0074	0071	0069	0068
00.008	1703	1346	1105	0799	0533	0494	0351	0308	0246	0147	0113	0101	0097	0092	0090	0088
00.010	1813	1460	1217	0906	0617	0584	0425	0375	0307	0183	0138	0124	0119	0113	0109	0108
00.020	2185	1857	1622	1305	0949	0949	0750	0681	0581	0371	0278	0245	0233	0219	0213	0209
00.030	2420	2115	1894	1588	1195	1225	1014	0938	0823	0560	0426	0373	0354	0333	0322	0318
00.040	2594	2309	2101	1809	1391	1456	1240	1163	1040	0745	0577	0505	0479	0451	0436	0429
00.050	2731	2463	2267	1991	1556	1651	1440	1360	1238	0923	0728	0642	0608	0571	0554	0542
00.060	2847	2595	2410	2148	1699	1822	1618	1540	1418	1097	0882	0781	0741	0694	0671	0658
00.070	2946	2708	2534	2284	1825	1973	1775	1702	1583	1260	1036	0924	0875	0820	0794	0777
00.080	3032	2808	2643	2405	1938	2110	1924	1850	1734	1419	1187	1064	1012	0951	0921	0900
00.090	3110	2897	2741	2518	2043	2235	2057	1987	1875	1568	1334	1207	1150	1082	1048	1025
00.100	3179	2977	2830	2618	2136	2352	2179	2114	2008	1708	1480	1349	1289	1215	1176	1152
00.130	3355	3181	3056	2876	2380	2648	2503	2448	2357	2099	1890	1764	1702	1619	1574	1544
00.170	3538	3395	3291	3147	2638	2969	2856	2810	2740	2540	2378	2276	2225	2152	2108	2079
00.200	3652	3528	3438	3315	2797	3169	3075	3041	2986	2829	2704	2625	2585	2529	2496	2472
00.250	3805	3708	3640	3551	3024	3446	3384	3363	3328	3234	3172	3136	3119	3096	3084	3074
00.300	3933	3858	3809	3746	3210	3678	3646	3632	3615	3580	3568	3568	3570	3579	3584	3591
00.400	4135	4096	4074	4055	3506	4046	4055	4059	4070	4123	4191	4243	4272	4318	4347	4368
00.500	4292	4281	4280	4294	3735	4334	4372	4387	4420	4536	4653	4733	4776	4838	4873	4897
00.600	4421	4433	4449	4491	3923	4564	4627	4656	4704	4862	5010	5105	5153	5215	5250	5272
00.700	4528	4559	4593	4655	4079	4758	4840	4875	4934	5125	5291	5393	5441	5502	5534	5554
00.800	4625	4671	4715	4793	4210	4924	5021	5064	5131	5343	5520	5623	5669	5727	5756	5773
00.900	4707	4767	4824	4920	4330	5069	5178	5224	5300	5527	5709	5811	5854	5909	5936	5952
01.000	4781	4855	4918	5030	4434	5197	5316	5365	5449	5685	5870	5968	6009	6059	6083	6097
01.200	4910	5004	5085	5219	4612	5413	5549	5603	5693	5945	6127	6216	6253	6295	6315	6328
01.400	5015	5127	5222	5375	4758	5591	5740	5798	5894	6149	6323	6405	6438	6475	6493	6504
01.600	5106	5233	5340	5508	4882	5741	5897	5957	6056	6315	6479	6554	6585	6618	6633	6643
01.800	5190	5330	5443	5624	4989	5869	6033	6095	6195	6451	6608	6678	6704	6734	6747	6756
02.000	5263	5414	5536	5727	5084	5983	6150	6213	6315	6565	6716	6779	6804	6831	6843	6852
04.000	5729	5945	6111	6355	5652	6655	6831	6891	6986	7187	7283	7321	7334	7351	7359	7363
06.000	5991	6238	6425	6686	5944	6992	7157	7214	7297	7461	7533	7560	7570	7583	7588	7592
08.000	6172	6437	6633	6903	6132	7203	7359	7411	7485	7624	7683	7706	7715	7725	7729	7732
10.000	6310	6588	6789	7062	6267	7354	7502	7549	7616	7736	7789	7808	7816	7825	7827	7831
20.000	6713	7018	7229	7495	6627	7751	7865	7899	7945	8027	8060	8074	8077	8084	8087	8089
30.000	6935	7248	7458	7709	6799	7939	8034	8063	8100	8162	8190	8201	8204	8210	8213	8214
40.000	7084	7399	7604	7846	6908	8056	8138	8164	8195	8249	8272	8282	8286	8290	8293	8292

$2\alpha/RT\beta = 3.40$

p/K'	γ-Values 1.00	1.50	2.00	3.00	4.00	5.00	7.00	8.00	10.0	20.0	40.0	70.0	100	200	400	1000
00.001	0880	0565	0386	0293	0125	0085	0050	0041	0034	0022	0019	0011	0019	0010	0019	0012
00.002	1119	0776	0567	0493	0222	0160	0098	0084	0064	0041	0033	0027	0030	0025	0029	0018
00.004	1399	1041	0810	0793	0380	0289	0190	0162	0127	0076	0060	0051	0052	0047	0048	0030
00.006	1581	1221	0982	0685	0511	0401	0276	0238	0188	0111	0086	0074	0074	0067	0069	0043
00.008	1719	1362	1120	0811	0624	0502	0356	0312	0249	0147	0113	0098	0097	0089	0090	0056
00.010	1831	1478	1234	0919	0725	0594	0432	0381	0310	0184	0138	0121	0119	0110	0109	0068
00.020	2207	1880	1644	1326	1113	0964	0763	0692	0591	0375	0280	0244	0233	0217	0213	0132
00.030	2442	2138	1917	1612	1399	1246	1032	0954	0837	0568	0430	0372	0356	0331	0323	0201
00.040	2617	2334	2127	1833	1632	1480	1261	1183	1060	0756	0582	0509	0482	0450	0438	0272
00.050	2755	2490	2296	2019	1825	1678	1464	1384	1259	0941	0740	0646	0613	0570	0557	0345
00.060	2871	2622	2438	2178	1991	1850	1647	1567	1444	1117	0897	0789	0748	0698	0676	0419
00.070	2971	2736	2562	2314	2138	2003	1806	1732	1611	1284	1053	0935	0885	0829	0803	0498
00.080	3058	2836	2673	2439	2273	2144	1957	1882	1766	1445	1208	1084	1027	0962	0930	0576
00.090	3135	2925	2772	2552	2391	2270	2092	2022	1910	1599	1359	1232	1168	1099	1061	0657
00.100	3206	3007	2860	2652	2503	2387	2216	2152	2043	1744	1509	1378	1313	1236	1194	0739
00.130	3382	3212	3090	2911	2786	2687	2544	2490	2399	2141	1930	1811	1738	1660	1604	0995
00.170	3567	3428	3327	3185	3088	3012	2901	2858	2790	2593	2432	2343	2279	2218	2159	1346
00.200	3680	3560	3474	3355	3275	3215	3124	3091	3037	2885	2766	2705	2650	2615	2562	1605
00.250	3834	3741	3678	3591	3535	3493	3437	3418	3385	3299	3244	3235	3199	3202	3171	2001
00.300	3962	3892	3846	3787	3752	3727	3699	3683	3673	3648	3647	3681	3660	3703	3684	2335
00.400	4164	4129	4114	4098	4094	4099	4112	4113	4132	4198	4276	4370	4370	4457	4456	2833
00.500	4323	4316	4320	4338	4361	4386	4429	4449	4485	4612	4740	4869	4874	4984	4978	3164
00.600	4452	4468	4489	4535	4579	4620	4687	4716	4769	4938	5097	5244	5249	5361	5351	3399
00.700	4560	4595	4632	4700	4760	4811	4901	4936	5000	5202	5377	5530	5534	5646	5628	3572
00.800	4655	4707	4755	4841	4915	4979	5082	5124	5198	5420	5604	5762	5758	5869	5846	3708
00.900	4738	4804	4864	4967	5052	5123	5238	5287	5365	5604	5792	5948	5939	6048	6021	3818
01.000	4813	4891	4960	5076	5173	5252	5377	5428	5514	5761	5951	6104	6093	6196	6164	3907
01.200	4941	5041	5126	5266	5374	5468	5609	5666	5759	6017	6203	6349	6330	6430	6391	4050
01.400	5048	5165	5262	5422	5545	5645	5798	5858	5956	6218	6395	6537	6509	6606	6563	4157
01.600	5140	5272	5381	5555	5689	5794	5955	6017	6118	6380	6548	6684	6651	6746	6700	4243
01.800	5221	5365	5485	5670	5811	5923	6089	6153	6256	6515	6673	6804	6768	6860	6811	4313
02.000	5293	5449	5576	5773	5919	6035	6207	6271	6374	6629	6777	6904	6865	6957	6905	4372
04.000	5759	5979	6150	6400	6574	6704	6881	6942	7036	7237	7331	7436	7381	7465	7404	4685
06.000	6023	6274	6462	6728	6908	7035	7203	7260	7342	7503	7574	7673	7610	7695	7626	4825
08.000	6203	6473	6670	6944	7122	7245	7402	7453	7526	7662	7721	7814	7751	7837	7765	4913
10.000	6339	6621	6825	7099	7275	7393	7540	7586	7652	7772	7821	7917	7849	7934	7861	4973
20.000	6743	7051	7261	7527	7685	7783	7898	7931	7975	8054	8088	8176	8104	8185	8114	5133
30.000	6962	7277	7487	7740	7883	7968	8062	8090	8125	8187	8215	8303	8228	8311	8237	5210
40.000	7111	7428	7634	7876	8005	8082	8165	8189	8219	8272	8295	8382	8308	8391	8314	5258

$2\alpha/RT\beta = 3.50$

p/K′	γ-Values 1.00	1.50	2.00	3.00	4.00	5.00	7.00	8.00	10.0	20.0	40.0	70.0	100	200	400	1000
00.001	0891	0573	0391	0207	0127	0086	0050	0042	0034	0022	0019	0019	0019	0019	0010	0019
00.002	1132	0787	0575	0341	0225	0162	0099	0084	0064	0041	0033	0031	0030	0029	0025	0029
00.004	1413	1054	0822	0543	0387	0293	0192	0163	0128	0076	0060	0054	0052	0049	0046	0048
00.006	1598	1237	0996	0695	0520	0407	0280	0241	0190	0112	0086	0077	0074	0071	0066	0068
00.008	1737	1379	1135	0826	0633	0510	0361	0316	0252	0148	0113	0101	0097	0092	0086	0088
00.010	1850	1496	1250	0936	0736	0605	0438	0387	0314	0185	0139	0124	0119	0113	0108	0108
00.020	2228	1901	1665	1350	1131	0980	0775	0703	0599	0379	0281	0246	0234	0219	0213	0209
00.030	2465	2162	1941	1641	1422	1267	1050	0971	0853	0576	0433	0377	0357	0335	0323	0320
00.040	2640	2358	2152	1868	1656	1503	1284	1204	1079	0770	0589	0512	0486	0455	0439	0433
00.050	2780	2516	2323	2057	1852	1706	1492	1410	1283	0956	0749	0655	0619	0579	0558	0548
00.060	2897	2650	2468	2222	2021	1880	1675	1595	1470	1137	0910	0800	0758	0705	0680	0668
00.070	2998	2766	2592	2360	2171	2036	1839	1764	1643	1309	1072	0948	0898	0837	0807	0790
00.080	3084	2865	2704	2487	2306	2179	1990	1915	1798	1474	1233	1098	1043	0973	0940	0919
00.090	3163	2956	2804	2602	2426	2306	2129	2059	1946	1630	1387	1251	1188	1112	1074	1049
00.100	3234	3038	2892	2702	2538	2424	2255	2190	2081	1780	1540	1400	1335	1254	1215	1183
00.130	3411	3244	3122	2966	2824	2730	2587	2533	2444	2186	1974	1843	1776	1686	1645	1602
00.170	3595	3460	3361	3246	3128	3055	2948	2907	2838	2648	2488	2389	2334	2261	2226	2181
00.200	3708	3592	3511	3418	3319	3261	3173	3141	3091	2947	2832	2756	2720	2664	2651	2607
00.250	3863	3775	3715	3660	3581	3540	3489	3471	3441	3366	3319	3295	3284	3269	3287	3258
00.300	3991	3925	3885	3857	3797	3776	3754	3744	3734	3719	3727	3743	3753	3776	3820	3803
00.400	4194	4164	4152	4174	4143	4151	4170	4179	4197	4274	4362	4433	4471	4529	4606	4590
00.500	4354	4353	4360	4415	4411	4439	4489	4511	4550	4691	4830	4927	4975	5046	5130	5111
00.600	4482	4503	4529	4617	4630	4674	4747	4779	4836	5017	5186	5292	5345	5414	5500	5474
00.700	4590	4631	4673	4780	4810	4867	4960	5000	5067	5280	5464	5575	5626	5689	5774	5743
00.800	4684	4741	4796	4925	4966	5035	5142	5188	5265	5496	5689	5798	5846	5904	5989	5952
00.900	4768	4839	4905	5053	5104	5178	5299	5350	5432	5679	5875	5979	6027	6078	6160	6121
01.000	4844	4928	4999	5164	5224	5305	5437	5491	5579	5836	6031	6131	6173	6221	6302	6259
01.200	4971	5076	5166	5355	5427	5523	5669	5727	5823	6089	6277	6367	6404	6446	6525	6477
01.400	5078	5200	5303	5512	5596	5699	5856	5919	6019	6288	6466	6549	6580	6615	6693	6642
01.600	5172	5309	5422	5648	5739	5848	6014	6078	6181	6447	6616	6689	6718	6750	6829	6772
01.800	5254	5404	5525	5762	5861	5976	6146	6211	6317	6581	6738	6804	6830	6859	6936	6880
02.000	5325	5486	5616	5865	5969	6088	6263	6328	6433	6691	6841	6900	6924	6949	7029	6971
04.000	5790	6015	6189	6499	6620	6752	6930	6994	7086	7287	7378	7413	7425	7441	7516	7452
06.000	6052	6308	6500	6829	6951	7080	7247	7305	7385	7545	7613	7640	7649	7661	7739	7669
08.000	6232	6506	6707	7046	7163	7286	7443	7495	7566	7699	7757	7777	7785	7795	7878	7802
10.000	6370	6655	6862	7202	7314	7433	7579	7625	7690	7806	7855	7873	7880	7889	7973	7894
20.000	6772	7082	7295	7631	7719	7816	7929	7963	8005	8081	8114	8127	8131	8137	8213	8141
30.000	6990	7307	7519	7841	7913	7997	8090	8116	8152	8213	8238	8247	8252	8256	8338	8261
40.000	7138	7457	7663	7977	8034	8110	8191	8212	8243	8293	8317	8325	8330	8333	8414	8336

$2\alpha/RT\beta = 3.60$

p/K'	1.00	1.50	2.00	3.00	4.00	5.00	7.00	8.00	10.0	20.0	40.0	70.0	100	200	400	1000
									γ-Values							
00.001	0901	0581	0398	0211	0128	0087	0050	0042	0034	0022	0019	0019	0019	0019	0019	0019
00.002	1144	0796	0583	0349	0229	0164	0100	0085	0064	0041	0033	0031	0030	0029	0029	0029
00.004	1429	1067	0834	0550	0393	0297	0194	0165	0129	0076	0060	0054	0052	0049	0048	0048
00.006	1614	1252	1010	0707	0529	0414	0283	0244	0192	0112	0086	0077	0074	0071	0069	0068
00.008	1755	1396	1151	0836	0644	0518	0366	0320	0255	0149	0113	0101	0097	0092	0090	0088
00.010	1869	1515	1268	0948	0748	0614	0445	0392	0318	0186	0139	0124	0119	0113	0109	0108
00.020	2249	1922	1687	1365	1150	0997	0789	0715	0610	0384	0283	0248	0234	0220	0214	0209
00.030	2488	2186	1966	1658	1444	1289	1069	0988	0867	0583	0437	0379	0360	0337	0325	0321
00.040	2665	2385	2179	1887	1683	1528	1307	1227	1099	0782	0595	0517	0488	0458	0442	0434
00.050	2805	2543	2350	2076	1881	1733	1518	1436	1307	0974	0759	0661	0624	0585	0564	0552
00.060	2923	2678	2497	2238	2053	1913	1705	1624	1498	1159	0926	0811	0765	0712	0687	0673
00.070	3023	2793	2622	2377	2204	2069	1872	1796	1673	1336	1090	0964	0909	0847	0818	0798
00.080	3111	2894	2735	2504	2341	2211	2026	1950	1834	1504	1254	1117	1058	0985	0951	0929
00.090	3189	2985	2835	2620	2461	2343	2165	2095	1983	1665	1415	1273	1209	1129	1088	1063
00.100	3260	3068	2924	2722	2576	2462	2295	2229	2121	1818	1573	1429	1361	1275	1228	1200
00.130	3439	3276	3157	2986	2864	2771	2631	2578	2489	2232	2020	1885	1817	1723	1670	1634
00.170	3623	3492	3396	3265	3172	3101	2996	2956	2890	2704	2549	2448	2396	2318	2271	2238
00.200	3738	3626	3546	3435	3362	3307	3225	3193	3145	3007	2900	2828	2792	2740	2709	2685
00.250	3893	3808	3751	3676	3628	3590	3542	3526	3501	3435	3396	3378	3371	3364	3360	3359
00.300	4021	3960	3921	3874	3844	3826	3810	3802	3795	3792	3811	3833	3850	3878	3899	3916
00.400	4226	4201	4190	4187	4193	4202	4228	4239	4260	4351	4454	4530	4573	4639	4680	4707
00.500	4384	4388	4400	4430	4461	4494	4549	4573	4617	4770	4921	5024	5077	5151	5194	5220
00.600	4513	4539	4569	4627	4681	4728	4806	4843	4903	5096	5275	5387	5443	5514	5551	5574
00.700	4622	4668	4713	4794	4862	4922	5021	5064	5135	5358	5551	5666	5719	5783	5817	5837
00.800	4718	4780	4836	4935	5018	5089	5204	5251	5332	5576	5775	5885	5935	5994	6022	6041
00.900	4801	4877	4945	5059	5154	5233	5360	5412	5500	5757	5958	6063	6109	6161	6187	6203
01.000	4875	4964	5040	5168	5275	5362	5498	5553	5647	5911	6110	6211	6253	6300	6323	6336
01.200	5003	5113	5207	5359	5477	5577	5728	5788	5889	6162	6352	6441	6477	6518	6538	6547
01.400	5109	5237	5343	5515	5647	5754	5917	5980	6083	6357	6537	6618	6647	6682	6699	6708
01.600	5203	5346	5462	5648	5789	5902	6071	6136	6242	6514	6682	6754	6781	6814	6826	6836
01.800	5285	5440	5565	5763	5911	6030	6204	6270	6377	6644	6801	6867	6891	6919	6931	6939
02.000	5357	5523	5658	5865	6020	6140	6319	6386	6492	6754	6901	6960	6983	7006	7020	7026
04.000	5823	6053	6228	6487	6666	6800	6979	7042	7137	7333	7422	7458	7471	7484	7490	7495
06.000	6083	6343	6537	6811	6994	7122	7292	7349	7429	7585	7652	7679	7687	7699	7703	7706
08.000	6264	6541	6743	7023	7202	7328	7484	7536	7605	7739	7792	7814	7820	7829	7834	7836
10.000	6399	6688	6896	7176	7352	7471	7618	7663	7727	7841	7887	7906	7911	7919	7925	7924
20.000	6798	7111	7327	7595	7751	7849	7959	7992	8036	8109	8140	8152	8156	8162	8165	8167
30.000	7018	7337	7547	7803	7943	8025	8117	8144	8178	8236	8262	8271	8275	8279	8282	8283
40.000	7165	7485	7691	7933	8062	8136	8216	8238	8267	8317	8337	8347	8351	8355	8354	8358

$2\alpha/RT\beta = 3.70$

p/K′	1.00	1.50	2.00	3.00	4.00	5.00	7.00	8.00	10.0	20.0	40.0	70.0	100	200	400	1000
	γ-Values															
00.001	0918	0590	0404	0214	0127	0085	0049	0039	0030	0016	0013	0011	0019	0019	0019	0019
00.002	1165	0809	0593	0352	0231	0164	0099	0081	0062	0036	0029	0027	0030	0029	0029	0029
00.004	1454	1086	0846	0561	0399	0301	0194	0165	0126	0073	0056	0051	0052	0049	0048	0048
00.006	1643	1273	1027	0717	0537	0420	0286	0244	0193	0109	0082	0074	0074	0071	0069	0068
00.008	1786	1422	1172	0852	0654	0526	0370	0321	0257	0147	0110	0098	0097	0092	0090	0088
00.010	1901	1541	1293	0965	0763	0623	0451	0397	0322	0186	0137	0121	0119	0113	0109	0108
00.020	2286	1955	1718	1391	1174	1016	0803	0729	0619	0386	0282	0247	0235	0221	0214	0210
00.030	2527	2225	2001	1690	1475	1314	1092	1009	0887	0592	0438	0378	0360	0338	0327	0322
00.040	2707	2424	2217	1921	1717	1562	1337	1251	1124	0797	0601	0521	0493	0460	0445	0437
00.050	2848	2585	2392	2117	1918	1769	1551	1471	1336	0993	0770	0668	0630	0588	0567	0556
00.060	2969	2724	2542	2282	2095	1953	1741	1662	1534	1185	0942	0822	0775	0718	0693	0678
00.070	3070	2839	2671	2426	2250	2115	1914	1835	1711	1368	1113	0978	0921	0855	0826	0805
00.080	3161	2944	2783	2555	2391	2261	2071	1998	1880	1543	1282	1141	1074	0998	0963	0939
00.090	3238	3034	2885	2669	2513	2394	2214	2145	2030	1706	1451	1302	1231	1146	1103	1076
00.100	3311	3118	2978	2771	2627	2515	2348	2282	2172	1864	1613	1463	1386	1297	1247	1219
00.130	3491	3330	3211	3042	2922	2831	2691	2639	2551	2292	2077	1941	1861	1762	1704	1667
00.170	3681	3547	3456	3325	3235	3165	3063	3025	2962	2780	2624	2524	2459	2381	2331	2298
00.200	3794	3684	3605	3502	3427	3374	3299	3268	3219	3091	2987	2919	2868	2819	2789	2764
00.250	3952	3870	3816	3746	3701	3666	3626	3608	3581	3530	3497	3487	3463	3461	3462	3466
00.300	4081	4022	3986	3945	3919	3904	3891	3887	3885	3893	3927	3957	3948	3988	4011	4034
00.400	4287	4268	4260	4263	4276	4288	4316	4330	4359	4463	4581	4668	4678	4750	4795	4824
00.500	4447	4456	4472	4508	4546	4581	4646	4672	4720	4889	5053	5164	5179	5258	5302	5327
00.600	4579	4610	4643	4708	4769	4817	4904	4944	5006	5218	5410	5531	5541	5612	5651	5674
00.700	4689	4739	4789	4874	4949	5014	5125	5167	5243	5482	5688	5809	5811	5877	5909	5929
00.800	4784	4852	4914	5022	5109	5189	5305	5357	5442	5699	5908	6025	6022	6081	6109	6126
00.900	4869	4950	5022	5147	5248	5330	5465	5522	5611	5882	6090	6199	6190	6243	6270	6284
01.000	4946	5039	5124	5258	5369	5460	5603	5663	5760	6039	6245	6346	6332	6379	6401	6415
01.200	5073	5188	5289	5450	5576	5678	5837	5900	6003	6288	6483	6572	6550	6589	6609	6619
01.400	5183	5317	5427	5607	5745	5854	6025	6090	6199	6481	6666	6746	6715	6748	6765	6774
01.600	5276	5427	5548	5742	5887	6007	6183	6252	6357	6639	6809	6882	6845	6876	6889	6897
01.800	5357	5521	5651	5855	6010	6133	6316	6383	6493	6770	6925	6991	6952	6978	6990	6998
02.000	5432	5606	5741	5958	6123	6245	6429	6500	6607	6875	7023	7084	7040	7063	7075	7083
04.000	5901	6138	6321	6585	6768	6904	7090	7155	7251	7450	7538	7573	7512	7527	7533	7537
06.000	6165	6434	6632	6912	7097	7230	7403	7459	7541	7697	7763	7786	7724	7734	7739	7742
08.000	6345	6629	6840	7124	7307	7435	7591	7642	7716	7846	7900	7921	7853	7861	7866	7870
10.000	6483	6777	6992	7279	7460	7580	7725	7769	7835	7949	7995	8013	7942	7951	7955	7954
20.000	6888	7205	7423	7697	7854	7954	8064	8096	8138	8212	8241	8255	8181	8187	8190	8191
30.000	7106	7430	7646	7902	8045	8131	8220	8246	8281	8339	8362	8371	8298	8302	8303	8305
40.000	7254	7579	7793	8033	8164	8238	8320	8340	8369	8415	8436	8446	8371	8374	8376	8377

$2\alpha/RT\beta = 3.80$

p/K'	γ-Values 1.00	1.50	2.00	3.00	4.00	5.00	7.00	8.00	10.0	20.0	40.0	70.0	100	200	400	1000
00.001	0928	0598	0409	0217	0129	0086	0049	0039	0030	0016	0013	0011	0019	0010	0019	0019
00.002	1178	0821	0602	0358	0235	0166	0100	0082	0062	0036	0029	0027	0030	0025	0029	0029
00.004	1470	1101	0859	0569	0405	0305	0196	0167	0127	0074	0056	0051	0052	0047	0048	0048
00.006	1660	1289	1042	0729	0546	0426	0291	0247	0194	0109	0082	0074	0074	0067	0069	0068
00.008	1804	1439	1190	0866	0665	0535	0377	0326	0261	0148	0110	0098	0097	0089	0090	0088
00.010	1920	1560	1310	0980	0775	0634	0459	0402	0326	0187	0137	0121	0119	0110	0109	0108
00.020	2308	1978	1741	1413	1194	1032	0817	0742	0630	0390	0283	0247	0236	0219	0214	0210
00.030	2552	2250	2029	1716	1500	1338	1112	1030	0903	0601	0442	0380	0362	0336	0328	0323
00.040	2733	2450	2245	1950	1744	1588	1363	1278	1144	0811	0609	0526	0495	0460	0446	0439
00.050	2875	2615	2421	2147	1949	1799	1579	1497	1363	1013	0782	0675	0635	0587	0571	0558
00.060	2993	2752	2570	2312	2127	1984	1775	1693	1565	1210	0959	0833	0782	0724	0699	0684
00.070	3097	2870	2703	2460	2285	2151	1948	1870	1745	1397	1135	0994	0934	0866	0834	0813
00.080	3188	2972	2816	2588	2426	2298	2108	2035	1915	1574	1310	1162	1094	1014	0973	0949
00.090	3266	3065	2920	2705	2551	2432	2254	2184	2070	1746	1482	1329	1253	1166	1119	1090
00.100	3340	3150	3011	2810	2667	2556	2390	2323	2214	1906	1652	1496	1417	1322	1268	1236
00.130	3520	3361	3246	3081	2963	2873	2737	2685	2601	2346	2128	1987	1906	1813	1744	1703
00.170	3706	3582	3493	3367	3278	3214	3113	3078	3015	2841	2689	2593	2527	2465	2399	2363
00.200	3823	3717	3645	3544	3473	3422	3349	3322	3278	3154	3058	2997	2949	2923	2876	2854
00.250	3979	3904	3854	3786	3748	3718	3680	3667	3644	3604	3582	3580	3558	3591	3571	3579
00.300	4111	4056	4027	3991	3970	3956	3949	3949	3948	3970	4014	4057	4053	4130	4130	4157
00.400	4318	4305	4302	4307	4326	4341	4377	4392	4427	4545	4675	4770	4786	4903	4911	4945
00.500	4479	4492	4511	4556	4598	4638	4701	4735	4788	4968	5146	5267	5282	5409	5409	5436
00.600	4610	4644	4682	4756	4820	4873	4968	5010	5076	5296	5501	5627	5637	5763	5750	5772
00.700	4721	4777	4830	4923	5004	5071	5185	5231	5313	5564	5777	5900	5902	6022	6000	6019
00.800	4818	4891	4954	5070	5165	5242	5369	5424	5510	5779	5996	6113	6108	6223	6195	6210
00.900	4902	4990	5061	5195	5300	5385	5528	5583	5679	5958	6174	6285	6274	6384	6349	6363
01.000	4976	5075	5163	5303	5420	5516	5666	5726	5826	6113	6322	6425	6408	6517	6477	6488
01.200	5105	5226	5332	5496	5630	5732	5900	5961	6069	6361	6557	6647	6621	6722	6677	6687
01.400	5216	5355	5468	5653	5796	5909	6083	6152	6261	6553	6734	6814	6781	6876	6829	6838
01.600	5308	5464	5587	5785	5939	6062	6241	6309	6422	6702	6877	6945	6907	7002	6948	6958
01.800	5390	5558	5692	5904	6065	6187	6372	6442	6552	6829	6990	7051	7009	7102	7048	7055
02.000	5465	5640	5783	6008	6172	6298	6485	6558	6668	6937	7082	7140	7096	7187	7131	7137
04.000	5935	6173	6362	6630	6820	6954	7140	7203	7299	7493	7583	7614	7554	7639	7574	7579
06.000	6197	6468	6667	6954	7141	7273	7444	7504	7585	7738	7801	7825	7759	7846	7775	7778
08.000	6376	6664	6875	7164	7348	7474	7636	7683	7753	7882	7934	7954	7886	7971	7899	7902
10.000	6512	6814	7030	7317	7497	7619	7762	7809	7872	7979	8025	8043	7973	8059	7985	7984
20.000	6915	7239	7455	7729	7885	7986	8096	8128	8168	8238	8269	8279	8207	8289	8214	8216
30.000	7134	7461	7678	7933	8072	8159	8247	8273	8303	8359	8385	8394	8320	8403	8324	8326
40.000	7281	7610	7820	8063	8189	8264	8340	8363	8390	8439	8457	8466	8391	8474	8397	8397

$2\alpha/RT\beta = 3.90$

γ-Values

p/K′	1.00	1.50	2.00	3.00	4.00	5.00	7.00	8.00	10.0	20.0	40.0	70.0	100	200	400	1000
00.001	0940	0607	0416	0221	0131	0087	0049	0039	0030	0016	0013	0011	0019	0019	0019	0019
00.002	1191	0832	0611	0363	0238	0169	0101	0083	0062	0036	0029	0027	0030	0029	0029	0029
00.004	1486	1115	0871	0579	0412	0310	0199	0169	0128	0074	0057	0051	0052	0049	0048	0048
00.006	1677	1305	1058	0741	0555	0434	0296	0251	0197	0110	0083	0074	0074	0071	0069	0068
00.008	1821	1458	1205	0879	0677	0544	0384	0331	0264	0149	0111	0098	0097	0092	0090	0088
00.010	1939	1579	1329	0996	0788	0645	0466	0409	0331	0188	0138	0122	0119	0113	0109	0108
00.020	2330	2000	1763	1435	1213	1051	0833	0755	0640	0395	0284	0249	0236	0222	0215	0211
00.030	2576	2275	2054	1742	1524	1361	1132	1049	0920	0611	0446	0383	0364	0340	0328	0324
00.040	2758	2476	2271	1978	1774	1617	1387	1302	1167	0825	0617	0530	0498	0466	0449	0440
00.050	2902	2643	2451	2177	1980	1830	1610	1527	1389	1033	0793	0683	0642	0597	0575	0562
00.060	3021	2781	2604	2346	2160	2019	1806	1725	1594	1235	0976	0843	0792	0734	0705	0689
00.070	3124	2898	2733	2494	2319	2186	1985	1905	1779	1426	1159	1013	0948	0876	0843	0821
00.080	3214	3004	2847	2625	2462	2335	2146	2073	1954	1610	1339	1183	1112	1027	0984	0961
00.090	3293	3096	2952	2741	2589	2470	2294	2225	2110	1785	1516	1356	1277	1184	1134	1105
00.100	3368	3180	3045	2847	2706	2595	2429	2366	2258	1947	1689	1528	1447	1346	1290	1256
00.130	3549	3395	3279	3121	3003	2916	2785	2736	2650	2399	2181	2039	1955	1849	1784	1742
00.170	3737	3616	3529	3409	3322	3261	3166	3132	3070	2905	2757	2664	2597	2521	2470	2433
00.200	3854	3753	3682	3587	3519	3473	3403	3377	3336	3222	3135	3079	3033	2992	2969	2948
00.250	4010	3938	3892	3830	3796	3766	3735	3724	3708	3677	3668	3674	3657	3673	3686	3697
00.300	4143	4093	4065	4033	4017	4009	4010	4010	4013	4048	4107	4158	4158	4214	4252	4284
00.400	4349	4340	4339	4354	4374	4395	4438	4456	4495	4626	4769	4876	4894	4977	5030	5063
00.500	4509	4528	4554	4602	4650	4695	4768	4798	4859	5052	5241	5367	5386	5469	5517	5544
00.600	4643	4684	4724	4803	4872	4930	5032	5074	5145	5379	5593	5724	5736	5809	5848	5871
00.700	4752	4816	4871	4970	5056	5130	5248	5296	5383	5643	5865	5990	5993	6058	6090	6109
00.800	4850	4926	4995	5118	5215	5299	5433	5487	5578	5857	6079	6200	6194	6250	6278	6294
00.900	4933	5025	5105	5243	5351	5443	5590	5647	5750	6038	6255	6367	6353	6403	6427	6442
01.000	5006	5111	5203	5355	5472	5573	5728	5792	5890	6187	6401	6504	6486	6529	6552	6564
01.200	5138	5262	5373	5546	5681	5787	5955	6021	6134	6430	6628	6718	6691	6728	6745	6755
01.400	5246	5392	5510	5701	5846	5968	6143	6215	6327	6619	6801	6879	6846	6878	6892	6901
01.600	5341	5501	5631	5833	5990	6114	6299	6368	6483	6769	6938	7006	6969	6995	7009	7014
01.800	5423	5595	5731	5952	6115	6240	6429	6504	6615	6890	7048	7110	7067	7092	7104	7111
02.000	5496	5676	5828	6052	6220	6352	6544	6615	6729	6998	7139	7196	7150	7173	7183	7191
04.000	5966	6208	6400	6671	6863	7000	7189	7251	7349	7541	7625	7657	7596	7609	7615	7618
06.000	6224	6504	6709	6995	7186	7319	7488	7546	7624	7776	7837	7861	7794	7806	7810	7813
08.000	6409	6699	6912	7204	7386	7516	7672	7723	7790	7918	7968	7988	7918	7926	7931	7934
10.000	6543	6848	7063	7353	7534	7655	7800	7846	7905	8014	8056	8075	8003	8012	8015	8016
20.000	6944	7271	7489	7762	7922	8017	8124	8155	8196	8266	8291	8304	8231	8235	8238	8239
30.000	7162	7490	7707	7964	8105	8186	8273	8296	8331	8385	8407	8416	8342	8345	8344	8348
40.000	7308	7639	7848	8091	8215	8292	8366	8385	8413	8457	8479	8485	8411	8414	8416	8417

$2\alpha/RT\beta = 4.00$

p/K'	1.00	1.50	2.00	3.00	4.00	5.00	7.00	8.00	10.0	20.0	40.0	70.0	100	200	400	1000
									γ-Values							
00.001	0951	0616	0423	0224	0133	0088	0050	0040	0030	0016	0013	0011	0010	0010	0010	0009
00.002	1205	0843	0620	0370	0242	0171	0102	0084	0063	0036	0029	0027	0026	0025	0025	0025
00.004	1502	1130	0883	0589	0419	0315	0202	0171	0129	0074	0057	0051	0049	0047	0046	0045
00.006	1694	1323	1073	0753	0565	0442	0300	0255	0199	0110	0083	0074	0072	0067	0066	0065
00.008	1842	1475	1223	0893	0689	0553	0390	0337	0268	0150	0111	0098	0094	0089	0086	0085
00.010	1961	1598	1347	1013	0802	0656	0474	0416	0335	0190	0138	0122	0116	0110	0108	0105
00.020	2354	2024	1786	1456	1233	1069	0848	0769	0653	0401	0286	0249	0235	0220	0215	0209
00.030	2600	2299	2080	1767	1548	1384	1152	1068	0937	0621	0450	0386	0363	0338	0328	0323
00.040	2782	2506	2299	2007	1802	1645	1412	1327	1190	0841	0625	0535	0501	0464	0450	0439
00.050	2926	2670	2480	2208	2011	1861	1639	1556	1418	1054	0807	0692	0648	0597	0576	0564
00.060	3047	2810	2635	2379	2195	2052	1840	1758	1627	1260	0996	0857	0803	0737	0709	0692
00.070	3149	2926	2766	2527	2354	2223	2021	1944	1813	1457	1180	1029	0965	0887	0849	0827
00.080	3243	3035	2883	2660	2498	2373	2184	2111	1994	1645	1367	1207	1131	1043	0999	0972
00.090	3323	3130	2984	2778	2629	2509	2336	2266	2150	1824	1552	1385	1307	1205	1152	1126
00.100	3395	3213	3078	2884	2745	2637	2473	2409	2303	1993	1730	1565	1484	1374	1317	1283
00.130	3577	3427	3316	3158	3048	2960	2833	2784	2699	2452	2236	2092	2015	1907	1838	1790
00.170	3768	3651	3567	3451	3368	3309	3216	3187	3128	2967	2827	2736	2690	2612	2566	2527
00.200	3883	3785	3719	3627	3567	3521	3457	3433	3394	3293	3214	3166	3142	3108	3088	3069
00.250	4039	3975	3930	3874	3841	3821	3795	3782	3768	3753	3756	3772	3786	3812	3832	3851
00.300	4171	4129	4103	4081	4071	4063	4069	4073	4080	4128	4200	4262	4300	4366	4412	4449
00.400	4381	4376	4381	4402	4428	4450	4503	4524	4564	4710	4866	4984	5045	5135	5191	5228
00.500	4539	4563	4593	4650	4702	4746	4832	4866	4926	5136	5334	5470	5535	5624	5673	5699
00.600	4674	4721	4768	4850	4924	4985	5090	5138	5216	5464	5684	5818	5883	5957	5997	6020
00.700	4785	4851	4911	5020	5111	5185	5311	5364	5450	5725	5954	6080	6138	6205	6237	6254
00.800	4882	4965	5037	5165	5268	5355	5494	5550	5647	5937	6163	6284	6331	6389	6418	6432
00.900	4966	5063	5146	5288	5404	5500	5651	5710	5813	6113	6336	6446	6492	6539	6564	6577
01.000	5040	5147	5247	5401	5527	5629	5789	5854	5960	6264	6477	6581	6620	6665	6686	6695
01.200	5170	5304	5414	5592	5732	5844	6017	6085	6196	6504	6702	6787	6821	6857	6876	6888
01.400	5278	5431	5553	5747	5898	6022	6206	6274	6391	6685	6870	6946	6973	7005	7020	7030
01.600	5371	5539	5670	5881	6045	6167	6359	6428	6543	6831	7002	7067	7092	7119	7133	7141
01.800	5454	5631	5774	6000	6163	6293	6487	6561	6675	6955	7110	7168	7189	7213	7225	7231
02.000	5528	5714	5870	6100	6270	6407	6599	6671	6784	7054	7198	7251	7271	7293	7303	7309
04.000	5995	6246	6438	6715	6909	7050	7235	7302	7394	7585	7667	7697	7709	7722	7726	7729
06.000	6257	6539	6746	7032	7228	7362	7531	7588	7665	7812	7875	7896	7908	7917	7924	7927
08.000	6440	6733	6950	7242	7430	7556	7710	7760	7829	7952	8000	8017	8026	8035	8040	8044
10.000	6574	6882	7098	7391	7570	7692	7836	7879	7940	8046	8086	8103	8108	8116	8119	8120
20.000	6972	7302	7522	7796	7952	8045	8152	8183	8223	8290	8317	8327	8331	8337	8340	8342
30.000	7189	7519	7735	7994	8132	8211	8298	8324	8354	8406	8427	8437	8441	8445	8444	8448
40.000	7334	7667	7875	8116	8244	8315	8389	8410	8436	8481	8497	8506	8510	8513	8513	8516

$2\alpha/RT\beta = 4.10$

γ-Values

p/K′	1.00	1.50	2.00	3.00	4.00	5.00	7.00	8.00	10.0	20.0	40.0	70.0	100	200	400	1000
00.001	0963	0624	0429	0228	0135	0089	0050	0040	0030	0016	0013	0011	0019	0019	0019	0019
00.002	1219	0855	0630	0376	0246	0174	0103	0084	0063	0036	0029	0027	0030	0029	0029	0029
00.004	1518	1144	0897	0599	0426	0321	0205	0173	0130	0074	0057	0051	0052	0049	0048	0048
00.006	1712	1339	1088	0766	0574	0449	0305	0259	0202	0110	0083	0074	0074	0071	0069	0068
00.008	1861	1495	1240	0908	0701	0563	0396	0342	0271	0150	0111	0098	0097	0092	0090	0088
00.010	1979	1617	1366	1029	0816	0668	0483	0422	0340	0191	0138	0122	0119	0113	0109	0108
00.020	2378	2047	1810	1480	1255	1088	0865	0784	0665	0406	0287	0250	0238	0222	0216	0211
00.030	2624	2326	2107	1793	1574	1410	1176	1089	0955	0632	0455	0388	0367	0343	0331	0327
00.040	2810	2534	2328	2037	1831	1673	1439	1353	1214	0856	0634	0541	0507	0472	0454	0445
00.050	2953	2697	2511	2239	2044	1892	1672	1587	1447	1078	0819	0700	0655	0606	0583	0569
00.060	3074	2839	2666	2413	2229	2086	1875	1792	1661	1291	1015	0870	0814	0747	0717	0700
00.070	3177	2959	2798	2563	2390	2260	2059	1981	1853	1490	1205	1046	0979	0898	0861	0837
00.080	3270	3067	2914	2696	2538	2411	2224	2152	2034	1683	1399	1232	1152	1057	1012	0983
00.090	3351	3161	3019	2815	2667	2551	2379	2312	2195	1868	1586	1417	1331	1224	1171	1135
00.100	3423	3243	3114	2924	2786	2680	2518	2453	2348	2038	1773	1604	1512	1400	1336	1296
00.130	3606	3459	3353	3200	3091	3007	2882	2833	2752	2510	2294	2149	2061	1947	1876	1825
00.170	3797	3686	3605	3492	3411	3359	3271	3240	3187	3032	2901	2813	2754	2681	2632	2595
00.200	3913	3821	3757	3670	3616	3575	3514	3491	3457	3363	3298	3255	3215	3187	3174	3158
00.250	4072	4008	3971	3921	3892	3875	3853	3844	3834	3830	3847	3873	3868	3903	3931	3955
00.300	4203	4167	4145	4124	4121	4120	4128	4135	4146	4208	4296	4369	4380	4457	4507	4547
00.400	4411	4413	4421	4449	4479	4508	4561	4588	4632	4794	4966	5090	5114	5208	5266	5301
00.500	4573	4601	4634	4697	4753	4806	4893	4931	4997	5219	5433	5572	5592	5680	5727	5753
00.600	4703	4758	4809	4899	4975	5046	5159	5203	5286	5543	5777	5915	5926	6001	6037	6059
00.700	4817	4887	4953	5068	5164	5242	5374	5430	5521	5806	6038	6169	6171	6235	6265	6282
00.800	4914	4999	5080	5211	5319	5411	5558	5615	5717	6015	6245	6367	6359	6414	6439	6454
00.900	4997	5099	5189	5339	5455	5559	5715	5779	5882	6188	6414	6526	6511	6557	6579	6593
01.000	5072	5187	5289	5450	5579	5684	5852	5918	6028	6338	6555	6655	6634	6675	6695	6706
01.200	5200	5340	5455	5642	5785	5899	6079	6150	6264	6573	6768	6858	6826	6861	6876	6886
01.400	5312	5467	5594	5795	5952	6071	6263	6335	6451	6755	6932	7007	6970	6999	7014	7021
01.600	5406	5574	5711	5929	6094	6223	6415	6490	6608	6897	7061	7127	7085	7111	7123	7128
01.800	5486	5667	5816	6046	6213	6350	6546	6615	6735	7013	7166	7224	7177	7200	7211	7218
02.000	5559	5751	5909	6144	6317	6455	6653	6727	6843	7114	7250	7304	7255	7277	7287	7293
04.000	6028	6284	6477	6759	6953	7095	7281	7347	7441	7629	7707	7737	7676	7687	7693	7696
06.000	6288	6572	6784	7075	7267	7401	7573	7628	7708	7851	7908	7931	7864	7874	7877	7880
08.000	6466	6770	6985	7283	7471	7595	7750	7799	7866	7982	8032	8051	7980	7988	7991	7994
10.000	6606	6914	7136	7427	7608	7730	7871	7914	7975	8076	8116	8132	8061	8067	8070	8073
20.000	7002	7331	7552	7828	7983	8077	8180	8211	8249	8313	8339	8352	8278	8281	8283	8285
30.000	7218	7551	7767	8021	8159	8238	8326	8347	8376	8426	8449	8457	8383	8386	8388	8389
40.000	7360	7694	7903	8144	8270	8340	8412	8431	8457	9501	8518	8526	8449	8453	8454	8455

$2\alpha/RT\beta = 4.20$

p/K′	γ-Values 1.00	1.50	2.00	3.00	4.00	5.00	7.00	8.00	10.0	20.0	40.0	70.0	100	200	400	1000
00.001	0975	0633	0436	0232	0138	0090	0050	0040	0030	0016	0013	0011	0019	0019	0019	0019
00.002	1233	0866	0639	0382	0250	0177	0104	0085	0063	0036	0029	0027	0030	0029	0029	0029
00.004	1536	1159	0911	0609	0434	0326	0209	0175	0132	0074	0057	0051	0052	0049	0048	0048
00.006	1730	1357	1104	0780	0584	0459	0310	0263	0204	0111	0083	0074	0074	0071	0069	0068
00.008	1881	1512	1258	0922	0713	0574	0403	0348	0275	0152	0111	0098	0097	0092	0090	0088
00.010	2000	1639	1386	1045	0830	0680	0492	0430	0346	0192	0139	0122	0120	0113	0109	0108
00.020	2400	2073	1833	1503	1277	1108	0882	0800	0678	0412	0290	0252	0238	0223	0217	0213
00.030	2648	2351	2134	1821	1601	1435	1197	1112	0974	0643	0461	0391	0371	0346	0333	0328
00.040	2835	2561	2357	2069	1863	1701	1468	1379	1239	0875	0643	0547	0511	0474	0456	0447
00.050	2977	2727	2541	2272	2077	1925	1704	1619	1476	1102	0836	0709	0662	0611	0587	0574
00.060	3099	2869	2698	2447	2264	2122	1909	1828	1694	1320	1036	0885	0826	0755	0724	0706
00.070	3206	2989	2830	2596	2428	2296	2098	2018	1891	1526	1234	1069	0996	0909	0871	0845
00.080	3298	3098	2949	2732	2576	2453	2266	2193	2075	1723	1433	1259	1176	1075	1027	0998
00.090	3378	3191	3055	2854	2707	2591	2422	2354	2241	1909	1628	1450	1360	1249	1190	1153
00.100	3451	3275	3150	2961	2827	2722	2563	2498	2395	2086	1816	1642	1549	1431	1362	1320
00.130	3632	3494	3390	3238	3136	3052	2933	2884	2806	2566	2354	2210	2120	2002	1926	1872
00.170	3829	3718	3642	3536	3461	3404	3324	3296	3244	3102	2979	2898	2837	2769	2721	2686
00.200	3944	3855	3795	3715	3660	3625	3570	3551	3518	3437	3381	3346	3313	3293	3285	3277
00.250	4101	4046	4008	3965	3940	3926	3911	3905	3898	3909	3940	3979	3978	4024	4062	4093
00.300	4236	4202	4185	4171	4171	4171	4189	4197	4213	4291	4393	4478	4497	4582	4639	4683
00.400	4444	4447	4462	4496	4533	4565	4626	4652	4703	4875	5062	5198	5225	5325	5385	5420
00.500	4604	4639	4678	4745	4804	4863	4956	4999	5066	5304	5528	5675	5695	5785	5831	5856
00.600	4737	4796	4849	4949	5031	5101	5222	5270	5357	5630	5868	6009	6021	6095	6131	6151
00.700	4850	4925	4994	5116	5216	5301	5437	5495	5591	5883	6128	6258	6259	6320	6348	6366
00.800	4946	5037	5120	5261	5373	5470	5619	5681	5785	6092	6330	6449	6440	6493	6516	6533
00.900	5030	5137	5233	5389	5513	5611	5776	5842	5951	6266	6495	6603	6587	6630	6652	6664
01.000	5104	5226	5332	5498	5631	5743	5911	5982	6096	6408	6630	6729	6705	6745	6764	6775
01.200	5234	5378	5496	5687	5836	5957	6141	6211	6329	6639	6841	6925	6890	6925	6939	6949
01.400	5343	5504	5636	5844	6006	6130	6323	6396	6514	6816	6997	7071	7031	7059	7072	7080
01.600	5438	5612	5755	5976	6144	6276	6473	6548	6666	6961	7121	7183	7142	7166	7176	7185
01.800	5520	5708	5858	6092	6264	6401	6600	6676	6791	7075	7224	7280	7233	7252	7263	7270
02.000	5591	5793	5952	6190	6374	6510	6709	6786	6900	7170	7308	7357	7306	7328	7336	7342
04.000	6060	6321	6516	6804	6996	7136	7329	7393	7485	7673	7749	7775	7714	7725	7731	7733
06.000	6321	6606	6822	7117	7307	7444	7614	7671	7748	7888	7944	7963	7896	7906	7909	7913
08.000	6499	6804	7022	7318	7506	7633	7789	7837	7903	8016	8062	8080	8009	8016	8020	8023
10.000	6635	6948	7170	7464	7647	7766	7906	7949	8007	8107	8145	8159	8089	8094	8099	8101
20.000	7030	7360	7584	7858	8011	8107	8210	8233	8277	8337	8364	8373	8300	8304	8304	8308
30.000	7244	7579	7797	8051	8185	8266	8347	8372	8402	8448	8468	8478	8402	8406	8408	8409
40.000	7388	7719	7933	8171	8295	8364	8435	8454	8478	8519	8537	8544	8468	8471	8473	8474

$2\alpha/RT\beta = 4.30$

p/K′	γ-Values 1.00	1.50	2.00	3.00	4.00	5.00	7.00	8.00	10.0	20.0	40.0	70.0	100	200	400	1000
00.001	0986	0642	0443	0236	0140	0092	0051	0040	0030	0016	0013	0011	0019	0019	0019	0019
00.002	1248	0879	0649	0388	0255	0180	0105	0086	0064	0036	0029	0027	0030	0029	0029	0029
00.004	1553	1175	0924	0619	0441	0332	0212	0178	0133	0074	0057	0051	0052	0049	0048	0048
00.006	1748	1375	1120	0792	0595	0466	0315	0267	0206	0111	0083	0074	0074	0071	0069	0068
00.008	1901	1531	1277	0938	0726	0584	0410	0354	0279	0152	0111	0098	0097	0092	0090	0088
00.010	2020	1659	1405	1063	0845	0693	0502	0438	0351	0193	0139	0122	0120	0113	0109	0108
00.020	2423	2096	1856	1526	1299	1128	0898	0816	0692	0417	0292	0252	0238	0223	0217	0213
00.030	2673	2376	2162	1848	1628	1461	1222	1133	0994	0654	0465	0392	0371	0346	0334	0329
00.040	2860	2588	2387	2097	1890	1733	1496	1406	1265	0894	0654	0551	0515	0477	0459	0450
00.050	3006	2755	2573	2305	2111	1959	1737	1649	1506	1125	0850	0718	0671	0615	0591	0577
00.060	3126	2900	2729	2482	2299	2159	1944	1864	1731	1348	1056	0899	0837	0764	0731	0711
00.070	3233	3022	2864	2634	2466	2334	2135	2058	1930	1562	1261	1090	1012	0921	0881	0853
00.080	3327	3130	2982	2769	2615	2493	2307	2234	2118	1763	1467	1286	1199	1092	1040	1009
00.090	3405	3223	3089	2893	2748	2634	2466	2399	2286	1956	1669	1485	1392	1272	1209	1170
00.100	3480	3307	3184	3000	2869	2765	2608	2547	2443	2137	1865	1686	1587	1462	1389	1344
00.130	3665	3529	3424	3280	3179	3100	2981	2937	2861	2627	2418	2275	2183	2062	1983	1924
00.170	3857	3752	3680	3578	3508	3458	3380	3351	3306	3172	3057	2984	2926	2863	2817	2783
00.200	3973	3890	3834	3759	3710	3676	3627	3609	3581	3513	3468	3445	3414	3404	3406	3403
00.250	4131	4083	4050	4011	3994	3981	3973	3968	3965	3991	4036	4085	4092	4150	4197	4236
00.300	4267	4239	4225	4218	4222	4227	4252	4263	4282	4375	4493	4587	4613	4710	4772	4822
00.400	4475	4483	4502	4544	4585	4619	4689	4718	4770	4962	5166	5307	5337	5441	5500	5536
00.500	4634	4677	4719	4794	4860	4919	5019	5065	5139	5390	5625	5774	5798	5888	5933	5957
00.600	4769	4833	4891	4997	5086	5160	5286	5339	5427	5712	5960	6103	6113	6186	6221	6241
00.700	4882	4961	5035	5163	5269	5358	5500	5560	5661	5968	6212	6347	6343	6403	6433	6449
00.800	4979	5075	5164	5310	5427	5528	5683	5748	5856	6171	6409	6531	6520	6572	6594	6607
00.900	5061	5176	5273	5437	5566	5669	5838	5906	6018	6340	6569	6678	6659	6703	6724	6737
01.000	5134	5264	5373	5547	5684	5798	5975	6046	6161	6484	6703	6798	6775	6813	6831	6842
01.200	5267	5415	5539	5737	5890	6010	6199	6273	6393	6710	6904	6987	6955	6986	7001	7009
01.400	5377	5539	5676	5890	6055	6186	6380	6457	6578	6884	7060	7130	7089	7117	7130	7136
01.600	5470	5648	5797	6024	6194	6332	6530	6606	6726	7019	7179	7243	7196	7219	7231	7236
01.800	5550	5744	5899	6137	6315	6452	6658	6732	6850	7134	7278	7332	7283	7305	7314	7320
02.000	5625	5830	5991	6238	6422	6561	6764	6841	6957	7224	7359	7409	7357	7376	7385	7391
04.000	6092	6356	6555	6846	7040	7185	7375	7439	7533	7714	7788	7815	7751	7762	7767	7770
06.000	6351	6642	6856	7157	7349	7486	7653	7710	7784	7924	7977	7998	7929	7937	7941	7944
08.000	6531	6837	7058	7357	7545	7671	7823	7873	7937	8049	8093	8109	8040	8045	8049	8051
10.000	6664	6979	7206	7501	7683	7803	7943	7986	8040	8135	8174	8187	8117	8123	8126	8127
20.000	7053	7389	7616	7888	8043	8136	8236	8265	8301	8362	8386	8396	8321	8325	8328	8329
30.000	7272	7607	7825	8078	8211	8294	8372	8394	8422	8469	8489	8497	8422	8425	8428	8428
40.000	7414	7747	7959	8198	8319	8388	8458	8473	8500	8538	8555	8561	8486	8489	8491	8492

$2\alpha/RT\beta = 4.40$

p/K'	1.00	1.50	2.00	3.00	4.00	5.00	7.00	8.00	10.0	20.0	40.0	70.0	100	200	400	1000
									γ-Values							
00.001	0998	0651	0450	0240	0142	0093	0051	0041	0030	0016	0013	0011	0019	0019	0019	0019
00.002	1263	0891	0659	0396	0260	0183	0107	0087	0064	0036	0029	0027	0030	0029	0029	0029
00.004	1571	1190	0938	0630	0450	0339	0215	0180	0135	0075	0057	0051	0052	0049	0048	0048
00.006	1767	1392	1137	0806	0606	0475	0320	0271	0209	0112	0083	0074	0074	0071	0069	0068
00.008	1921	1552	1295	0953	0739	0596	0418	0360	0283	0153	0111	0099	0097	0092	0090	0088
00.010	2042	1681	1424	1082	0860	0706	0511	0446	0357	0195	0140	0122	0120	0113	0109	0108
00.020	2447	2120	1883	1550	1322	1150	0916	0832	0705	0423	0294	0254	0240	0224	0219	0215
00.030	2698	2406	2189	1876	1656	1487	1246	1157	1016	0667	0471	0396	0374	0348	0335	0330
00.040	2887	2616	2415	2129	1922	1763	1525	1436	1292	0912	0663	0557	0519	0480	0462	0452
00.050	3031	2785	2603	2337	2144	1993	1769	1684	1540	1150	0866	0730	0677	0621	0596	0581
00.060	3155	2931	2763	2515	2335	2195	1983	1901	1766	1380	1080	0916	0849	0772	0737	0718
00.070	3261	3052	2897	2671	2504	2375	2178	2098	1972	1598	1293	1113	1031	0935	0892	0863
00.080	3356	3159	3017	2807	2656	2534	2351	2280	2160	1807	1506	1318	1226	1112	1056	1023
00.090	3435	3254	3125	2932	2788	2677	2512	2444	2334	2004	1713	1525	1425	1299	1233	1191
00.100	3509	3339	3219	3039	2912	2810	2655	2595	2493	2189	1915	1732	1631	1499	1420	1371
00.130	3695	3563	3462	3322	3222	3148	3033	2990	2916	2691	2485	2341	2251	2127	2043	1981
00.170	3887	3785	3718	3622	3556	3510	3436	3411	3365	3244	3141	3075	3021	2964	2923	2892
00.200	4005	3923	3872	3804	3759	3729	3685	3669	3649	3589	3558	3547	3523	3525	3533	3539
00.250	4163	4120	4086	4055	4045	4035	4030	4032	4035	4075	4135	4200	4209	4281	4338	4383
00.300	4300	4272	4266	4267	4273	4284	4315	4328	4352	4462	4593	4706	4732	4839	4908	4959
00.400	4506	4522	4545	4589	4637	4676	4754	4785	4844	5050	5264	5418	5448	5557	5617	5652
00.500	4665	4716	4762	4840	4913	4978	5085	5131	5211	5472	5718	5878	5899	5989	6031	6056
00.600	4804	4871	4933	5046	5139	5216	5348	5404	5498	5791	6050	6198	6205	6276	6312	6328
00.700	4916	4999	5078	5211	5322	5416	5564	5626	5728	6046	6299	6430	6428	6486	6513	6527
00.800	5010	5113	5204	5359	5484	5583	5747	5814	5924	6250	6494	6612	6597	6647	6669	6682
00.900	5094	5218	5317	5484	5617	5729	5901	5971	6083	6416	6650	6755	6732	6775	6794	6806
01.000	5165	5301	5416	5594	5735	5855	6036	6108	6229	6555	6774	6873	6843	6880	6897	6908
01.200	5299	5451	5580	5787	5943	6064	6258	6335	6456	6778	6975	7052	7017	7046	7060	7069
01.400	5408	5577	5721	5939	6106	6240	6441	6516	6638	6949	7121	7187	7147	7173	7185	7192
01.600	5505	5688	5839	6071	6245	6383	6586	6665	6789	7082	7237	7298	7251	7272	7282	7290
01.800	5584	5785	5940	6184	6369	6509	6714	6790	6910	7187	7332	7386	7334	7355	7364	7370
02.000	5654	5867	6032	6286	6470	6614	6818	6895	7011	7283	7412	7460	7405	7424	7432	7438
04.000	6122	6393	6594	6889	7088	7232	7420	7484	7578	7757	7826	7852	7787	7799	7803	7806
06.000	6383	6677	6897	7196	7393	7526	7694	7748	7826	7960	8009	8030	7960	7969	7972	7974
08.000	6561	6871	7095	7396	7580	7709	7861	7907	7973	8081	8122	8138	8068	8074	8077	8079
10.000	6692	7011	7239	7540	7718	7835	7973	8017	8072	8167	8201	8215	8143	8149	8152	8154
20.000	7083	7419	7648	7920	8073	8166	8263	8291	8327	8385	8409	8417	8343	8346	8351	8351
30.000	7300	7637	7853	8105	8240	8315	8397	8418	8445	8491	8508	8516	8441	8445	8447	8447
40.000	7437	7773	7986	8225	8343	8412	8479	8497	8519	8557	8575	8580	8504	8507	8509	8510

$2\alpha/RT\beta = 4.50$

p/K′	1.00	1.50	2.00	3.00	4.00	5.00	7.00	8.00	10.0	20.0	40.0	70.0	100	200	400	1000
	γ-Values															
00.001	1003	0659	0457	0245	0147	0097	0054	0044	0035	0022	0019	0019	0019	0019	0019	0019
00.002	1267	0900	0668	0404	0265	0188	0110	0092	0067	0041	0033	0031	0030	0029	0029	0029
00.004	1575	1200	0950	0638	0457	0345	0221	0185	0140	0078	0060	0054	0052	0049	0048	0048
00.006	1776	1405	1148	0818	0617	0483	0327	0279	0213	0116	0086	0077	0074	0071	0069	0068
00.008	1926	1561	1307	0967	0753	0607	0426	0370	0288	0156	0114	0101	0097	0092	0090	0088
00.010	2047	1691	1437	1095	0873	0720	0521	0456	0364	0197	0141	0125	0120	0113	0109	0108
00.020	2453	2135	1897	1566	1339	1171	0933	0846	0719	0430	0299	0256	0241	0224	0219	0215
00.030	2703	2416	2203	1896	1675	1508	1268	1176	1034	0678	0479	0403	0376	0349	0337	0331
00.040	2891	2630	2433	2150	1946	1786	1550	1460	1315	0929	0677	0563	0524	0483	0464	0455
00.050	3037	2798	2621	2358	2167	2021	1796	1708	1567	1172	0885	0741	0685	0626	0600	0585
00.060	3161	2941	2776	2535	2360	2221	2014	1932	1795	1408	1101	0933	0864	0782	0744	0725
00.070	3265	3063	2910	2689	2528	2398	2205	2130	2003	1629	1319	1136	1051	0948	0903	0872
00.080	3358	3171	3032	2827	2679	2560	2383	2312	2193	1843	1538	1345	1253	1131	1071	1036
00.090	3439	3267	3139	2952	2812	2704	2543	2478	2368	2043	1750	1560	1462	1328	1254	1209
00.100	3511	3351	3235	3062	2935	2839	2689	2629	2529	2231	1959	1773	1675	1535	1450	1397
00.130	3696	3569	3478	3344	3251	3176	3066	3024	2954	2738	2542	2401	2322	2196	2109	2042
00.170	3890	3797	3731	3643	3580	3537	3472	3445	3406	3297	3207	3150	3120	3070	3035	3012
00.200	4005	3934	3885	3822	3785	3758	3720	3707	3690	3646	3631	3629	3634	3648	3669	3684
00.250	4165	4124	4100	4075	4064	4060	4061	4067	4077	4126	4206	4282	4331	4416	4482	4534
00.300	4297	4280	4277	4282	4294	4310	4346	4360	4393	4515	4664	4783	4854	4970	5044	5097
00.400	4504	4525	4554	4607	4654	4699	4784	4817	4879	5099	5322	5478	5560	5670	5728	5763
00.500	4667	4717	4767	4852	4931	4998	5111	5158	5244	5517	5769	5925	5998	6087	6131	6153
00.600	4798	4873	4938	5055	5151	5236	5370	5429	5528	5831	6089	6234	6296	6363	6397	6414
00.700	4907	5000	5084	5222	5334	5431	5584	5646	5756	6075	6328	6456	6509	6565	6590	6606
00.800	5002	5112	5208	5363	5490	5596	5763	5831	5945	6272	6512	6628	6672	6720	6741	6753
00.900	5087	5212	5318	5488	5626	5740	5916	5986	6104	6433	6662	6764	6803	6844	6862	6873
01.000	5161	5298	5414	5597	5746	5864	6049	6121	6242	6568	6783	6876	6910	6945	6962	6971
01.200	5292	5450	5579	5785	5945	6072	6268	6342	6467	6783	6973	7049	7076	7105	7120	7126
01.400	5397	5572	5714	5939	6108	6244	6443	6520	6643	6946	7116	7179	7203	7228	7239	7246
01.600	5490	5678	5832	6068	6247	6385	6588	6664	6787	7075	7228	7283	7302	7323	7334	7340
01.800	5573	5774	5932	6179	6362	6505	6711	6785	6907	7180	7317	7367	7384	7404	7412	7417
02.000	5644	5856	6024	6278	6465	6608	6816	6889	7006	7269	7393	7437	7453	7469	7479	7483
04.000	6105	6375	6578	6868	7068	7211	7399	7462	7551	7723	7789	7812	7823	7833	7838	7841
06.000	6359	6655	6873	7172	7366	7498	7665	7717	7790	7918	7965	7984	7991	8000	8003	8004
08.000	6536	6846	7067	7367	7553	7678	7825	7870	7930	8036	8074	8089	8096	8103	8106	8107
10.000	6667	6985	7212	7505	7684	7800	7933	7976	8027	8117	8151	8165	8169	8175	8177	8179
20.000	7053	7386	7610	7879	8027	8118	8212	8239	8272	8329	8352	8360	8364	8369	8372	8372
30.000	7263	7597	7812	8059	8189	8265	8340	8362	8389	8432	8450	8459	8459	8464	8465	8466
40.000	7401	7732	7940	8172	8291	8354	8420	8438	8458	8497	8513	8521	8522	8525	8526	8528

$2\alpha/RT\beta = 4.60$

p/K'	γ-Values 1.00	1.50	2.00	3.00	4.00	5.00	7.00	8.00	10.0	20.0	40.0	70.0	100	200	400	1000
00.001	1022	0672	0464	0249	0147	0096	0052	0041	0031	0016	0013	0011	0019	0019	0019	0019
00.002	1292	0916	0681	0410	0269	0189	0110	0088	0065	0036	0029	0027	0030	0029	0029	0029
00.004	1606	1224	0967	0652	0466	0351	0222	0186	0138	0075	0057	0051	0052	0049	0048	0048
00.006	1804	1428	1170	0835	0628	0493	0332	0280	0216	0113	0083	0074	0074	0071	0069	0068
00.008	1962	1594	1335	0987	0768	0619	0433	0374	0293	0155	0112	0099	0097	0092	0090	0088
00.010	2083	1721	1466	1118	0893	0733	0530	0464	0370	0199	0140	0122	0120	0113	0109	0108
00.020	2493	2169	1932	1599	1369	1193	0954	0867	0734	0437	0298	0256	0242	0226	0220	0216
00.030	2750	2463	2247	1934	1713	1542	1297	1206	1061	0694	0483	0403	0379	0351	0338	0332
00.040	2939	2673	2478	2194	1987	1829	1587	1497	1352	0953	0687	0571	0530	0487	0466	0457
00.050	3082	2846	2669	2406	2214	2064	1841	1754	1609	1206	0904	0752	0695	0633	0605	0590
00.060	3211	2995	2827	2588	2411	2273	2064	1980	1842	1449	1131	0953	0877	0791	0752	0732
00.070	3320	3115	2965	2744	2583	2456	2261	2183	2057	1678	1357	1163	1073	0965	0915	0884
00.080	3413	3223	3087	2885	2737	2619	2440	2370	2253	1897	1586	1386	1284	1155	1091	1052
00.090	3491	3320	3196	3010	2874	2765	2605	2540	2431	2104	1809	1609	1500	1359	1281	1231
00.100	3564	3408	3290	3123	2999	2903	2755	2694	2595	2298	2024	1834	1725	1577	1485	1427
00.130	3754	3631	3539	3408	3317	3245	3140	3100	3031	2821	2629	2491	2399	2272	2182	2110
00.170	3948	3854	3797	3712	3651	3614	3549	3532	3491	3397	3314	3267	3224	3184	3158	3142
00.200	4062	3997	3951	3894	3859	3837	3806	3796	3778	3750	3750	3762	3750	3780	3812	3837
00.250	4225	4190	4170	4151	4145	4145	4158	4162	4172	4243	4339	4429	4455	4554	4630	4688
00.300	4364	4344	4348	4358	4379	4398	4441	4461	4495	4634	4803	4936	4977	5101	5179	5234
00.400	4570	4600	4627	4688	4743	4791	4882	4920	4992	5225	5466	5635	5672	5783	5841	5874
00.500	4728	4790	4845	4939	5023	5095	5216	5268	5358	5647	5910	6075	6096	6184	6224	6246
00.600	4869	4944	5015	5143	5246	5333	5477	5539	5639	5961	6231	6378	6382	6449	6481	6499
00.700	4980	5077	5164	5312	5433	5531	5695	5760	5873	6207	6468	6597	6589	6642	6667	6682
00.800	5075	5194	5290	5456	5588	5699	5873	5944	6059	6403	6650	6764	6746	6792	6812	6824
00.900	5156	5293	5400	5580	5724	5844	6028	6100	6224	6562	6796	6897	6871	6910	6927	6938
01.000	5231	5376	5499	5690	5845	5966	6160	6235	6358	6699	6916	7005	6973	7008	7024	7034
01.200	5365	5525	5665	5880	6047	6178	6382	6458	6584	6912	7101	7175	7136	7162	7175	7184
01.400	5476	5655	5803	6033	6209	6349	6555	6636	6763	7074	7240	7302	7257	7280	7291	7299
01.600	5568	5766	5921	6164	6350	6492	6702	6779	6905	7197	7348	7402	7353	7374	7384	7389
01.800	5645	5859	6021	6280	6466	6614	6821	6900	7025	7301	7437	7485	7431	7451	7460	7464
02.000	5717	5937	6116	6376	6568	6715	6927	7006	7122	7389	7512	7553	7499	7515	7523	7528
04.000	6183	6460	6672	6971	7174	7318	7510	7572	7665	7834	7898	7925	7857	7867	7871	7875
06.000	6445	6749	6968	7273	7470	7605	7773	7826	7900	8026	8073	8088	8021	8030	8033	8033
08.000	6617	6934	7165	7469	7659	7783	7933	7978	8038	8141	8180	8192	8123	8130	8132	8134
10.000	6753	7080	7309	7606	7788	7907	8041	8083	8134	8221	8254	8267	8194	8200	8203	8205
20.000	7140	7481	7708	7982	8131	8222	8316	8342	8374	8430	8452	8460	8384	8389	8392	8392
30.000	7354	7689	7911	8163	8292	8366	8442	8462	8488	8531	8547	8556	8478	8482	8483	8485
40.000	7491	7829	8040	8272	8392	8456	8518	8537	8559	8595	8610	8617	8539	8542	8543	8544

$2\alpha/RT\beta = 4.70$

p/K'	γ-Values 1.00	1.50	2.00	3.00	4.00	5.00	7.00	8.00	10.0	20.0	40.0	70.0	100	200	400	1000
00.001	1034	0682	0472	0253	0150	0097	0053	0042	0031	0016	0013	0011	0019	0019	0019	0019
00.002	1307	0931	0692	0417	0274	0192	0111	0089	0065	0036	0029	0027	0030	0029	0029	0029
00.004	1624	1241	0982	0664	0475	0357	0226	0189	0140	0075	0057	0051	0052	0049	0048	0048
00.006	1824	1446	1187	0849	0640	0502	0338	0285	0219	0113	0083	0074	0074	0071	0069	0068
00.008	1984	1614	1352	1004	0782	0632	0442	0380	0298	0156	0112	0099	0097	0092	0090	0088
00.010	2102	1741	1487	1136	0909	0748	0542	0473	0377	0201	0140	0122	0120	0113	0109	0108
00.020	2516	2194	1960	1626	1394	1216	0973	0885	0750	0444	0300	0256	0243	0226	0220	0216
00.030	2777	2491	2275	1963	1743	1572	1324	1232	1086	0709	0491	0405	0380	0353	0339	0334
00.040	2965	2702	2509	2226	2022	1861	1619	1528	1381	0974	0699	0577	0534	0489	0470	0460
00.050	3110	2877	2701	2439	2251	2103	1878	1789	1645	1235	0924	0763	0703	0637	0610	0593
00.060	3239	3026	2860	2627	2450	2310	2102	2020	1885	1486	1158	0972	0892	0800	0759	0738
00.070	3351	3147	2999	2784	2623	2499	2304	2228	2101	1723	1395	1192	1094	0979	0927	0894
00.080	3441	3254	3122	2924	2778	2663	2486	2417	2301	1946	1630	1423	1316	1177	1108	1066
00.090	3518	3354	3231	3052	2917	2810	2654	2589	2483	2158	1861	1654	1544	1392	1307	1254
00.100	3592	3443	3327	3163	3043	2948	2804	2747	2649	2356	2082	1890	1777	1621	1523	1458
00.130	3787	3663	3576	3451	3364	3296	3194	3158	3092	2890	2704	2572	2481	2355	2261	2185
00.170	3979	3891	3836	3756	3702	3666	3609	3592	3559	3475	3407	3371	3333	3307	3291	3283
00.200	4094	4037	3991	3940	3911	3890	3865	3858	3845	3832	3850	3872	3872	3916	3961	4000
00.250	4256	4225	4210	4201	4198	4203	4220	4227	4243	4332	4442	4549	4582	4695	4779	4842
00.300	4396	4381	4388	4405	4433	4454	4504	4525	4567	4725	4908	5052	5101	5235	5315	5370
00.400	4601	4639	4668	4739	4797	4853	4947	4988	5064	5313	5571	5743	5782	5894	5950	5980
00.500	4764	4827	4886	4987	5076	5155	5279	5336	5429	5732	6008	6175	6192	6278	6317	6339
00.600	4901	4981	5060	5194	5300	5390	5545	5606	5712	6042	6318	6467	6469	6532	6564	6578
00.700	5012	5116	5206	5360	5486	5592	5758	5825	5941	6284	6547	6675	6667	6717	6741	6754
00.800	5105	5233	5332	5504	5641	5756	5938	6010	6131	6476	6724	6840	6819	6861	6881	6892
00.900	5189	5329	5443	5632	5777	5898	6089	6165	6291	6637	6866	6967	6938	6975	6993	7002
01.000	5262	5414	5540	5742	5897	6026	6219	6300	6426	6766	6983	7072	7038	7071	7086	7093
01.200	5398	5563	5707	5930	6097	6235	6440	6521	6645	6974	7161	7237	7192	7219	7231	7237
01.400	5510	5695	5846	6080	6262	6404	6612	6695	6821	7131	7299	7357	7310	7332	7343	7348
01.600	5600	5805	5961	6213	6399	6545	6758	6841	6960	7256	7403	7454	7404	7423	7431	7437
01.800	5679	5894	6064	6323	6514	6663	6877	6957	7080	7354	7488	7533	7480	7498	7505	7510
02.000	5751	5975	6155	6421	6619	6767	6981	7058	7176	7438	7559	7600	7542	7559	7568	7572
04.000	6215	6495	6712	7015	7217	7362	7553	7617	7706	7873	7937	7959	7891	7900	7906	7909
06.000	6479	6785	7007	7314	7511	7644	7811	7864	7934	8059	8101	8121	8050	8058	8062	8062
08.000	6649	6968	7201	7506	7694	7818	7966	8011	8071	8171	8206	8221	8149	8156	8158	8160
10.000	6782	7112	7342	7642	7824	7941	8074	8113	8163	8248	8280	8291	8219	8224	8226	8229
20.000	7169	7512	7738	8013	8159	8249	8341	8366	8399	8451	8472	8481	8404	8409	8411	8412
30.000	7379	7717	7938	8189	8316	8391	8464	8484	8508	8549	8567	8572	8498	8500	8502	8503
40.000	7515	7858	8066	8299	8415	8478	8540	8558	8577	8614	8629	8633	8557	8559	8560	8559

$2\alpha/RT\beta = 4.80$

γ-Values

p/K'	1.00	1.50	2.00	3.00	4.00	5.00	7.00	8.00	10.0	20.0	40.0	70.0	100	200	400	1000
00.001	1046	0693	0480	0257	0152	0099	0053	0042	0031	0016	0013	0011	0019	0019	0019	0019
00.002	1321	0945	0703	0425	0279	0196	0113	0091	0066	0036	0029	0027	0030	0029	0029	0029
00.004	1642	1258	0998	0676	0484	0364	0230	0192	0141	0075	0057	0051	0052	0049	0048	0048
00.006	1844	1466	1205	0864	0652	0513	0344	0290	0222	0114	0083	0074	0074	0071	0069	0068
00.008	2003	1635	1373	1022	0797	0644	0451	0388	0303	0157	0112	0099	0097	0092	0090	0088
00.010	2124	1763	1507	1156	0926	0763	0552	0483	0385	0202	0141	0122	0120	0113	0109	0108
00.020	2540	2218	1986	1653	1419	1240	0995	0904	0766	0452	0303	0258	0244	0226	0220	0216
00.030	2804	2518	2304	1994	1772	1601	1352	1258	1110	0724	0496	0408	0383	0354	0340	0335
00.040	2990	2729	2540	2259	2057	1894	1653	1561	1412	0998	0712	0584	0537	0493	0471	0462
00.050	3135	2910	2733	2475	2287	2140	1915	1829	1682	1267	0944	0778	0714	0645	0614	0597
00.060	3269	3056	2895	2663	2489	2351	2143	2061	1927	1524	1189	0993	0908	0812	0768	0744
00.070	3379	3177	3037	2821	2664	2541	2349	2272	2148	1767	1433	1222	1119	0997	0940	0905
00.080	3468	3288	3159	2964	2821	2708	2535	2465	2351	1996	1677	1462	1350	1204	1128	1083
00.090	3549	3388	3269	3092	2961	2856	2704	2643	2534	2213	1915	1707	1588	1429	1336	1279
00.100	3623	3478	3364	3205	3089	2998	2854	2800	2705	2418	2144	1951	1834	1670	1564	1494
00.130	3817	3698	3614	3493	3412	3346	3249	3214	3155	2961	2785	2657	2569	2443	2351	2269
00.170	4008	3929	3875	3802	3753	3716	3669	3656	3626	3556	3502	3480	3448	3436	3433	3436
00.200	4124	4073	4031	3987	3962	3945	3929	3921	3915	3918	3953	3991	3998	4058	4118	4172
00.250	4289	4261	4254	4246	4251	4259	4283	4293	4315	4418	4551	4672	4710	4838	4930	4996
00.300	4429	4419	4430	4452	4486	4513	4573	4595	4641	4812	5017	5175	5227	5365	5448	5503
00.400	4632	4677	4713	4787	4853	4909	5015	5057	5136	5403	5670	5850	5891	6002	6057	6086
00.500	4796	4863	4928	5036	5128	5213	5346	5401	5502	5819	6100	6270	6286	6370	6408	6427
00.600	4938	5019	5103	5242	5356	5449	5608	5674	5784	6127	6405	6552	6552	6614	6642	6657
00.700	5046	5156	5249	5410	5540	5650	5822	5893	6010	6366	6631	6757	6743	6792	6814	6827
00.800	5139	5270	5377	5552	5696	5815	6000	6075	6200	6552	6801	6912	6888	6929	6947	6958
00.900	5221	5364	5484	5680	5833	5958	6151	6228	6355	6706	6938	7035	7004	7039	7054	7064
01.000	5295	5451	5584	5790	5949	6082	6283	6361	6492	6839	7051	7137	7100	7131	7143	7154
01.200	5432	5603	5750	5976	6149	6287	6496	6579	6711	7040	7223	7292	7248	7274	7284	7292
01.400	5540	5733	5885	6130	6312	6458	6675	6752	6881	7190	7352	7409	7362	7382	7393	7398
01.600	5629	5841	6003	6260	6449	6598	6812	6892	7020	7312	7453	7504	7451	7470	7478	7484
01.800	5708	5931	6105	6371	6565	6715	6930	7010	7134	7409	7535	7582	7526	7542	7550	7555
02.000	5784	6009	6196	6465	6667	6821	7034	7113	7229	7491	7604	7645	7587	7601	7610	7614
04.000	6246	6531	6752	7055	7262	7407	7594	7658	7746	7908	7970	7992	7925	7933	7938	7942
06.000	6508	6819	7042	7352	7550	7684	7848	7899	7972	8087	8135	8149	8079	8085	8090	8091
08.000	6678	7001	7233	7541	7730	7853	8000	8043	8103	8199	8236	8247	8176	8180	8184	8187
10.000	6813	7145	7377	7678	7859	7974	8104	8143	8192	8273	8306	8317	8243	8249	8250	8253
20.000	7195	7544	7768	8040	8187	8275	8364	8392	8422	8473	8493	8501	8425	8429	8431	8432
30.000	7403	7747	7966	8215	8341	8414	8486	8506	8530	8569	8585	8592	8515	8518	8519	8521
40.000	7541	7886	8092	8323	8436	8500	8560	8577	8596	8630	8645	8651	8574	8576	8576	8579

$2\alpha/RT\beta = 4.90$

p/K'	1.00	1.50	2.00	3.00	4.00	5.00	7.00	8.00	10.0	20.0	40.0	70.0	100	200	400	1000
									γ-Values							
00.001	1059	0704	0488	0262	0155	0100	0054	0042	0031	0016	0013	0011	0019	0019	0019	0019
00.002	1336	0959	0715	0433	0285	0199	0114	0092	0067	0036	0029	0027	0030	0029	0029	0029
00.004	1658	1276	1013	0687	0494	0371	0234	0195	0143	0076	0057	0051	0052	0049	0048	0048
00.006	1866	1485	1223	0879	0665	0523	0351	0296	0226	0114	0083	0074	0074	0071	0069	0068
00.008	2025	1655	1394	1040	0812	0657	0460	0396	0309	0159	0113	0099	0097	0092	0090	0088
00.010	2145	1785	1529	1175	0944	0779	0565	0493	0392	0204	0141	0123	0120	0113	0110	0108
00.020	2565	2244	2012	1680	1445	1265	1016	0923	0783	0461	0307	0259	0245	0227	0221	0218
00.030	2832	2546	2334	2025	1805	1633	1382	1286	1135	0741	0503	0412	0386	0355	0342	0336
00.040	3018	2759	2572	2292	2091	1927	1688	1594	1445	1023	0728	0593	0544	0497	0475	0463
00.050	3164	2942	2767	2513	2324	2179	1955	1868	1720	1300	0967	0793	0724	0651	0620	0602
00.060	3300	3085	2928	2701	2529	2393	2187	2104	1970	1565	1219	1016	0927	0822	0777	0751
00.070	3407	3208	3070	2860	2706	2585	2395	2320	2194	1814	1473	1254	1147	1015	0954	0915
00.080	3499	3321	3197	3006	2865	2752	2583	2514	2403	2050	1728	1506	1387	1231	1148	1100
00.090	3575	3424	3305	3134	3005	2905	2752	2694	2588	2272	1973	1760	1638	1469	1367	1303
00.100	3650	3514	3402	3247	3134	3046	2906	2854	2761	2481	2210	2012	1894	1724	1608	1530
00.130	3851	3732	3654	3540	3460	3400	3308	3274	3217	3035	2869	2748	2661	2540	2448	2364
00.170	4040	3966	3913	3848	3802	3773	3729	3717	3692	3640	3600	3591	3569	3573	3584	3602
00.200	4153	4110	4071	4034	4013	4003	3991	3988	3986	4005	4059	4110	4128	4206	4282	4347
00.250	4323	4298	4295	4295	4304	4316	4346	4361	4389	4510	4661	4793	4843	4982	5081	5150
00.300	4462	4455	4470	4504	4539	4572	4638	4664	4712	4904	5125	5296	5349	5496	5580	5632
00.400	4663	4714	4752	4837	4907	4970	5080	5127	5212	5490	5773	5959	5998	6107	6159	6188
00.500	4830	4899	4973	5087	5186	5270	5412	5473	5574	5904	6197	6364	6379	6459	6495	6513
00.600	4970	5060	5143	5292	5408	5510	5672	5741	5855	6207	6490	6637	6633	6693	6719	6734
00.700	5075	5195	5291	5458	5593	5705	5890	5959	6082	6440	6707	6834	6816	6864	6884	6897
00.800	5170	5307	5420	5603	5753	5871	6062	6140	6266	6628	6874	6980	6956	6994	7012	7022
00.900	5251	5404	5527	5731	5885	6015	6215	6293	6424	6777	7006	7100	7067	7101	7115	7123
01.000	5329	5486	5625	5840	6002	6135	6344	6425	6553	6902	7113	7197	7161	7189	7202	7208
01.200	5467	5642	5791	6023	6206	6346	6555	6640	6772	7100	7284	7346	7304	7327	7337	7344
01.400	5570	5772	5931	6177	6364	6513	6730	6812	6939	7250	7407	7463	7412	7432	7441	7447
01.600	5663	5876	6047	6306	6499	6649	6869	6950	7078	7367	7505	7552	7499	7517	7523	7530
01.800	5740	5966	6145	6417	6616	6768	6988	7069	7186	7461	7584	7626	7569	7586	7592	7597
02.000	5814	6047	6235	6513	6714	6872	7085	7162	7285	7540	7652	7688	7630	7643	7652	7655
04.000	6274	6566	6788	7097	7304	7451	7636	7700	7787	7947	8004	8024	7957	7966	7970	7973
06.000	6539	6849	7077	7393	7590	7725	7887	7938	8005	8118	8161	8179	8108	8114	8117	8119
08.000	6706	7034	7272	7576	7765	7892	8033	8079	8134	8225	8260	8275	8202	8207	8210	8211
10.000	6843	7180	7411	7712	7890	8004	8135	8171	8222	8300	8329	8341	8266	8273	8275	8276
20.000	7225	7572	7796	8066	8215	8302	8390	8415	8444	8492	8514	8522	8445	8449	8450	8452
30.000	7432	7772	7994	8240	8367	8436	8507	8526	8550	8588	8604	8610	8533	8536	8537	8539
40.000	7567	7910	8117	8350	8457	8521	8580	8596	8615	8648	8662	8668	8589	8592	8593	8594

$2\alpha/RT\beta = 5.00$

p/K'	1.00	1.50	2.00	3.00	4.00	5.00	7.00	8.00	10.0	20.0	40.0	70.0	100	200	400	1000
									γ-Values							
00.001	1071	0715	0496	0267	0158	0102	0054	0042	0031	0016	0013	0011	0019	0019	0019	0019
00.002	1351	0972	0726	0441	0290	0203	0116	0093	0067	0036	0029	0027	0030	0029	0029	0029
00.004	1677	1292	1031	0700	0504	0379	0239	0198	0145	0076	0057	0051	0052	0049	0048	0048
00.006	1886	1505	1242	0894	0679	0533	0358	0302	0230	0115	0084	0074	0074	0071	0069	0068
00.008	2045	1676	1415	1059	0828	0670	0470	0404	0315	0160	0113	0099	0097	0092	0090	0088
00.010	2165	1807	1550	1196	0963	0795	0576	0503	0400	0206	0142	0123	0120	0113	0110	0108
00.020	2587	2270	2040	1708	1471	1291	1037	0945	0801	0470	0309	0260	0245	0228	0221	0218
00.030	2859	2574	2362	2058	1835	1664	1412	1315	1161	0758	0511	0416	0388	0358	0343	0338
00.040	3043	2790	2604	2324	2126	1965	1725	1629	1478	1050	0742	0602	0550	0500	0478	0467
00.050	3193	2974	2798	2549	2361	2216	1993	1908	1760	1334	0991	0808	0736	0657	0624	0607
00.060	3330	3118	2963	2741	2568	2438	2229	2149	2014	1606	1254	1041	0946	0834	0785	0759
00.070	3434	3240	3108	2900	2747	2630	2441	2367	2241	1864	1516	1290	1178	1034	0967	0926
00.080	3527	3356	3234	3047	2909	2796	2633	2564	2455	2105	1779	1553	1429	1261	1171	1118
00.090	3603	3460	3341	3176	3049	2953	2806	2747	2643	2333	2034	1817	1691	1512	1401	1330
00.100	3680	3549	3438	3289	3183	3096	2963	2908	2819	2546	2279	2081	1960	1784	1658	1571
00.130	3882	3766	3696	3584	3508	3452	3364	3335	3280	3112	2956	2843	2760	2645	2554	2471
00.170	4070	4003	3953	3893	3852	3826	3791	3781	3760	3723	3703	3710	3694	3716	3743	3778
00.200	4185	4147	4111	4083	4067	4059	4055	4056	4056	4093	4161	4232	4260	4358	4449	4525
00.250	4356	4333	4337	4343	4358	4373	4412	4429	4462	4602	4771	4921	4976	5126	5229	5301
00.300	4492	4495	4512	4551	4594	4631	4701	4733	4787	4998	5234	5418	5474	5625	5708	5759
00.400	4691	4751	4797	4886	4962	5030	5146	5197	5286	5582	5874	6065	6103	6210	6260	6286
00.500	4865	4937	5015	5135	5243	5328	5478	5541	5648	5991	6285	6458	6471	6545	6580	6596
00.600	5003	5098	5185	5341	5464	5570	5736	5809	5926	6286	6574	6718	6714	6769	6794	6807
00.700	5110	5235	5335	5507	5647	5763	5952	6027	6152	6520	6785	6905	6888	6932	6952	6963
00.800	5200	5345	5461	5652	5806	5930	6126	6206	6334	6698	6946	7053	7021	7059	7074	7085
00.900	5283	5440	5568	5778	5937	6072	6278	6356	6490	6849	7073	7165	7129	7161	7174	7184
01.000	5365	5524	5668	5887	6056	6193	6403	6487	6619	6971	7179	7256	7217	7244	7257	7265
01.200	5499	5683	5833	6074	6256	6402	6617	6697	6831	7163	7339	7402	7356	7378	7388	7394
01.400	5605	5811	5970	6225	6416	6565	6785	6865	7000	7305	7461	7512	7460	7480	7488	7494
01.600	5693	5913	6088	6353	6549	6705	6922	7004	7131	7420	7554	7600	7545	7562	7568	7573
01.800	5773	6003	6185	6460	6665	6820	7040	7120	7243	7511	7631	7670	7613	7628	7634	7639
02.000	5848	6085	6277	6559	6763	6921	7137	7216	7335	7586	7693	7729	7670	7684	7692	7695
04.000	6308	6603	6827	7137	7345	7493	7679	7744	7826	7981	8037	8057	7989	7997	8001	8004
06.000	6567	6883	7114	7428	7627	7760	7921	7971	8037	8150	8193	8205	8135	8142	8145	8147
08.000	6735	7068	7304	7614	7801	7926	8066	8107	8164	8252	8286	8299	8227	8232	8234	8236
10.000	6875	7212	7442	7746	7923	8037	8166	8203	8249	8325	8355	8365	8289	8295	8299	8300
20.000	7254	7600	7826	8096	8242	8325	8416	8437	8467	8512	8533	8539	8464	8468	8469	8470
30.000	7457	7801	8021	8264	8391	8459	8527	8546	8568	8606	8622	8628	8549	8553	8554	8554
40.000	7594	7935	8142	8372	8481	8540	8600	8614	8633	8665	8680	8684	8605	8608	8609	8609

$2\alpha/RT\beta = 5.10$

p/K'	1.00	1.50	2.00	3.00	4.00	5.00	7.00	8.00	10.0	20.0	40.0	70.0	100	200	400	1000
	γ-Values															
00.001	1083	0725	0504	0272	0161	0104	0055	0043	0031	0016	0013	0011	0019	0019	0019	0019
00.002	1366	0986	0738	0450	0296	0207	0118	0094	0068	0036	0029	0027	0030	0029	0029	0029
00.004	1694	1310	1047	0713	0514	0387	0243	0202	0147	0076	0057	0051	0052	0049	0048	0048
00.006	1908	1527	1261	0910	0692	0544	0365	0308	0234	0116	0084	0074	0074	0071	0069	0068
00.008	2067	1697	1435	1077	0844	0684	0480	0412	0321	0161	0113	0099	0097	0092	0090	0088
00.010	2187	1829	1572	1216	0980	0811	0588	0514	0408	0208	0142	0123	0120	0113	0110	0108
00.020	2610	2296	2068	1737	1498	1316	1061	0967	0820	0478	0312	0261	0245	0228	0221	0218
00.030	2886	2603	2391	2091	1868	1696	1443	1345	1189	0776	0520	0420	0390	0359	0344	0338
00.040	3069	2823	2640	2361	2163	2002	1761	1665	1513	1076	0758	0610	0556	0504	0480	0469
00.050	3222	3007	2833	2586	2400	2256	2036	1948	1801	1370	1018	0824	0746	0665	0630	0611
00.060	3359	3150	2998	2775	2613	2480	2275	2195	2060	1650	1290	1068	0966	0847	0793	0768
00.070	3465	3273	3146	2941	2791	2675	2491	2415	2292	1913	1563	1328	1209	1055	0982	0939
00.080	3554	3391	3270	3088	2954	2845	2681	2618	2508	2163	1835	1603	1474	1294	1196	1138
00.090	3631	3496	3377	3219	3099	3002	2857	2800	2700	2395	2098	1879	1748	1559	1439	1362
00.100	3710	3584	3474	3332	3227	3146	3017	2965	2878	2612	2351	2153	2031	1848	1714	1617
00.130	3913	3800	3738	3630	3560	3506	3426	3395	3345	3189	3046	2943	2867	2758	2673	2589
00.170	4099	4041	3992	3941	3904	3886	3854	3845	3831	3809	3809	3829	3826	3868	3914	3969
00.200	4215	4184	4153	4131	4119	4117	4117	4122	4127	4184	4273	4364	4399	4515	4620	4708
00.250	4391	4370	4379	4387	4411	4435	4480	4498	4536	4694	4881	5046	5109	5271	5379	5449
00.300	4526	4534	4551	4603	4649	4689	4767	4802	4863	5089	5344	5536	5600	5751	5833	5881
00.400	4724	4790	4842	4931	5017	5088	5214	5264	5359	5674	5976	6169	6205	6311	6357	6382
00.500	4897	4973	5056	5186	5295	5390	5542	5612	5722	6076	6378	6547	6557	6630	6661	6676
00.600	5033	5138	5226	5392	5520	5628	5805	5878	5999	6368	6657	6801	6791	6843	6867	6878
00.700	5141	5273	5379	5556	5706	5821	6015	6092	6220	6596	6861	6979	6958	6999	7017	7028
00.800	5232	5381	5506	5701	5859	5990	6188	6271	6403	6772	7016	7117	7085	7121	7135	7143
00.900	5315	5476	5611	5827	5992	6129	6338	6421	6553	6914	7139	7227	7188	7218	7232	7238
01.000	5398	5562	5709	5935	6109	6246	6465	6549	6684	7037	7240	7317	7273	7298	7311	7318
01.200	5533	5722	5875	6122	6308	6456	6675	6760	6893	7224	7395	7456	7406	7427	7438	7443
01.400	5638	5849	6015	6274	6465	6617	6843	6925	7056	7364	7512	7561	7507	7527	7533	7540
01.600	5724	5951	6128	6399	6603	6755	6978	7063	7186	7472	7602	7645	7588	7605	7611	7617
01.800	5802	6041	6226	6508	6714	6871	7090	7172	7296	7560	7677	7715	7654	7670	7675	7680
02.000	5882	6121	6314	6606	6811	6967	7188	7268	7383	7635	7737	7772	7712	7723	7732	7734
04.000	6338	6637	6865	7178	7386	7533	7722	7781	7868	8015	8068	8087	8019	8028	8030	8033
06.000	6600	6917	7150	7465	7663	7795	7955	8006	8072	8180	8217	8233	8163	8169	8171	8173
08.000	6765	7104	7341	7650	7836	7958	8100	8139	8194	8279	8312	8324	8251	8256	8258	8260
10.000	6904	7241	7476	7778	7957	8070	8195	8230	8277	8350	8376	8389	8313	8319	8322	8322
20.000	7281	7630	7853	8125	8268	8352	8437	8458	8489	8534	8551	8558	8484	8487	8487	8489
30.000	7482	7829	8045	8291	8414	8483	8549	8567	8589	8624	8639	8643	8568	8570	8571	8573
40.000	7621	7959	8168	8394	8503	8560	8618	8633	8651	8682	8696	8701	8620	8623	8624	8625

$2\alpha/RT\beta = 5.20$

p/K′	1.00	1.50	2.00	3.00	4.00	5.00	7.00	8.00	10.0	20.0	40.0	70.0	100	200	400	1000
									γ-Values							
00.001	1096	0736	0514	0278	0164	0106	0056	0043	0031	0016	0013	0011	0019	0019	0019	0019
00.002	1380	1000	0752	0458	0302	0211	0119	0096	0068	0036	0029	0027	0030	0029	0029	0029
00.004	1711	1327	1065	0727	0524	0395	0248	0205	0149	0076	0057	0051	0052	0049	0048	0048
00.006	1931	1549	1282	0927	0705	0556	0375	0314	0238	0116	0084	0074	0074	0071	0069	0068
00.008	2087	1717	1456	1096	0862	0698	0491	0421	0328	0163	0113	0099	0097	0092	0090	0088
00.010	2208	1852	1594	1237	1000	0829	0603	0525	0417	0211	0143	0124	0120	0113	0110	0108
00.020	2632	2324	2098	1766	1526	1344	1086	0989	0841	0489	0315	0263	0247	0230	0223	0219
00.030	2912	2630	2422	2123	1900	1730	1473	1376	1218	0794	0529	0423	0393	0360	0347	0340
00.040	3094	2855	2675	2396	2200	2041	1797	1703	1550	1106	0776	0620	0563	0507	0482	0471
00.050	3252	3040	2866	2624	2442	2295	2078	1991	1844	1410	1045	0842	0760	0672	0634	0616
00.060	3389	3181	3035	2817	2655	2524	2319	2242	2109	1699	1330	1097	0989	0861	0804	0775
00.070	3494	3305	3187	2981	2833	2720	2540	2467	2344	1967	1612	1369	1243	1079	1000	0953
00.080	3580	3428	3307	3131	2999	2893	2733	2669	2563	2221	1893	1656	1523	1331	1224	1159
00.090	3659	3531	3414	3259	3145	3052	2914	2856	2757	2459	2167	1944	1812	1613	1483	1394
00.100	3743	3619	3511	3374	3274	3194	3072	3024	2940	2681	2428	2233	2108	1920	1776	1668
00.130	3944	3835	3779	3678	3611	3559	3486	3460	3413	3269	3143	3050	2978	2881	2804	2727
00.170	4128	4080	4031	3987	3957	3941	3918	3913	3904	3898	3918	3955	3960	4026	4092	4169
00.200	4247	4221	4196	4179	4173	4176	4182	4189	4202	4273	4388	4492	4538	4675	4793	4891
00.250	4425	4406	4419	4439	4464	4494	4547	4570	4609	4785	4996	5175	5243	5415	5523	5592
00.300	4558	4574	4592	4652	4705	4745	4835	4873	4934	5181	5454	5658	5723	5875	5955	5999
00.400	4755	4827	4889	4985	5073	5149	5279	5336	5436	5761	6075	6271	6306	6405	6451	6474
00.500	4934	5011	5099	5236	5352	5451	5611	5678	5792	6158	6468	6636	6642	6712	6739	6755
00.600	5069	5179	5272	5439	5575	5685	5870	5946	6070	6449	6739	6876	6865	6915	6936	6947
00.700	5171	5312	5424	5606	5759	5882	6077	6158	6290	6669	6935	7049	7025	7064	7082	7092
00.800	5262	5421	5547	5750	5912	6047	6253	6337	6469	6843	7084	7182	7148	7181	7194	7204
00.900	5349	5513	5652	5876	6044	6182	6397	6481	6620	6985	7203	7285	7247	7274	7287	7295
01.000	5432	5601	5747	5985	6165	6307	6523	6611	6748	7100	7300	7372	7328	7352	7364	7370
01.200	5565	5762	5919	6169	6359	6508	6731	6814	6954	7282	7450	7506	7455	7476	7486	7491
01.400	5668	5886	6056	6320	6516	6673	6896	6981	7110	7418	7561	7608	7554	7572	7578	7584
01.600	5755	5986	6169	6445	6651	6808	7029	7112	7241	7524	7648	7690	7631	7647	7654	7658
01.800	5835	6077	6262	6555	6761	6920	7144	7223	7345	7610	7721	7757	7695	7710	7715	7721
02.000	5916	6158	6357	6650	6860	7019	7239	7318	7436	7681	7781	7813	7750	7762	7770	7772
04.000	6372	6675	6905	7217	7426	7573	7762	7821	7903	8050	8099	8119	8049	8055	8060	8062
06.000	6627	6949	7183	7502	7700	7835	7991	8039	8104	8209	8246	8258	8189	8194	8197	8198
08.000	6797	7136	7375	7684	7870	7991	8129	8169	8225	8305	8335	8347	8275	8279	8281	8283
10.000	6934	7273	7508	7811	7990	8101	8223	8258	8305	8375	8399	8411	8336	8341	8343	8345
20.000	7310	7656	7883	8154	8294	8378	8459	8484	8510	8552	8571	8577	8502	8504	8506	8507
30.000	7507	7856	8072	8313	8436	8504	8569	8586	8606	8643	8656	8660	8583	8587	8588	8589
40.000	7644	7985	8194	8416	8523	8582	8637	8651	8668	8700	8711	8716	8636	8640	8639	8640

$2\alpha/RT\beta = 5.30$

p/K'	γ-Values 1.00	1.50	2.00	3.00	4.00	5.00	7.00	8.00	10.0	20.0	40.0	70.0	100	200	400	1000
00.001	1107	0747	0523	0284	0167	0108	0056	0044	0032	0016	0013	0011	0019	0019	0019	0019
00.002	1396	1014	0766	0467	0308	0215	0122	0097	0069	0036	0029	0027	0030	0029	0029	0029
00.004	1728	1343	1081	0741	0534	0403	0253	0209	0152	0076	0057	0051	0052	0049	0048	0048
00.006	1954	1572	1303	0943	0719	0568	0383	0321	0243	0117	0084	0074	0074	0071	0069	0068
00.008	2106	1738	1476	1115	0878	0712	0502	0431	0334	0164	0113	0099	0097	0092	0090	0088
00.010	2230	1876	1619	1259	1021	0847	0616	0538	0426	0213	0143	0124	0120	0113	0110	0108
00.020	2657	2354	2131	1793	1553	1371	1110	1012	0861	0498	0318	0263	0247	0230	0223	0219
00.030	2942	2658	2453	2159	1933	1764	1506	1407	1248	0815	0539	0427	0396	0361	0347	0340
00.040	3119	2888	2708	2431	2237	2078	1836	1742	1586	1137	0796	0630	0569	0511	0487	0474
00.050	3283	3071	2897	2665	2482	2339	2121	2034	1887	1447	1075	0863	0774	0680	0642	0620
00.060	3422	3212	3074	2854	2697	2568	2367	2288	2157	1744	1371	1128	1014	0875	0813	0784
00.070	3521	3340	3223	3024	2877	2767	2589	2518	2398	2024	1665	1414	1281	1104	1017	0966
00.080	3609	3464	3343	3176	3049	2943	2785	2723	2620	2282	1957	1714	1576	1371	1253	1180
00.090	3689	3567	3449	3306	3192	3103	2969	2912	2817	2526	2238	2017	1880	1672	1530	1430
00.100	3774	3653	3551	3417	3321	3248	3131	3083	3000	2755	2508	2316	2190	1998	1846	1725
00.130	3978	3869	3819	3724	3662	3616	3544	3523	3481	3350	3241	3159	3097	3013	2950	2881
00.170	4157	4120	4071	4035	4010	4001	3982	3982	3975	3989	4026	4086	4102	4190	4278	4374
00.200	4279	4256	4242	4231	4226	4230	4246	4257	4276	4366	4499	4625	4682	4837	4967	5070
00.250	4457	4441	4461	4484	4521	4552	4615	4637	4685	4882	5111	5303	5378	5555	5665	5730
00.300	4589	4614	4632	4703	4761	4807	4905	4945	5015	5276	5564	5776	5843	5996	6071	6114
00.400	4788	4861	4931	5032	5129	5209	5350	5406	5511	5851	6174	6371	6404	6500	6540	6562
00.500	4969	5050	5138	5288	5404	5509	5674	5748	5868	6244	6554	6722	6724	6790	6816	6831
00.600	5101	5218	5316	5492	5630	5742	5932	6010	6137	6527	6816	6952	6937	6984	7005	7015
00.700	5202	5350	5471	5654	5814	5940	6141	6225	6360	6743	7007	7120	7090	7127	7144	7152
00.800	5293	5457	5590	5802	5965	6105	6315	6398	6534	6913	7152	7242	7207	7238	7252	7261
00.900	5384	5551	5693	5922	6100	6241	6460	6542	6686	7050	7265	7343	7302	7328	7341	7348
01.000	5468	5639	5791	6030	6216	6361	6583	6673	6808	7164	7358	7427	7379	7404	7414	7420
01.200	5600	5801	5965	6217	6409	6561	6789	6874	7013	7339	7503	7556	7505	7523	7532	7537
01.400	5698	5924	6098	6368	6569	6724	6954	7040	7167	7472	7609	7655	7599	7615	7621	7627
01.600	5785	6023	6207	6490	6699	6857	7085	7167	7294	7573	7695	7735	7673	7689	7694	7700
01.800	5867	6113	6305	6599	6810	6967	7194	7280	7397	7658	7762	7799	7736	7750	7755	7760
02.000	5950	6196	6400	6696	6906	7069	7288	7365	7486	7724	7820	7852	7787	7800	7807	7810
04.000	6402	6711	6945	7260	7468	7618	7799	7858	7943	8081	8131	8148	8078	8085	8089	8091
06.000	6656	6982	7217	7537	7738	7870	8025	8075	8137	8235	8273	8285	8214	8219	8222	8223
08.000	6825	7169	7408	7718	7905	8025	8161	8202	8249	8331	8359	8372	8298	8303	8303	8307
10.000	6964	7305	7539	7844	8021	8131	8252	8287	8330	8396	8423	8431	8359	8361	8364	8366
20.000	7337	7688	7912	8181	8321	8401	8483	8504	8530	8572	8589	8595	8520	8522	8524	8524
30.000	7530	7882	8097	8340	8460	8526	8588	8604	8624	8658	8674	8678	8600	8602	8603	8604
40.000	7670	8011	8218	8437	8543	8601	8654	8668	8684	8715	8728	8732	8652	8653	8655	8656

$2\alpha/RT\beta = 5.40$

γ-Values

p/K′	1.00	1.50	2.00	3.00	4.00	5.00	7.00	8.00	10.0	20.0	40.0	70.0	100	200	400	1000
00.001	1120	0753	0532	0290	0173	0113	0059	0047	0036	0022	0013	0011	0019	0019	0019	0019
00.002	1410	1025	0778	0476	0316	0222	0125	0103	0073	0041	0029	0027	0030	0029	0029	0029
00.004	1746	1360	1095	0752	0545	0412	0260	0216	0158	0080	0057	0051	0052	0049	0048	0048
00.006	1978	1572	1320	0959	0734	0580	0392	0330	0249	0121	0084	0074	0074	0071	0069	0068
00.008	2126	1759	1487	1131	0894	0728	0514	0443	0342	0168	0113	0099	0097	0092	0090	0088
00.010	2254	1894	1635	1278	1038	0866	0630	0551	0437	0217	0144	0124	0121	0113	0110	0108
00.020	2681	2368	2147	1812	1573	1399	1134	1033	0881	0510	0322	0266	0248	0231	0224	0220
00.030	2969	2683	2467	2183	1956	1792	1534	1433	1273	0835	0550	0433	0399	0365	0349	0343
00.040	3146	2894	2726	2456	2261	2108	1867	1775	1616	1163	0818	0642	0578	0516	0490	0478
00.050	3314	3076	2912	2690	2513	2369	2156	2068	1921	1482	1108	0886	0792	0689	0647	0626
00.060	3452	3240	3093	2876	2723	2601	2406	2324	2196	1788	1415	1163	1041	0892	0825	0792
00.070	3546	3357	3236	3047	2905	2794	2627	2558	2440	2070	1721	1463	1323	1134	1037	0981
00.080	3636	3459	3353	3196	3076	2974	2824	2762	2663	2338	2023	1778	1633	1416	1287	1205
00.090	3719	3561	3463	3324	3217	3135	3008	2951	2863	2584	2315	2093	1955	1736	1582	1470
00.100	3808	3662	3569	3438	3348	3276	3169	3124	3047	2812	2591	2405	2280	2086	1925	1789
00.130	4009	3890	3831	3748	3691	3648	3585	3564	3526	3417	3343	3276	3222	3156	3107	3057
00.170	4184	4110	4083	4056	4042	4032	4024	4021	4021	4050	4141	4216	4247	4361	4468	4584
00.200	4314	4271	4259	4250	4255	4264	4285	4297	4329	4434	4617	4760	4828	4999	5141	5246
00.250	4494	4460	4468	4501	4543	4578	4643	4675	4734	4939	5226	5432	5512	5694	5802	5863
00.300	4618	4605	4644	4722	4780	4832	4938	4978	5057	5332	5676	5895	5961	6113	6185	6222
00.400	4821	4873	4938	5043	5145	5231	5377	5437	5541	5893	6273	6470	6499	6591	6627	6648
00.500	5006	5059	5139	5299	5418	5528	5697	5772	5895	6276	6642	6803	6803	6864	6889	6903
00.600	5132	5203	5322	5496	5643	5761	5952	6030	6165	6552	6894	7026	7006	7050	7068	7080
00.700	5231	5344	5471	5661	5820	5950	6155	6239	6375	6759	7077	7183	7155	7188	7202	7212
00.800	5327	5464	5586	5806	5969	6112	6323	6409	6549	6922	7217	7306	7266	7294	7308	7315
00.900	5419	5559	5692	5926	6105	6244	6466	6551	6690	7053	7326	7400	7357	7381	7393	7400
01.000	5503	5636	5790	6028	6220	6364	6587	6676	6815	7160	7414	7481	7432	7455	7465	7470
01.200	5634	5782	5961	6215	6407	6562	6788	6874	7010	7327	7553	7605	7550	7570	7576	7582
01.400	5728	5914	6085	6362	6566	6719	6947	7032	7161	7454	7657	7699	7643	7658	7664	7669
01.600	5817	6022	6195	6482	6692	6851	7076	7157	7283	7553	7737	7774	7715	7729	7734	7739
01.800	5902	6108	6298	6589	6799	6959	7181	7264	7383	7632	7804	7838	7775	7788	7793	7797
02.000	5983	6182	6392	6687	6895	7052	7274	7350	7467	7698	7861	7890	7826	7837	7844	7846
04.000	6435	6684	6921	7232	7441	7586	7768	7826	7907	8037	8160	8177	8108	8114	8117	8119
06.000	6686	6964	7195	7509	7704	7832	7986	8032	8093	8187	8298	8311	8238	8243	8246	8248
08.000	6857	7133	7374	7684	7867	7984	8116	8153	8202	8277	8384	8394	8320	8325	8328	8329
10.000	6995	7274	7506	7806	7979	8086	8203	8235	8277	8341	8444	8454	8380	8384	8384	8386
20.000	7367	7648	7872	8132	8269	8347	8423	8445	8470	8511	8607	8613	8537	8539	8542	8543
30.000	7556	7834	8048	8288	8403	8465	8525	8543	8561	8594	8689	8694	8615	8619	8619	8619
40.000	7697	7966	8167	8382	8486	8539	8590	8603	8618	8647	8743	8748	8666	8670	8669	8670

$2\alpha/RT\beta = 5.50$

p/K′	1.00	1.50	2.00	3.00	4.00	5.00	7.00	8.00	10.0	20.0	40.0	70.0	100	200	400	1000
									γ-Values							
00.001	1133	0766	0543	0296	0177	0115	0060	0047	0036	0022	0019	0019	0019	0019	0019	0019
00.002	1427	1042	0791	0485	0322	0227	0128	0105	0074	0041	0033	0031	0030	0029	0029	0029
00.004	1765	1382	1111	0767	0557	0422	0265	0220	0161	0080	0060	0054	0052	0049	0048	0048
00.006	1967	1592	1342	0977	0749	0594	0400	0337	0254	0122	0087	0078	0074	0071	0069	0068
00.008	2146	1784	1507	1149	0911	0743	0526	0452	0349	0169	0116	0102	0097	0092	0090	0088
00.010	2267	1916	1662	1301	1060	0883	0644	0563	0448	0220	0145	0126	0121	0113	0110	0108
00.020	2694	2392	2180	1840	1601	1427	1161	1058	0904	0522	0327	0268	0249	0231	0224	0220
00.030	2976	2716	2500	2216	1992	1825	1567	1466	1303	0858	0564	0440	0401	0366	0349	0343
00.040	3154	2922	2760	2494	2298	2148	1907	1814	1657	1197	0838	0653	0585	0519	0492	0479
00.050	3316	3112	2947	2726	2555	2411	2201	2114	1967	1525	1141	0908	0807	0697	0653	0631
00.060	3455	3276	3135	2916	2765	2646	2457	2374	2246	1839	1455	1198	1071	0908	0835	0801
00.070	3553	3390	3272	3091	2952	2840	2678	2611	2497	2130	1772	1510	1369	1164	1056	0995
00.080	3638	3491	3388	3237	3125	3027	2879	2818	2723	2404	2085	1836	1696	1467	1323	1231
00.090	3727	3596	3500	3365	3262	3186	3065	3012	2923	2654	2378	2166	2035	1810	1643	1515
00.100	3815	3702	3611	3485	3396	3329	3224	3184	3111	2890	2666	2489	2376	2183	2014	1865
00.130	4003	3924	3869	3793	3741	3703	3646	3629	3595	3504	3430	3382	3352	3308	3283	3253
00.170	4192	4149	4125	4106	4097	4092	4090	4089	4095	4143	4233	4329	4394	4533	4663	4793
00.200	4328	4311	4304	4297	4307	4323	4351	4368	4403	4529	4707	4866	4974	5162	5311	5415
00.250	4483	4496	4507	4551	4599	4639	4709	4746	4810	5034	5301	5520	5644	5829	5934	5990
00.300	4603	4639	4688	4774	4834	4891	5006	5048	5132	5424	5738	5963	6079	6227	6293	6328
00.400	4836	4915	4980	5093	5202	5289	5446	5506	5616	5981	6314	6509	6592	6676	6710	6731
00.500	4988	5094	5182	5352	5472	5589	5762	5841	5965	6357	6670	6825	6881	6936	6960	6973
00.600	5111	5240	5369	5543	5699	5820	6017	6097	6235	6627	6909	7032	7074	7114	7133	7143
00.700	5236	5386	5515	5712	5871	6009	6217	6304	6442	6828	7080	7181	7213	7247	7261	7268
00.800	5336	5503	5625	5855	6023	6168	6385	6471	6613	6988	7214	7296	7321	7349	7361	7369
00.900	5415	5594	5731	5970	6159	6300	6525	6615	6754	7115	7317	7387	7409	7433	7443	7450
01.000	5482	5672	5835	6073	6271	6420	6646	6734	6875	7221	7401	7461	7482	7502	7512	7518
01.200	5609	5819	6004	6265	6456	6613	6843	6932	7067	7383	7532	7580	7597	7614	7620	7626
01.400	5727	5955	6125	6407	6615	6771	7001	7086	7216	7504	7632	7671	7683	7700	7705	7710
01.600	5819	6060	6234	6526	6739	6902	7127	7210	7335	7600	7708	7743	7755	7768	7773	7778
01.800	5893	6144	6340	6636	6846	7008	7233	7311	7431	7676	7771	7801	7813	7825	7830	7834
02.000	5956	6216	6435	6732	6943	7100	7322	7399	7514	7740	7825	7851	7862	7873	7878	7882
04.000	6406	6715	6957	7270	7481	7627	7806	7863	7940	8069	8112	8128	8135	8142	8145	8147
06.000	6667	6996	7233	7542	7739	7867	8019	8062	8120	8214	8247	8258	8263	8268	8270	8272
08.000	6826	7164	7406	7717	7898	8014	8142	8181	8230	8302	8328	8339	8342	8347	8351	8352
10.000	6964	7308	7538	7836	8010	8115	8230	8262	8302	8364	8386	8397	8401	8404	8404	8407
20.000	7329	7676	7901	8159	8293	8370	8446	8466	8489	8528	8545	8552	8554	8557	8558	8559
30.000	7515	7858	8073	8311	8425	8486	8543	8560	8577	8609	8623	8628	8630	8633	8635	8635
40.000	7656	7991	8191	8405	8506	8557	8606	8619	8636	8663	8674	8678	8681	8683	8684	8684

$2\alpha/RT\beta = 5.60$

p/K'	γ-Values 1.00	1.50	2.00	3.00	4.00	5.00	7.00	8.00	10.0	20.0	40.0	70.0	100	200	400	1000
00.001	1146	0778	0556	0302	0177	0114	0059	0045	0032	0016	0013	0011	0010	0010	0010	0009
00.002	1440	1054	0807	0494	0329	0230	0129	0102	0071	0037	0029	0027	0026	0025	0025	0025
00.004	1780	1394	1129	0787	0572	0431	0270	0222	0160	0077	0057	0051	0049	0047	0046	0045
00.006	2024	1640	1369	1000	0768	0609	0408	0343	0258	0120	0084	0075	0072	0067	0066	0065
00.008	2165	1800	1536	1174	0930	0761	0538	0463	0357	0169	0114	0100	0095	0089	0086	0085
00.010	2305	1958	1703	1333	1086	0902	0661	0577	0457	0222	0144	0124	0118	0111	0108	0105
00.020	2732	2449	2223	1880	1646	1459	1190	1089	0928	0535	0331	0269	0248	0230	0224	0218
00.030	3020	2740	2557	2262	2041	1868	1609	1503	1345	0886	0574	0445	0402	0365	0351	0343
00.040	3199	2997	2806	2548	2349	2202	1956	1863	1709	1239	0864	0669	0591	0520	0495	0479
00.050	3385	3167	3002	2778	2616	2469	2260	2177	2025	1581	1180	0935	0827	0701	0657	0634
00.060	3507	3303	3195	2978	2825	2711	2519	2442	2311	1905	1511	1243	1105	0928	0847	0809
00.070	3602	3450	3333	3160	3020	2911	2746	2678	2569	2204	1842	1570	1425	1203	1079	1014
00.080	3691	3580	3448	3301	3193	3100	2953	2895	2798	2486	2167	1918	1768	1529	1367	1265
00.090	3781	3674	3565	3433	3333	3258	3140	3095	3006	2743	2478	2262	2130	1897	1710	1573
00.100	3881	3754	3681	3555	3468	3403	3302	3264	3196	2985	2771	2600	2495	2289	2125	1959
00.130	4068	3981	3931	3861	3816	3782	3733	3716	3692	3611	3559	3524	3506	3494	3487	3486
00.170	4244	4236	4205	4186	4177	4178	4181	4187	4197	4269	4374	4493	4575	4738	4890	5035
00.200	4386	4362	4377	4378	4388	4405	4449	4469	4507	4655	4857	5034	5156	5368	5520	5626
00.250	4560	4557	4579	4633	4692	4737	4815	4853	4921	5170	5456	5686	5818	6009	6113	6164
00.300	4676	4741	4768	4860	4926	4985	5111	5157	5240	5560	5892	6126	6245	6392	6454	6484
00.400	4889	4969	5058	5182	5302	5389	5554	5616	5735	6117	6461	6657	6739	6821	6856	6872
00.500	5074	5168	5261	5443	5568	5694	5874	5956	6082	6491	6809	6962	7020	7073	7096	7109
00.600	5193	5344	5457	5636	5799	5918	6131	6212	6351	6757	7043	7166	7205	7243	7260	7270
00.700	5292	5465	5599	5808	5973	6115	6331	6420	6564	6961	7213	7310	7340	7371	7383	7394
00.800	5391	5563	5709	5953	6126	6277	6501	6590	6733	7117	7340	7419	7446	7470	7482	7489
00.900	5493	5660	5814	6067	6263	6407	6638	6730	6873	7239	7440	7506	7530	7552	7562	7567
01.000	5578	5759	5928	6171	6372	6527	6762	6849	6993	7344	7525	7583	7600	7619	7628	7633
01.200	5693	5919	6094	6365	6563	6725	6955	7046	7188	7504	7651	7698	7712	7730	7737	7741
01.400	5787	6033	6212	6506	6720	6878	7115	7201	7331	7624	7745	7786	7799	7816	7821	7827
01.600	5878	6129	6324	6624	6843	7013	7240	7323	7451	7715	7822	7855	7868	7882	7891	7894
01.800	5969	6221	6435	6738	6952	7115	7346	7424	7545	7789	7885	7914	7925	7938	7945	7949
02.000	6058	6315	6530	6833	7049	7209	7433	7512	7629	7854	7934	7963	7972	7985	7992	7996
04.000	6508	6820	7055	7374	7584	7733	7912	7971	8048	8175	8217	8233	8239	8244	8248	8247
06.000	6739	7077	7330	7646	7842	7972	8124	8172	8227	8318	8349	8360	8364	8371	8374	8376
08.000	6919	7274	7502	7820	8004	8120	8248	8287	8332	8403	8430	8439	8442	8449	8452	8452
10.000	7053	7398	7638	7937	8115	8220	8332	8364	8405	8465	8486	8496	8500	8504	8504	8507
20.000	7424	7767	8000	8259	8395	8472	8546	8567	8590	8627	8642	8649	8651	8652	8651	8649
30.000	7601	7965	8172	8409	8525	8586	8644	8661	8678	8711	8723	8727	8729	8733	8733	8736
40.000	7751	8084	8292	8505	8605	8656	8705	8720	8733	8760	8774	8779	8782	8786	8788	8789

$2\alpha/RT\beta = 5.70$

p/K′	γ-Values 1.00	1.50	2.00	3.00	4.00	5.00	7.00	8.00	10.0	20.0	40.0	70.0	100	200	400	1000
00.001	1168	0795	0565	0309	0184	0120	0062	0048	0036	0022	0019	0019	0019	0019	0019	0019
00.002	1468	1080	0818	0504	0336	0237	0133	0108	0075	0042	0033	0031	0030	0029	0029	0029
00.004	1810	1425	1143	0799	0584	0442	0278	0229	0167	0081	0060	0054	0052	0049	0048	0048
00.006	2007	1632	1386	1018	0785	0622	0418	0354	0264	0124	0087	0078	0074	0071	0069	0068
00.008	2194	1834	1549	1189	0948	0778	0550	0475	0365	0173	0117	0103	0097	0092	0090	0088
00.010	2308	1958	1718	1353	1104	0920	0676	0591	0469	0227	0146	0127	0121	0113	0110	0108
00.020	2740	2442	2239	1898	1668	1486	1213	1113	0951	0548	0337	0271	0251	0232	0226	0221
00.030	3033	2779	2576	2282	2069	1895	1639	1535	1374	0908	0589	0452	0408	0369	0353	0346
00.040	3205	2979	2821	2578	2374	2230	1991	1896	1743	1270	0888	0681	0603	0528	0499	0485
00.050	3385	3191	3023	2801	2644	2503	2293	2210	2066	1619	1216	0962	0846	0718	0666	0641
00.060	3512	3340	3213	3003	2851	2737	2561	2481	2351	1950	1555	1282	1142	0950	0860	0819
00.070	3605	3449	3342	3181	3050	2942	2782	2718	2615	2259	1901	1628	1475	1238	1105	1030
00.080	3695	3556	3458	3319	3219	3132	2996	2937	2842	2546	2236	1985	1840	1588	1413	1292
00.090	3796	3676	3583	3453	3357	3287	3181	3134	3055	2807	2549	2347	2216	1982	1792	1627
00.100	3885	3783	3703	3584	3499	3439	3342	3307	3242	3051	2852	2694	2591	2407	2233	2057
00.130	4059	3989	3943	3882	3840	3811	3770	3759	3737	3684	3650	3641	3637	3646	3672	3713
00.170	4262	4231	4220	4212	4210	4210	4221	4225	4243	4334	4470	4608	4699	4889	5054	5202
00.200	4398	4394	4388	4391	4411	4438	4489	4513	4559	4723	4947	5139	5267	5484	5640	5733
00.250	4539	4562	4585	4655	4716	4759	4844	4889	4966	5226	5529	5770	5902	6088	6181	6227
00.300	4667	4713	4785	4875	4940	5013	5144	5188	5284	5612	5952	6186	6302	6440	6495	6525
00.400	4904	4995	5058	5198	5315	5407	5582	5644	5765	6155	6500	6689	6766	6839	6871	6884
00.500	5046	5163	5269	5449	5583	5710	5895	5978	6110	6517	6832	6977	7027	7077	7096	7107
00.600	5178	5318	5461	5639	5810	5932	6142	6231	6371	6775	7052	7166	7203	7237	7253	7263
00.700	5309	5471	5594	5817	5975	6127	6345	6430	6576	6967	7212	7301	7330	7357	7371	7379
00.800	5402	5580	5703	5951	6134	6278	6505	6598	6740	7118	7332	7407	7429	7453	7463	7471
00.900	5474	5663	5818	6062	6264	6410	6643	6735	6879	7237	7428	7490	7508	7531	7540	7546
01.000	5539	5739	5928	6170	6372	6532	6762	6850	6994	7335	7505	7558	7576	7595	7605	7608
01.200	5681	5902	6084	6359	6560	6718	6952	7043	7180	7487	7626	7669	7681	7699	7704	7710
01.400	5797	6035	6200	6494	6713	6876	7106	7191	7321	7601	7717	7752	7765	7779	7784	7789
01.600	5880	6130	6316	6616	6831	6999	7228	7310	7434	7689	7788	7820	7831	7843	7848	7852
01.800	5951	6211	6429	6729	6941	7099	7330	7408	7526	7761	7849	7877	7885	7896	7902	7904
02.000	6016	6284	6517	6819	7037	7195	7417	7492	7606	7820	7898	7923	7931	7941	7947	7950
04.000	6467	6783	7028	7344	7558	7703	7879	7934	8008	8128	8169	8184	8189	8196	8198	8200
06.000	6723	7059	7304	7611	7808	7936	8081	8124	8179	8265	8294	8305	8309	8315	8318	8319
08.000	6883	7227	7467	7782	7961	8075	8200	8236	8281	8347	8372	8382	8386	8390	8393	8394
10.000	7024	7372	7606	7895	8070	8172	8281	8311	8350	8407	8428	8437	8441	8445	8447	8448
20.000	7382	7732	7958	8212	8339	8415	8487	8504	8529	8563	8579	8585	8588	8591	8592	8594
30.000	7568	7912	8124	8358	8466	8526	8580	8595	8613	8641	8655	8659	8662	8663	8665	8666
40.000	7702	8037	8237	8445	8544	8594	8641	8651	8666	8692	8701	8707	8709	8711	8712	8713

$2\alpha/RT\beta = 5.80$

p/K′	γ-Values															
	1.00	1.50	2.00	3.00	4.00	5.00	7.00	8.00	10.0	20.0	40.0	70.0	100	200	400	1000
00.001	1186	0811	0577	0315	0188	0122	0063	0049	0037	0022	0019	0019	0019	0019	0019	0019
00.002	1488	1098	0830	0514	0342	0243	0135	0110	0076	0042	0033	0031	0030	0029	0029	0029
00.004	1831	1446	1159	0816	0597	0454	0284	0234	0170	0081	0060	0054	0052	0049	0048	0048
00.006	2029	1654	1408	1041	0803	0636	0428	0362	0270	0125	0087	0078	0074	0071	0069	0068
00.008	2216	1857	1570	1209	0969	0798	0563	0488	0375	0175	0117	0103	0097	0092	0090	0088
00.010	2328	1979	1746	1379	1124	0940	0691	0606	0481	0230	0147	0127	0121	0113	0110	0108
00.020	2763	2467	2267	1924	1704	1513	1241	1141	0974	0563	0342	0273	0252	0233	0226	0221
00.030	3060	2809	2616	2314	2111	1930	1679	1571	1412	0936	0605	0458	0413	0371	0355	0348
00.040	3232	3009	2852	2622	2414	2269	2035	1938	1788	1310	0916	0697	0614	0532	0502	0488
00.050	3419	3231	3065	2836	2688	2552	2338	2259	2118	1668	1260	0994	0869	0728	0673	0647
00.060	3539	3372	3251	3053	2893	2781	2615	2537	2405	2008	1613	1331	1181	0973	0875	0830
00.070	3630	3479	3376	3224	3102	2996	2834	2773	2673	2325	1971	1694	1537	1282	1134	1049
00.080	3726	3591	3492	3359	3265	3184	3057	2999	2908	2619	2317	2069	1922	1659	1467	1327
00.090	3835	3722	3628	3499	3403	3338	3239	3194	3121	2886	2643	2446	2318	2083	1883	1697
00.100	3919	3822	3748	3636	3556	3496	3403	3370	3308	3133	2951	2804	2710	2537	2366	2181
00.130	4085	4020	3978	3925	3891	3865	3832	3827	3807	3777	3766	3778	3788	3828	3888	3974
00.170	4301	4277	4270	4264	4264	4267	4284	4292	4320	4432	4593	4751	4855	5066	5245	5394
00.200	4430	4431	4430	4436	4462	4500	4562	4586	4635	4823	5066	5278	5414	5640	5794	5879
00.250	4567	4595	4623	4711	4773	4818	4912	4965	5044	5322	5645	5895	6027	6210	6296	6337
00.300	4700	4753	4835	4922	4994	5077	5212	5258	5364	5705	6058	6295	6409	6538	6590	6617
00.400	4936	5032	5098	5254	5369	5463	5648	5712	5843	6242	6589	6775	6849	6917	6944	6959
00.500	5075	5197	5317	5496	5641	5767	5963	6045	6182	6596	6908	7050	7097	7141	7161	7169
00.600	5216	5362	5505	5691	5862	5992	6205	6296	6439	6848	7121	7229	7263	7295	7311	7318
00.700	5345	5514	5633	5870	6026	6185	6408	6493	6641	7035	7273	7359	7385	7412	7424	7430
00.800	5432	5615	5742	5996	6190	6331	6563	6659	6800	7180	7388	7459	7481	7503	7513	7518
00.900	5503	5697	5866	6108	6315	6467	6704	6794	6939	7296	7479	7539	7557	7578	7586	7591
01.000	5569	5773	5974	6221	6418	6586	6821	6906	7052	7390	7555	7605	7622	7639	7648	7652
01.200	5720	5947	6124	6406	6613	6766	7007	7097	7233	7538	7670	7711	7724	7739	7745	7750
01.400	5829	6072	6237	6536	6759	6928	7158	7243	7373	7647	7758	7792	7804	7817	7821	7826
01.600	5909	6164	6360	6664	6877	7047	7277	7359	7483	7732	7826	7857	7868	7878	7884	7887
01.800	5979	6242	6474	6776	6990	7148	7378	7454	7572	7803	7885	7911	7919	7930	7936	7939
02.000	6046	6318	6554	6860	7080	7243	7460	7537	7648	7859	7932	7955	7965	7974	7979	7982
04.000	6497	6817	7060	7384	7598	7741	7912	7968	8041	8158	8196	8208	8214	8220	8223	8225
06.000	6750	7088	7339	7644	7844	7968	8112	8153	8207	8289	8317	8328	8331	8338	8341	8342
08.000	6915	7261	7497	7815	7990	8105	8227	8262	8306	8371	8394	8403	8407	8412	8413	8415
10.000	7051	7401	7639	7926	8100	8200	8307	8335	8372	8427	8448	8457	8460	8463	8466	8466
20.000	7409	7759	7982	8235	8364	8436	8507	8523	8547	8582	8596	8601	8604	8608	8608	8609
30.000	7595	7938	8150	8379	8486	8545	8598	8611	8628	8658	8669	8673	8676	8680	8679	8680
40.000	7725	8060	8258	8465	8562	8612	8655	8666	8682	8705	8717	8720	8723	8725	8726	8727

$2\alpha/RT\beta = 5.90$

p/K′	γ-Values 1.00	1.50	2.00	3.00	4.00	5.00	7.00	8.00	10.0	20.0	40.0	70.0	100	200	400	1000
00.001	1203	0826	0587	0321	0192	0124	0064	0050	0037	0022	0019	0019	0019	0019	0019	0019
00.002	1506	1115	0842	0526	0349	0247	0138	0112	0077	0042	0033	0031	0030	0029	0029	0029
00.004	1852	1466	1173	0830	0611	0464	0290	0239	0173	0081	0060	0054	0052	0049	0048	0048
00.006	2054	1680	1428	1062	0821	0650	0439	0372	0277	0127	0088	0078	0074	0071	0069	0068
00.008	2238	1879	1594	1230	0992	0818	0577	0501	0385	0178	0118	0103	0097	0092	0090	0088
00.010	2348	2000	1774	1404	1145	0960	0708	0624	0494	0234	0148	0127	0121	0113	0110	0108
00.020	2786	2492	2295	1954	1740	1543	1270	1169	1000	0578	0347	0274	0253	0233	0227	0221
00.030	3084	2837	2655	2345	2153	1966	1720	1608	1449	0965	0620	0466	0416	0373	0356	0349
00.040	3259	3040	2882	2664	2457	2309	2085	1982	1832	1350	0947	0715	0626	0538	0506	0493
00.050	3453	3269	3111	2872	2729	2605	2385	2305	2174	1722	1305	1027	0895	0741	0680	0653
00.060	3567	3403	3287	3103	2942	2826	2668	2596	2463	2071	1675	1383	1226	1002	0890	0841
00.070	3656	3508	3409	3265	3152	3055	2891	2828	2732	2394	2044	1765	1603	1334	1168	1068
00.080	3758	3629	3530	3399	3308	3234	3116	3064	2976	2695	2400	2159	2012	1740	1531	1369
00.090	3873	3767	3679	3550	3454	3390	3295	3254	3185	2970	2737	2551	2429	2194	1989	1781
00.100	3952	3861	3792	3689	3613	3560	3471	3439	3380	3218	3055	2922	2836	2677	2517	2330
00.130	4113	4052	4014	3969	3941	3920	3898	3897	3885	3873	3885	3919	3944	4018	4113	4245
00.170	4341	4325	4318	4314	4317	4322	4350	4363	4399	4528	4716	4895	5012	5244	5432	5576
00.200	4461	4468	4469	4479	4516	4567	4631	4656	4710	4919	5187	5414	5559	5792	5941	6019
00.250	4596	4629	4664	4770	4829	4875	4984	5041	5121	5418	5757	6015	6148	6328	6405	6441
00.300	4738	4797	4886	4968	5048	5144	5277	5328	5445	5797	6160	6401	6513	6635	6680	6705
00.400	4966	5068	5136	5313	5421	5524	5713	5780	5918	6326	6675	6859	6928	6991	7015	7030
00.500	5104	5230	5369	5541	5703	5823	6031	6110	6255	6674	6984	7119	7162	7205	7220	7232
00.600	5257	5410	5547	5746	5913	6056	6268	6364	6506	6919	7187	7289	7321	7352	7366	7373
00.700	5380	5555	5670	5919	6080	6240	6470	6555	6706	7100	7333	7414	7437	7464	7474	7481
00.800	5462	5650	5781	6039	6247	6385	6620	6721	6865	7241	7444	7509	7529	7551	7559	7566
00.900	5531	5731	5915	6154	6365	6526	6764	6850	6998	7352	7531	7586	7604	7623	7630	7636
01.000	5599	5808	6017	6275	6468	6639	6875	6966	7108	7446	7602	7650	7666	7682	7691	7695
01.200	5760	5993	6161	6450	6666	6818	7063	7149	7287	7588	7713	7751	7765	7779	7784	7790
01.400	5860	6109	6274	6579	6804	6979	7206	7297	7423	7692	7798	7831	7841	7854	7859	7863
01.600	5937	6196	6408	6714	6925	7091	7326	7406	7530	7774	7865	7892	7903	7913	7919	7921
01.800	6007	6275	6514	6820	7039	7196	7424	7501	7619	7842	7921	7945	7953	7963	7969	7972
02.000	6077	6353	6593	6902	7125	7289	7504	7582	7691	7897	7967	7988	7997	8006	8010	8014
04.000	6532	6856	7093	7422	7637	7779	7947	8002	8074	8185	8221	8236	8239	8246	8248	8250
06.000	6776	7117	7373	7677	7879	8001	8140	8182	8236	8312	8340	8350	8354	8359	8363	8363
08.000	6950	7298	7528	7846	8022	8135	8255	8288	8329	8392	8415	8424	8428	8432	8432	8434
10.000	7078	7430	7672	7956	8127	8226	8331	8360	8396	8448	8468	8476	8479	8483	8485	8486
20.000	7432	7784	8010	8260	8386	8457	8525	8543	8563	8597	8611	8617	8619	8623	8624	8624
30.000	7623	7966	8177	8402	8507	8563	8616	8628	8644	8673	8684	8689	8690	8692	8694	8694
40.000	7747	8082	8279	8486	8580	8628	8672	8682	8697	8720	8731	8734	8736	8738	8739	8739

$2\alpha/RT\beta = 6.00$

γ-Values

p/K'	1.00	1.50	2.00	3.00	4.00	5.00	7.00	8.00	10.0	20.0	40.0	70.0	100	200	400	1000
00.001	1220	0840	0597	0327	0198	0127	0065	0050	0037	0022	0019	0019	0010	0019	0019	0019
00.002	1524	1131	0854	0539	0355	0253	0141	0114	0078	0042	0033	0031	0026	0029	0029	0029
00.004	1871	1485	1189	0845	0624	0474	0297	0245	0177	0082	0060	0054	0049	0049	0048	0048
00.006	2082	1709	1447	1082	0838	0664	0452	0382	0285	0128	0088	0078	0072	0071	0069	0068
00.008	2258	1900	1622	1254	1019	0837	0593	0515	0394	0181	0118	0103	0095	0092	0090	0088
00.010	2369	2021	1798	1428	1165	0983	0725	0641	0509	0239	0148	0127	0118	0114	0110	0108
00.020	2810	2518	2321	1984	1773	1573	1302	1196	1028	0596	0354	0275	0251	0234	0227	0223
00.030	3108	2863	2694	2376	2193	2007	1764	1649	1487	0997	0638	0474	0418	0376	0358	0352
00.040	3291	3077	2911	2704	2505	2347	2136	2031	1878	1393	0980	0736	0637	0542	0510	0494
00.050	3484	3305	3159	2909	2769	2654	2437	2353	2224	1779	1356	1064	0924	0754	0688	0658
00.060	3591	3431	3322	3154	2996	2873	2718	2652	2527	2138	1737	1441	1278	1032	0907	0851
00.070	3682	3538	3440	3306	3201	3111	2953	2886	2791	2468	2122	1843	1685	1389	1206	1091
00.080	3796	3674	3570	3441	3352	3283	3174	3127	3048	2772	2490	2255	2111	1831	1606	1418
00.090	3910	3811	3730	3608	3513	3447	3354	3314	3251	3055	2837	2663	2554	2320	2112	1883
00.100	3982	3896	3833	3738	3671	3622	3541	3512	3455	3305	3161	3046	2994	2833	2690	2511
00.130	4140	4083	4049	4014	3995	3981	3968	3974	3966	3971	4005	4064	4118	4216	4347	4524
00.170	4380	4372	4365	4362	4367	4376	4418	4438	4483	4626	4839	5039	5207	5421	5613	5747
00.200	4492	4504	4506	4524	4574	4635	4699	4725	4784	5017	5309	5550	5748	5939	6081	6149
00.250	4623	4660	4707	4826	4881	4930	5060	5115	5195	5514	5869	6135	6320	6438	6509	6542
00.300	4781	4849	4934	5013	5108	5211	5341	5401	5523	5890	6264	6503	6671	6727	6768	6789
00.400	4995	5102	5172	5368	5471	5590	5775	5851	5992	6409	6761	6939	7066	7062	7084	7097
00.500	5133	5265	5424	5585	5765	5879	6096	6172	6327	6749	7057	7184	7292	7264	7281	7291
00.600	5300	5461	5588	5805	5963	6120	6332	6430	6574	6989	7253	7349	7446	7407	7419	7427
00.700	5412	5593	5707	5966	6138	6291	6529	6616	6768	7165	7390	7466	7561	7513	7523	7530
00.800	5490	5683	5823	6082	6300	6441	6679	6783	6928	7301	7496	7559	7649	7597	7605	7611
00.900	5558	5762	5967	6203	6412	6584	6823	6907	7056	7408	7581	7633	7721	7667	7674	7680
01.000	5630	5845	6062	6328	6517	6691	6929	7024	7163	7498	7648	7693	7782	7723	7732	7736
01.200	5797	6037	6196	6493	6716	6873	7117	7201	7339	7634	7755	7793	7879	7818	7823	7827
01.400	5892	6145	6313	6621	6848	7027	7256	7346	7473	7736	7837	7866	7953	7889	7895	7898
01.600	5965	6228	6458	6764	6974	7137	7376	7453	7575	7818	7901	7929	8013	7948	7953	7956
01.800	6035	6307	6555	6862	7084	7247	7470	7549	7664	7880	7956	7978	8063	7996	8000	8004
02.000	6111	6392	6628	6940	7166	7333	7548	7625	7733	7933	7999	8021	8104	8038	8041	8044
04.000	6570	6898	7126	7465	7676	7816	7984	8035	8105	8211	8248	8259	8343	8270	8273	8274
06.000	6801	7145	7405	7712	7912	8032	8170	8211	8260	8336	8362	8373	8456	8381	8383	8385
08.000	6986	7336	7558	7876	8052	8164	8281	8314	8353	8413	8436	8444	8526	8451	8451	8454
10.000	7104	7458	7702	7987	8155	8252	8356	8384	8417	8467	8487	8495	8576	8501	8503	8504
20.000	7455	7808	8034	8284	8410	8479	8545	8561	8582	8615	8628	8634	8717	8639	8638	8640
30.000	7652	7995	8201	8422	8527	8581	8631	8645	8661	8686	8698	8703	8791	8707	8707	8708
40.000	7767	8103	8300	8507	8597	8646	8687	8698	8710	8734	8742	8749	8839	8751	8753	8753

$2\alpha/RT\beta = 6.10$

p/K'	γ-Values 1.00	1.50	2.00	3.00	4.00	5.00	7.00	8.00	10.0	20.0	40.0	70.0	100	200	400	1000
00.001	1235	0853	0606	0332	0204	0129	0066	0049	0038	0022	0019	0019	0019	0019	0019	0019
00.002	1541	1147	0866	0555	0363	0258	0144	0111	0080	0042	0033	0031	0030	0029	0029	0029
00.004	1888	1502	1204	0859	0636	0485	0304	0251	0180	0083	0061	0054	0052	0049	0048	0048
00.006	2117	1746	1466	1101	0854	0677	0466	0389	0292	0129	0088	0078	0074	0071	0069	0068
00.008	2279	1921	1656	1284	1045	0855	0612	0525	0405	0184	0118	0103	0098	0092	0090	0088
00.010	2389	2042	1822	1450	1185	1010	0742	0661	0523	0244	0149	0128	0122	0114	0110	0108
00.020	2836	2547	2347	2020	1805	1607	1339	1227	1060	0615	0360	0278	0255	0236	0228	0223
00.030	3132	2890	2729	2408	2230	2055	1805	1705	1521	1028	0657	0482	0426	0378	0360	0353
00.040	3327	3118	2941	2741	2561	2387	2186	2098	1922	1442	1015	0758	0654	0550	0513	0497
00.050	3513	3338	3207	2950	2808	2700	2496	2420	2275	1840	1406	1106	0956	0772	0696	0666
00.060	3616	3459	3356	3200	3058	2922	2768	2721	2594	2209	1807	1504	1331	1069	0926	0864
00.070	3707	3566	3473	3343	3247	3166	3022	2975	2852	2549	2206	1926	1758	1457	1250	1115
00.080	3840	3726	3619	3484	3398	3329	3231	3204	3116	2853	2587	2357	2214	1933	1696	1476
00.090	3945	3852	3777	3667	3578	3513	3418	3401	3315	3139	2943	2782	2672	2457	2255	2008
00.100	4011	3929	3871	3786	3725	3680	3610	3606	3536	3402	3272	3174	3115	3000	2883	2726
00.130	4166	4114	4085	4064	4055	4050	4046	4076	4048	4066	4130	4212	4271	4419	4586	4799
00.170	4417	4415	4409	4406	4415	4432	4490	4557	4568	4732	4966	5184	5328	5592	5784	5906
00.200	4521	4539	4541	4569	4641	4703	4764	4826	4865	5118	5433	5685	5843	6081	6214	6271
00.250	4651	4692	4756	4878	4933	4987	5136	5224	5270	5615	5983	6250	6383	6545	6608	6637
00.300	4826	4902	4978	5056	5172	5274	5407	5519	5598	5984	6366	6603	6707	6813	6851	6871
00.400	5023	5135	5209	5420	5521	5660	5839	5972	6061	6491	6845	7016	7076	7129	7149	7163
00.500	5163	5300	5476	5629	5824	5932	6159	6287	6395	6824	7128	7250	7288	7324	7339	7346
00.600	5341	5509	5626	5863	6013	6182	6401	6543	6646	7056	7313	7405	7433	7458	7472	7477
00.700	5442	5627	5743	6011	6200	6345	6586	6734	6830	7226	7446	7518	7539	7561	7571	7576
00.800	5517	5714	5870	6126	6351	6499	6740	6898	6990	7359	7548	7605	7624	7643	7650	7656
00.900	5586	5794	6017	6255	6458	6638	6879	7018	7111	7463	7629	7677	7692	7709	7714	7720
01.000	5663	5883	6101	6380	6569	6739	6979	7144	7222	7548	7694	7734	7751	7763	7773	7776
01.200	5833	6078	6231	6533	6763	6928	7171	7312	7388	7680	7797	7830	7842	7855	7860	7864
01.400	5921	6178	6355	6668	6892	7072	7302	7461	7521	7779	7875	7903	7914	7924	7929	7932
01.600	5993	6260	6504	6812	7027	7183	7425	7565	7622	7856	7938	7962	7971	7981	7985	7989
01.800	6061	6337	6593	6902	7129	7295	7513	7664	7708	7917	7988	8011	8019	8028	8031	8034
02.000	6154	6441	6664	6978	7208	7376	7592	7735	7774	7968	8032	8051	8059	8069	8072	8074
04.000	6610	6942	7156	7506	7714	7851	8018	8142	8133	8238	8272	8285	8289	8294	8298	8299
06.000	6826	7173	7436	7748	7945	8063	8198	8315	8286	8359	8384	8395	8399	8402	8403	8406
08.000	7020	7373	7591	7905	8083	8192	8305	8417	8377	8434	8455	8464	8467	8471	8473	8474
10.000	7129	7485	7731	8021	8181	8276	8380	8486	8439	8487	8505	8512	8516	8520	8522	8523
20.000	7479	7832	8057	8306	8433	8499	8563	8660	8599	8630	8643	8649	8651	8653	8655	8655
30.000	7679	8021	8228	8445	8547	8598	8648	8745	8675	8701	8712	8717	8718	8721	8722	8723
40.000	7789	8125	8320	8527	8616	8661	8701	8797	8726	8748	8758	8761	8763	8765	8765	8766

$2\alpha/RT\beta = 6.20$

p/K′	1.00	1.50	2.00	3.00	4.00	5.00	7.00	8.00	10.0	20.0	40.0	70.0	100	200	400	1000
									γ-Values							
00.001	1249	0865	0615	0338	0210	0132	0067	0052	0038	0022	0019	0019	0019	0019	0019	0019
00.002	1556	1161	0876	0572	0370	0263	0146	0118	0081	0042	0033	0031	0030	0029	0029	0029
00.004	1906	1519	1222	0872	0647	0494	0312	0260	0185	0083	0061	0054	0052	0049	0048	0048
00.006	2152	1782	1483	1119	0869	0689	0479	0399	0298	0130	0088	0078	0074	0071	0069	0068
00.008	2297	1940	1698	1321	1070	0872	0631	0539	0414	0187	0119	0103	0098	0092	0090	0088
00.010	2413	2067	1845	1471	1206	1042	0762	0678	0536	0249	0150	0128	0122	0114	0110	0108
00.020	2868	2583	2371	2065	1834	1650	1380	1251	1095	0633	0366	0279	0256	0236	0228	0223
00.030	3155	2915	2762	2441	2265	2111	1842	1746	1558	1059	0678	0492	0431	0380	0360	0353
00.040	3374	3173	2972	2776	2616	2429	2232	2144	1967	1495	1056	0782	0669	0555	0517	0500
00.050	3541	3370	3248	2996	2845	2743	2562	2458	2325	1898	1462	1149	0991	0789	0705	0671
00.060	3639	3485	3387	3243	3118	2981	2817	2757	2653	2276	1881	1569	1391	1110	0948	0876
00.070	3735	3599	3504	3379	3289	3215	3091	3026	2922	2629	2295	2019	1845	1533	1304	1145
00.080	3886	3780	3681	3534	3444	3379	3284	3243	3181	2943	2687	2463	2328	2048	1802	1549
00.090	3977	3889	3821	3723	3643	3586	3494	3456	3389	3220	3054	2907	2804	2609	2420	2165
00.100	4038	3962	3907	3829	3775	3735	3676	3653	3609	3500	3392	3310	3269	3178	3098	2983
00.130	4193	4145	4123	4124	4127	4127	4123	4128	4123	4158	4261	4361	4442	4625	4825	5067
00.170	4451	4455	4449	4449	4463	4487	4576	4600	4645	4840	5095	5326	5483	5759	5945	6052
00.200	4547	4569	4577	4614	4715	4768	4825	4859	4953	5224	5552	5819	5979	6215	6336	6388
00.250	4677	4723	4818	4927	4981	5048	5207	5251	5346	5713	6093	6361	6493	6646	6702	6728
00.300	4870	4954	5020	5099	5246	5332	5476	5560	5669	6077	6463	6698	6799	6897	6930	6948
00.400	5051	5168	5245	5469	5571	5731	5903	6003	6130	6570	6924	7091	7147	7196	7215	7227
00.500	5197	5339	5524	5675	5879	5985	6218	6303	6462	6901	7196	7312	7347	7380	7393	7403
00.600	5377	5552	5663	5918	6061	6241	6470	6550	6713	7125	7373	7460	7484	7511	7520	7527
00.700	5471	5661	5779	6054	6262	6394	6641	6749	6889	7287	7500	7568	7587	7608	7617	7622
00.800	5544	5746	5928	6168	6398	6565	6804	6894	7049	7415	7599	7652	7669	7687	7692	7699
00.900	5614	5827	6061	6319	6503	6691	6933	7024	7166	7517	7675	7720	7733	7750	7755	7761
01.000	5706	5932	6140	6427	6628	6785	7030	7137	7279	7599	7738	7777	7790	7802	7811	7815
01.200	5866	6116	6266	6572	6808	6985	7221	7307	7438	7726	7838	7868	7879	7890	7895	7900
01.400	5948	6209	6404	6720	6934	7116	7349	7441	7567	7821	7911	7939	7948	7958	7963	7967
01.600	6019	6290	6547	6855	7078	7233	7472	7546	7667	7894	7972	7995	8004	8013	8017	8022
01.800	6092	6372	6627	6940	7172	7342	7556	7638	7750	7954	8021	8042	8050	8060	8063	8064
02.000	6200	6492	6695	7016	7245	7417	7637	7708	7813	8003	8062	8083	8089	8098	8102	8103
04.000	6648	6982	7186	7545	7747	7885	8050	8101	8165	8264	8297	8307	8313	8319	8320	8322
06.000	6851	7200	7464	7786	7977	8092	8226	8265	8314	8383	8405	8415	8419	8423	8423	8427
08.000	7051	7406	7628	7933	8115	8220	8331	8361	8398	8454	8476	8482	8487	8489	8491	8493
10.000	7153	7511	7760	8052	8207	8303	8401	8430	8461	8506	8525	8532	8534	8537	8539	8540
20.000	7500	7854	8080	8329	8455	8517	8581	8596	8615	8646	8657	8664	8666	8670	8669	8670
30.000	7705	8047	8250	8464	8565	8617	8663	8676	8690	8714	8726	8730	8732	8735	8735	8735
40.000	7807	8143	8340	8546	8633	8677	8716	8726	8739	8761	8770	8774	8776	8778	8778	8779

$2\alpha/RT\beta = 6.30$

p/K'	γ-Values 1.00	1.50	2.00	3.00	4.00	5.00	7.00	8.00	10.0	20.0	40.0	70.0	100	200	400	1000
00.001	1262	0876	0623	0343	0215	0135	0068	0053	0038	0022	0019	0019	0019	0019	0019	0019
00.002	1571	1175	0888	0585	0380	0268	0149	0120	0082	0042	0033	0031	0030	0029	0029	0029
00.004	1922	1534	1242	0885	0659	0503	0324	0266	0190	0083	0061	0054	0052	0049	0048	0048
00.006	2181	1814	1500	1136	0884	0703	0491	0408	0304	0132	0088	0078	0074	0071	0069	0068
00.008	2314	1958	1736	1356	1091	0889	0648	0550	0424	0190	0119	0103	0098	0092	0090	0088
00.010	2439	2095	1866	1491	1227	1072	0788	0694	0548	0253	0151	0128	0122	0114	0110	0108
00.020	2912	2633	2395	2121	1863	1702	1417	1281	1130	0652	0375	0282	0257	0236	0228	0223
00.030	3175	2937	2792	2480	2297	2161	1878	1792	1595	1091	0702	0503	0436	0382	0363	0355
00.040	3423	3230	3005	2808	2663	2479	2274	2198	2020	1544	1098	0808	0689	0561	0521	0504
00.050	3566	3399	3286	3057	2883	2785	2624	2529	2374	1954	1526	1200	1031	0808	0715	0678
00.060	3663	3512	3416	3283	3171	3055	2867	2808	2712	2341	1957	1644	1461	1158	0974	0891
00.070	3765	3634	3538	3416	3330	3263	3151	3098	3008	2705	2378	2114	1939	1617	1367	1177
00.080	3929	3831	3744	3601	3498	3430	3336	3299	3242	3037	2784	2579	2448	2177	1924	1637
00.090	4007	3924	3862	3773	3703	3655	3576	3542	3478	3307	3161	3043	2948	2779	2604	2359
00.100	4065	3992	3942	3870	3821	3787	3735	3717	3681	3590	3512	3455	3431	3372	3334	3280
00.130	4222	4179	4168	4199	4202	4197	4192	4195	4197	4258	4387	4517	4614	4836	5063	5319
00.170	4482	4492	4488	4491	4509	4547	4656	4677	4718	4942	5218	5471	5637	5919	6099	6191
00.200	4573	4599	4611	4668	4782	4827	4886	4926	5047	5328	5666	5948	6114	6344	6452	6497
00.250	4707	4758	4886	4972	5028	5117	5272	5315	5435	5806	6200	6469	6599	6741	6792	6814
00.300	4909	5000	5060	5142	5314	5389	5562	5637	5737	6165	6559	6791	6886	6976	7005	7022
00.400	5076	5198	5284	5515	5621	5791	5975	6076	6197	6655	7002	7162	7214	7257	7277	7284
00.500	5238	5387	5567	5721	5929	6039	6275	6379	6524	6974	7261	7371	7403	7435	7448	7455
00.600	5413	5594	5698	5968	6109	6297	6534	6608	6779	7191	7432	7513	7535	7560	7569	7574
00.700	5499	5694	5814	6096	6321	6444	6694	6814	6952	7350	7553	7616	7634	7653	7661	7667
00.800	5567	5773	5991	6211	6444	6627	6867	6948	7106	7469	7646	7697	7711	7729	7734	7740
00.900	5642	5859	6104	6377	6549	6739	6984	7086	7222	7568	7719	7762	7774	7790	7795	7801
01.000	5760	5994	6175	6473	6690	6833	7082	7188	7333	7648	7780	7817	7829	7842	7848	7852
01.200	5897	6152	6300	6611	6852	7034	7269	7362	7492	7770	7875	7905	7915	7925	7931	7935
01.400	5975	6240	6465	6781	6979	7159	7402	7486	7614	7860	7948	7972	7982	7992	7995	8000
01.600	6044	6318	6588	6899	7125	7288	7515	7595	7713	7932	8005	8028	8036	8045	8049	8053
01.800	6125	6409	6662	6978	7211	7386	7598	7679	7788	7989	8053	8073	8080	8090	8093	8093
02.000	6243	6541	6728	7053	7283	7457	7684	7746	7855	8036	8094	8111	8119	8125	8130	8132
04.000	6683	7021	7217	7561	7781	7917	8082	8132	8193	8289	8320	8331	8336	8341	8342	8345
06.000	6874	7226	7494	7824	8007	8122	8253	8290	8337	8404	8427	8435	8440	8443	8445	8446
08.000	7080	7437	7670	7960	8146	8246	8354	8385	8421	8474	8495	8502	8505	8508	8510	8512
10.000	7176	7536	7786	8080	8232	8329	8425	8451	8479	8525	8542	8550	8552	8555	8557	8558
20.000	7521	7877	8102	8348	8475	8538	8597	8614	8633	8661	8674	8678	8681	8683	8684	8685
30.000	7728	8070	8272	8483	8584	8634	8680	8691	8705	8730	8741	8745	8746	8748	8749	8749
40.000	7828	8164	8359	8566	8651	8692	8730	8741	8752	8773	8781	8787	8788	8789	8791	8791

$2\alpha/RT\beta = 6.40$

p/K'	1.00	1.50	2.00	3.00	4.00	5.00	7.00	8.00	10.0	20.0	40.0	70.0	100	200	400	1000
	γ-Values															
00.001	1275	0887	0631	0349	0220	0139	0069	0053	0038	0022	0019	0019	0019	0019	0019	0019
00.002	1584	1187	0898	0598	0396	0274	0153	0124	0083	0042	0033	0031	0030	0029	0029	0029
00.004	1937	1550	1270	0898	0669	0514	0336	0273	0197	0084	0061	0054	0052	0049	0048	0048
00.006	2208	1841	1516	1152	0898	0718	0502	0417	0311	0134	0088	0078	0074	0071	0069	0068
00.008	2333	1976	1767	1386	1112	0905	0663	0563	0435	0193	0119	0103	0098	0092	0090	0088
00.010	2479	2140	1885	1512	1255	1097	0820	0708	0561	0260	0152	0128	0122	0114	0110	0108
00.020	2967	2696	2419	2170	1890	1747	1451	1324	1159	0669	0384	0285	0259	0237	0230	0225
00.030	3197	2962	2820	2537	2328	2205	1912	1833	1643	1133	0727	0516	0444	0384	0364	0356
00.040	3463	3277	3048	2841	2708	2556	2314	2245	2094	1591	1137	0842	0711	0571	0525	0507
00.050	3591	3427	3321	3133	2923	2823	2679	2604	2431	2011	1594	1255	1080	0834	0727	0685
00.060	3685	3538	3446	3320	3220	3126	2926	2859	2766	2407	2031	1728	1538	1214	1006	0906
00.070	3807	3683	3577	3450	3368	3307	3206	3163	3088	2778	2469	2227	2048	1716	1445	1219
00.080	3964	3872	3797	3680	3577	3494	3392	3355	3300	3123	2894	2705	2575	2319	2069	1754
00.090	4035	3956	3900	3818	3755	3715	3649	3621	3572	3415	3273	3173	3100	2965	2815	2601
00.100	4090	4022	3973	3909	3865	3835	3791	3775	3746	3677	3629	3597	3593	3580	3589	3616
00.130	4254	4218	4232	4271	4265	4259	4256	4260	4265	4379	4508	4674	4791	5046	5296	5550
00.170	4511	4526	4523	4531	4558	4634	4727	4745	4788	5035	5336	5618	5790	6073	6241	6318
00.200	4598	4629	4643	4746	4844	4880	4947	5013	5131	5424	5790	6075	6244	6465	6561	6600
00.250	4740	4798	4945	5015	5076	5209	5333	5378	5533	5894	6298	6576	6700	6833	6876	6897
00.300	4945	5042	5096	5187	5375	5441	5645	5704	5807	6248	6657	6880	6970	7051	7077	7093
00.400	5102	5228	5328	5558	5678	5851	6062	6141	6271	6738	7080	7230	7278	7319	7336	7345
00.500	5299	5460	5607	5779	5976	6103	6330	6456	6584	7042	7326	7429	7460	7487	7499	7506
00.600	5444	5630	5732	6015	6164	6346	6591	6664	6838	7252	7486	7563	7585	7607	7616	7621
00.700	5526	5726	5853	6134	6371	6495	6751	6872	7024	7407	7603	7661	7680	7697	7704	7710
00.800	5593	5804	6046	6262	6487	6684	6924	7000	7160	7524	7691	7739	7753	7769	7775	7781
00.900	5674	5896	6142	6430	6598	6786	7032	7146	7285	7616	7762	7803	7813	7829	7834	7839
01.000	5810	6052	6210	6514	6747	6880	7145	7236	7384	7695	7821	7854	7866	7879	7885	7889
01.200	5926	6186	6334	6650	6893	7084	7314	7416	7545	7813	7912	7940	7949	7959	7965	7969
01.400	6000	6269	6521	6835	7035	7200	7458	7532	7656	7899	7982	8006	8014	8024	8028	8032
01.600	6069	6348	6624	6937	7169	7341	7557	7643	7758	7968	8039	8059	8066	8073	8079	8082
01.800	6171	6462	6694	7012	7251	7425	7642	7720	7828	8022	8083	8103	8110	8118	8122	8122
02.000	6280	6583	6758	7091	7323	7494	7725	7791	7896	8069	8123	8141	8148	8154	8158	8160
04.000	6714	7055	7251	7615	7813	7949	8114	8161	8220	8313	8343	8354	8359	8363	8363	8366
06.000	6899	7252	7521	7857	8035	8147	8279	8315	8361	8425	8446	8456	8459	8463	8465	8466
08.000	7109	7467	7708	7987	8175	8272	8379	8408	8444	8493	8514	8521	8523	8526	8528	8529
10.000	7199	7561	7813	8110	8256	8357	8446	8471	8499	8544	8560	8568	8569	8572	8573	8574
20.000	7541	7897	8124	8371	8495	8557	8616	8632	8648	8676	8688	8693	8695	8698	8698	8699
30.000	7751	8092	8293	8503	8602	8653	8696	8706	8719	8743	8752	8757	8759	8761	8762	8763
40.000	7847	8184	8378	8586	8667	8708	8745	8754	8767	8787	8796	8799	8800	8802	8803	8804

$2\alpha/RT\beta = 6.50$

p/K'	1.00	1.50	2.00	3.00	4.00	5.00	7.00	8.00	10.0	20.0	40.0	70.0	100	200	400	1000
									γ-Values							
00.001	1286	0897	0639	0355	0224	0146	0070	0054	0039	0022	0019	0019	0019	0019	0019	0019
00.002	1598	1199	0910	0608	0414	0283	0160	0129	0084	0042	0033	0031	0030	0029	0029	0029
00.004	1953	1565	1323	0911	0680	0525	0345	0278	0201	0085	0061	0054	0052	0049	0048	0048
00.006	2231	1865	1532	1167	0912	0739	0512	0428	0318	0136	0089	0078	0074	0071	0069	0068
00.008	2348	1992	1794	1411	1130	0921	0677	0577	0452	0197	0120	0103	0098	0092	0090	0088
00.010	2540	2209	1905	1530	1307	1119	0845	0723	0575	0268	0153	0129	0122	0114	0110	0108
00.020	3008	2744	2442	2210	1918	1786	1479	1383	1185	0692	0393	0289	0261	0238	0230	0226
00.030	3217	2984	2846	2618	2359	2244	1947	1869	1717	1188	0748	0529	0450	0386	0366	0358
00.040	3498	3317	3129	2871	2748	2626	2352	2289	2165	1636	1182	0874	0734	0580	0529	0510
00.050	3613	3453	3354	3192	2972	2864	2726	2665	2521	2086	1651	1310	1127	0863	0739	0692
00.060	3707	3563	3474	3354	3263	3185	3019	2915	2817	2492	2121	1805	1614	1279	1042	0921
00.070	3880	3770	3628	3487	3407	3350	3256	3218	3155	2853	2594	2342	2151	1830	1535	1268
00.080	3997	3911	3843	3744	3666	3595	3460	3416	3363	3200	3029	2822	2724	2480	2236	1900
00.090	4061	3987	3933	3860	3804	3766	3711	3686	3646	3531	3420	3314	3248	3157	3048	2889
00.100	4113	4048	4005	3947	3906	3880	3845	3832	3808	3760	3759	3764	3780	3802	3856	3977
00.130	4310	4283	4323	4328	4322	4315	4312	4318	4329	4492	4642	4823	4970	5254	5516	5762
00.170	4538	4558	4556	4569	4615	4729	4791	4808	4854	5124	5476	5753	5931	6215	6372	6436
00.200	4622	4657	4675	4836	4898	4930	5013	5121	5202	5511	5916	6204	6364	6578	6664	6697
00.250	4796	4864	4995	5056	5126	5286	5391	5442	5617	5981	6402	6676	6796	6918	6958	6975
00.300	4976	5079	5132	5238	5425	5490	5719	5767	5884	6331	6746	6965	7048	7123	7148	7160
00.400	5125	5254	5396	5598	5766	5902	6139	6201	6366	6811	7150	7295	7340	7378	7392	7401
00.500	5355	5526	5647	5869	6022	6191	6387	6524	6648	7105	7389	7483	7511	7537	7549	7555
00.600	5474	5665	5764	6058	6246	6394	6647	6720	6894	7310	7541	7611	7632	7652	7661	7666
00.700	5551	5754	5898	6172	6420	6552	6817	6927	7088	7462	7652	7707	7722	7739	7746	7752
00.800	5617	5831	6094	6338	6530	6735	6977	7052	7211	7579	7735	7780	7794	7810	7815	7820
00.900	5716	5945	6178	6477	6661	6829	7080	7198	7344	7665	7804	7841	7853	7866	7872	7877
01.000	5850	6099	6244	6555	6796	6933	7210	7284	7430	7741	7860	7892	7902	7915	7921	7924
01.200	5953	6217	6369	6687	6932	7126	7357	7462	7592	7854	7947	7974	7983	7993	7999	8002
01.400	6024	6298	6569	6883	7102	7243	7506	7573	7698	7938	8015	8038	8046	8056	8059	8063
01.600	6094	6376	6659	6974	7211	7387	7597	7685	7797	8002	8069	8089	8097	8104	8109	8111
01.800	6235	6533	6727	7047	7287	7464	7694	7757	7866	8057	8115	8131	8138	8145	8150	8152
02.000	6315	6622	6788	7135	7364	7532	7765	7837	7935	8100	8153	8168	8175	8182	8185	8188
04.000	6743	7087	7291	7646	7843	7979	8141	8190	8249	8338	8365	8376	8380	8385	8384	8389
06.000	6924	7280	7548	7889	8062	8175	8303	8339	8384	8447	8467	8476	8479	8483	8484	8486
08.000	7133	7493	7742	8015	8200	8295	8402	8429	8463	8514	8531	8539	8541	8545	8546	8547
10.000	7221	7583	7837	8136	8280	8381	8469	8490	8520	8561	8577	8582	8586	8589	8591	8591
20.000	7562	7919	8145	8391	8515	8577	8633	8648	8663	8691	8701	8708	8709	8712	8713	8713
30.000	7773	8115	8314	8520	8619	8669	8710	8721	8733	8757	8768	8770	8772	8774	8775	8776
40.000	7866	8203	8398	8603	8683	8723	8760	8768	8779	8798	8808	8811	8812	8814	8814	8816

$2\alpha/RT\beta = 6.60$

p/K'	1.00	1.50	2.00	3.00	4.00	5.00	7.00	8.00	10.0	20.0	40.0	70.0	100	200	400	1000
	γ-Values															
00.001	1297	0906	0647	0363	0228	0151	0071	0055	0040	0022	0019	0019	0019	0019	0019	0019
00.002	1610	1211	0923	0619	0426	0301	0167	0132	0085	0042	0033	0031	0030	0029	0029	0029
00.004	1967	1578	1363	0925	0692	0540	0353	0283	0205	0086	0061	0054	0052	0049	0048	0048
00.006	2251	1886	1548	1181	0927	0784	0522	0453	0331	0138	0089	0078	0074	0071	0069	0068
00.008	2365	2010	1819	1432	1146	0937	0691	0611	0477	0204	0120	0103	0098	0092	0090	0088
00.010	2582	2255	1923	1550	1359	1139	0865	0738	0599	0276	0154	0129	0122	0114	0110	0108
00.020	3043	2783	2468	2245	1943	1818	1506	1425	1210	0731	0402	0291	0261	0239	0231	0226
00.030	3236	3006	2872	2676	2388	2278	1985	1904	1775	1231	0773	0542	0458	0389	0368	0361
00.040	3528	3352	3209	2901	2782	2680	2391	2329	2218	1688	1255	0907	0766	0589	0534	0515
00.050	3636	3478	3383	3239	3078	2902	2770	2717	2610	2191	1709	1395	1183	0895	0751	0700
00.060	3731	3590	3502	3388	3302	3234	3112	3030	2871	2609	2237	1921	1721	1362	1088	0941
00.070	3936	3837	3741	3528	3446	3391	3304	3269	3213	2982	2695	2441	2292	1948	1650	1340
00.080	4026	3945	3883	3794	3730	3679	3587	3538	3455	3272	3133	2984	2858	2672	2420	2099
00.090	4086	4015	3966	3897	3848	3814	3765	3744	3710	3621	3540	3483	3440	3379	3293	3230
00.100	4137	4076	4034	3980	3945	3924	3896	3890	3872	3871	3920	3925	3944	4047	4146	4340
00.130	4391	4382	4386	4379	4370	4364	4368	4374	4396	4586	4803	5008	5142	5447	5726	5951
00.170	4564	4588	4589	4608	4729	4800	4848	4867	4926	5218	5608	5890	6084	6355	6493	6547
00.200	4645	4684	4709	4898	4946	4979	5132	5202	5271	5596	6020	6318	6489	6687	6761	6787
00.250	4878	4962	5038	5096	5186	5349	5446	5532	5686	6100	6515	6773	6888	7000	7035	7050
00.300	5005	5114	5165	5343	5476	5536	5782	5825	6000	6439	6830	7044	7125	7192	7213	7227
00.400	5148	5282	5494	5638	5854	5951	6202	6258	6446	6882	7219	7358	7398	7433	7447	7454
00.500	5400	5579	5681	5941	6064	6267	6463	6583	6739	7166	7446	7538	7563	7586	7597	7603
00.600	5502	5698	5795	6098	6326	6441	6697	6810	6948	7367	7594	7659	7677	7697	7705	7710
00.700	5576	5783	5993	6210	6465	6647	6901	6976	7145	7514	7699	7749	7764	7780	7787	7792
00.800	5642	5860	6134	6415	6572	6781	7024	7130	7261	7627	7779	7821	7834	7848	7853	7857
00.900	5799	6041	6213	6518	6745	6871	7130	7247	7396	7715	7845	7879	7890	7903	7908	7912
01.000	5885	6140	6275	6590	6841	7016	7262	7329	7475	7783	7897	7928	7937	7950	7956	7959
01.200	5979	6247	6408	6732	6970	7168	7400	7507	7637	7892	7983	8007	8015	8025	8030	8034
01.400	6048	6325	6609	6925	7156	7290	7551	7622	7751	7974	8048	8070	8077	8084	8090	8092
01.600	6122	6408	6692	7008	7249	7428	7635	7726	7836	8038	8100	8119	8126	8134	8137	8140
01.800	6281	6585	6755	7082	7322	7502	7739	7795	7905	8090	8143	8162	8166	8174	8178	8180
02.000	6344	6655	6818	7211	7429	7570	7802	7877	7969	8131	8181	8197	8202	8209	8211	8214
04.000	6769	7116	7361	7678	7872	8008	8170	8218	8278	8361	8388	8397	8401	8405	8409	8409
06.000	6950	7308	7574	7918	8089	8201	8333	8366	8406	8465	8486	8494	8498	8501	8503	8504
08.000	7158	7520	7772	8044	8226	8319	8424	8453	8483	8531	8550	8556	8558	8562	8563	8564
10.000	7242	7606	7861	8159	8305	8402	8489	8511	8538	8579	8594	8600	8603	8606	8608	8609
20.000	7582	7940	8166	8412	8533	8597	8649	8663	8680	8705	8718	8722	8724	8726	8727	8727
30.000	7792	8134	8333	8538	8635	8683	8726	8736	8748	8770	8780	8783	8785	8786	8787	8788
40.000	7884	8220	8425	8621	8698	8739	8773	8782	8792	8812	8819	8823	8822	8826	8828	8828

$2\alpha/RT\beta = 6.70$

p/K′	γ-Values 1.00	1.50	2.00	3.00	4.00	5.00	7.00	8.00	10.0	20.0	40.0	70.0	100	200	400	1000
00.001	1308	0915	0654	0393	0232	0155	0074	0056	0040	0022	0019	0019	0019	0019	0019	0019
00.002	1622	1222	0938	0627	0436	0312	0172	0134	0087	0042	0033	0031	0030	0029	0029	0029
00.004	1981	1592	1390	0942	0706	0583	0360	0289	0209	0086	0061	0054	0052	0049	0048	0048
00.006	2271	1906	1565	1196	0945	0812	0531	0473	0352	0139	0089	0078	0074	0071	0069	0068
00.008	2380	2025	1841	1453	1163	0956	0703	0639	0491	0208	0121	0103	0098	0092	0090	0088
00.010	2614	2290	1940	1570	1391	1158	0882	0755	0639	0281	0155	0130	0123	0114	0110	0108
00.020	3072	2815	2495	2275	1970	1847	1531	1458	1235	0758	0421	0294	0263	0240	0231	0226
00.030	3257	3029	2896	2720	2419	2310	2044	1937	1819	1264	0825	0565	0471	0392	0369	0361
00.040	3554	3382	3259	2930	2816	2722	2435	2367	2264	1802	1303	0966	0791	0600	0538	0518
00.050	3656	3501	3410	3280	3165	2947	2810	2762	2672	2257	1828	1453	1256	0935	0769	0708
00.060	3754	3616	3530	3420	3340	3278	3175	3124	2951	2685	2316	2017	1804	1439	1148	0963
00.070	3977	3886	3809	3596	3487	3433	3349	3317	3265	3090	2779	2611	2413	2112	1791	1429
00.080	4051	3974	3919	3838	3782	3737	3668	3637	3594	3350	3225	3120	3059	2840	2641	2338
00.090	4109	4042	3995	3933	3889	3859	3815	3798	3769	3696	3639	3612	3587	3596	3572	3643
00.100	4157	4100	4063	4016	3985	3968	3953	3955	3961	4027	4032	4060	4168	4255	4426	4730
00.130	4441	4441	4435	4422	4414	4411	4420	4428	4486	4666	4921	5146	5333	5656	5924	6118
00.170	4589	4617	4618	4647	4822	4857	4901	4923	5077	5382	5715	6042	6211	6479	6607	6652
00.200	4667	4710	4742	4950	4991	5024	5232	5271	5332	5696	6117	6429	6599	6785	6851	6875
00.250	4927	5020	5078	5133	5326	5405	5504	5659	5752	6204	6608	6872	6978	7076	7107	7122
00.300	5032	5145	5196	5432	5521	5584	5838	5882	6083	6534	6908	7128	7197	7258	7277	7289
00.400	5173	5311	5555	5675	5914	5997	6259	6313	6513	6946	7295	7420	7456	7486	7500	7507
00.500	5435	5620	5714	5995	6106	6325	6566	6636	6818	7247	7500	7587	7611	7633	7643	7649
00.600	5526	5726	5827	6135	6384	6484	6745	6885	7002	7436	7643	7705	7721	7739	7748	7752
00.700	5598	5809	6070	6247	6507	6712	6962	7025	7197	7564	7743	7791	7805	7820	7826	7832
00.800	5667	5890	6173	6468	6613	6825	7072	7197	7319	7674	7822	7860	7873	7885	7890	7895
00.900	5851	6100	6245	6557	6802	6915	7214	7293	7443	7761	7884	7916	7926	7938	7943	7948
01.000	5917	6176	6304	6625	6884	7078	7310	7374	7522	7823	7935	7963	7972	7984	7989	7992
01.200	6003	6275	6516	6830	7009	7207	7445	7549	7677	7929	8015	8040	8048	8058	8061	8064
01.400	6071	6351	6646	6961	7201	7371	7590	7681	7798	8008	8079	8098	8107	8114	8119	8122
01.600	6152	6442	6723	7042	7285	7468	7674	7764	7872	8071	8130	8148	8154	8162	8165	8168
01.800	6319	6628	6784	7114	7357	7537	7780	7832	7952	8121	8172	8188	8194	8202	8204	8207
02.000	6372	6687	6849	7264	7485	7638	7835	7912	8003	8161	8208	8223	8228	8235	8238	8240
04.000	6795	7145	7408	7706	7902	8039	8197	8245	8304	8382	8409	8419	8422	8427	8430	8430
06.000	6992	7351	7598	7944	8114	8240	8360	8391	8429	8485	8505	8512	8516	8520	8522	8522
08.000	7179	7543	7799	8094	8251	8343	8449	8474	8506	8549	8567	8573	8576	8579	8581	8581
10.000	7262	7628	7885	8182	8339	8426	8508	8531	8557	8594	8610	8617	8619	8622	8623	8624
20.000	7601	7959	8187	8432	8551	8614	8667	8677	8694	8721	8731	8734	8738	8739	8740	8741
30.000	7812	8154	8352	8555	8652	8700	8739	8749	8761	8782	8790	8796	8798	8799	8800	8800
40.000	7904	8240	8456	8637	8714	8753	8786	8794	8805	8823	8832	8835	8836	8838	8839	8838

$2\alpha/RT\beta = 6.80$

p/K'	γ-Values 1.00	1.50	2.00	3.00	4.00	5.00	7.00	8.00	10.0	20.0	40.0	70.0	100	200	400	1000
00.001	1316	0913	0688	0379	0230	0155	0072	0055	0035	0017	0013	0011	0011	0010	0010	0010
00.002	1623	1267	1007	0653	0440	0301	0172	0128	0086	0035	0026	0023	0022	0021	0020	0020
00.004	2074	1610	1338	1011	0733	0577	0372	0305	0212	0080	0054	0047	0045	0043	0041	0041
00.006	2229	1872	1599	1260	1014	0824	0568	0478	0346	0137	0083	0073	0069	0065	0063	0062
00.008	2471	2112	1827	1484	1225	1025	0735	0629	0499	0205	0117	0099	0093	0088	0085	0084
00.010	2557	2216	2044	1613	1356	1209	0900	0782	0634	0285	0153	0126	0119	0111	0108	0106
00.020	3000	2807	2544	2321	2066	1844	1571	1496	1300	0790	0430	0299	0264	0239	0230	0225
00.030	3311	3157	2911	2730	2481	2374	2104	2037	1832	1332	0838	0587	0482	0394	0371	0361
00.040	3589	3318	3215	3063	2833	2741	2504	2434	2335	1837	1376	1013	0836	0614	0543	0520
00.050	3696	3582	3490	3239	3143	3068	2920	2791	2708	2340	1891	1535	1343	0990	0789	0715
00.060	3785	3684	3612	3507	3428	3358	3155	3114	3049	2734	2416	2141	1944	1543	1238	0994
00.070	4001	3902	3811	3634	3568	3519	3445	3415	3365	3109	2937	2710	2547	2278	1965	1573
00.080	4071	3995	3942	3867	3815	3775	3716	3692	3653	3459	3349	3271	3221	3045	2939	2663
00.090	4125	4062	4019	3961	3924	3896	3858	3844	3823	3771	3817	3805	3800	3806	3904	4031
00.100	4176	4126	4095	4067	4132	4120	4100	4092	4081	4066	4177	4216	4331	4498	4739	5075
00.130	4454	4423	4406	4393	4409	4514	4531	4539	4559	4753	5064	5286	5456	5809	6100	6262
00.170	4583	4602	4722	4741	4774	4909	4961	4992	5135	5413	5851	6150	6337	6596	6705	6742
00.200	4695	4819	4833	4989	5038	5079	5258	5290	5446	5832	6219	6538	6702	6873	6929	6951
00.250	4921	4950	5112	5179	5356	5411	5612	5656	5823	6245	6699	6954	7054	7143	7170	7183
00.300	5007	5191	5237	5445	5526	5695	5904	5954	6124	6600	6995	7190	7260	7313	7331	7341
00.400	5290	5361	5554	5779	5884	6061	6286	6418	6578	7022	7346	7470	7503	7531	7543	7550
00.500	5398	5620	5703	5953	6164	6349	6578	6706	6866	7299	7552	7630	7652	7672	7681	7687
00.600	5493	5722	5934	6189	6403	6581	6810	6930	7083	7475	7685	7741	7757	7773	7781	7786
00.700	5693	5909	6029	6304	6525	6709	6999	7068	7254	7618	7780	7824	7837	7851	7858	7862
00.800	5759	6007	6209	6488	6705	6883	7109	7220	7363	7719	7855	7890	7902	7914	7920	7923
00.900	5812	6071	6294	6568	6791	6972	7247	7350	7482	7798	7914	7945	7955	7966	7971	7974
01.000	5862	6128	6356	6654	6935	7108	7327	7427	7582	7860	7964	7990	7999	8010	8015	8017
01.200	6070	6337	6553	6843	7062	7238	7491	7584	7722	7963	8042	8064	8072	8081	8085	8088
01.400	6145	6421	6653	6997	7225	7389	7625	7709	7828	8038	8103	8122	8129	8137	8141	8143
01.600	6203	6490	6727	7096	7308	7507	7731	7806	7903	8096	8152	8169	8175	8182	8186	8188
01.800	6258	6641	6873	7165	7431	7586	7798	7872	7976	8144	8193	8208	8214	8221	8224	8226
02.000	6404	6706	6939	7286	7493	7647	7877	7943	8035	8184	8227	8242	8247	8253	8256	8258
04.000	6825	7127	7424	7749	7950	8072	8225	8267	8321	8399	8423	8432	8436	8440	8442	8443
06.000	7000	7385	7665	7966	8147	8261	8376	8406	8445	8498	8517	8525	8528	8531	8533	8534
08.000	7195	7574	7789	8110	8270	8367	8462	8489	8518	8561	8577	8584	8586	8589	8591	8591
10.000	7279	7659	7925	8195	8357	8441	8522	8545	8569	8605	8620	8626	8628	8631	8632	8633
20.000	7672	8016	8232	8461	8566	8623	8675	8688	8703	8727	8738	8742	8744	8746	8747	8748
30.000	7806	9182	8378	8576	8666	8709	8747	8757	8769	8789	8798	8801	8803	8804	8805	8806
40.000	7957	8266	8456	8649	8725	8761	8793	8801	8811	8828	8837	8840	8841	8843	8844	8844

$2\alpha/RT\beta = 6.90$

p/K'	γ-Values 1.00	1.50	2.00	3.00	4.00	5.00	7.00	8.00	10.0	20.0	40.0	70.0	100	200	400	1000
00.001	1326	0928	0696	0412	0235	0161	0073	0056	0035	0017	0013	0011	0011	0010	0010	0010
00.002	1638	1280	1022	0669	0448	0306	0177	0131	0087	0035	0026	0023	0022	0021	0020	0020
00.004	2097	1629	1356	1028	0744	0587	0380	0311	0216	0081	0054	0047	0045	0043	0041	0041
00.006	2245	1889	1620	1277	1033	0845	0582	0488	0353	0139	0084	0073	0069	0065	0063	0062
00.008	2489	2131	1845	1505	1247	1045	0749	0641	0510	0216	0117	0099	0094	0088	0085	0084
00.010	2573	2237	2070	1714	1447	1238	0916	0800	0646	0290	0156	0126	0119	0111	0108	0106
00.020	3125	2831	2571	2360	2093	1872	1598	1525	1326	0815	0439	0302	0266	0240	0231	0226
00.030	3330	3185	2942	2762	2514	2408	2137	2074	1866	1367	0899	0603	0491	0397	0373	0362
00.040	3617	3343	3241	3101	2868	2776	2639	2473	2379	1881	1427	1047	0864	0633	0548	0524
00.050	3717	3609	3526	3275	3177	3107	2995	2841	2753	2396	1948	1656	1400	1027	0813	0724
00.060	3808	3708	3639	3541	3469	3410	3304	3161	3098	2797	2487	2219	2023	1669	1333	1033
00.070	4031	3943	3878	3676	3607	3559	3490	3463	3419	3277	3021	2899	2723	2473	2148	1759
00.080	4095	4024	3974	3905	3859	3823	3772	3752	3719	3626	3541	3476	3340	3275	3213	3035
00.090	4147	4088	4048	3995	3962	3938	3907	3897	3882	3949	3926	3922	3931	4057	4200	4463
00.100	4197	4153	4130	4212	4192	4176	4156	4149	4140	4145	4278	4429	4469	4752	5028	5409
00.130	4481	4452	4436	4429	4558	4569	4583	4593	4720	4927	5171	5478	5647	5991	6263	6398
00.170	4605	4740	4759	4778	4922	4962	5013	5148	5199	5574	5952	6299	6476	6710	6803	6835
00.200	4824	4849	4863	5040	5081	5129	5312	5346	5513	5914	6357	6657	6812	6963	7010	7029
00.250	4946	4977	5152	5217	5403	5457	5664	5718	5882	6376	6778	7034	7130	7212	7236	7249
00.300	5029	5225	5268	5485	5664	5742	5958	6009	6183	6669	7066	7264	7326	7373	7390	7399
00.400	5319	5389	5589	5823	6015	6106	6336	6476	6643	7082	7415	7524	7555	7581	7592	7599
00.500	5420	5652	5739	5995	6205	6394	6626	6759	6920	7354	7604	7677	7697	7716	7725	7730
00.600	5522	5749	5968	6227	6445	6630	6856	6982	7134	7542	7731	7784	7798	7814	7821	7825
00.700	5723	5958	6060	6418	6636	6751	7047	7160	7304	7667	7823	7863	7876	7889	7895	7899
00.800	5784	6038	6254	6527	6746	6927	7200	7264	7441	7763	7893	7927	7937	7949	7955	7958
00.900	5835	6098	6327	6603	6829	7012	7290	7396	7525	7841	7951	7979	7989	7999	8004	8007
01.000	5885	6156	6386	6759	6978	7152	7408	7469	7626	7904	7999	8023	8032	8042	8046	8049
01.200	6100	6371	6595	6877	7099	7323	7531	7623	7763	7999	8074	8095	8102	8111	8115	8117
01.400	6170	6448	6683	7049	7262	7426	7662	7746	7868	8070	8133	8151	8157	8165	8169	8171
01.600	6226	6517	6756	7130	7343	7553	7770	7845	7938	8127	8180	8196	8202	8209	8212	8214
01.800	6281	6680	6912	7198	7468	7621	7833	7906	8010	8173	8220	8234	8240	8246	8249	8251
02.000	6443	6736	6969	7325	7526	7714	7912	7978	8069	8212	8253	8267	8272	8278	8281	8283
04.000	6851	7152	7460	7780	7984	8099	8252	8292	8348	8421	8444	8453	8456	8460	8462	8463
06.000	7096	7408	7697	7992	8172	8286	8399	8428	8466	8517	8535	8543	8545	8549	8550	8551
08.000	7217	7599	7813	8134	8293	8389	8483	8509	8537	8578	8594	8600	8603	8605	8607	8608
10.000	7300	7681	7949	8237	8379	8462	8541	8563	8587	8622	8636	8642	8644	8646	8648	8648
20.000	7698	8038	8256	8482	8583	8643	8691	8703	8718	8741	8751	8755	8757	8759	8760	8761
30.000	7824	8204	8399	8593	8682	8723	8761	8771	8782	8801	8810	8813	8815	8816	8817	8818
40.000	7976	8284	8473	8665	8740	8774	8806	8814	8823	8840	8848	8851	8853	8854	8855	8855

$2\alpha/RT\beta = 7.00$

p/K'	γ-Values 1.00	1.50	2.00	3.00	4.00	5.00	7.00	8.00	10.0	20.0	40.0	70.0	100	200	400	1000
00.001	1336	1001	0704	0422	0259	0164	0079	0056	0037	0017	0013	0011	0011	0010	0010	0010
00.002	1746	1292	1036	0680	0455	0336	0181	0142	0088	0035	0026	0023	0022	0021	0020	0020
00.004	2116	1739	1456	1043	0756	0596	0387	0317	0219	0082	0054	0047	0045	0043	0041	0041
00.006	2262	1905	1731	1292	1049	0861	0593	0497	0386	0146	0084	0073	0069	0065	0063	0062
00.008	2507	2149	1862	1523	1266	1062	0761	0700	0519	0220	0118	0099	0094	0088	0085	0084
00.010	2589	2370	2093	1744	1476	1260	0931	0867	0659	0312	0157	0127	0119	0112	0108	0106
00.020	3170	2852	2714	2391	2119	2002	1718	1551	1350	0834	0468	0311	0269	0241	0231	0226
00.030	3349	3210	3091	2790	2663	2438	2170	2107	2002	1398	0924	0642	0512	0402	0375	0364
00.040	3642	3496	3266	3134	3025	2809	2686	2629	2419	2023	1467	1132	0926	0657	0556	0528
00.050	3736	3635	3558	3435	3210	3144	3043	3000	2796	2446	2099	1715	1520	1096	0844	0735
00.060	3835	3732	3666	3573	3507	3454	3370	3333	3145	2970	2663	2396	2191	1819	1444	1091
00.070	4058	3977	3919	3835	3774	3601	3534	3509	3468	3346	3219	2998	2936	2694	2352	2013
00.080	4117	4050	4004	3940	3898	3866	3820	3803	3775	3697	3633	3590	3566	3527	3496	3465
00.090	4167	4112	4075	4029	4000	3981	4084	4075	4061	4027	4014	4150	4161	4321	4493	4883
00.100	4218	4182	4291	4260	4238	4223	4206	4200	4195	4334	4370	4542	4701	5008	5294	5666
00.130	4507	4479	4465	4468	4610	4617	4632	4645	4798	5015	5265	5594	5839	6166	6411	6523
00.170	4626	4783	4793	4813	4982	5010	5180	5214	5257	5655	6042	6402	6578	6817	6895	6922
00.200	4862	4878	4893	5083	5121	5291	5362	5511	5572	5987	6442	6745	6901	7047	7087	7104
00.250	4970	5006	5187	5255	5446	5504	5714	5867	5939	6449	6891	7123	7207	7279	7300	7312
00.300	5051	5257	5298	5523	5722	5786	6009	6155	6238	6734	7158	7329	7387	7431	7446	7455
00.400	5346	5419	5621	5862	6070	6149	6386	6528	6698	7178	7473	7578	7605	7629	7640	7646
00.500	5443	5681	5882	6134	6245	6436	6747	6807	6970	7406	7652	7723	7741	7759	7767	7772
00.600	5666	5776	5999	6263	6485	6675	6901	7029	7181	7592	7774	7825	7838	7852	7859	7863
00.700	5750	5995	6090	6473	6690	6867	7091	7209	7350	7712	7863	7901	7913	7925	7931	7934
00.800	5807	6067	6291	6563	6784	6968	7250	7307	7488	7805	7932	7962	7972	7983	7989	7992
00.900	5857	6125	6358	6637	6941	7116	7331	7438	7567	7880	7986	8013	8022	8032	8037	8039
01.000	5909	6183	6415	6801	7018	7192	7453	7548	7667	7942	8032	8055	8064	8073	8077	8080
01.200	6128	6401	6631	6910	7199	7369	7608	7662	7801	8033	8105	8125	8132	8140	8144	8146
01.400	6193	6474	6712	7088	7297	7461	7698	7782	7905	8104	8162	8179	8185	8192	8196	8198
01.600	6247	6544	6786	7162	7433	7591	7805	7881	7987	8159	8208	8223	8229	8235	8238	8240
01.800	6306	6713	6946	7230	7502	7654	7866	7939	8041	8202	8246	8260	8265	8271	8274	8276
02.000	6474	6763	6998	7359	7557	7753	7945	8010	8100	8239	8278	8292	8297	8302	8305	8307
04.000	6874	7176	7490	7808	8014	8125	8277	8317	8372	8442	8464	8472	8476	8479	8481	8483
06.000	7129	7431	7725	8016	8196	8310	8421	8449	8487	8536	8553	8560	8563	8566	8568	8569
08.000	7237	7622	7836	8158	8315	8411	8503	8529	8557	8595	8610	8616	8619	8622	8623	8624
10.000	7322	7703	7971	8261	8401	8483	8560	8581	8605	8638	8651	8657	8659	8662	8663	8664
20.000	7720	8058	8277	8501	8600	8659	8707	8718	8732	8754	8764	8768	8770	8772	8773	8773
30.000	7842	8224	8419	8610	8698	8738	8775	8784	8795	8813	8822	8825	8827	8828	8829	8830
40.000	7994	8301	8490	8681	8754	8788	8818	8826	8835	8852	8859	8862	8864	8865	8866	8866

$2\alpha/RT\beta = 7.10$

p/K'	1.00	1.50	2.00	3.00	4.00	5.00	7.00	8.00	10.0	20.0	40.0	70.0	100	200	400	1000
									γ-Values							
00.001	1345	1017	0711	0429	0265	0167	0081	0057	0037	0017	0013	0011	0011	0010	0010	0010
00.002	1772	1303	1048	0690	0461	0345	0184	0145	0095	0036	0026	0023	0022	0021	0020	0020
00.004	2134	1764	1479	1056	0830	0605	0393	0322	0241	0084	0054	0048	0045	0043	0041	0041
00.006	2405	1923	1757	1307	1064	0875	0603	0505	0394	0148	0084	0073	0069	0065	0063	0062
00.008	2523	2165	1878	1541	1283	1077	0774	0717	0528	0223	0119	0099	0094	0088	0085	0084
00.010	2605	2403	2113	1768	1498	1279	0947	0885	0724	0318	0158	0127	0119	0112	0108	0106
00.020	3200	2873	2749	2418	2143	2038	1753	1575	1374	0852	0479	0315	0271	0242	0232	0227
00.030	3367	3233	3128	2817	2704	2467	2327	2138	2045	1522	0947	0658	0521	0405	0377	0366
00.040	3665	3539	3290	3164	3069	2841	2726	2678	2456	2074	1599	1164	0952	0689	0564	0532
00.050	3755	3658	3586	3477	3243	3178	3085	3047	2979	2627	2155	1871	1570	1182	0904	0747
00.060	4004	3757	3691	3603	3541	3493	3418	3388	3335	3030	2730	2468	2386	1993	1574	1177
00.070	4082	4007	3953	3878	3825	3785	3724	3699	3516	3407	3305	3224	3032	2938	2578	2227
00.080	4138	4075	4031	3973	3934	3905	3864	3849	3825	3761	3712	3684	3815	3797	3789	3934
00.090	4187	4136	4102	4062	4185	4168	4144	4135	4121	4094	4239	4254	4411	4592	4780	5189
00.100	4240	4357	4333	4300	4280	4266	4251	4248	4247	4406	4589	4776	4942	5156	5543	5925
00.130	4530	4504	4493	4647	4655	4660	4679	4836	4859	5089	5466	5791	5956	6330	6546	6637
00.170	4647	4818	4825	4847	5029	5053	5246	5270	5312	5726	6208	6546	6707	6916	6981	7004
00.200	4893	4904	4922	5122	5159	5344	5410	5574	5627	6055	6583	6862	7001	7126	7160	7175
00.250	4992	5176	5219	5426	5485	5672	5763	5924	6099	6515	6964	7193	7281	7342	7361	7372
00.300	5073	5285	5327	5558	5768	5828	6056	6214	6387	6858	7223	7397	7448	7486	7500	7508
00.400	5371	5582	5651	5899	6116	6191	6527	6577	6750	7236	7526	7628	7654	7675	7685	7691
00.500	5464	5708	5924	6178	6391	6477	6801	6853	7018	7485	7705	7767	7783	7800	7807	7812
00.600	5702	5802	6029	6296	6522	6716	7018	7074	7227	7638	7819	7864	7877	7890	7897	7900
00.700	5775	6028	6121	6514	6733	6916	7133	7254	7394	7754	7902	7938	7948	7960	7966	7969
00.800	5829	6094	6325	6596	6820	7006	7293	7401	7530	7845	7968	7997	8006	8016	8022	8024
00.900	5878	6150	6387	6671	6987	7163	7370	7478	7642	7918	8020	8045	8054	8063	8068	8071
01.000	5934	6327	6444	6838	7054	7230	7493	7590	7705	7978	8065	8087	8094	8103	8107	8110
01.200	6154	6429	6663	6941	7243	7408	7648	7734	7837	8070	8135	8154	8160	8168	8172	8174
01.400	6215	6499	6740	7123	7330	7495	7732	7816	7939	8135	8190	8206	8212	8219	8222	8224
01.600	6268	6678	6911	7193	7472	7626	7839	7914	8021	8188	8234	8249	8254	8261	8264	8266
01.800	6448	6742	6976	7334	7534	7686	7927	7994	8072	8231	8271	8285	8290	8296	8299	8300
02.000	6501	6789	7026	7391	7588	7786	7976	8040	8130	8267	8303	8316	8320	8326	8329	8330
04.000	6897	7201	7518	7835	8041	8174	8301	8349	8395	8462	8483	8492	8495	8499	8500	8501
06.000	7156	7452	7751	8040	8220	8332	8442	8474	8507	8554	8571	8577	8580	8583	8585	8586
08.000	7256	7645	7913	8180	8337	8432	8526	8547	8575	8612	8626	8632	8635	8637	8639	8640
10.000	7428	7724	7993	8284	8421	8502	8581	8599	8622	8654	8667	8672	8674	8677	8678	8678
20.000	7741	8078	8297	8519	8625	8675	8721	8733	8746	8768	8777	8781	8783	8785	8786	8786
30.000	7860	8243	8437	8626	8713	8752	8788	8797	8807	8825	8834	8837	8838	8840	8841	8841
40.000	8011	8318	8530	8696	8768	8802	8831	8838	8847	8863	8870	8873	8875	8876	8877	8877

$2\alpha/RT\beta = 7.20$

p/K′	γ-Values 1.00	1.50	2.00	3.00	4.00	5.00	7.00	8.00	10.0	20.0	40.0	70.0	100	200	400	1000
00.001	1355	1029	0718	0436	0269	0169	0082	0062	0038	0017	0013	0011	0011	0010	0010	0010
00.002	1792	1313	1058	0699	0467	0352	0187	0147	0096	0036	0026	0023	0022	0021	0020	0020
00.004	2150	1784	1497	1068	0845	0614	0399	0327	0247	0085	0054	0048	0045	0043	0041	0041
00.006	2435	2068	1778	1320	1077	0887	0612	0513	0402	0150	0085	0073	0069	0065	0063	0062
00.008	2539	2181	1895	1557	1299	1092	0856	0730	0587	0227	0120	0099	0094	0088	0086	0084
00.010	2622	2428	2132	1789	1518	1296	1042	0900	0739	0323	0163	0127	0120	0112	0108	0106
00.020	3226	2892	2777	2443	2166	2068	1781	1599	1501	0938	0488	0327	0275	0243	0233	0227
00.030	3384	3255	3158	2842	2738	2495	2365	2168	2080	1559	1037	0713	0550	0411	0379	0368
00.040	3687	3572	3314	3192	3105	3029	2761	2718	2640	2116	1640	1275	1034	0703	0574	0536
00.050	3774	3681	3613	3513	3435	3211	3123	3089	3030	2683	2343	1927	1719	1286	0959	0764
00.060	4036	3945	3717	3632	3574	3529	3461	3434	3389	3083	2945	2680	2462	2194	1726	1305
00.070	4105	4034	3984	3915	3867	3831	3778	3757	3724	3463	3376	3313	3275	3044	2979	2598
00.080	4158	4098	4058	4004	3968	3942	3906	3893	3872	3983	3947	3928	3921	4078	4087	4415
00.090	4207	4159	4130	4258	4234	4217	4193	4185	4174	4156	4324	4502	4523	4864	5176	5546
00.100	4426	4393	4369	4336	4317	4306	4295	4294	4457	4471	4673	4877	5060	5400	5774	6146
00.130	4553	4528	4520	4691	4695	4700	4881	4894	4914	5156	5553	5893	6141	6483	6668	6742
00.170	4668	4849	4854	4881	5071	5094	5300	5321	5504	5792	6290	6636	6796	7001	7062	7082
00.200	4921	4930	4950	5158	5197	5390	5597	5628	5679	6224	6663	6940	7078	7198	7229	7243
00.250	5013	5216	5250	5470	5523	5721	5938	5976	6161	6577	7032	7275	7345	7403	7420	7429
00.300	5094	5312	5356	5592	5810	5869	6102	6266	6446	6920	7284	7455	7504	7539	7552	7559
00.400	5394	5621	5680	5933	6157	6350	6577	6622	6798	7289	7589	7677	7700	7720	7729	7735
00.500	5484	5734	5960	6217	6437	6516	6849	6897	7134	7534	7750	7809	7824	7840	7847	7851
00.600	5732	5828	6058	6329	6558	6754	7067	7116	7329	7682	7860	7902	7914	7927	7933	7936
00.700	5798	6057	6280	6552	6772	6958	7173	7295	7435	7810	7941	7973	7983	7994	8000	8003
00.800	5850	6119	6356	6629	6855	7043	7333	7444	7571	7895	8003	8030	8039	8049	8053	8056
00.900	5899	6174	6415	6809	7027	7203	7467	7516	7683	7963	8054	8077	8085	8094	8098	8101
01.000	6096	6366	6472	6872	7088	7265	7531	7629	7742	8012	8096	8117	8124	8133	8137	8139
01.200	6178	6455	6693	6972	7280	7444	7684	7771	7871	8102	8164	8182	8188	8195	8199	8201
01.400	6236	6523	6767	7155	7362	7587	7806	7882	7972	8165	8217	8233	8238	8245	8248	8250
01.600	6289	6714	6947	7222	7507	7658	7871	7946	8052	8216	8260	8274	8279	8285	8288	8290
01.800	6480	6769	7005	7369	7565	7766	7960	8026	8117	8257	8296	8309	8314	8320	8322	8324
02.000	6526	6814	7052	7420	7677	7818	8006	8070	8159	8292	8327	8339	8344	8349	8351	8353
04.000	6918	7318	7543	7860	8067	8202	8337	8372	8417	8482	8502	8511	8514	8517	8519	8520
06.000	7181	7474	7775	8063	8242	8355	8463	8495	8526	8572	8588	8594	8597	8600	8601	8602
08.000	7275	7666	7940	8202	8358	8452	8546	8566	8592	8628	8642	8648	8650	8653	8654	8655
10.000	7454	7815	8014	8305	8441	8521	8599	8616	8638	8669	8682	8687	8689	8691	8692	8693
20.000	7761	8097	8317	8537	8642	8691	8736	8747	8760	8780	8790	8794	8795	8797	8798	8799
30.000	7948	8261	8455	8642	8727	8768	8801	8810	8819	8837	8845	8848	8850	8851	8852	8852
40.000	8028	8375	8548	8710	8782	8815	8843	8850	8858	8874	8881	8884	8885	8887	8888	8888

$2\alpha/RT\beta$ = 7.30

p/K′	1.00	1.50	2.00	3.00	4.00	5.00	7.00	8.00	10.0	20.0	40.0	70.0	100	200	400	1000
	γ-Values															
00.001	1365	1040	0725	0441	0273	0172	0083	0063	0038	0017	0013	0011	0011	0010	0010	0010
00.002	1808	1324	1069	0707	0473	0357	0189	0149	0098	0036	0026	0023	0022	0021	0020	0020
00.004	2165	1801	1514	1080	0857	0686	0449	0367	0251	0086	0054	0048	0045	0043	0041	0041
00.006	2458	2091	1796	1334	1090	0899	0621	0575	0408	0151	0085	0073	0069	0065	0063	0062
00.008	2553	2196	2053	1572	1314	1105	0872	0742	0599	0247	0122	0100	0094	0088	0086	0084
00.010	2801	2450	2150	1808	1536	1312	1060	0914	0752	0328	0164	0128	0120	0112	0109	0106
00.020	3248	2910	2803	2465	2188	2094	1806	1750	1528	0959	0532	0331	0277	0243	0233	0228
00.030	3402	3276	3185	2866	2768	2684	2399	2348	2111	1589	1060	0729	0588	0418	0381	0369
00.040	3707	3600	3513	3219	3138	3071	2794	2754	2686	2153	1676	1308	1136	0744	0586	0540
00.050	3792	3702	3638	3545	3475	3418	3158	3126	3073	2731	2400	2116	1895	1413	1028	0781
00.060	4063	3981	3921	3660	3604	3562	3500	3476	3435	3308	3012	2748	2690	2267	1900	1490
00.070	4126	4059	4012	3948	3904	3871	3824	3806	3776	3692	3619	3567	3360	3309	3265	3040
00.080	4177	4121	4083	4033	4001	3977	3946	4112	4095	4052	4024	4014	4189	4366	4549	4880
00.090	4225	4182	4334	4299	4275	4259	4238	4231	4222	4388	4400	4597	4783	4992	5442	5855
00.100	4461	4425	4400	4369	4352	4343	4512	4513	4514	4702	4912	5121	5303	5636	5988	6331
00.130	4574	4551	4546	4729	4732	4739	4935	4945	4964	5371	5760	6086	6242	6624	6788	6840
00.170	4864	4877	4881	5082	5109	5133	5348	5368	5564	5983	6454	6771	6913	7088	7139	7156
00.200	4946	4954	5148	5191	5395	5432	5650	5677	5870	6290	6734	7048	7168	7269	7296	7308
00.250	5034	5249	5278	5509	5559	5766	5990	6023	6216	6723	7139	7338	7411	7460	7476	7485
00.300	5285	5338	5547	5624	5849	6049	6274	6314	6499	6977	7369	7518	7559	7590	7602	7609
00.400	5416	5653	5707	5966	6195	6396	6622	6768	6933	7340	7638	7723	7745	7763	7772	7777
00.500	5505	5759	5992	6253	6477	6670	6892	7029	7182	7581	7793	7850	7864	7878	7885	7889
00.600	5759	6007	6085	6360	6593	6791	7110	7233	7375	7746	7901	7939	7950	7962	7968	7971
00.700	5820	6085	6316	6586	6809	6997	7288	7335	7529	7851	7977	8007	8017	8027	8032	8035
00.800	5871	6144	6385	6659	6888	7078	7371	7484	7609	7932	8037	8062	8071	8080	8085	8087
00.900	5919	6199	6442	6847	7063	7240	7507	7606	7721	7998	8086	8108	8115	8124	8128	8130
01.000	6126	6399	6628	6904	7121	7299	7566	7666	7813	8052	8127	8146	8153	8161	8165	8167
01.200	6200	6480	6721	7105	7314	7478	7719	7806	7933	8133	8193	8209	8215	8222	8226	8228
01.400	6256	6547	6793	7185	7393	7624	7840	7917	8003	8194	8244	8258	8264	8270	8273	8275
01.600	6309	6744	6979	7250	7539	7689	7902	7977	8083	8243	8286	8299	8304	8310	8312	8314
01.800	6508	6795	7032	7401	7594	7800	7991	8056	8147	8283	8321	8333	8337	8343	8345	8347
02.000	6549	6837	7077	7449	7710	7848	8035	8120	8187	8317	8350	8362	8366	8371	8374	8375
04.000	6938	7346	7568	7885	8092	8227	8361	8395	8439	8502	8521	8529	8532	8535	8537	8538
06.000	7203	7585	7798	8085	8293	8376	8490	8514	8547	8589	8605	8611	8613	8616	8618	8618
08.000	7293	7686	7964	8223	8400	8472	8564	8584	8610	8644	8658	8663	8665	8668	8669	8670
10.000	7477	7841	8034	8326	8460	8540	8616	8635	8654	8684	8696	8701	8703	8705	8707	8707
20.000	7779	8115	8335	8554	8659	8706	8750	8761	8773	8793	8802	8806	8808	8809	8810	8811
30.000	7970	8279	8471	8657	8742	8782	8814	8822	8832	8849	8856	8860	8861	8862	8863	8864
40.000	8044	8394	8565	8724	8795	8828	8855	8861	8870	8885	8892	8895	8896	8897	8898	8898

$2\alpha/RT\beta = 7.40$

p/K'	γ-Values 1.00	1.50	2.00	3.00	4.00	5.00	7.00	8.00	10.0	20.0	40.0	70.0	100	200	400	1000
00.001	1499	1050	0732	0446	0277	0174	0084	0064	0038	0017	0013	0011	0011	0010	0010	0010
00.002	1823	1333	1078	0715	0536	0362	0192	0151	0099	0036	0026	0023	0022	0021	0020	0020
00.004	2179	1817	1528	1091	0868	0697	0458	0374	0255	0090	0054	0048	0045	0043	0042	0041
00.006	2478	2111	1813	1472	1102	0909	0699	0585	0414	0163	0086	0073	0069	0065	0063	0062
00.008	2567	2211	2078	1587	1328	1118	0885	0753	0609	0250	0123	0100	0094	0088	0086	0084
00.010	2827	2470	2167	1826	1553	1327	1076	0927	0763	0363	0165	0128	0120	0112	0109	0107
00.020	3269	2928	2825	2487	2374	2118	1829	1778	1551	0976	0542	0348	0282	0244	0234	0229
00.030	3608	3295	3210	2889	2795	2721	2428	2383	2140	1617	1177	0801	0599	0427	0384	0371
00.040	3725	3625	3547	3244	3167	3106	2825	2787	2725	2353	1848	1447	1165	0796	0600	0545
00.050	3809	3723	3661	3574	3510	3460	3192	3162	3113	2775	2450	2169	1947	1562	1111	0812
00.060	4087	4011	3956	3879	3634	3594	3536	3514	3477	3366	3071	2991	2759	2495	2250	1638
00.070	4146	4082	4038	3979	3938	3908	3865	3849	3822	3749	3689	3649	3627	3592	3564	3536
00.080	4195	4142	4107	4061	4032	4203	4175	4165	4149	4111	4092	4279	4283	4483	4849	5185
00.090	4244	4397	4370	4335	4313	4297	4279	4273	4267	4451	4652	4856	4881	5255	5691	6115
00.100	4489	4453	4429	4400	4385	4378	4564	4564	4564	4770	4993	5213	5405	5861	6242	6489
00.130	4594	4574	4761	4763	4766	4775	4982	4992	5012	5439	5843	6176	6415	6723	6889	6931
00.170	4898	4903	4908	5122	5145	5170	5392	5413	5617	6048	6531	6850	7021	7169	7211	7226
00.200	4969	4977	5186	5222	5439	5471	5698	5723	5925	6350	6871	7118	7237	7335	7359	7370
00.250	5053	5279	5305	5544	5594	5806	6037	6068	6266	6783	7201	7412	7470	7516	7530	7538
00.300	5319	5362	5584	5655	5885	6094	6323	6358	6548	7031	7425	7571	7612	7639	7650	7657
00.400	5437	5683	5734	5997	6230	6437	6665	6818	6984	7435	7686	7769	7788	7805	7813	7818
00.500	5525	5783	6022	6286	6514	6712	6933	7074	7227	7624	7840	7889	7902	7916	7922	7926
00.600	5784	6039	6111	6530	6751	6827	7150	7277	7417	7788	7939	7975	7985	7996	8002	8005
00.700	5841	6110	6349	6618	6843	7033	7329	7372	7570	7889	8013	8040	8049	8059	8064	8067
00.800	5890	6167	6413	6689	7031	7208	7407	7521	7693	7968	8070	8093	8101	8110	8115	8117
00.900	5940	6223	6468	6881	7097	7275	7544	7645	7757	8032	8117	8137	8145	8153	8157	8159
01.000	6153	6428	6663	6935	7153	7332	7600	7700	7849	8085	8157	8175	8182	8189	8193	8195
01.200	6222	6504	6748	7140	7346	7510	7751	7839	7966	8162	8220	8236	8242	8248	8251	8253
01.400	6276	6569	6818	7214	7505	7657	7873	7949	8058	8225	8270	8284	8289	8295	8298	8300
01.600	6481	6772	7008	7374	7569	7719	7970	8037	8112	8271	8310	8323	8328	8333	8336	8338
01.800	6533	6819	7058	7430	7623	7831	8020	8085	8175	8310	8344	8356	8360	8366	8368	8370
02.000	6571	6860	7102	7476	7741	7876	8093	8150	8227	8341	8373	8384	8388	8393	8396	8397
04.000	6957	7372	7591	7908	8116	8251	8384	8417	8460	8521	8540	8547	8550	8553	8555	8556
06.000	7224	7612	7820	8155	8316	8397	8510	8534	8565	8606	8621	8627	8629	8632	8634	8634
08.000	7311	7706	7986	8243	8421	8505	8582	8604	8627	8660	8673	8678	8680	8683	8684	8685
10.000	7499	7863	8053	8345	8479	8568	8633	8651	8671	8699	8711	8715	8717	8720	8721	8721
20.000	7797	8132	8352	8570	8674	8721	8764	8774	8786	8806	8814	8818	8820	8821	8822	8823
30.000	7989	8295	8488	8672	8756	8795	8827	8834	8843	8860	8868	8871	8872	8873	8874	8875
40.000	8059	8412	8581	8738	8808	8840	8867	8873	8881	8896	8902	8905	8906	8908	8908	8909

$2\alpha/RT\beta = 7.50$

p/K'	γ-Values 1.00	1.50	2.00	3.00	4.00	5.00	7.00	8.00	10.0	20.0	40.0	70.0	100	200	400	1000
00.001	1515	1060	0826	0451	0280	0176	0085	0065	0041	0017	0013	0011	0011	0010	0010	0010
00.002	1836	1343	1087	0722	0546	0367	0220	0153	0100	0036	0026	0023	0022	0021	0020	0020
00.004	2192	1831	1542	1101	0879	0707	0465	0380	0258	0091	0054	0048	0045	0043	0042	0041
00.006	2495	2129	1828	1492	1113	0920	0711	0594	0470	0165	0086	0073	0069	0065	0063	0062
00.008	2581	2225	2099	1601	1341	1248	0898	0763	0618	0254	0123	0100	0094	0088	0086	0084
00.010	2849	2489	2183	1842	1568	1342	1090	0939	0774	0369	0172	0129	0120	0112	0109	0107
00.020	3288	2945	2847	2507	2402	2140	1851	1803	1573	0992	0551	0353	0290	0245	0235	0229
00.030	3636	3314	3233	3106	2821	2752	2456	2414	2167	1790	1202	0818	0649	0439	0386	0373
00.040	3743	3648	3577	3268	3195	3138	3049	2818	2760	2394	1885	1479	1292	0862	0634	0550
00.050	3826	3742	3684	3601	3542	3496	3424	3195	3149	3009	2680	2390	2154	1604	1212	0860
00.060	4109	4037	3987	3916	3866	3828	3571	3550	3516	3417	3325	3056	3015	2747	2495	1940
00.070	4165	4105	4063	4007	3970	3942	3903	3888	3865	3801	3752	3924	3909	3886	3872	4063
00.080	4213	4163	4130	4292	4266	4247	4220	4211	4197	4165	4355	4359	4561	4770	5139	5584
00.090	4465	4428	4402	4368	4347	4333	4317	4313	4509	4508	4725	4943	5144	5511	5922	6330
00.100	4515	4479	4456	4429	4417	4612	4610	4609	4610	4830	5066	5457	5644	6073	6414	6625
00.130	4613	4595	4796	4795	4799	5005	5026	5035	5242	5500	5917	6360	6503	6846	6983	7017
00.170	4926	4928	4933	5157	5179	5391	5434	5630	5666	6107	6601	6974	7096	7245	7280	7293
00.200	4991	4999	5219	5253	5478	5508	5742	5766	5974	6524	6938	7216	7317	7399	7419	7429
00.250	5072	5306	5331	5577	5801	5844	6080	6111	6313	6838	7299	7470	7529	7569	7582	7589
00.300	5347	5385	5617	5860	5919	6135	6367	6401	6593	7155	7503	7627	7661	7686	7697	7702
00.400	5457	5710	5759	6027	6264	6475	6705	6863	7030	7483	7741	7812	7829	7845	7853	7857
00.500	5730	5805	6050	6317	6549	6751	6971	7115	7269	7699	7880	7927	7939	7952	7958	7961
00.600	5807	6069	6137	6567	6790	6978	7188	7318	7457	7827	7975	8009	8019	8030	8035	8038
00.700	5862	6135	6378	6648	6875	7067	7368	7482	7608	7925	8047	8073	8081	8091	8095	8098
00.800	5909	6190	6439	6850	7067	7245	7442	7557	7731	8003	8102	8124	8131	8140	8144	8146
00.900	6134	6407	6493	6913	7129	7308	7579	7681	7791	8065	8147	8166	8173	8181	8185	8187
01.000	6178	6454	6694	6963	7183	7451	7695	7734	7884	8116	8185	8203	8209	8216	8220	8222
01.200	6242	6527	6773	7172	7376	7541	7783	7871	7998	8195	8247	8262	8267	8274	8277	8279
01.400	6295	6739	6975	7242	7538	7689	7903	7979	8088	8252	8295	8308	8313	8319	8322	8324
01.600	6511	6797	7035	7406	7598	7748	8001	8067	8159	8297	8334	8346	8351	8356	8359	8360
01.800	6556	6842	7082	7458	7724	7861	8048	8113	8202	8334	8367	8378	8383	8388	8390	8392
02.000	6592	6882	7125	7502	7769	7904	8121	8177	8254	8366	8395	8406	8410	8415	8417	8418
04.000	6976	7395	7614	7931	8139	8275	8406	8438	8485	8539	8558	8565	8567	8571	8572	8573
06.000	7244	7636	7841	8178	8338	8436	8529	8552	8583	8622	8637	8643	8645	8648	8649	8650
08.000	7328	7725	8007	8303	8442	8524	8600	8621	8644	8675	8688	8693	8695	8697	8699	8699
10.000	7519	7884	8072	8364	8516	8586	8650	8667	8686	8713	8725	8729	8731	8733	8734	8735
20.000	7814	8149	8369	8586	8690	8735	8777	8787	8799	8818	8826	8830	8831	8833	8834	8834
30.000	8007	8311	8503	8701	8769	8808	8839	8846	8855	8871	8879	8882	8883	8884	8885	8885
40.000	8074	8428	8596	8752	8820	8852	8878	8884	8892	8906	8913	8916	8917	8918	8919	8919

$2a/Kr_p = 7.00$

p/K′	γ-Values 1.00	1.50	2.00	3.00	4.00	5.00	7.00	8.00	10.0	20.0	40.0	70.0	100	200	400	1000
00.001	1528	1068	0837	0456	0283	0205	0097	0066	0042	0017	0013	0011	0011	0010	0010	0010
00.002	1849	1353	1096	0729	0553	0371	0223	0175	0113	0037	0026	0023	0022	0021	0020	0020
00.004	2204	1844	1554	1111	0888	0716	0471	0385	0261	0092	0054	0048	0045	0043	0042	0041
00.006	2512	2145	1842	1509	1248	1037	0721	0602	0478	0167	0086	0073	0069	0065	0063	0062
00.008	2594	2239	2118	1770	1494	1265	0909	0864	0626	0257	0127	0100	0094	0089	0086	0084
00.010	2868	2506	2198	1858	1583	1495	1104	1058	0877	0375	0174	0129	0120	0113	0109	0107
00.020	3305	2962	2866	2527	2427	2161	1871	1825	1593	1115	0612	0377	0293	0246	0236	0230
00.030	3661	3332	3254	3137	2845	2780	2482	2442	2376	1821	1224	0910	0660	0455	0389	0375
00.040	3760	3670	3603	3501	3221	3168	3086	3052	2793	2431	2089	1648	1321	0944	0663	0556
00.050	4058	3761	3705	3627	3571	3529	3464	3438	3394	3056	2730	2442	2203	1781	1441	0937
00.060	4130	4062	4014	3948	3903	3868	3818	3799	3767	3463	3383	3325	3080	3020	2763	2326
00.070	4183	4126	4087	4034	3999	3974	3938	3925	3904	4064	4024	4001	3991	4189	4186	4590
00.080	4230	4183	4153	4329	4303	4285	4261	4253	4240	4427	4422	4636	4841	5055	5416	5926
00.090	4495	4456	4430	4398	4379	4367	4354	4562	4559	4766	4987	5206	5403	5755	6134	6507
00.100	4538	4503	4481	4457	4447	4655	4651	4651	4654	4886	5317	5541	5735	6270	6569	6744
00.130	4631	4616	4827	4825	4830	5047	5066	5076	5293	5557	6125	6442	6662	6958	7070	7098
00.170	4951	4951	4957	5189	5211	5433	5473	5678	5712	6302	6758	7045	7193	7316	7345	7357
00.200	5011	5021	5249	5281	5514	5544	5783	5978	6021	6584	7000	7279	7391	7459	7476	7486
00.250	5091	5331	5356	5608	5841	5880	6121	6307	6357	6982	7357	7536	7585	7619	7631	7638
00.300	5373	5408	5647	5897	5952	6173	6408	6580	6637	7207	7555	7676	7709	7732	7741	7747
00.400	5476	5735	5968	6226	6296	6512	6743	6904	7073	7528	7785	7854	7869	7884	7891	7895
00.500	5759	5827	6077	6347	6582	6787	7008	7155	7309	7740	7922	7964	7975	7987	7992	7996
00.600	5828	6096	6331	6601	6825	7016	7224	7356	7495	7865	8013	8043	8052	8062	8067	8070
00.700	5881	6158	6406	6677	6906	7100	7404	7520	7645	7976	8079	8104	8112	8121	8125	8128
00.800	5928	6212	6464	6885	7101	7280	7553	7591	7767	8047	8133	8153	8161	8169	8173	8175
00.900	6161	6436	6672	6942	7159	7339	7613	7715	7867	8105	8177	8195	8201	8209	8212	8214
01.000	6200	6479	6722	6991	7320	7485	7729	7819	7916	8146	8213	8230	8236	8243	8246	8248
01.200	6261	6549	6798	7201	7405	7571	7866	7901	8029	8223	8273	8287	8292	8298	8301	8303
01.400	6313	6768	7005	7268	7569	7718	7933	8009	8117	8278	8319	8332	8337	8342	8345	8347
01.600	6536	6821	7060	7436	7626	7840	8030	8096	8187	8322	8358	8369	8374	8379	8381	8383
01.800	6578	6864	7106	7484	7754	7889	8075	8139	8229	8358	8390	8400	8405	8409	8412	8413
02.000	6612	6903	7271	7527	7796	7930	8148	8204	8280	8389	8417	8427	8431	8436	8438	8439
04.000	6994	7418	7635	8018	8161	8297	8427	8459	8505	8558	8575	8582	8585	8588	8589	8590
06.000	7263	7658	7861	8200	8359	8457	8548	8570	8601	8639	8653	8658	8661	8663	8664	8665
08.000	7345	7744	8028	8325	8461	8542	8617	8638	8660	8690	8702	8707	8709	8712	8713	8714
10.000	7537	7904	8155	8383	8535	8604	8666	8683	8701	8727	8738	8743	8745	8747	8748	8749
20.000	7830	8165	8386	8602	8705	8749	8792	8800	8811	8830	8838	8842	8843	8845	8845	8846
30.000	8024	8327	8518	8716	8789	8821	8851	8858	8867	8882	8889	8892	8893	8895	8896	8896
40.000	8088	8444	8611	8777	8837	8864	8889	8895	8903	8916	8923	8926	8927	8928	8929	8929

$2\alpha/RT\beta = 7.70$

p/K'	γ-Values 1.00	1.50	2.00	3.00	4.00	5.00	7.00	8.00	10.0	20.0	40.0	70.0	100	200	400	1000
00.001	1540	1076	0846	0460	0286	0208	0099	0066	0042	0017	0013	0011	0011	0010	0010	0010
00.002	1860	1508	1104	0736	0560	0375	0226	0178	0114	0037	0026	0023	0022	0021	0020	0020
00.004	2216	1857	1566	1250	0897	0724	0477	0389	0300	0092	0054	0048	0045	0043	0042	0041
00.006	2527	2160	1856	1524	1263	1051	0730	0609	0485	0169	0087	0073	0069	0065	0063	0062
00.008	2606	2445	2135	1789	1511	1280	0919	0877	0715	0288	0128	0100	0095	0089	0086	0084
00.010	2886	2521	2212	1872	1760	1514	1116	1072	0890	0380	0175	0130	0121	0113	0109	0107
00.020	3322	3192	2885	2545	2450	2180	2069	1846	1774	1133	0622	0382	0304	0247	0236	0230
00.030	3682	3349	3274	3165	2868	2806	2711	2468	2407	1848	1373	0928	0726	0460	0393	0377
00.040	3777	3690	3627	3534	3245	3195	3120	3089	2823	2465	2128	1681	1475	1044	0700	0562
00.050	4084	3780	3726	3651	3599	3559	3499	3476	3436	3099	2775	2693	2440	1984	1606	1056
00.060	4150	4085	4040	3978	3936	3904	3858	3840	3812	3731	3437	3388	3360	3311	3050	2792
00.070	4200	4146	4109	4060	4028	4004	3972	3960	4166	4119	4086	4070	4285	4281	4704	5089
00.080	4247	4427	4399	4362	4337	4321	4299	4291	4281	4484	4698	4711	4928	5331	5678	6210
00.090	4520	4482	4457	4426	4409	4398	4610	4607	4604	4825	5058	5286	5491	5985	6328	6656
00.100	4559	4525	4505	4484	4697	4693	4690	4691	4909	5143	5387	5781	5963	6370	6708	6849
00.130	4648	4858	4855	4853	5075	5086	5104	5320	5340	5789	6195	6613	6808	7061	7151	7174
00.170	4974	4973	4981	5220	5242	5471	5704	5723	5755	6361	6823	7156	7260	7382	7407	7418
00.200	5031	5042	5277	5513	5548	5770	5822	6025	6064	6639	7126	7367	7450	7516	7531	7540
00.250	5108	5355	5380	5637	5877	5914	6159	6351	6545	7035	7410	7588	7636	7668	7679	7685
00.300	5396	5429	5675	5931	5983	6209	6447	6625	6807	7255	7604	7727	7754	7775	7784	7789
00.400	5494	5759	6001	6262	6327	6546	6908	6943	7113	7571	7827	7893	7908	7922	7929	7932
00.500	5784	5849	6102	6375	6614	6821	7156	7192	7346	7780	7960	7999	8009	8021	8026	8029
00.600	5849	6121	6363	6632	6858	7051	7259	7392	7599	7923	8047	8075	8084	8093	8098	8101
00.700	5900	6180	6432	6705	6936	7132	7438	7556	7680	8010	8112	8134	8142	8150	8154	8157
00.800	5946	6233	6488	6917	7133	7313	7588	7691	7801	8080	8163	8182	8189	8197	8200	8203
00.900	6185	6462	6703	6971	7189	7370	7645	7748	7901	8136	8205	8222	8228	8235	8239	8241
01.000	6221	6503	6749	7018	7352	7517	7762	7852	7947	8182	8241	8256	8262	8268	8272	8273
01.200	6280	6570	6821	7229	7433	7681	7898	7975	8058	8251	8298	8312	8317	8322	8325	8327
01.400	6331	6795	7032	7294	7598	7747	7961	8037	8145	8304	8343	8355	8360	8365	8368	8370
01.600	6560	6844	7084	7463	7652	7870	8058	8123	8214	8346	8380	8392	8396	8401	8403	8404
01.800	6598	6885	7128	7510	7782	7916	8137	8193	8270	8382	8412	8422	8426	8430	8433	8434
02.000	6631	6923	7299	7647	7822	8011	8174	8229	8304	8411	8438	8448	8452	8456	8458	8459
04.000	7011	7439	7656	8043	8228	8318	8448	8488	8524	8576	8592	8599	8602	8604	8606	8607
06.000	7281	7679	7881	8221	8379	8476	8566	8593	8618	8654	8668	8674	8676	8678	8679	8680
08.000	7491	7762	8047	8345	8480	8560	8638	8654	8676	8705	8717	8722	8723	8726	8727	8727
10.000	7555	7923	8176	8401	8553	8620	8685	8698	8716	8741	8752	8756	8758	8760	8761	8762
20.000	7846	8181	8402	8638	8719	8768	8805	8814	8824	8842	8850	8853	8854	8856	8857	8857
30.000	8040	8342	8533	8730	8802	8834	8863	8870	8878	8893	8900	8903	8904	8905	8906	8906
40.000	8102	8459	8625	8790	8850	8876	8900	8906	8913	8927	8933	8936	8937	8938	8938	8939

$2\alpha/RT\beta = 7.80$

p/K′	1.00	1.50	2.00	3.00	4.00	5.00	7.00	8.00	10.0	20.0	40.0	70.0	100	200	400	1000
	γ-Values															
00.001	1551	1084	0854	0464	0335	0210	0100	0076	0043	0017	0013	0011	0011	0010	0010	0010
00.002	1871	1522	1112	0843	0566	0437	0229	0180	0115	0037	0026	0023	0022	0021	0020	0020
00.004	2227	1869	1577	1264	0905	0732	0483	0394	0304	0100	0054	0048	0045	0043	0042	0041
00.006	2541	2174	1869	1538	1277	1063	0739	0616	0491	0187	0088	0073	0070	0065	0063	0062
00.008	2618	2465	2151	1806	1527	1294	1046	0889	0726	0292	0128	0100	0095	0089	0086	0084
00.010	2903	2537	2226	1887	1780	1530	1263	1086	0902	0431	0186	0130	0121	0113	0109	0107
00.020	3338	3217	2903	2776	2471	2400	2093	1865	1798	1150	0701	0415	0306	0249	0237	0231
00.030	3702	3366	3293	3191	2889	2831	2742	2493	2435	1874	1397	1044	0738	0482	0397	0379
00.040	3792	3710	3650	3563	3499	3221	3150	3122	3075	2709	2163	1880	1656	1065	0748	0571
00.050	4106	4032	3746	3674	3625	3587	3532	3510	3475	3137	3040	2744	2488	2213	1796	1234
00.060	4168	4106	4064	4005	3966	3936	3894	3878	3852	3780	3719	3678	3655	3384	3354	3323
00.070	4217	4165	4130	4084	4054	4267	4238	4227	4210	4169	4143	4363	4361	4587	5011	5387
00.080	4263	4456	4428	4392	4369	4353	4334	4327	4319	4535	4763	4992	5207	5597	5921	6441
00.090	4543	4506	4482	4453	4437	4658	4651	4648	4646	4879	5123	5545	5741	6199	6561	6775
00.100	4579	4547	4528	4509	4733	4728	4726	4728	4956	5200	5451	5858	6180	6546	6834	6948
00.130	4665	4886	4882	4880	5112	5122	5140	5365	5383	5846	6260	6687	6885	7154	7227	7246
00.170	4996	4994	5226	5249	5485	5507	5747	5764	5982	6416	6884	7220	7345	7443	7466	7476
00.200	5050	5285	5303	5549	5580	5810	6045	6068	6275	6690	7183	7425	7516	7570	7584	7592
00.250	5126	5378	5403	5665	5911	6133	6365	6393	6591	7085	7498	7648	7687	7715	7725	7731
00.300	5418	5450	5701	5963	6199	6243	6484	6666	6851	7301	7672	7772	7799	7817	7826	7831
00.400	5512	5782	6030	6295	6527	6578	6948	6981	7152	7658	7876	7932	7945	7958	7965	7968
00.500	5808	6070	6126	6403	6644	6854	7194	7227	7468	7847	7995	8033	8043	8053	8059	8062
00.600	5868	6144	6393	6661	6889	7085	7392	7426	7636	7958	8079	8107	8115	8124	8128	8131
00.700	5917	6202	6457	6732	7099	7279	7471	7590	7771	8044	8142	8164	8171	8179	8183	8185
00.800	6164	6439	6511	6946	7163	7344	7621	7725	7834	8111	8192	8210	8217	8224	8227	8229
00.900	6207	6487	6731	6998	7217	7499	7675	7779	7933	8166	8233	8249	8255	8261	8265	8267
01.000	6242	6525	6774	7178	7382	7547	7793	7883	8014	8211	8267	8282	8287	8293	8296	8298
01.200	6298	6590	6844	7256	7460	7711	7928	8005	8116	8277	8323	8336	8340	8346	8349	8350
01.400	6534	6819	7058	7436	7625	7774	8035	8101	8172	8331	8367	8378	8383	8388	8390	8392
01.600	6581	6866	7108	7490	7678	7898	8085	8150	8240	8371	8403	8413	8417	8422	8424	8426
01.800	6618	6906	7150	7534	7809	7941	8164	8219	8296	8405	8433	8443	8447	8451	8453	8454
02.000	6649	7095	7325	7675	7847	8038	8199	8254	8328	8433	8459	8468	8472	8476	8478	8479
04.000	7028	7459	7676	8066	8250	8339	8468	8508	8543	8593	8609	8615	8618	8621	8622	8623
06.000	7299	7699	7899	8241	8398	8495	8584	8610	8635	8670	8683	8688	8690	8693	8694	8695
08.000	7512	7779	8066	8364	8498	8578	8655	8670	8691	8719	8731	8735	8737	8739	8741	8741
10.000	7572	7941	8196	8455	8570	8637	8700	8713	8730	8755	8765	8770	8771	8773	8774	8775
20.000	7861	8196	8417	8654	8733	8782	8817	8826	8836	8853	8861	8864	8866	8867	8868	8868
30.000	8056	8356	8547	8744	8815	8846	8874	8881	8889	8904	8910	8913	8914	8916	8916	8917
40.000	8116	8474	8639	8803	8862	8887	8911	8916	8923	8937	8943	8945	8946	8947	8948	8948

$2\alpha/RT\beta = 7.90$

p/K′	1.00	1.50	2.00	3.00	4.00	5.00	7.00	8.00	10.0	20.0	40.0	70.0	100	200	400	1000
	γ-Values															
00.001	1561	1091	0862	0542	0340	0213	0101	0077	0043	0017	0013	0011	0011	0010	0010	0010
00.002	1882	1534	1259	0852	0572	0443	0231	0182	0117	0037	0026	0023	0022	0021	0020	0020
00.004	2238	1880	1588	1276	1035	0739	0488	0458	0308	0101	0055	0048	0045	0043	0042	0041
00.006	2554	2187	1881	1552	1290	1075	0747	0711	0497	0189	0090	0074	0070	0065	0064	0062
00.008	2630	2483	2166	1822	1542	1307	1060	0900	0736	0295	0134	0101	0095	0089	0086	0085
00.010	2918	2551	2448	2093	1799	1546	1279	1098	0914	0437	0187	0132	0121	0113	0109	0107
00.020	3353	3239	2920	2800	2491	2424	2116	2076	1820	1165	0712	0420	0322	0250	0238	0232
00.030	3721	3621	3312	3214	3142	2854	2771	2737	2461	2090	1419	1062	0822	0510	0402	0381
00.040	3807	3728	3671	3589	3530	3245	3179	3153	3109	2746	2403	1914	1688	1188	0808	0585
00.050	4127	4057	4007	3696	3650	3614	3562	3542	3510	3412	3086	2790	2753	2466	2012	1486
00.060	4185	4126	4086	4031	3994	3967	3928	3913	3889	3825	3773	3739	3720	3692	3671	3893
00.070	4233	4183	4151	4350	4323	4303	4276	4266	4251	4215	4434	4429	4661	4890	5306	5803
00.080	4521	4481	4455	4420	4399	4384	4367	4362	4593	4583	4823	5063	5287	5691	6250	6627
00.090	4564	4528	4505	4478	4465	4696	4688	4686	4685	4929	5390	5620	5822	6292	6710	6888
00.100	4598	4567	4549	4772	4765	4761	4761	4994	5000	5252	5704	6087	6260	6706	6948	7038
00.130	4682	4912	4907	4906	5147	5155	5396	5406	5424	5898	6460	6843	7015	7240	7299	7314
00.170	5016	5014	5256	5276	5520	5540	5787	5803	6027	6467	7029	7281	7423	7503	7522	7531
00.200	5068	5313	5328	5581	5610	5847	6087	6108	6320	6860	7297	7503	7570	7622	7635	7642
00.250	5374	5400	5646	5692	5942	6171	6407	6432	6633	7132	7548	7696	7734	7760	7769	7775
00.300	5438	5470	5726	5993	6234	6275	6677	6705	6892	7418	7718	7818	7841	7858	7866	7870
00.400	5529	5804	6058	6325	6561	6767	6986	7016	7293	7699	7915	7969	7981	7994	7999	8003
00.500	5829	6098	6149	6429	6672	6885	7229	7365	7505	7884	8034	8066	8075	8085	8090	8093
00.600	5887	6167	6420	6689	6919	7116	7427	7547	7672	7992	8113	8137	8145	8153	8158	8160
00.700	5935	6222	6481	6915	7131	7312	7502	7622	7805	8075	8172	8192	8199	8207	8211	8213
00.800	6189	6466	6533	6974	7192	7373	7652	7757	7865	8141	8220	8237	8243	8250	8254	8256
00.900	6228	6510	6757	7023	7365	7531	7778	7868	7964	8194	8259	8275	8280	8287	8290	8292
01.000	6261	6547	6798	7208	7411	7576	7823	7913	8045	8238	8293	8307	8312	8318	8321	8322
01.200	6316	6610	6866	7281	7590	7740	7956	8033	8144	8306	8347	8359	8364	8369	8372	8373
01.400	6558	6842	7083	7464	7652	7800	8063	8129	8221	8355	8389	8400	8405	8409	8412	8413
01.600	6602	6887	7130	7515	7790	7924	8110	8175	8266	8394	8424	8435	8438	8443	8445	8446
01.800	6637	6925	7171	7557	7834	7966	8189	8244	8320	8427	8454	8463	8467	8471	8473	8474
02.000	6667	7120	7349	7700	7871	8063	8223	8277	8364	8455	8479	8488	8491	8495	8497	8499
04.000	7044	7479	7695	8087	8272	8390	8500	8527	8562	8610	8625	8632	8634	8637	8638	8639
06.000	7315	7718	7918	8261	8417	8514	8602	8627	8653	8685	8698	8703	8705	8707	8709	8709
08.000	7532	7900	8084	8382	8515	8595	8671	8686	8706	8733	8745	8749	8751	8753	8754	8755
10.000	7589	7959	8215	8473	8587	8653	8715	8728	8745	8768	8778	8782	8784	8786	8787	8788
20.000	7876	8284	8432	8669	8747	8795	8830	8838	8848	8865	8872	8875	8877	8878	8879	8879
30.000	8071	8370	8561	8757	8827	8858	8886	8892	8900	8914	8921	8923	8924	8926	8926	8927
40.000	8129	8488	8652	8816	8873	8900	8921	8927	8934	8946	8952	8955	8956	8957	8958	8958

$2\alpha/kT\beta = 8.00$

p/K'	γ-Values 1.00	1.50	2.00	3.00	4.00	5.00	7.00	8.00	10.0	20.0	40.0	70.0	100	200	400	1000
00.001	1571	1098	0869	0548	0344	0215	0102	0078	0043	0017	0013	0011	0011	0010	0010	0010
00.002	1891	1546	1271	0861	0578	0448	0234	0184	0118	0038	0026	0023	0022	0021	0020	0020
00.004	2248	1891	1598	1288	1047	0853	0568	0464	0311	0101	0055	0048	0045	0043	0042	0041
00.006	2567	2200	1893	1564	1301	1085	0754	0720	0503	0191	0090	0074	0070	0066	0064	0062
00.008	2870	2500	2180	1837	1556	1319	1072	0909	0745	0338	0135	0101	0095	0089	0086	0085
00.010	2933	2564	2467	2112	1816	1560	1293	1110	0924	0442	0202	0132	0121	0113	0109	0107
00.020	3367	3259	2936	2823	2510	2447	2136	2100	1840	1323	0722	0465	0324	0252	0238	0232
00.030	3738	3645	3329	3236	3168	2875	2797	2766	2486	2118	1602	1079	0926	0547	0408	0383
00.040	3822	3746	3691	3614	3559	3516	3206	3181	3141	2780	2440	2143	1898	1335	0882	0608
00.050	4147	4080	4033	3968	3923	3640	3591	3573	3542	3454	3128	3072	3039	2743	2253	1664
00.060	4202	4146	4107	4055	4021	3995	3959	3946	3924	3866	3822	4044	4030	4009	4242	4232
00.070	4248	4201	4170	4381	4355	4336	4311	4302	4288	4504	4490	4728	4731	5188	5587	6153
00.080	4544	4505	4479	4446	4426	4413	4645	4641	4635	4867	5107	5343	5558	5939	6443	6760
00.090	4584	4549	4527	4502	4737	4731	4724	4722	4961	5204	5453	5870	6059	6486	6844	6989
00.100	4617	4587	4570	4803	4796	4792	5033	5035	5040	5301	5766	6159	6460	6851	7065	7121
00.130	4698	4935	4930	4931	5179	5187	5435	5445	5677	5948	6521	6911	7084	7309	7366	7379
00.170	5035	5034	5284	5302	5553	5572	5824	5840	6069	6659	7085	7378	7481	7560	7576	7585
00.200	5085	5338	5351	5612	5639	5882	6126	6146	6363	6911	7351	7555	7629	7671	7683	7690
00.250	5398	5420	5674	5718	5973	6207	6445	6469	6673	7268	7595	7749	7781	7803	7812	7817
00.300	5458	5718	5749	6021	6267	6487	6716	6742	6931	7461	7761	7860	7881	7897	7905	7909
00.400	5546	5824	6084	6354	6593	6803	7021	7176	7332	7738	7959	8005	8016	8028	8033	8036
00.500	5850	6123	6171	6636	6863	7058	7263	7401	7541	7920	8067	8098	8107	8116	8121	8124
00.600	5905	6188	6445	6716	6947	7146	7461	7582	7705	8025	8144	8167	8174	8182	8186	8188
00.700	5951	6242	6503	6945	7162	7343	7623	7653	7838	8120	8202	8220	8227	8234	8237	8239
00.800	6211	6491	6736	7001	7219	7402	7682	7788	7945	8180	8247	8263	8269	8276	8279	8281
00.900	6248	6532	6782	7048	7395	7561	7808	7899	7993	8229	8285	8300	8305	8311	8314	8316
01.000	6279	6567	6820	7236	7438	7604	7851	7942	8074	8265	8318	8331	8336	8342	8344	8346
01.200	6332	6811	7050	7306	7619	7768	7984	8061	8171	8331	8370	8382	8386	8391	8394	8395
01.400	6580	6864	7106	7490	7677	7901	8090	8155	8247	8379	8411	8422	8426	8431	8433	8434
01.600	6621	6907	7151	7538	7817	7949	8135	8231	8308	8417	8445	8455	8459	8463	8465	8467
01.800	6654	6944	7336	7580	7858	8052	8214	8268	8343	8449	8474	8483	8487	8491	8493	8494
02.000	6684	7143	7372	7725	7894	8087	8278	8324	8387	8476	8499	8507	8511	8514	8516	8518
04.000	7060	7497	7822	8108	8293	8410	8520	8546	8583	8627	8642	8648	8650	8653	8654	8655
06.000	7332	7736	8027	8280	8436	8532	8626	8644	8669	8700	8712	8717	8719	8722	8723	8723
08.000	7550	7920	8101	8400	8532	8611	8686	8704	8721	8747	8758	8762	8764	8766	8767	8768
10.000	7605	7975	8233	8491	8603	8668	8730	8744	8759	8781	8791	8795	8797	8799	8799	8800
20.000	7890	8301	8497	8684	8770	8808	8842	8850	8860	8876	8883	8886	8887	8889	8890	8890
30.000	8085	8384	8615	8770	8840	8872	8897	8903	8910	8924	8931	8933	8934	8936	8936	8937
40.000	8233	8501	8665	8828	8885	8911	8932	8937	8944	8956	8962	8964	8965	8966	8967	8967

$2\alpha/RT\beta = 8.10$

p/K'	γ-Values 1.00	1.50	2.00	3.00	4.00	5.00	7.00	8.00	10.0	20.0	40.0	70.0	100	200	400	1000
00.001	1580	1104	0876	0553	0347	0217	0103	0079	0049	0017	0013	0011	0011	0010	0010	0010
00.002	1901	1556	1282	0869	0583	0453	0277	0186	0137	0039	0026	0023	0022	0021	0020	0020
00.004	2258	1901	1608	1298	1057	0863	0575	0469	0315	0102	0055	0048	0045	0043	0042	0041
00.006	2579	2212	1904	1575	1312	1095	0872	0728	0586	0218	0090	0074	0070	0066	0064	0062
00.008	2888	2515	2193	1850	1568	1330	1084	0919	0753	0342	0135	0102	0095	0089	0086	0085
00.010	2946	2577	2485	2129	1831	1574	1306	1121	0934	0511	0204	0134	0122	0113	0110	0107
00.020	3381	3277	2951	2843	2528	2468	2156	2121	1858	1341	0825	0471	0346	0255	0239	0233
00.030	3755	3666	3346	3257	3193	3142	2822	2793	2744	2144	1625	1224	0940	0594	0416	0385
00.040	3836	3762	3710	3637	3585	3545	3231	3208	3170	2811	2473	2178	1930	1506	0974	0647
00.050	4165	4101	4057	3996	3954	3922	3876	3601	3573	3492	3422	3118	3089	2795	2519	2053
00.060	4218	4164	4127	4078	4046	4022	3989	3976	3957	4163	4125	4103	4092	4332	4571	4804
00.070	4263	4219	4447	4409	4384	4367	4344	4336	4324	4552	4788	4791	5032	5474	5850	6436
00.080	4566	4527	4502	4471	4453	4441	4682	4678	4673	4916	5166	5409	5631	6171	6617	6893
00.090	4603	4569	4548	4526	4769	4763	4757	4756	5003	5255	5511	5939	6132	6662	6964	7079
00.100	4634	4605	4590	4831	4824	4822	5071	5073	5078	5568	6010	6374	6533	6981	7155	7199
00.130	4713	4958	4952	5201	5209	5217	5472	5481	5720	6181	6578	7051	7199	7384	7429	7441
00.170	5053	5053	5309	5327	5584	5603	5860	6080	6109	6710	7139	7433	7549	7613	7628	7636
00.200	5101	5361	5374	5640	5888	5914	6163	6182	6403	6958	7402	7624	7684	7719	7730	7736
00.250	5420	5440	5701	5964	6001	6240	6481	6504	6711	7313	7673	7793	7824	7845	7853	7858
00.300	5476	5743	5772	6047	6298	6522	6752	6777	6968	7503	7821	7902	7921	7935	7942	7946
00.400	5562	5844	6108	6381	6623	6836	7054	7212	7368	7816	7995	8039	8050	8061	8066	8069
00.500	5869	6147	6193	6665	6894	7092	7295	7435	7575	7954	8099	8129	8137	8146	8151	8153
00.600	5922	6209	6469	6741	6975	7175	7492	7615	7738	8075	8173	8195	8202	8210	8214	8216
00.700	5968	6261	6525	6973	7190	7373	7655	7761	7869	8150	8229	8247	8253	8260	8264	8266
00.800	6232	6514	6762	7026	7245	7429	7711	7817	7976	8209	8274	8289	8295	8301	8304	8306
00.900	6267	6553	6806	7072	7423	7589	7837	7929	8021	8257	8311	8324	8330	8335	8338	8340
01.000	6297	6587	6842	7262	7464	7630	7878	7969	8102	8296	8342	8355	8359	8365	8367	8369
01.200	6348	6835	7075	7329	7645	7794	8010	8087	8197	8355	8393	8404	8408	8413	8415	8417
01.400	6601	6885	7128	7515	7701	7928	8115	8181	8272	8401	8433	8443	8447	8451	8454	8455
01.600	6640	6926	7172	7562	7842	7974	8200	8256	8332	8439	8466	8475	8479	8483	8485	8486
01.800	6672	6963	7360	7713	7882	8077	8237	8292	8366	8470	8494	8503	8506	8510	8512	8513
02.000	6700	7165	7394	7748	7995	8111	8301	8347	8409	8496	8518	8526	8529	8533	8535	8536
04.000	7243	7515	7843	8129	8313	8430	8539	8564	8601	8643	8657	8663	8665	8668	8669	8670
06.000	7347	7754	8047	8298	8453	8549	8642	8660	8684	8715	8727	8731	8733	8735	8737	8737
08.000	7568	7938	8118	8418	8549	8641	8702	8719	8736	8761	8771	8775	8777	8779	8780	8781
10.000	7620	7992	8250	8508	8619	8683	8744	8758	8773	8794	8804	8808	8809	8811	8812	8812
20.000	7903	8318	8513	8698	8784	8821	8854	8862	8871	8887	8894	8897	8898	8900	8900	8901
30.000	8099	8461	8629	8783	8852	8883	8908	8914	8921	8934	8941	8943	8944	8945	8946	8946
40.000	8248	8514	8678	8840	8896	8922	8942	8947	8954	8966	8971	8974	8975	8976	8976	8977

$2\alpha/RT\beta = 8.20$

p/K'	1.00	1.50	2.00	3.00	4.00	5.00	7.00	8.00	10.0	20.0	40.0	70.0	100	200	400	1000
									γ-Values							
00.001	1588	1111	0882	0558	0351	0219	0104	0079	0049	0017	0013	0011	0011	0010	0010	0010
00.002	1910	1566	1291	0876	0587	0457	0280	0220	0139	0039	0026	0023	0022	0021	0020	0020
00.004	2268	1911	1809	1308	1067	0872	0581	0474	0370	0113	0055	0048	0045	0043	0042	0041
00.006	2590	2223	2125	1586	1323	1104	0882	0736	0593	0220	0093	0074	0070	0066	0064	0063
00.008	2904	2530	2206	1863	1580	1513	1094	1060	0761	0346	0144	0102	0095	0089	0086	0085
00.010	2959	2590	2501	2146	1846	1586	1319	1283	1075	0517	0205	0135	0122	0113	0110	0108
00.020	3394	3295	3219	2862	2786	2487	2174	2141	2086	1357	0837	0476	0374	0256	0240	0234
00.030	3770	3686	3623	3276	3216	3169	2845	2818	2772	2168	1838	1242	1067	0601	0427	0387
00.040	3849	3778	3728	3659	3610	3572	3516	3494	3198	3094	2746	2436	2170	1704	1199	0713
00.050	4182	4121	4079	4021	3982	3952	3909	3893	3867	3528	3465	3421	3397	3095	2808	2533
00.060	4233	4181	4147	4100	4070	4048	4281	4270	4253	4207	4174	4417	4410	4655	4895	5336
00.070	4277	4500	4472	4436	4412	4396	4375	4367	4357	4597	4844	5091	5325	5746	6094	6661
00.080	4585	4548	4524	4495	4478	4728	4717	4714	4709	4962	5220	5681	5891	6385	6773	7005
00.090	4620	4588	4568	4809	4799	4793	4788	5041	5043	5302	5775	6176	6353	6821	7072	7162
00.100	4650	4623	4871	4858	4852	4850	5106	5108	5114	5617	6069	6440	6715	7057	7238	7272
00.130	4989	4979	4974	5230	5237	5246	5507	5516	5761	6230	6765	7113	7303	7453	7489	7499
00.170	5071	5071	5333	5592	5613	5859	5893	6119	6146	6757	7270	7519	7602	7664	7677	7684
00.200	5117	5383	5395	5667	5921	5945	6198	6408	6441	7003	7502	7672	7731	7765	7775	7780
00.250	5440	5459	5725	5993	6029	6272	6516	6710	6900	7356	7717	7841	7867	7885	7893	7897
00.300	5494	5766	5794	6073	6327	6556	6787	6963	7136	7608	7861	7941	7958	7972	7978	7982
00.400	5819	5863	6131	6407	6652	6867	7086	7246	7403	7853	8030	8073	8082	8093	8098	8101
00.500	5888	6169	6424	6693	6923	7123	7326	7468	7607	8012	8133	8159	8167	8175	8180	8182
00.600	5939	6229	6492	6765	7001	7203	7522	7646	7769	8106	8204	8223	8230	8237	8241	8243
00.700	6210	6490	6546	7000	7218	7401	7685	7792	7899	8179	8257	8273	8279	8286	8289	8291
00.800	6252	6536	6787	7051	7271	7569	7819	7846	8005	8236	8300	8314	8319	8325	8328	8330
00.900	6285	6573	6828	7248	7450	7616	7865	7957	8091	8283	8335	8348	8353	8359	8361	8363
01.000	6314	6606	6863	7287	7490	7656	7970	8049	8129	8321	8365	8378	8382	8387	8390	8391
01.200	6364	6857	7099	7352	7671	7819	8035	8113	8223	8379	8415	8426	8430	8434	8437	8438
01.400	6621	6905	7149	7539	7725	7953	8140	8205	8297	8425	8454	8464	8467	8472	8474	8475
01.600	6657	6945	7192	7584	7866	7997	8225	8280	8355	8461	8486	8495	8499	8503	8505	8506
01.800	6688	7153	7383	7737	7905	8101	8260	8314	8403	8491	8513	8522	8525	8529	8531	8532
02.000	6716	7186	7415	7770	8018	8133	8324	8369	8431	8516	8537	8545	8548	8551	8553	8554
04.000	7263	7532	7863	8148	8332	8449	8557	8582	8618	8659	8673	8678	8681	8683	8684	8685
06.000	7362	7771	8066	8316	8503	8585	8659	8676	8700	8729	8740	8745	8747	8749	8750	8751
08.000	7585	7956	8135	8434	8590	8657	8717	8733	8750	8774	8784	8788	8790	8792	8793	8793
10.000	7635	8007	8266	8524	8634	8709	8758	8772	8786	8807	8816	8820	8821	8823	8824	8824
20.000	7917	8333	8528	8712	8797	8833	8866	8874	8882	8898	8905	8907	8909	8910	8911	8911
30.000	8112	8476	8643	8796	8864	8895	8918	8924	8931	8944	8950	8953	8954	8955	8956	8956
40.000	8263	8527	8691	8852	8907	8932	8952	8957	8963	8975	8980	8983	8984	8985	8985	8986

$2\alpha/RT\beta = 8.30$

p/K′	1.00	1.50	2.00	3.00	4.00	5.00	7.00	8.00	10.0	20.0	40.0	70.0	100	200	400	1000
	γ-Values															
00.001	1596	1117	0888	0563	0354	0221	0124	0080	0049	0017	0013	0011	0011	0010	0010	0010
00.002	1918	1576	1301	0883	0691	0461	0283	0223	0140	0039	0026	0023	0022	0021	0020	0020
00.004	2277	1921	1823	1318	1076	0880	0587	0479	0375	0114	0055	0048	0046	0043	0042	0041
00.006	2601	2234	2141	1597	1333	1269	0891	0743	0599	0222	0093	0074	0070	0066	0064	0063
00.008	2919	2543	2218	1876	1592	1528	1104	1072	0885	0350	0145	0102	0095	0089	0086	0085
00.010	2972	2602	2517	2161	1859	1795	1330	1297	1087	0523	0227	0138	0122	0114	0110	0108
00.020	3407	3311	3240	2880	2809	2506	2424	2160	2108	1372	0847	0535	0377	0260	0241	0234
00.030	3785	3705	3645	3295	3238	3193	2867	2841	2798	2424	1864	1415	1083	0662	0455	0389
00.040	4134	3794	3746	3679	3633	3598	3545	3525	3491	3127	2781	2471	2203	1732	1355	0765
00.050	4199	4140	4100	4045	4008	3980	3940	3925	3901	3833	3776	3467	3446	3411	3384	3090
00.060	4248	4198	4165	4121	4093	4343	4315	4305	4289	4248	4487	4474	4469	4972	5208	5806
00.070	4563	4523	4496	4460	4438	4423	4404	4397	4656	4638	4895	5151	5392	5826	6318	6804
00.080	4604	4568	4545	4517	4769	4760	4750	4747	4743	5005	5504	5744	6135	6580	6912	7101
00.090	4637	4606	4588	4837	4827	4822	5077	5078	5080	5347	5831	6241	6557	6963	7169	7239
00.100	4666	4641	4896	4883	4877	5136	5140	5142	5396	5663	6125	6638	6782	7170	7315	7342
00.130	5010	4999	4994	5257	5264	5521	5539	5781	5799	6277	6819	7171	7363	7517	7546	7555
00.170	5087	5348	5356	5621	5641	5892	6141	6156	6379	6801	7320	7569	7663	7713	7725	7731
00.200	5132	5404	5416	5692	5951	5975	6231	6446	6477	7162	7550	7733	7781	7808	7818	7823
00.250	5459	5477	5748	6021	6271	6302	6548	6746	6938	7397	7759	7882	7907	7924	7931	7935
00.300	5511	5788	5814	6097	6355	6587	6820	6999	7174	7648	7900	7980	7995	8007	8013	8017
00.400	5841	5882	6154	6432	6679	6897	7117	7279	7436	7888	8069	8105	8114	8124	8128	8131
00.500	5905	6190	6450	6719	6951	7153	7472	7499	7720	8045	8164	8189	8196	8204	8208	8210
00.600	5954	6248	6514	6789	7182	7366	7551	7676	7866	8136	8232	8250	8257	8264	8267	8269
00.700	6232	6513	6567	7025	7244	7428	7714	7821	7928	8207	8283	8299	8304	8311	8314	8316
00.800	6271	6556	6810	7074	7431	7598	7848	7940	8033	8263	8325	8338	8343	8349	8352	8354
00.900	6303	6592	6849	7274	7476	7642	7892	7984	8118	8308	8359	8372	8376	8382	8384	8386
01.000	6330	6624	6884	7311	7514	7779	7997	8076	8154	8346	8388	8400	8404	8409	8412	8413
01.200	6595	6878	7121	7510	7695	7844	8114	8180	8247	8404	8436	8447	8451	8455	8457	8459
01.400	6639	6924	7169	7562	7846	7978	8164	8229	8320	8447	8475	8484	8487	8491	8494	8495
01.600	6674	6963	7367	7605	7890	8020	8248	8303	8378	8482	8506	8514	8518	8522	8523	8525
01.800	6704	7175	7404	7760	7926	8124	8282	8362	8424	8511	8532	8540	8543	8547	8549	8550
02.000	6731	7205	7435	7792	8041	8155	8346	8391	8452	8535	8555	8563	8566	8569	8571	8572
04.000	7281	7549	7883	8167	8351	8468	8575	8607	8635	8675	8688	8693	8695	8698	8699	8700
06.000	7377	7787	8084	8333	8521	8602	8674	8695	8715	8743	8754	8759	8760	8762	8763	8764
08.000	7601	7973	8233	8451	8606	8673	8731	8748	8764	8787	8797	8801	8802	8804	8805	8806
10.000	7649	8023	8282	8540	8649	8724	8772	8785	8799	8819	8828	8832	8833	8835	8836	8836
20.000	7930	8348	8542	8725	8810	8845	8878	8885	8893	8908	8915	8918	8919	8920	8921	8921
30.000	8125	8490	8656	8808	8875	8906	8929	8934	8941	8954	8960	8962	8963	8964	8965	8965
40.000	8277	8540	8736	8863	8918	8943	8962	8967	8973	8984	8990	8992	8993	8994	8994	8994

$2\alpha/RT\beta = 8.40$

p/K'	1.00	1.50	2.00	3.00	4.00	5.00	7.00	8.00	10.0	20.0	40.0	70.0	100	200	400	1000
									γ-Values							
00.001	1604	1283	0894	0568	0357	0267	0126	0096	0050	0017	0013	0011	0011	0010	0010	0010
00.002	1926	1585	1309	0889	0697	0465	0286	0225	0142	0041	0026	0023	0022	0021	0020	0020
00.004	2524	2150	1836	1326	1085	0888	0593	0483	0379	0115	0056	0048	0046	0043	0042	0041
00.006	2612	2245	2155	1808	1522	1281	0899	0750	0605	0224	0094	0074	0070	0066	0064	0063
00.008	2933	2556	2230	1887	1804	1541	1114	1082	0894	0408	0156	0103	0095	0089	0087	0085
00.010	2984	2867	2531	2175	1872	1811	1341	1309	1098	0529	0228	0138	0123	0114	0110	0108
00.020	3419	3327	3260	2897	2830	2523	2446	2177	2129	1567	0976	0541	0415	0265	0241	0235
00.030	3799	3722	3666	3313	3258	3216	3153	2863	2823	2451	1887	1434	1236	0739	0479	0391
00.040	4153	3809	3763	3699	3655	3622	3573	3554	3522	3159	2813	2757	2471	1959	1536	0899
00.050	4214	4158	4120	4068	4032	4006	3969	3955	3933	3871	3820	3787	3769	3741	3719	3421
00.060	4262	4214	4183	4419	4392	4373	4347	4337	4322	4285	4536	4792	4789	5044	5509	6201
00.070	4583	4544	4517	4484	4463	4449	4705	4700	4692	4942	5197	5445	5675	6080	6521	6958
00.080	4621	4586	4565	4539	4799	4790	4780	4778	5040	5298	5557	6004	6203	6656	7037	7187
00.090	4653	4623	4606	4863	4854	4849	5112	5113	5115	5626	5884	6461	6623	7035	7258	7311
00.100	4681	4657	4920	4907	4902	5167	5171	5174	5434	5706	6357	6699	6945	7271	7386	7407
00.130	5029	5018	5014	5283	5290	5553	5571	5818	5835	6320	6990	7294	7454	7577	7600	7609
00.170	5103	5371	5378	5649	5667	5924	6177	6191	6417	6983	7436	7644	7711	7760	7770	7776
00.200	5147	5424	5435	5716	5981	6003	6263	6482	6692	7205	7594	7778	7824	7851	7859	7864
00.250	5478	5495	5771	6047	6301	6330	6580	6781	6975	7520	7825	7925	7946	7961	7968	7972
00.300	5527	5809	5834	6120	6381	6617	6851	7033	7209	7686	7952	8015	8030	8041	8047	8050
00.400	5861	5900	6175	6456	6705	6926	7286	7310	7569	7922	8102	8136	8145	8154	8158	8161
00.500	5922	6210	6473	6744	6978	7181	7504	7528	7752	8076	8193	8217	8224	8231	8235	8237
00.600	5970	6266	6535	6992	7210	7394	7680	7705	7896	8165	8260	8277	8283	8289	8292	8294
00.700	6252	6535	6787	7049	7269	7454	7741	7850	8011	8245	8309	8324	8329	8335	8338	8340
00.800	6289	6576	6832	7096	7458	7624	7875	7968	8060	8297	8349	8362	8367	8372	8375	8377
00.900	6319	6611	6870	7298	7500	7667	7918	8010	8145	8339	8382	8394	8399	8404	8406	8408
01.000	6346	6642	6903	7334	7537	7805	8023	8102	8214	8370	8411	8422	8426	8431	8433	8435
01.200	6615	6899	7143	7534	7719	7867	8139	8205	8298	8427	8457	8467	8471	8475	8477	8479
01.400	6657	6942	7189	7584	7870	8001	8187	8286	8343	8468	8495	8503	8507	8511	8513	8514
01.600	6690	6980	7389	7745	7912	8111	8271	8325	8400	8502	8525	8533	8536	8540	8542	8543
01.800	6719	7195	7425	7782	8031	8146	8338	8384	8446	8530	8551	8559	8562	8565	8567	8568
02.000	6746	7224	7455	7813	8063	8176	8367	8412	8472	8554	8573	8580	8583	8586	8588	8589
04.000	7299	7566	7901	8185	8369	8486	8592	8624	8652	8690	8703	8708	8710	8712	8713	8714
06.000	7391	7803	8101	8405	8538	8618	8690	8710	8730	8757	8767	8772	8773	8775	8776	8777
08.000	7616	7989	8250	8466	8622	8688	8749	8762	8778	8800	8809	8813	8815	8817	8817	8818
10.000	7663	8037	8298	8555	8684	8738	8788	8798	8812	8831	8840	8843	8845	8846	8847	8848
20.000	8062	8362	8557	8738	8822	8861	8889	8896	8904	8919	8925	8928	8929	8930	8931	8931
30.000	8138	8504	8670	8820	8886	8917	8940	8945	8951	8964	8969	8972	8973	8974	8974	8975
40.000	8291	8552	8749	8874	8928	8953	8972	8976	8982	8993	8998	9001	9002	9003	9003	9003

$2\alpha/RT\beta = 8.50$

p/K′	1.00	1.50	2.00	3.00	4.00	5.00	7.00	8.00	10.0	20.0	40.0	70.0	100	200	400	1000
									γ-Values							
00.001	1611	1292	0899	0572	0360	0269	0127	0097	0050	0017	0013	0011	0011	0010	0010	0010
00.002	1934	1593	1317	0895	0704	0469	0288	0227	0143	0041	0026	0023	0022	0021	0020	0020
00.004	2538	2164	1848	1335	1093	0895	0598	0573	0382	0116	0056	0048	0046	0043	0042	0041
00.006	2622	2255	2169	1822	1536	1292	0907	0878	0610	0261	0098	0074	0070	0066	0064	0063
00.008	2946	2569	2241	1898	1819	1554	1286	1092	0903	0413	0158	0103	0096	0089	0087	0085
00.010	2995	2884	2545	2188	1884	1826	1537	1321	1109	0620	0230	0143	0123	0114	0110	0108
00.020	3431	3342	3278	2914	2849	2540	2467	2437	2148	1584	0988	0617	0419	0272	0242	0236
00.030	3813	3739	3685	3609	3277	3238	3178	3154	2846	2475	2133	1636	1252	0833	0511	0395
00.040	4171	4107	3779	3718	3676	3644	3598	3580	3551	3188	3115	2792	2505	2213	1745	1100
00.050	4229	4175	4139	4089	4056	4031	3996	3983	3963	3906	3861	3833	3818	4078	4062	4047
00.060	4275	4230	4483	4445	4419	4401	4376	4367	4354	4601	4582	4847	5105	5352	5792	6418
00.070	4602	4563	4538	4506	4487	4474	4738	4733	4726	4985	5248	5503	5738	6315	6790	7081
00.080	4638	4604	4583	4559	4826	4818	4810	4807	5077	5342	5607	6063	6429	6829	7148	7265
00.090	4668	4640	4903	4888	4879	4875	5144	5145	5148	5671	6139	6521	6807	7160	7338	7379
00.100	4695	4953	4942	4930	5196	5197	5201	5463	5470	5973	6411	6879	7092	7363	7452	7469
00.130	5047	5036	5032	5308	5314	5583	5844	5853	5869	6546	7041	7347	7508	7630	7652	7660
00.170	5118	5392	5398	5675	5693	5954	6210	6224	6454	7026	7482	7691	7765	7804	7814	7820
00.200	5161	5443	5713	5740	6008	6255	6499	6515	6728	7246	7682	7832	7869	7891	7899	7904
00.250	5495	5771	5792	6072	6330	6358	6794	6814	7010	7559	7865	7963	7983	7997	8004	8007
00.300	5542	5829	6097	6368	6407	6645	6881	7066	7243	7722	7988	8051	8063	8075	8080	8083
00.400	5880	6160	6195	6478	6730	6953	7317	7340	7602	7989	8133	8167	8174	8183	8187	8190
00.500	5938	6230	6496	6768	7004	7208	7534	7660	7783	8106	8224	8244	8251	8258	8262	8264
00.600	5985	6283	6555	7018	7236	7422	7709	7733	7925	8209	8286	8302	8308	8314	8317	8319
00.700	6271	6556	6811	7073	7293	7479	7768	7877	8039	8272	8334	8348	8353	8359	8361	8363
00.800	6306	6595	6854	7281	7483	7650	7902	7995	8131	8322	8373	8385	8390	8395	8397	8399
00.900	6335	6629	6890	7322	7524	7691	8013	8035	8170	8363	8405	8416	8421	8425	8428	8429
01.000	6361	6659	7107	7356	7682	7830	8048	8127	8239	8397	8433	8443	8447	8452	8454	8455
01.200	6634	6918	7163	7558	7741	7976	8163	8229	8321	8448	8478	8487	8491	8495	8497	8498
01.400	6674	6960	7208	7605	7893	8024	8254	8309	8385	8489	8514	8523	8526	8530	8532	8533
01.600	6706	7181	7411	7768	7933	8134	8292	8346	8421	8521	8544	8552	8555	8558	8560	8561
01.800	6734	7214	7444	7803	8054	8168	8360	8405	8466	8549	8569	8576	8579	8583	8584	8585
02.000	6966	7243	7473	7833	8084	8253	8387	8432	8503	8572	8591	8598	8600	8603	8605	8606
04.000	7316	7581	7919	8203	8387	8503	8609	8641	8668	8705	8717	8722	8724	8726	8728	8728
06.000	7405	7819	8118	8422	8555	8634	8705	8725	8745	8770	8781	8785	8786	8788	8789	8790
08.000	7631	8005	8267	8527	8638	8703	8763	8775	8791	8812	8822	8825	8827	8829	8829	8830
10.000	7677	8052	8313	8570	8698	8752	8801	8811	8825	8843	8852	8855	8856	8858	8859	8859
20.000	8077	8376	8570	8751	8835	8873	8900	8907	8915	8929	8935	8938	8939	8940	8941	8941
30.000	8150	8517	8682	8847	8903	8927	8950	8955	8961	8973	8979	8981	8982	8983	8983	8984
40.000	8304	8564	8761	8885	8943	8963	8981	8986	8991	9002	9007	9009	9010	9011	9012	9012

$2\alpha/RT\beta = 8.60$

p/K′	γ-Values 1.00	1.50	2.00	3.00	4.00	5.00	7.00	8.00	10.0	20.0	40.0	70.0	100	200	400	1000
00.001	1618	1300	0904	0576	0362	0272	0128	0098	0059	0017	0013	0011	0011	0010	0010	0010
00.002	1942	1601	1325	0901	0709	0560	0291	0229	0144	0041	0026	0023	0022	0021	0020	0020
00.004	2551	2177	1859	1343	1100	0902	0602	0579	0386	0132	0057	0048	0046	0043	0042	0041
00.006	2632	2513	2181	1835	1548	1303	0914	0886	0721	0264	0099	0074	0070	0066	0064	0063
00.008	2959	2581	2500	2139	1833	1566	1298	1102	0912	0417	0158	0105	0096	0090	0087	0085
00.010	3006	2899	2558	2201	2126	1840	1550	1332	1118	0626	0260	0143	0125	0114	0110	0108
00.020	3442	3356	3295	2929	2868	2820	2486	2458	2166	1600	0999	0624	0469	0282	0243	0236
00.030	3826	3755	3704	3631	3296	3258	3202	3180	2867	2498	2158	1656	1434	0948	0551	0400
00.040	4188	4127	4084	3736	3696	3666	3622	3606	3579	3501	3149	2826	2799	2492	1980	1385
00.050	4243	4192	4156	4109	4078	4055	4022	4010	3991	3939	4189	4164	4152	4134	4406	4673
00.060	4288	4534	4506	4468	4444	4427	4404	4396	4384	4640	4901	4898	5163	5648	6057	6700
00.070	4619	4582	4558	4528	4509	4782	4768	4764	4757	5025	5295	5786	6007	6530	6940	7183
00.080	4654	4621	4601	4864	4853	4845	4837	5112	5111	5383	5884	6307	6491	6985	7247	7338
00.090	4683	4656	4926	4911	4903	5175	5175	5176	5443	5714	6189	6578	6868	7270	7418	7444
00.100	4709	4974	4963	4951	5225	5225	5229	5497	5504	6016	6461	6936	7151	7444	7514	7528
00.130	5065	5053	5326	5331	5602	5611	5877	5886	6131	6589	7089	7456	7587	7684	7702	7709
00.170	5133	5412	5418	5700	5717	5983	6243	6256	6489	7067	7527	7756	7810	7847	7856	7861
00.200	5452	5461	5737	5762	6034	6286	6532	6548	6763	7285	7724	7873	7909	7930	7938	7942
00.250	5511	5793	5812	6096	6357	6593	6827	6845	7043	7596	7902	8000	8019	8032	8038	8042
00.300	5557	5848	6121	6395	6431	6672	7076	7097	7275	7814	8034	8085	8096	8107	8112	8115
00.400	5898	6182	6215	6501	6755	6979	7347	7493	7633	8022	8168	8196	8203	8212	8215	8218
00.500	5954	6248	6518	6791	7028	7234	7562	7689	7812	8156	8251	8271	8277	8284	8287	8289
00.600	5999	6300	6575	7043	7262	7448	7737	7847	7953	8236	8312	8327	8332	8338	8341	8343
00.700	6289	6576	6833	7095	7316	7628	7881	7903	8066	8297	8358	8371	8376	8382	8384	8386
00.800	6322	6614	6874	7306	7508	7675	7927	8021	8157	8346	8396	8407	8412	8417	8419	8421
00.900	6351	6646	6909	7344	7547	7820	8038	8117	8195	8386	8427	8438	8442	8447	8449	8450
01.000	6376	6885	7129	7378	7706	7854	8072	8151	8263	8419	8454	8464	8468	8472	8474	8476
01.200	6652	6936	7183	7580	7763	7999	8186	8252	8344	8469	8498	8507	8510	8514	8516	8518
01.400	6690	6977	7226	7626	7915	8046	8277	8332	8407	8510	8533	8541	8544	8548	8550	8551
01.600	6721	7201	7431	7789	7954	8156	8314	8367	8457	8541	8562	8570	8573	8576	8578	8579
01.800	6749	7233	7463	7823	8075	8188	8381	8425	8486	8568	8587	8594	8596	8600	8601	8602
02.000	6984	7260	7491	7852	8104	8274	8407	8472	8523	8590	8608	8614	8617	8620	8622	8623
04.000	7332	7742	7937	8291	8450	8521	8637	8657	8686	8719	8731	8736	8738	8740	8741	8742
06.000	7585	7834	8134	8439	8571	8650	8725	8740	8759	8783	8793	8797	8799	8801	8802	8802
08.000	7645	8020	8283	8543	8653	8717	8777	8789	8805	8825	8834	8837	8839	8840	8841	8842
10.000	7690	8171	8328	8585	8713	8765	8814	8825	8837	8855	8863	8866	8868	8869	8870	8870
20.000	8092	8390	8584	8763	8846	8884	8911	8918	8925	8939	8945	8948	8949	8950	8951	8951
30.000	8162	8530	8695	8858	8914	8938	8960	8964	8971	8982	8988	8990	8991	8992	8992	8993
40.000	8316	8633	8773	8908	8953	8973	8991	8995	9000	9011	9016	9018	9019	9020	9020	9021

$2\alpha/RT\beta = 8.70$

p/K′	1.00	1.50	2.00	3.00	4.00	5.00	7.00	8.00	10.0	20.0	40.0	70.0	100	200	400	1000
	γ-Values															
00.001	1625	1308	0909	0579	0438	0274	0129	0098	0059	0017	0013	0011	0011	0010	0010	0010
00.002	1949	1609	1332	0907	0715	0565	0293	0231	0173	0041	0026	0023	0022	0021	0020	0020
00.004	2563	2189	1870	1542	1108	0908	0714	0584	0389	0133	0057	0048	0046	0043	0042	0041
00.006	2641	2527	2193	1848	1559	1313	1068	0894	0728	0266	0099	0074	0070	0066	0064	0063
00.008	2971	2592	2514	2154	1846	1577	1308	1111	0919	0421	0175	0105	0096	0090	0087	0085
00.010	3017	2914	2570	2213	2142	1854	1563	1342	1297	0633	0262	0150	0125	0114	0110	0108
00.020	3743	3370	3311	2944	2885	2840	2504	2478	2182	1615	1157	0719	0473	0283	0244	0237
00.030	3838	3770	3721	3652	3314	3277	3224	3203	3169	2789	2181	1888	1451	0960	0603	0406
00.040	4204	4145	4104	4048	4009	3686	3645	3630	3604	3532	3180	3138	3114	2794	2242	1575
00.050	4257	4207	4174	4129	4099	4077	4047	4035	4017	4265	4230	4209	4199	4475	4749	5265
00.060	4301	4554	4527	4491	4468	4452	4431	4423	4412	4677	4947	5215	5470	5926	6300	6917
00.070	4636	4600	4577	4548	4531	4810	4797	4793	4788	5063	5592	5841	6066	6598	7074	7266
00.080	4669	4638	4619	4889	4877	4870	5145	5144	5144	5422	5933	6363	6697	7123	7337	7406
00.090	4697	4672	4948	4933	4926	5204	5204	5205	5478	5754	6237	6779	7033	7369	7485	7505
00.100	4723	4994	4983	4972	5252	5252	5525	5529	5536	6057	6677	6989	7280	7503	7573	7584
00.130	5081	5070	5349	5353	5630	5638	5909	5917	6167	6630	7242	7505	7637	7734	7749	7756
00.170	5147	5432	5437	5723	5993	6010	6273	6286	6522	7106	7628	7800	7858	7888	7897	7902
00.200	5470	5479	5759	5783	6060	6315	6563	6578	6796	7429	7764	7921	7950	7968	7975	7979
00.250	5527	5813	5831	6119	6384	6623	6857	6875	7074	7632	7960	8038	8054	8066	8071	8075
00.300	5572	5867	6143	6420	6668	6698	7107	7126	7306	7849	8067	8118	8128	8138	8142	8145
00.400	5914	6203	6234	6735	6970	7174	7375	7523	7663	8052	8197	8224	8231	8239	8243	8245
00.500	5969	6266	6538	6813	7052	7259	7589	7718	7840	8185	8279	8297	8303	8309	8313	8314
00.600	6266	6551	6593	7066	7286	7473	7764	7874	7980	8262	8337	8351	8356	8362	8365	8367
00.700	6306	6595	6854	7116	7486	7654	7907	8001	8092	8322	8382	8394	8399	8404	8407	8408
00.800	6338	6631	6894	7329	7531	7699	7952	8045	8182	8370	8418	8429	8433	8438	8441	8442
00.900	6365	6663	6927	7366	7569	7844	8063	8142	8255	8409	8449	8459	8463	8467	8469	8471
01.000	6390	6905	7150	7398	7729	7877	8096	8174	8286	8441	8475	8484	8488	8492	8494	8496
01.200	6669	6954	7202	7601	7784	8022	8209	8274	8366	8492	8518	8526	8529	8533	8535	8536
01.400	6706	6994	7244	7646	7937	8067	8299	8353	8428	8529	8552	8560	8563	8566	8568	8569
01.600	6736	7221	7451	7810	7974	8177	8371	8416	8477	8560	8580	8587	8590	8593	8595	8596
01.800	6763	7251	7482	7843	8096	8208	8401	8445	8505	8586	8604	8611	8613	8616	8618	8619
02.000	7001	7277	7509	7871	8124	8295	8455	8492	8541	8608	8624	8631	8634	8636	8638	8639
04.000	7347	7759	7953	8309	8467	8565	8653	8673	8702	8734	8745	8750	8752	8754	8755	8756
06.000	7601	7848	8150	8455	8586	8665	8740	8754	8773	8796	8806	8810	8811	8813	8814	8815
08.000	7659	8035	8299	8558	8667	8731	8790	8804	8817	8837	8845	8849	8850	8852	8853	8853
10.000	7702	8186	8342	8599	8727	8779	8827	8837	8849	8866	8874	8877	8879	8880	8881	8881
20.000	8105	8403	8597	8776	8858	8896	8922	8928	8936	8949	8955	8957	8958	8959	8960	8960
30.000	8174	8543	8707	8870	8925	8948	8969	8974	8980	8991	8997	8999	9000	9001	9001	9002
40.000	8328	8646	8785	8918	8963	8984	9000	9004	9009	9020	9024	9026	9027	9028	9029	9029

$2\alpha/RT\beta = 8.80$

p/K'	γ-Values 1.00	1.50	2.00	3.00	4.00	5.00	7.00	8.00	10.0	20.0	40.0	70.0	100	200	400	1000
00.001	1632	1315	1064	0583	0441	0276	0130	0099	0060	0017	0013	0011	0011	0010	0010	0010
00.002	2195	1616	1339	1062	0720	0569	0355	0280	0175	0044	0026	0023	0022	0021	0020	0020
00.004	2575	2200	1880	1553	1287	1064	0720	0589	0466	0134	0057	0048	0046	0043	0042	0041
00.006	2650	2540	2205	1859	1570	1322	1078	0902	0734	0268	0105	0074	0070	0066	0064	0063
00.008	2983	2603	2528	2167	1858	1588	1319	1119	0927	0424	0176	0107	0096	0090	0087	0085
00.010	3027	2928	2582	2225	2156	1866	1575	1547	1308	0638	0264	0151	0126	0114	0111	0108
00.020	3759	3383	3326	3244	2902	2858	2521	2496	2455	1846	1169	0726	0537	0297	0246	0238
00.030	3851	3784	3738	3672	3626	3296	3245	3226	3194	2814	2460	1910	1663	1100	0669	0419
00.040	4219	4162	4123	4069	4033	4006	3667	3652	3629	3562	3505	3172	3150	2830	2529	2009
00.050	4270	4222	4190	4147	4119	4098	4369	4359	4343	4300	4269	4547	4539	4815	5083	5578
00.060	4613	4573	4547	4513	4491	4476	4456	4449	4734	4712	4990	5264	5525	5991	6522	7047
00.070	4652	4617	4594	4567	4847	4837	4825	4821	5102	5372	5639	6109	6316	6791	7192	7345
00.080	4683	4653	4636	4912	4901	4894	5175	5175	5174	5715	5978	6588	6755	7246	7429	7470
00.090	4711	4981	4969	4954	4948	5232	5232	5233	5511	6029	6477	6832	7089	7428	7547	7562
00.100	4736	5013	5002	4992	5277	5278	5556	5560	5567	6096	6725	7149	7334	7574	7628	7638
00.130	5097	5086	5371	5375	5656	5664	5939	5947	6200	6669	7288	7599	7706	7782	7795	7801
00.170	5161	5450	5455	5746	6020	6036	6302	6532	6554	7272	7670	7857	7903	7928	7936	7940
00.200	5488	5496	5781	6061	6084	6342	6593	6608	6828	7467	7839	7959	7987	8004	8011	8015
00.250	5543	5833	5850	6141	6408	6651	6887	7076	7255	7740	7995	8072	8087	8099	8104	8107
00.300	5586	5884	6165	6444	6695	6723	7137	7155	7465	7882	8100	8149	8158	8168	8172	8175
00.400	5931	6222	6252	6760	6996	7201	7403	7551	7692	8082	8226	8252	8258	8266	8269	8272
00.500	5983	6283	6558	6834	7074	7283	7616	7745	7867	8212	8305	8322	8328	8334	8337	8339
00.600	6284	6571	6612	7089	7309	7497	7790	7900	8006	8288	8361	8375	8380	8385	8388	8389
00.700	6322	6613	6874	7137	7510	7678	7932	8027	8117	8355	8405	8416	8421	8426	8428	8430
00.800	6353	6648	6912	7351	7553	7721	7976	8070	8206	8399	8440	8451	8455	8459	8461	8463
00.900	6380	6679	6945	7387	7719	7868	8087	8166	8279	8435	8470	8479	8483	8487	8490	8491
01.000	6404	6925	7170	7418	7751	7900	8118	8197	8309	8462	8495	8504	8508	8512	8514	8515
01.200	6686	6971	7220	7622	7914	8044	8230	8296	8387	8512	8537	8545	8548	8552	8554	8555
01.400	6721	7010	7434	7665	7958	8161	8320	8374	8449	8548	8570	8577	8580	8584	8585	8586
01.600	6750	7239	7469	7830	8084	8197	8391	8436	8497	8578	8597	8604	8607	8610	8612	8613
01.800	6776	7268	7499	7862	8116	8288	8420	8465	8536	8603	8621	8627	8630	8633	8634	8635
02.000	7018	7293	7526	7998	8143	8314	8474	8511	8560	8625	8641	8647	8650	8652	8654	8655
04.000	7362	7776	7970	8327	8484	8582	8669	8688	8717	8748	8759	8763	8765	8767	8768	8769
06.000	7616	7991	8165	8471	8601	8680	8754	8767	8786	8809	8818	8822	8824	8825	8826	8827
08.000	7673	8049	8314	8573	8681	8745	8803	8816	8830	8849	8857	8860	8862	8863	8864	8865
10.000	7865	8202	8356	8612	8740	8792	8839	8849	8861	8878	8885	8888	8890	8891	8892	8892
20.000	8118	8415	8609	8809	8870	8907	8933	8939	8946	8958	8964	8967	8968	8969	8970	8970
30.000	8185	8555	8718	8881	8935	8960	8979	8983	8989	9000	9005	9008	9009	9010	9010	9010
40.000	8340	8658	8796	8929	8973	8993	9009	9013	9018	9028	9033	9035	9036	9036	9037	9037

$2\alpha/RT\beta$ = 8.90

p/K'	1.00	1.50	2.00	3.00	4.00	5.00	7.00	8.00	10.0	20.0	40.0	70.0	100	200	400	1000
	γ-Values															
00.001	1638	1322	1071	0697	0445	0278	0131	0100	0060	0017	0013	0011	0011	0010	0010	0010
00.002	2205	1624	1346	1070	0725	0574	0358	0282	0176	0045	0026	0023	0022	0021	0020	0020
00.004	2585	2211	1889	1563	1296	1073	0726	0594	0470	0135	0059	0048	0046	0043	0042	0041
00.006	2937	2553	2216	1870	1580	1331	1086	0909	0740	0271	0106	0075	0070	0066	0064	0063
00.008	2994	2613	2542	2180	1869	1598	1328	1302	1086	0503	0178	0108	0096	0090	0087	0085
00.010	3037	2942	2594	2497	2170	1878	1586	1559	1319	0644	0303	0151	0127	0115	0111	0109
00.020	3774	3396	3341	3263	2917	2876	2537	2513	2475	1863	1180	0844	0542	0315	0247	0239
00.030	3862	3798	3753	3691	3647	3613	3265	3247	3217	2838	2484	2171	1901	1264	0750	0442
00.040	4233	4179	4141	4090	4055	4030	3993	3979	3956	3590	3538	3503	3484	3154	2840	2540
00.050	4283	4237	4206	4165	4442	4423	4397	4387	4372	4333	4605	4591	4875	5146	5406	6069
00.060	4630	4591	4566	4533	4512	4498	4779	4773	4764	5036	5308	5572	5819	6250	6723	7191
00.070	4667	4633	4612	4586	4872	4862	4851	4848	5134	5410	5683	6162	6372	6964	7297	7417
00.080	4697	4669	4952	4933	4923	4918	5204	5203	5204	5755	6242	6641	6940	7306	7499	7530
00.090	4724	5001	4989	4975	5260	5258	5259	5538	5542	6069	6523	7013	7234	7513	7605	7617
00.100	4749	5031	5020	5302	5301	5302	5586	5589	5858	6354	6771	7199	7446	7639	7680	7689
00.130	5112	5102	5392	5395	5681	5689	5968	5975	6232	6881	7331	7645	7751	7827	7838	7844
00.170	5174	5468	5473	5767	6045	6061	6330	6563	6782	7310	7710	7897	7942	7966	7973	7978
00.200	5505	5512	5801	6085	6107	6369	6622	6834	7034	7503	7877	8002	8025	8040	8046	8049
00.250	5558	5852	5868	6162	6432	6677	6915	7107	7287	7775	8029	8107	8120	8130	8135	8138
00.300	5599	5901	6185	6466	6720	6946	7165	7335	7495	7914	8140	8179	8188	8197	8201	8204
00.400	5947	6241	6512	6783	7020	7228	7429	7579	7720	8111	8256	8279	8285	8292	8295	8297
00.500	5997	6300	6577	6854	7271	7459	7641	7771	7969	8239	8331	8347	8352	8358	8361	8362
00.600	6301	6590	6850	7110	7332	7521	7814	7926	8092	8325	8385	8398	8402	8408	8410	8412
00.700	6338	6631	6894	7157	7534	7702	7957	8051	8141	8379	8427	8438	8442	8447	8449	8451
00.800	6368	6665	6930	7373	7575	7744	8074	8153	8230	8422	8461	8471	8475	8480	8482	8483
00.900	6394	6915	7161	7407	7742	7890	8109	8189	8302	8457	8490	8499	8503	8507	8509	8510
01.000	6659	6943	7190	7590	7773	7921	8201	8267	8331	8486	8515	8523	8527	8531	8533	8534
01.200	6702	6988	7238	7642	7935	8065	8251	8354	8408	8531	8555	8563	8566	8570	8571	8573
01.400	6735	7025	7453	7814	7978	8182	8340	8394	8468	8567	8588	8595	8598	8601	8603	8604
01.600	6764	7257	7487	7850	8105	8217	8411	8456	8516	8596	8614	8621	8624	8627	8628	8629
01.800	7010	7285	7516	7880	8135	8307	8439	8505	8555	8621	8637	8644	8646	8649	8650	8651
02.000	7034	7309	7695	8017	8161	8333	8493	8529	8578	8642	8657	8663	8665	8668	8669	8670
04.000	7377	7793	7985	8344	8501	8598	8684	8710	8731	8761	8772	8777	8778	8780	8781	8782
06.000	7631	8007	8180	8486	8616	8694	8768	8784	8799	8821	8830	8834	8836	8837	8838	8839
08.000	7686	8063	8328	8587	8695	8770	8816	8829	8842	8860	8868	8872	8873	8875	8875	8876
10.000	7879	8216	8445	8664	8753	8804	8851	8861	8872	8889	8896	8899	8900	8902	8902	8903
20.000	8131	8428	8621	8821	8889	8918	8943	8949	8956	8968	8974	8976	8977	8978	8979	8979
30.000	8196	8567	8730	8892	8946	8970	8988	8993	8998	9009	9014	9016	9017	9018	9019	9019
40.000	8352	8669	8807	8939	8983	9002	9018	9021	9027	9036	9041	9043	9044	9045	9045	9045

$2\alpha/RT\beta = 9.00$

p/K'	1.00	1.50	2.00	3.00	4.00	5.00	7.00	8.00	10.0	20.0	40.0	70.0	100	200	400	1000
									γ-Values							
00.001	1866	1329	1078	0702	0448	0280	0161	0123	0060	0017	0013	0011	0011	0010	0010	0010
00.002	2216	1853	1552	1077	0729	0578	0361	0285	0177	0045	0026	0023	0022	0021	0020	0020
00.004	2596	2221	1899	1573	1305	1081	0731	0710	0474	0158	0059	0048	0046	0043	0042	0041
00.006	2950	2565	2226	1880	1589	1539	1094	1069	0746	0321	0115	0075	0070	0066	0064	0063
00.008	3004	2903	2554	2192	1880	1831	1338	1312	1095	0508	0201	0111	0097	0090	0087	0085
00.010	3341	2954	2885	2512	2183	2133	1819	1571	1330	0761	0354	0173	0129	0115	0111	0109
00.020	3788	3408	3355	3280	3226	2893	2834	2810	2493	1878	1369	0984	0716	0369	0252	0239
00.030	3873	3812	3768	3708	3667	3635	3588	3570	3238	3155	2788	2458	2167	1656	1096	0552
00.040	4247	4195	4158	4109	4077	4052	4017	4004	3983	3925	3877	3846	3829	3801	3781	3761
00.050	4295	4251	4529	4491	4466	4448	4423	4414	4399	4667	4644	4928	5204	5466	6136	6719
00.060	4646	4609	4584	4553	4533	4823	4807	4802	4793	5072	5351	5865	6095	6642	7071	7324
00.070	4681	4649	4628	4908	4895	4886	5170	5168	5165	5447	5970	6412	6756	7194	7413	7483
00.080	4711	4683	4972	4954	4945	5235	5231	5231	5513	5792	6487	6845	7106	7451	7568	7587
00.090	4737	5020	5007	4994	5284	5283	5566	5568	5572	6107	6745	7175	7362	7606	7660	7670
00.100	5065	5048	5038	5325	5324	5606	5613	5880	5890	6393	6969	7340	7545	7704	7730	7738
00.130	5127	5410	5411	5696	5705	5977	5995	6248	6486	6919	7466	7726	7824	7871	7880	7886
00.170	5481	5485	5772	5788	6070	6330	6581	6593	6814	7347	7797	7948	7983	8003	8010	8014
00.200	5521	5527	5821	6108	6374	6394	6650	6864	7066	7630	7942	8042	8060	8073	8079	8082
00.250	5572	5870	6149	6427	6455	6703	6942	7136	7317	7807	8078	8140	8151	8161	8165	8168
00.300	5893	5917	6205	6488	6744	6972	7192	7364	7525	7993	8170	8210	8217	8225	8229	8232
00.400	5962	6259	6532	6805	7044	7253	7587	7606	7840	8168	8285	8305	8310	8317	8320	8322
00.500	6011	6316	6596	7075	7295	7484	7778	7889	7995	8281	8356	8370	8375	8381	8384	8386
00.600	6318	6609	6870	7131	7353	7676	7931	7950	8117	8349	8408	8420	8424	8429	8432	8433
00.700	6353	6648	6913	7354	7556	7725	7980	8075	8213	8401	8449	8459	8463	8468	8470	8471
00.800	6382	6681	6948	7393	7728	7877	8097	8177	8291	8443	8482	8492	8495	8499	8502	8503
00.900	6407	6934	7181	7427	7763	7912	8132	8211	8324	8478	8510	8519	8522	8526	8528	8529
01.000	6676	6960	7208	7611	7794	8035	8223	8289	8382	8506	8534	8542	8546	8549	8551	8552
01.200	6717	7004	7255	7661	7956	8086	8321	8375	8451	8552	8573	8581	8584	8587	8589	8590
01.400	6749	7242	7472	7834	7997	8203	8398	8443	8504	8586	8605	8612	8615	8618	8619	8620
01.600	6777	7274	7505	7868	8124	8236	8431	8475	8535	8614	8631	8638	8640	8643	8645	8645
01.800	7026	7301	7533	8009	8154	8327	8487	8524	8573	8638	8653	8659	8662	8665	8666	8667
02.000	7049	7324	7713	8036	8255	8352	8511	8547	8604	8658	8673	8678	8681	8683	8685	8685
04.000	7390	7808	8001	8360	8517	8614	8699	8724	8748	8775	8785	8790	8791	8793	8794	8795
06.000	7646	8022	8195	8501	8660	8725	8781	8797	8813	8834	8842	8846	8847	8849	8850	8850
08.000	7699	8077	8343	8601	8731	8784	8832	8841	8854	8872	8880	8883	8884	8886	8886	8887
10.000	7893	8230	8459	8677	8766	8826	8865	8873	8884	8899	8907	8910	8911	8912	8913	8913
20.000	8144	8440	8633	8833	8901	8928	8953	8959	8965	8977	8983	8985	8986	8987	8988	8988
30.000	8319	8578	8741	8902	8956	8980	8998	9002	9007	9018	9023	9025	9026	9027	9027	9028
40.000	8363	8681	8818	8949	8992	9011	9026	9030	9035	9045	9049	9051	9052	9053	9053	9055

$2\alpha/RT\beta = 9.10$

p/K'	γ-Values 1.00	1.50	2.00	3.00	4.00	5.00	7.00	8.00	10.0	20.0	40.0	70.0	100	200	400	1000
00.001	1876	1335	1084	0707	0451	0344	0162	0124	0073	0017	0013	0011	0011	0010	0010	0010
00.002	2225	1864	1562	1084	0868	0582	0364	0287	0179	0049	0026	0023	0022	0021	0020	0020
00.004	2606	2230	1907	1582	1314	1088	0737	0716	0478	0160	0061	0048	0046	0043	0042	0041
00.006	2962	2576	2236	1890	1598	1551	1102	1078	0886	0324	0116	0076	0070	0066	0064	0063
00.008	3014	2918	2566	2204	1891	1844	1550	1322	1104	0606	0202	0116	0097	0090	0087	0085
00.010	3355	2966	2901	2526	2196	2149	1834	1582	1339	0768	0358	0174	0133	0115	0111	0109
00.020	3802	3727	3368	3297	3246	2909	2853	2831	2511	2142	1585	0995	0723	0371	0260	0240
00.030	4197	3825	3783	3725	3686	3655	3611	3594	3566	3181	2815	2484	2190	1902	1265	0606
00.040	4260	4210	4175	4128	4097	4074	4041	4028	4009	3955	3911	3883	3868	4157	4140	4124
00.050	4307	4577	4549	4512	4488	4471	4447	4439	4425	4702	4980	5256	5254	5770	6384	6969
00.060	4661	4625	4601	4571	4860	4849	4834	4828	4821	5106	5659	5915	6351	6838	7201	7404
00.070	4695	4664	4644	4930	4917	4909	5199	5197	5194	5749	6239	6641	6811	7316	7503	7543
00.080	4724	5005	4991	4974	5264	5260	5257	5542	5544	6076	6533	7029	7253	7536	7627	7642
00.090	4749	5037	5025	5311	5308	5307	5594	5596	5869	6370	6789	7224	7475	7670	7712	7720
00.100	5082	5065	5055	5347	5346	5633	5640	5912	5920	6429	7012	7464	7632	7753	7778	7784
00.130	5141	5431	5430	5719	5728	6004	6270	6278	6519	6955	7507	7798	7866	7912	7921	7926
00.170	5499	5501	5792	5808	6093	6357	6610	6621	6845	7495	7834	7994	8022	8039	8045	8048
00.200	5537	5827	5840	6130	6399	6418	6879	6893	7096	7665	7977	8076	8095	8106	8112	8115
00.250	5585	5887	6171	6451	6704	6728	7148	7164	7346	7901	8122	8171	8181	8190	8195	8197
00.300	5911	5933	6224	6509	6767	6997	7218	7391	7553	8023	8206	8238	8245	8253	8257	8259
00.400	5976	6276	6552	6826	7067	7277	7614	7745	7867	8195	8311	8330	8335	8342	8345	8347
00.500	6024	6331	6613	7097	7318	7507	7803	7915	8021	8306	8380	8394	8398	8404	8406	8408
00.600	6334	6626	6890	7151	7531	7700	7956	8051	8141	8372	8431	8442	8446	8451	8453	8455
00.700	6367	6664	6931	7376	7578	7747	8003	8098	8237	8430	8470	8480	8484	8488	8490	8492
00.800	6395	6922	7168	7413	7751	7900	8120	8199	8313	8469	8502	8511	8515	8519	8521	8522
00.900	6668	6952	7200	7602	7785	7933	8215	8282	8345	8501	8529	8538	8541	8545	8547	8548
01.000	6692	6977	7226	7631	7813	8057	8244	8310	8403	8528	8553	8561	8564	8567	8569	8570
01.200	6731	7019	7451	7679	7976	8106	8341	8396	8471	8570	8591	8598	8601	8605	8606	8607
01.400	6763	7259	7490	7853	8110	8222	8418	8463	8523	8604	8622	8629	8631	8634	8636	8637
01.600	7016	7290	7522	7886	8144	8317	8449	8493	8566	8631	8647	8654	8656	8659	8660	8661
01.800	7042	7316	7706	8029	8172	8345	8506	8542	8590	8654	8669	8675	8677	8680	8681	8682
02.000	7064	7339	7731	8054	8274	8369	8529	8581	8621	8674	8688	8693	8696	8698	8699	8700
04.000	7404	7823	8129	8376	8532	8629	8723	8739	8762	8788	8798	8802	8804	8806	8807	8807
06.000	7660	8037	8303	8515	8674	8739	8794	8810	8826	8846	8854	8858	8859	8861	8861	8862
08.000	7711	8090	8356	8615	8744	8797	8844	8853	8866	8883	8891	8894	8895	8896	8897	8897
10.000	7907	8244	8473	8691	8795	8838	8877	8885	8895	8910	8917	8920	8921	8923	8923	8924
20.000	8156	8452	8694	8844	8911	8942	8963	8969	8975	8987	8992	8994	8995	8996	8997	8997
30.000	8331	8589	8790	8913	8966	8989	9007	9011	9016	9026	9031	9033	9034	9035	9035	9036
40.000	8373	8692	8828	8959	9001	9020	9035	9039	9044	9053	9057	9059	9060	9061	9061	9061

$2\alpha/RT\beta = 9.20$

p/K'	γ-Values 1.00	1.50	2.00	3.00	4.00	5.00	7.00	8.00	10.0	20.0	40.0	70.0	100	200	400	1000
00.001	1885	1341	1090	0711	0454	0347	0164	0125	0074	0017	0013	0011	0011	0010	0010	0010
00.002	2234	1874	1571	1090	0874	0585	0366	0289	0219	0050	0026	0023	0022	0021	0020	0020
00.004	2615	2240	1916	1591	1322	1095	0877	0721	0577	0161	0061	0048	0046	0043	0042	0041
00.006	2973	2587	2245	1900	1836	1561	1109	1087	0893	0390	0129	0077	0071	0066	0064	0063
00.008	3024	2932	2577	2215	2152	1856	1561	1331	1112	0611	0234	0116	0099	0090	0087	0086
00.010	3368	2978	2915	2540	2207	2163	1847	1593	1555	0910	0360	0175	0137	0115	0111	0109
00.020	3815	3743	3381	3313	3264	3226	2871	2850	2527	2160	1600	1163	0845	0413	0268	0241
00.030	4212	4153	3797	3742	3704	3675	3633	3617	3591	3205	2840	2796	2484	1923	1461	0753
00.040	4273	4224	4191	4146	4116	4094	4063	4051	4033	3983	4260	4235	4222	4202	4500	4790
00.050	4318	4595	4568	4532	4509	4493	4471	4463	4762	4734	5020	5302	5572	6056	6609	7157
00.060	4676	4641	4618	4589	4884	4873	4859	4854	5149	5427	5701	6190	6404	7012	7315	7475
00.070	4709	4678	4971	4951	4939	4931	5226	5224	5222	5786	6284	6691	6998	7373	7570	7601
00.080	4736	5024	5010	4993	5288	5285	5570	5572	5574	6113	6576	7078	7304	7612	7682	7694
00.090	5073	5054	5042	5334	5330	5329	5621	5623	5900	6408	6990	7366	7524	7728	7761	7768
00.100	5097	5080	5071	5368	5367	5659	5937	5941	5949	6671	7189	7510	7679	7803	7823	7829
00.130	5154	5449	5448	5742	5749	6030	6298	6307	6550	7149	7625	7839	7915	7952	7959	7964
00.170	5515	5517	5812	6099	6115	6382	6638	6649	6875	7530	7909	8029	8057	8073	8079	8082
00.200	5551	5847	5858	6152	6424	6671	6908	6921	7125	7698	8034	8112	8127	8138	8143	8146
00.250	5599	5904	6191	6473	6729	6751	7176	7191	7510	7933	8153	8202	8210	8219	8223	8226
00.300	5928	5948	6242	6530	6789	7021	7402	7418	7695	8053	8234	8265	8272	8279	8283	8285
00.400	5990	6293	6572	6847	7089	7300	7639	7771	7894	8245	8337	8354	8360	8366	8368	8370
00.500	6309	6598	6630	7118	7340	7530	7827	7940	8045	8331	8404	8416	8421	8426	8428	8430
00.600	6349	6643	6909	7170	7554	7722	7980	8075	8165	8404	8452	8463	8467	8472	8474	8475
00.700	6381	6680	6948	7396	7598	7768	8104	8184	8259	8452	8490	8500	8504	8508	8510	8511
00.800	6409	6941	7188	7433	7772	7921	8142	8221	8335	8490	8522	8531	8534	8538	8540	8541
00.900	6685	6969	7218	7623	7805	7954	8237	8303	8397	8521	8548	8556	8560	8563	8565	8566
01.000	6708	6993	7244	7651	7948	8078	8265	8331	8423	8547	8571	8579	8582	8585	8587	8588
01.200	6746	7034	7470	7833	7996	8203	8361	8415	8490	8588	8608	8615	8618	8621	8623	8624
01.400	6776	7276	7507	7872	8130	8241	8437	8482	8542	8621	8638	8645	8647	8650	8652	8653
01.600	7033	7306	7538	7904	8162	8336	8468	8534	8583	8648	8663	8669	8672	8675	8676	8677
01.800	7057	7331	7724	8047	8268	8363	8524	8560	8608	8670	8685	8690	8692	8695	8696	8697
02.000	7078	7536	7748	8071	8292	8437	8546	8599	8638	8689	8703	8708	8710	8713	8714	8715
04.000	7417	7838	8145	8391	8547	8644	8737	8753	8775	8801	8811	8815	8816	8818	8819	8820
06.000	7673	8051	8318	8529	8689	8753	8812	8823	8838	8857	8866	8869	8870	8872	8873	8873
08.000	7723	8218	8370	8629	8758	8809	8856	8867	8878	8894	8901	8904	8906	8907	8908	8908
10.000	7921	8257	8486	8703	8808	8850	8888	8897	8906	8921	8927	8930	8931	8933	8933	8934
20.000	8168	8541	8707	8855	8922	8952	8973	8978	8984	8996	9001	9003	9004	9005	9006	9006
30.000	8344	8600	8802	8923	8975	8999	9016	9020	9025	9035	9039	9041	9042	9043	9044	9044
40.000	8384	8703	8838	8968	9011	9029	9043	9047	9052	9061	9065	9067	9067	9068	9069	9069

$2\alpha/RT\beta = 9.30$

p/K'	1.00	1.50	2.00	3.00	4.00	5.00	7.00	8.00	10.0	20.0	40.0	70.0	100	200	400	1000
								γ-Values								
00.001	1894	1347	1095	0716	0457	0350	0165	0126	0074	0018	0013	0011	0011	0010	0010	0010
00.002	2243	1883	1580	1096	0880	0705	0448	0355	0221	0050	0026	0023	0022	0021	0020	0020
00.004	2624	2248	2178	1599	1330	1102	0884	0727	0581	0192	0061	0048	0046	0043	0042	0041
00.006	2984	2597	2529	2163	1848	1571	1301	1095	0900	0393	0130	0077	0071	0066	0064	0063
00.008	3033	2944	2588	2225	2166	1867	1572	1340	1304	0616	0235	0123	0099	0090	0087	0086
00.010	3381	2989	2929	2552	2219	2176	1859	1835	1566	0918	0427	0193	0138	0115	0111	0109
00.020	3827	3758	3709	3328	3281	3245	2888	2868	2835	2177	1614	1174	0853	0467	0278	0241
00.030	4227	4170	4130	3757	3721	3693	3654	3639	3614	3228	3169	2823	2800	2198	1686	0974
00.040	4285	4238	4206	4163	4134	4113	4084	4073	4376	4330	4294	4272	4577	4560	4855	5414
00.050	4650	4612	4585	4551	4529	4514	4493	4801	4790	4764	5349	5620	5874	6320	6811	7298
00.060	4690	4656	4634	4922	4907	4896	4883	4878	5179	5462	5995	6238	6639	7070	7446	7539
00.070	4721	4692	4991	4971	4959	5258	5252	5250	5541	5822	6326	6898	7049	7478	7632	7655
00.080	4748	5041	5027	5012	5311	5308	5599	5600	5602	6149	6801	7241	7432	7680	7734	7743
00.090	5089	5070	5059	5355	5352	5643	5647	5649	5930	6444	7032	7412	7621	7776	7807	7813
00.100	5112	5096	5393	5388	5680	5683	5965	5969	6232	6707	7231	7617	7754	7851	7866	7872
00.130	5167	5467	5466	5763	5770	6055	6326	6334	6580	7184	7663	7902	7954	7990	7997	8001
00.170	5531	5532	5831	6121	6137	6407	6664	6884	7087	7563	7944	8070	8092	8106	8111	8115
00.200	5566	5865	5875	6172	6447	6696	6935	6947	7154	7730	8067	8144	8159	8169	8173	8176
00.250	5611	5920	6211	6495	6752	6982	7203	7377	7539	7963	8182	8230	8239	8247	8251	8253
00.300	5943	6238	6259	6549	6811	7044	7428	7443	7723	8120	8262	8292	8298	8305	8309	8311
00.400	6004	6309	6590	6866	7110	7322	7664	7796	7919	8270	8362	8378	8383	8389	8392	8393
00.500	6325	6616	6879	7139	7362	7552	7851	7964	8133	8367	8426	8438	8442	8447	8450	8451
00.600	6364	6660	6927	7373	7575	7745	8003	8098	8238	8427	8473	8484	8487	8492	8494	8495
00.700	6395	6695	6965	7416	7619	7905	8126	8206	8281	8473	8510	8520	8523	8527	8529	8530
00.800	6422	6958	7207	7451	7793	7942	8163	8243	8357	8510	8541	8549	8553	8556	8558	8559
00.900	6701	6986	7236	7643	7825	8070	8258	8324	8417	8541	8567	8575	8578	8581	8583	8584
01.000	6723	7009	7261	7670	7968	8098	8285	8351	8443	8566	8589	8596	8599	8602	8604	8605
01.200	6759	7257	7488	7853	8014	8223	8380	8434	8509	8606	8625	8632	8635	8638	8639	8640
01.400	6789	7292	7524	7890	8149	8260	8456	8500	8560	8639	8655	8661	8663	8666	8668	8668
01.600	7048	7321	7554	8037	8180	8355	8516	8552	8601	8664	8679	8685	8687	8690	8691	8692
01.800	7072	7346	7741	8065	8286	8381	8541	8577	8634	8686	8700	8705	8707	8710	8711	8712
02.000	7092	7552	7764	8088	8309	8455	8586	8615	8654	8704	8718	8723	8725	8727	8728	8729
04.000	7429	7852	8160	8406	8602	8659	8751	8767	8789	8814	8823	8827	8829	8830	8831	8832
06.000	7686	8065	8333	8594	8702	8766	8825	8835	8850	8869	8877	8880	8881	8883	8884	8884
08.000	7734	8232	8383	8642	8770	8822	8868	8878	8889	8905	8912	8915	8916	8917	8918	8919
10.000	7933	8270	8499	8716	8820	8862	8899	8908	8917	8931	8938	8940	8941	8943	8943	8944
20.000	8179	8554	8718	8866	8932	8962	8983	8988	8994	9005	9010	9012	9013	9014	9014	9015
30.000	8355	8611	8813	8933	8990	9008	9025	9028	9034	9043	9048	9050	9050	9051	9052	9052
40.000	8394	8713	8848	8978	9023	9039	9052	9055	9060	9069	9073	9075	9075	9076	9076	9077

$2\alpha/RT\beta = 9.40$

p/K'	γ-Values 1.00	1.50	2.00	3.00	4.00	5.00	7.00	8.00	10.0	20.0	40.0	70.0	100	200	400	1000
00.001	1902	1353	1101	0720	0459	0352	0166	0126	0075	0018	0013	0011	0011	0010	0010	0010
00.002	2252	1891	1588	1102	0886	0710	0451	0357	0222	0050	0026	0023	0022	0021	0020	0020
00.004	2633	2257	2189	1607	1337	1108	0890	0732	0586	0194	0065	0048	0046	0043	0042	004
00.006	2995	2607	2542	2176	1859	1581	1310	1102	0906	0396	0130	0078	0071	0066	0064	0063
00.008	3042	2957	2599	2235	2178	1878	1582	1559	1313	0621	0277	0124	0100	0090	0087	0086
00.010	3392	2999	2942	2564	2508	2188	1871	1848	1578	0925	0431	0194	0145	0115	0111	0109
00.020	3839	3773	3725	3342	3298	3263	2904	2885	2854	2470	1862	1371	1000	0535	0292	0242
00.030	4241	4186	4148	4096	3737	3711	3673	3659	3636	3570	3195	3158	2828	2499	1939	1286
00.040	4297	4251	4220	4179	4152	4132	4428	4418	4402	4360	4327	4627	4618	4914	5202	5967
00.050	4666	4628	4602	4570	4549	4534	4834	4827	4817	5104	5388	5664	5923	6562	6990	7390
00.060	4703	4670	4649	4943	4928	4918	4905	5211	5207	5496	6037	6493	6688	7223	7528	7599
00.070	4734	4705	5009	4990	4979	5283	5277	5275	5570	5855	6577	6945	7215	7570	7689	7707
00.080	4760	5058	5044	5029	5334	5331	5625	5626	5907	6182	6843	7287	7479	7729	7783	7790
00.090	5104	5086	5075	5376	5373	5668	5672	5953	5959	6478	7073	7535	7666	7827	7851	7857
00.100	5127	5110	5412	5408	5704	5707	5992	5996	6262	6742	7388	7659	7797	7895	7908	7913
00.130	5180	5483	5482	5784	6069	6078	6353	6360	6609	7218	7765	7939	7998	8026	8033	8037
00.170	5546	5547	5850	6143	6415	6430	6690	6912	7117	7595	7978	8104	8126	8138	8143	8146
00.200	5579	5882	5892	6191	6469	6721	6961	7160	7343	7841	8098	8177	8189	8198	8203	8205
00.250	5624	5935	6229	6515	6775	7007	7228	7404	7567	7992	8220	8259	8266	8274	8278	8280
00.300	5959	6256	6276	6803	7042	7067	7454	7608	7750	8148	8293	8318	8324	8330	8334	8336
00.400	6017	6324	6608	6885	7317	7506	7687	7821	8024	8295	8386	8401	8406	8411	8414	8416
00.500	6341	6634	6898	7159	7382	7574	7874	7988	8157	8390	8448	8459	8464	8468	8471	8472
00.600	6378	6675	6944	7394	7596	7766	8025	8121	8261	8448	8494	8504	8507	8512	8514	8515
00.700	6408	6710	6981	7435	7777	7926	8148	8228	8343	8493	8530	8539	8542	8546	8548	8549
00.800	6434	6975	7225	7631	7813	7962	8248	8315	8377	8533	8560	8568	8571	8574	8576	8577
00.900	6716	7002	7253	7662	7844	8091	8278	8345	8437	8559	8585	8592	8595	8599	8600	8601
01.000	6737	7024	7277	7688	7988	8117	8356	8410	8486	8584	8606	8613	8616	8619	8621	8622
01.200	6772	7275	7506	7871	8033	8242	8399	8484	8545	8624	8642	8648	8651	8654	8655	8656
01.400	6801	7308	7540	7907	8167	8278	8474	8518	8578	8655	8671	8677	8679	8682	8683	8684
01.600	7063	7336	7569	8055	8197	8373	8534	8570	8618	8680	8694	8700	8702	8705	8706	8707
01.800	7086	7360	7758	8082	8303	8398	8558	8611	8650	8701	8714	8720	8722	8724	8725	8726
02.000	7105	7568	7780	8105	8326	8472	8603	8632	8670	8719	8732	8737	8739	8741	8742	8743
04.000	7441	7866	8176	8421	8617	8696	8765	8785	8802	8826	8835	8839	8841	8842	8843	8844
06.000	7699	8079	8347	8608	8716	8779	8837	8850	8863	8880	8888	8891	8892	8894	8895	8895
08.000	7909	8246	8396	8694	8783	8834	8880	8890	8900	8915	8922	8925	8926	8928	8928	8929
10.000	7945	8282	8512	8729	8832	8873	8910	8918	8927	8941	8947	8950	8951	8952	8953	8953
20.000	8190	8566	8730	8877	8943	8972	8992	8997	9003	9013	9018	9021	9021	9022	9023	9023
30.000	8366	8686	8823	8956	8999	9018	9033	9037	9042	9051	9056	9058	9058	9059	9060	9060
40.000	8404	8723	8858	8987	9032	9047	9060	9063	9068	9076	9080	9082	9083	9084	9084	9084

$2\alpha/RT\beta = 9.50$

p/K'	1.00	1.50	2.00	3.00	4.00	5.00	7.00	8.00	10.0	20.0	40.0	70.0	100	200	400	1000
	γ-Values															
00.001	1909	1358	1106	0723	0561	0354	0207	0127	0075	0018	0013	0011	0011	0010	0010	0010
00.002	2260	1899	1595	1107	0891	0715	0454	0360	0224	0057	0026	0023	0022	0021	0020	0020
00.004	2641	2265	2200	1614	1344	1303	0896	0737	0590	0195	0065	0048	0046	0043	0042	0041
00.006	3005	2616	2554	2187	1870	1590	1319	1109	0913	0399	0149	0079	0071	0066	0064	0063
00.008	3051	2968	2609	2527	2190	1888	1592	1570	1323	0745	0279	0133	0103	0090	0087	0086
00.010	3404	3010	2954	2576	2522	2200	1882	1860	1588	0932	0434	0219	0145	0116	0112	0109
00.020	3850	3787	3741	3356	3313	3281	3232	2902	2872	2489	1878	1383	1173	0620	0311	0243
00.030	4254	4201	4165	4115	4081	3728	3692	3679	3657	3595	3543	3185	3166	2822	2500	1701
00.040	4308	4264	4234	4195	4497	4478	4452	4442	4427	4388	4681	4665	4970	5258	5794	6432
00.050	4680	4643	4619	4587	4567	4876	4859	4852	4843	5135	5425	5967	6206	6780	7149	7484
00.060	4716	4684	4664	4963	4949	4939	5241	5238	5233	5527	6076	6538	6900	7357	7602	7654
00.070	4746	5042	5027	5008	5311	5306	5300	5598	5598	6147	6617	7130	7361	7651	7743	7756
00.080	4771	5074	5061	5360	5355	5352	5651	5652	5936	6452	6882	7429	7590	7788	7829	7835
00.090	5119	5101	5090	5395	5392	5692	5977	5980	5986	6724	7252	7578	7750	7875	7894	7899
00.100	5141	5125	5431	5426	5726	5729	6018	6022	6291	6775	7427	7752	7861	7937	7948	7953
00.130	5192	5500	5498	5803	6092	6101	6378	6623	6637	7392	7801	7993	8039	8061	8068	8071
00.170	5560	5561	5867	6163	6438	6453	6714	6938	7145	7725	8042	8140	8158	8169	8174	8176
00.200	5592	5899	5908	6210	6490	6745	6986	7187	7371	7872	8147	8207	8218	8227	8231	8234
00.250	5935	5950	6247	6535	6797	7031	7253	7430	7594	8072	8248	8286	8293	8300	8304	8306
00.300	5973	6273	6292	6824	7065	7277	7478	7633	7776	8175	8319	8343	8349	8355	8358	8360
00.400	6029	6339	6625	6904	7339	7530	7710	7844	8048	8320	8409	8424	8428	8433	8436	8438
00.500	6356	6650	6917	7178	7402	7735	7896	8010	8180	8413	8470	8480	8484	8489	8491	8492
00.600	6392	6691	6961	7414	7616	7786	8046	8142	8283	8476	8514	8523	8527	8531	8533	8534
00.700	6421	6961	7210	7453	7798	7947	8169	8249	8364	8518	8549	8557	8561	8564	8566	8567
00.800	6707	6992	7242	7651	7833	7982	8269	8335	8429	8552	8578	8586	8589	8592	8594	8595
00.900	6731	7017	7269	7680	7981	8111	8298	8364	8457	8580	8602	8610	8612	8616	8617	8618
01.000	6751	7039	7293	7705	8007	8136	8375	8430	8505	8604	8623	8630	8633	8636	8637	8638
01.200	6785	7291	7523	7889	8149	8261	8458	8502	8563	8641	8658	8664	8667	8670	8671	8672
01.400	7051	7323	7556	7924	8185	8361	8492	8535	8608	8671	8686	8692	8694	8697	8698	8699
01.600	7077	7350	7749	8073	8295	8390	8551	8587	8634	8696	8709	8715	8717	8719	8720	8721
01.800	7099	7562	7774	8098	8321	8467	8574	8628	8666	8716	8729	8734	8736	8738	8740	8740
02.000	7118	7584	7796	8120	8342	8488	8619	8648	8692	8734	8746	8751	8753	8755	8756	8757
04.000	7454	7879	8190	8500	8631	8710	8778	8798	8815	8838	8847	8851	8852	8854	8855	8855
06.000	7711	8092	8361	8622	8729	8792	8849	8862	8874	8891	8899	8902	8903	8904	8905	8906
08.000	7922	8259	8489	8707	8795	8846	8891	8901	8911	8926	8932	8935	8936	8938	8938	8939
10.000	7957	8294	8525	8740	8844	8885	8921	8929	8938	8951	8957	8960	8961	8962	8963	8963
20.000	8201	8577	8741	8887	8953	8981	9002	9006	9012	9022	9027	9029	9030	9031	9031	9032
30.000	8377	8697	8834	8965	9008	9027	9042	9045	9050	9059	9064	9065	9066	9067	9067	9068
40.000	8414	8733	8899	9006	9040	9056	9068	9071	9076	9084	9088	9090	9090	9091	9091	9092

$2\alpha/RT\beta = 9.60$

p/K'	γ-Values 1.00	1.50	2.00	3.00	4.00	5.00	7.00	8.00	10.0	20.0	40.0	70.0	100	200	400	1000
00.001	1917	1363	1111	0727	0565	0356	0209	0160	0094	0019	0013	0011	0011	0010	0010	0010
00.002	2267	1907	1602	1113	0896	0719	0457	0362	0225	0057	0026	0023	0022	0021	0020	0020
00.004	2649	2273	2210	1862	1566	1311	0902	0883	0594	0196	0070	0048	0046	0043	0042	0041
00.006	3015	2625	2565	2198	1879	1598	1327	1306	1084	0483	0149	0081	0071	0066	0064	0063
00.008	3376	2979	2618	2540	2201	1897	1601	1580	1331	0750	0281	0134	0103	0091	0088	0086
00.010	3415	3019	2966	2587	2535	2212	1892	1872	1598	1105	0519	0220	0156	0116	0112	0109
00.020	3861	3800	3756	3369	3328	3297	3251	3233	2889	2507	1893	1610	1183	0725	0337	0245
00.030	4267	4216	4180	4133	4100	4076	4041	4028	3676	3618	3570	3537	3194	3165	2828	2222
00.040	4319	4276	4248	4542	4517	4499	4475	4465	4451	4740	4715	5017	5010	5588	6089	6804
00.050	4694	4658	4634	4604	4585	4899	4882	4876	4867	5166	5745	6010	6253	6834	7355	7561
00.060	4729	4698	4678	4982	4968	4959	5266	5263	5259	5842	6353	6772	7089	7475	7668	7707
00.070	4757	5058	5043	5026	5333	5328	5625	5625	5625	6181	6656	7175	7407	7723	7793	7803
00.080	5109	5089	5076	5380	5375	5372	5675	5676	5964	6486	7086	7472	7686	7842	7873	7879
00.090	5133	5115	5105	5414	5714	5715	6003	6006	6275	6758	7291	7685	7823	7920	7935	7940
00.100	5154	5138	5449	5444	5748	5750	6043	6310	6319	6998	7464	7791	7919	7977	7986	7991
00.130	5204	5515	5514	5822	6114	6123	6402	6650	6879	7425	7836	8029	8074	8095	8101	8105
00.170	5574	5877	5884	6183	6461	6474	6954	6964	7173	7756	8073	8171	8189	8199	8203	8206
00.200	5605	5915	5924	6228	6511	6767	7010	7213	7399	7901	8176	8237	8247	8255	8259	8261
00.250	5950	5964	6264	6554	6818	7054	7276	7455	7619	8100	8282	8313	8319	8326	8329	8331
00.300	5987	6290	6308	6845	7087	7300	7501	7658	7800	8201	8347	8367	8373	8379	8382	8383
00.400	6042	6353	6641	7137	7360	7552	7852	7867	8072	8361	8433	8446	8450	8455	8457	8459
00.500	6370	6666	6935	7196	7587	7757	8017	8114	8203	8435	8491	8501	8504	8509	8511	8512
00.600	6405	6705	6977	7433	7636	7806	8067	8163	8304	8497	8534	8542	8546	8550	8552	8553
00.700	6433	6978	7228	7471	7818	7967	8189	8270	8384	8538	8568	8576	8579	8582	8584	8585
00.800	6722	7007	7259	7670	7851	8101	8289	8355	8449	8571	8596	8603	8606	8609	8611	8612
00.900	6745	7032	7285	7698	8000	8130	8317	8383	8476	8598	8619	8626	8629	8632	8634	8635
01.000	6764	7053	7308	7722	8026	8155	8394	8449	8524	8621	8640	8647	8649	8652	8654	8655
01.200	6798	7307	7539	7907	8168	8279	8476	8520	8580	8658	8674	8680	8682	8685	8686	8687
01.400	7066	7338	7571	7941	8202	8379	8509	8576	8625	8687	8701	8707	8709	8711	8713	8713
01.600	7091	7364	7765	8090	8312	8407	8568	8603	8660	8711	8724	8729	8731	8733	8735	8735
01.800	7112	7578	7789	8115	8337	8483	8615	8644	8682	8731	8743	8748	8750	8752	8753	8754
02.000	7130	7599	7810	8136	8358	8504	8634	8663	8707	8748	8760	8765	8766	8769	8770	8770
04.000	7465	7892	8204	8515	8645	8724	8792	8811	8829	8850	8859	8862	8864	8865	8866	8867
06.000	7723	8104	8374	8635	8741	8804	8861	8873	8886	8902	8909	8912	8914	8915	8916	8916
08.000	7934	8272	8502	8720	8807	8867	8904	8912	8922	8936	8943	8945	8946	8948	8948	8949
10.000	7969	8306	8537	8752	8856	8896	8933	8939	8948	8961	8967	8969	8970	8971	8972	8973
20.000	8212	8589	8752	8916	8969	8991	9011	9015	9020	9031	9035	9037	9038	9039	9040	9040
30.000	8387	8708	8844	8975	9017	9036	9050	9054	9058	9067	9071	9073	9074	9075	9075	9076
40.000	8424	8743	8909	9015	9049	9064	9076	9079	9083	9092	9095	9097	9098	9098	9099	9099

$2\alpha/RT\beta = 9.70$

p/K′	γ-Values 1.00	1.50	2.00	3.00	4.00	5.00	7.00	8.00	10.0	20.0	40.0	70.0	100	200	400	1000
00.001	1924	1586	1115	0731	0568	0358	0210	0161	0094	0019	0013	0011	0011	0010	0010	0010
00.002	2275	1915	1609	1310	0901	0724	0460	0365	0280	0057	0026	0023	0022	0021	0020	0020
00.004	2657	2568	2220	1871	1575	1319	0907	0889	0598	0239	0070	0048	0046	0043	0042	0041
00.006	3024	2634	2576	2209	1889	1607	1335	1315	1092	0487	0150	0081	0071	0066	0064	0063
00.008	3388	2990	2933	2552	2212	2173	1852	1589	1340	0756	0336	0148	0107	0091	0088	0086
00.010	3425	3348	2977	2597	2548	2222	2168	1882	1850	1113	0523	0253	0169	0116	0112	0110
00.020	3872	3812	3770	3712	3342	3313	3269	3252	2905	2524	2173	1624	1386	0731	0412	0248
00.030	4279	4230	4196	4150	4119	4096	4063	4050	4030	3975	3596	3566	3550	3194	3175	2837
00.040	4330	4288	4596	4560	4536	4519	4496	4487	4474	4770	5065	5054	5354	5901	6360	7090
00.050	4707	4672	4649	4620	4932	4920	4904	4899	4891	5499	5782	6293	6514	7028	7464	7627
00.060	4741	4710	5022	5000	4987	4979	5290	5287	5588	5874	6392	6815	7135	7577	7728	7757
00.070	4768	5074	5060	5042	5354	5349	5650	5650	5650	6213	6886	7338	7534	7787	7841	7849
00.080	5124	5104	5091	5400	5395	5698	5699	5987	5991	6518	7124	7596	7728	7891	7915	7920
00.090	5147	5129	5439	5432	5736	5737	6029	6032	6303	6791	7449	7724	7863	7962	7974	7979
00.100	5167	5151	5466	5461	5768	5771	6066	6337	6345	7031	7601	7871	7957	8015	8023	8027
00.130	5215	5530	5529	5840	6136	6409	6668	6676	6907	7456	7923	8076	8111	8128	8134	8137
00.170	5587	5894	5900	6202	6482	6738	6979	6989	7199	7786	8128	8205	8218	8228	8232	8234
00.200	5618	5930	6226	6512	6530	6789	7226	7238	7425	7929	8205	8265	8274	8282	8286	8288
00.250	5965	5978	6280	6573	6838	7076	7467	7480	7644	8126	8308	8338	8344	8350	8354	8356
00.300	6000	6306	6589	6864	7108	7323	7524	7681	7825	8258	8372	8391	8396	8402	8405	8406
00.400	6054	6367	6657	7157	7381	7573	7875	7990	8095	8384	8455	8467	8471	8476	8478	8480
00.500	6384	6682	6952	7214	7608	7778	8039	8136	8224	8465	8511	8520	8524	8528	8530	8531
00.600	6418	6720	6993	7452	7655	7947	8170	8251	8325	8517	8553	8561	8564	8568	8570	8571
00.700	6445	6994	7245	7489	7837	7987	8209	8289	8404	8557	8586	8593	8596	8600	8602	8603
00.800	6736	7023	7275	7688	7870	8120	8308	8375	8468	8589	8613	8620	8623	8626	8628	8629
00.900	6758	7046	7300	7715	8019	8148	8389	8443	8519	8616	8636	8643	8646	8649	8650	8651
01.000	6777	7067	7515	7739	8043	8255	8413	8467	8542	8638	8656	8663	8665	8668	8669	8670
01.200	6810	7322	7554	7924	8185	8296	8494	8538	8597	8674	8689	8695	8697	8700	8701	8702
01.400	7080	7352	7586	8078	8219	8396	8557	8593	8641	8703	8716	8721	8724	8726	8727	8728
01.600	7104	7378	7781	8106	8329	8423	8584	8619	8676	8726	8738	8743	8745	8747	8749	8749
01.800	7125	7593	7805	8130	8353	8500	8630	8659	8697	8745	8757	8762	8764	8766	8767	8768
02.000	7143	7613	7825	8252	8441	8520	8650	8691	8722	8762	8773	8778	8780	8782	8783	8783
04.000	7476	7905	8218	8529	8659	8738	8804	8824	8841	8862	8870	8874	8875	8877	8877	8878
06.000	7735	8117	8387	8648	8754	8830	8873	8885	8897	8913	8920	8923	8924	8925	8926	8926
08.000	7947	8284	8515	8732	8819	8878	8915	8922	8932	8946	8952	8955	8956	8957	8958	8958
10.000	7980	8318	8548	8764	8867	8913	8943	8950	8958	8970	8976	8979	8980	8981	8981	8982
20.000	8222	8600	8763	8926	8979	9000	9020	9024	9029	9039	9044	9046	9046	9047	9048	9048
30.000	8397	8718	8854	8984	9026	9044	9058	9062	9066	9075	9079	9081	9082	9082	9083	9083
40.000	8534	8753	8919	9024	9057	9072	9084	9087	9091	9099	9103	9104	9105	9106	9106	9106

$2\alpha/RT\beta = 9.80$

p/K'	1.00	1.50	2.00	3.00	4.00	5.00	7.00	8.00	10.0	20.0	40.0	70.0	100	200	400	1000
								γ-Values								
00.001	1930	1593	1120	0879	0571	0444	0212	0162	0095	0019	0013	0011	0011	0010	0010	0010
00.002	2282	1922	1616	1317	0905	0728	0463	0367	0282	0058	0026	0023	0022	0021	0020	0020
00.004	2664	2578	2229	1881	1583	1326	0912	0895	0724	0240	0070	0048	0046	0043	0042	0041
00.006	3033	2642	2586	2219	1898	1859	1342	1323	1099	0490	0175	0085	0071	0066	0064	0063
00.008	3399	3000	2946	2563	2222	2185	1863	1598	1347	0761	0338	0148	0107	0091	0088	0086
00.010	3435	3361	2988	2916	2560	2523	2181	1893	1862	1121	0526	0255	0170	0117	0112	0110
00.020	3882	3824	3784	3728	3356	3328	3286	3270	3243	2848	2189	1881	1397	0860	0469	0255
00.030	4291	4243	4210	4166	4136	4114	4083	4071	4052	4000	3958	3931	3578	3555	3537	3188
00.040	4340	4638	4612	4578	4555	4539	4516	4508	4496	4797	5099	5397	5683	5948	6608	7235
00.050	4720	4686	4664	4636	4952	4941	4926	4921	5236	5530	5817	6334	6558	7201	7559	7685
00.060	4752	4723	5039	5018	5005	5320	5312	5310	5615	5905	6428	7025	7302	7667	7783	7805
00.070	4779	5089	5075	5059	5374	5370	5675	5674	5964	6244	6923	7379	7576	7831	7885	7892
00.080	5137	5118	5106	5418	5413	5720	5721	6013	6016	6549	7305	7636	7811	7937	7955	7960
00.090	5160	5142	5457	5450	5757	5758	6053	6056	6330	6822	7486	7816	7926	8002	8012	8016
00.100	5179	5487	5482	5786	5788	6081	6089	6362	6615	7062	7637	7907	8007	8052	8059	8063
00.130	5549	5544	5852	5858	6156	6432	6693	6701	6934	7486	7956	8109	8143	8160	8165	8168
00.170	5600	5910	5916	6220	6503	6761	7004	7013	7224	7815	8158	8234	8247	8256	8260	8262
00.200	5630	5945	6244	6532	6794	6810	7251	7262	7450	8024	8246	8292	8301	8308	8312	8314
00.250	5979	6281	6296	6591	6858	7097	7491	7503	7791	8152	8334	8363	8368	8374	8378	8379
00.300	6014	6321	6607	6883	7129	7344	7691	7704	7950	8282	8395	8414	8419	8424	8427	8429
00.400	6355	6649	6673	7176	7400	7594	7897	8013	8118	8406	8476	8488	8492	8496	8498	8500
00.500	6397	6697	6968	7231	7628	7798	8060	8157	8245	8486	8531	8540	8543	8547	8549	8550
00.600	6430	6734	7008	7470	7673	7967	8190	8271	8345	8536	8571	8579	8582	8586	8588	8589
00.700	6457	7010	7262	7674	7856	8005	8295	8362	8424	8578	8603	8611	8614	8617	8619	8620
00.800	6750	7037	7291	7706	7887	8139	8327	8393	8487	8609	8630	8637	8640	8643	8644	8645
00.900	6772	7060	7315	7732	8037	8166	8407	8462	8537	8634	8653	8659	8662	8665	8666	8667
01.000	6790	7081	7532	7900	8061	8273	8431	8485	8559	8655	8672	8678	8681	8683	8685	8686
01.200	6821	7337	7570	7940	8203	8313	8511	8555	8614	8690	8705	8710	8712	8715	8716	8717
01.400	7094	7366	7600	8095	8235	8413	8574	8610	8657	8718	8731	8736	8738	8740	8741	8742
01.600	7117	7391	7796	8122	8346	8493	8600	8653	8691	8740	8752	8757	8759	8761	8762	8763
01.800	7137	7608	7819	8145	8369	8516	8646	8674	8719	8759	8771	8775	8777	8779	8780	8781
02.000	7154	7627	7839	8268	8456	8535	8665	8706	8737	8776	8787	8791	8793	8795	8796	8796
04.000	7682	7917	8232	8543	8672	8751	8824	8836	8853	8873	8882	8885	8886	8888	8888	8889
06.000	7746	8129	8400	8661	8791	8842	8888	8896	8908	8923	8930	8933	8934	8935	8936	8937
08.000	7958	8296	8527	8744	8849	8889	8926	8934	8942	8956	8962	8965	8966	8967	8967	8968
10.000	7991	8329	8560	8807	8878	8924	8954	8960	8967	8980	8985	8988	8989	8990	8990	8991
20.000	8232	8610	8773	8936	8989	9009	9029	9032	9038	9047	9052	9054	9054	9055	9056	9056
30.000	8408	8729	8864	8993	9035	9053	9066	9070	9074	9083	9087	9088	9089	9090	9090	9090
40.000	8544	8762	8928	9032	9065	9080	9092	9094	9099	9106	9110	9111	9112	9113	9113	9113

$2\alpha/RT\beta = 9.90$

p/K'	γ-Values 1.00	1.50	2.00	3.00	4.00	5.00	7.00	8.00	10.0	20.0	40.0	70.0	100	200	400	1000
00.001	1937	1600	1320	0884	0574	0447	0213	0163	0095	0019	0013	0011	0011	0010	0010	0010
00.002	2289	1929	1622	1324	0910	0731	0466	0453	0284	0068	0026	0023	0022	0021	0020	0020
00.004	2672	2589	2238	1890	1592	1333	1088	0900	0729	0242	0077	0048	0046	0043	0042	0041
00.006	3042	2962	2596	2228	1907	1870	1571	1331	1106	0494	0176	0085	0072	0067	0064	0063
00.008	3410	3010	2957	2574	2232	2196	1874	1607	1577	0912	0340	0167	0112	0091	0088	0086
00.010	3445	3374	2998	2929	2571	2536	2193	2173	1873	1128	0633	0298	0189	0119	0112	0110
00.020	3892	3836	3797	3743	3705	3342	3302	3287	3262	2866	2497	1896	1630	1013	0540	0267
00.030	4302	4256	4224	4181	4153	4132	4102	4091	4073	4024	3985	3960	3946	3925	3910	3895
00.040	4691	4653	4628	4594	4573	4557	4536	4528	4852	4824	5131	5434	5723	6240	6831	7416
00.050	4732	4699	4678	4987	4971	4960	4946	5267	5261	5560	6122	6595	6796	7352	7643	7739
00.060	4763	5071	5055	5035	5023	5342	5334	5332	5641	5934	6685	7066	7345	7713	7834	7850
00.070	4790	5104	5090	5400	5393	5389	5698	5698	5991	6520	6959	7522	7685	7887	7928	7933
00.080	5151	5131	5120	5436	5431	5742	6035	6037	6041	6801	7342	7674	7850	7976	7994	7998
00.090	5172	5155	5473	5467	5777	5778	6076	6079	6356	7048	7521	7853	7963	8039	8048	8052
00.100	5191	5503	5498	5806	5807	6103	6382	6387	6642	7092	7671	7975	8043	8087	8093	8097
00.130	5563	5558	5869	5875	6176	6455	6718	6725	6959	7639	7987	8150	8177	8190	8195	8198
00.170	5613	5926	5931	6238	6523	6783	7027	7232	7420	7927	8186	8264	8275	8283	8287	8289
00.200	5641	5960	6261	6551	6815	6831	7275	7285	7620	8051	8273	8319	8327	8334	8337	8339
00.250	5993	6298	6312	6608	6877	7117	7514	7672	7815	8219	8363	8387	8392	8398	8401	8402
00.300	6026	6336	6624	6902	7148	7365	7714	7726	7973	8306	8421	8436	8441	8446	8449	8450
00.400	6369	6665	6688	7194	7420	7614	7918	8034	8139	8428	8498	8508	8512	8516	8518	8520
00.500	6410	6711	6984	7444	7647	7818	8080	8178	8320	8506	8550	8559	8562	8566	8568	8569
00.600	6442	6747	7023	7487	7837	7987	8210	8291	8407	8560	8589	8597	8600	8603	8605	8606
00.700	6739	7025	7278	7693	7874	8023	8315	8381	8475	8597	8621	8628	8631	8634	8635	8636
00.800	6764	7051	7306	7723	8028	8157	8345	8412	8505	8626	8647	8653	8656	8659	8660	8661
00.900	6784	7074	7330	7748	8055	8184	8425	8480	8555	8651	8669	8675	8677	8680	8682	8682
01.000	6802	7316	7548	7918	8078	8291	8448	8502	8595	8672	8688	8694	8696	8699	8700	8701
01.200	6833	7351	7585	7956	8219	8397	8528	8571	8644	8705	8719	8725	8727	8729	8731	8731
01.400	7107	7379	7785	8111	8335	8429	8590	8626	8673	8732	8745	8750	8752	8754	8755	8756
01.600	7130	7600	7811	8138	8361	8509	8640	8669	8706	8754	8766	8771	8772	8775	8776	8776
01.800	7149	7622	7834	8160	8384	8531	8661	8689	8733	8773	8784	8788	8790	8792	8793	8794
02.000	7166	7641	7853	8283	8472	8592	8698	8721	8751	8789	8800	8804	8805	8807	8808	8809
04.000	7695	8076	8245	8556	8686	8764	8836	8848	8865	8885	8893	8896	8897	8899	8899	8900
06.000	7757	8140	8412	8673	8803	8853	8899	8907	8919	8934	8940	8943	8944	8945	8946	8946
08.000	7970	8308	8539	8756	8860	8900	8936	8944	8953	8966	8972	8974	8975	8976	8977	8977
10.000	8002	8443	8571	8819	8889	8935	8964	8970	8977	8989	8995	8997	8998	8999	8999	9000
20.000	8242	8621	8784	8946	8998	9021	9037	9041	9046	9055	9060	9062	9062	9063	9064	9064
30.000	8417	8738	8873	9003	9043	9061	9074	9078	9082	9090	9094	9096	9096	9097	9098	9098
40.000	8554	8771	8937	9041	9074	9088	9099	9102	9106	9113	9117	9119	9119	9120	9120	9121

$2\alpha/RT\beta = 10.00$

p/K′	γ-Values 1.00	1.50	2.00	3.00	4.00	5.00	7.00	8.00	10.0	20.0	40.0	70.0	100	200	400	1000
00.001	1943	1607	1327	0889	0577	0449	0214	0164	0096	0021	0013	0011	0011	0010	0010	0010
00.002	2295	1936	1628	1330	1086	0735	0572	0456	0285	0068	0026	0023	0022	0021	0020	0020
00.004	2992	2598	2247	1898	1599	1340	1095	0906	0733	0243	0078	0049	0046	0043	0042	0041
00.006	3050	2973	2606	2238	1915	1879	1580	1338	1112	0600	0177	0090	0072	0067	0065	0063
00.008	3421	3019	2969	2585	2242	2207	1884	1865	1587	0919	0411	0168	0112	0091	0088	0086
00.010	3454	3385	3008	2942	2582	2548	2204	2186	1883	1333	0637	0300	0190	0120	0112	0110
00.020	3901	3847	3809	3757	3722	3694	3317	3303	3279	2884	2514	2183	1642	1193	0629	0278
00.030	4313	4268	4237	4196	4169	4149	4120	4110	4093	4047	4011	4331	4319	4300	4287	4613
00.040	4704	4668	4643	4611	4590	4575	4893	4886	4876	5177	5475	5764	6035	6508	7030	7549
00.050	4744	4712	4691	5005	4990	4979	5294	5291	5285	5588	6156	6635	7010	7484	7717	7788
00.060	4774	5086	5070	5051	5368	5362	5355	5667	5665	6235	6721	7252	7490	7789	7884	7894
00.070	5139	5118	5105	5419	5412	5408	5720	5720	6016	6551	7166	7561	7780	7938	7968	7973
00.080	5164	5145	5462	5453	5763	5763	6059	6061	6338	6832	7376	7780	7920	8018	8031	8035
00.090	5184	5168	5490	5483	5797	5798	6098	6374	6381	7079	7658	7931	8018	8075	8083	8086
00.100	5202	5518	5513	5824	5826	6125	6406	6411	6668	7294	7789	8010	8087	8120	8126	8130
00.130	5577	5572	5886	6186	6195	6476	6741	6748	6984	7669	8060	8181	8207	8220	8224	8227
00.170	5625	5940	5946	6255	6542	6804	7050	7257	7445	7954	8233	8291	8302	8309	8313	8315
00.200	5652	5973	6278	6570	6836	7074	7297	7480	7645	8077	8299	8345	8352	8358	8362	8363
00.250	6006	6313	6327	6874	7119	7137	7536	7695	7839	8244	8387	8410	8415	8421	8423	8425
00.300	6038	6351	6641	6919	7167	7385	7736	7873	7996	8329	8443	8458	8463	8468	8470	8471
00.400	6383	6681	6951	7212	7438	7633	7939	8055	8229	8462	8518	8528	8531	8535	8538	8539
00.500	6423	6725	7000	7462	7665	7837	8100	8198	8340	8526	8569	8577	8580	8584	8586	8587
00.600	6454	7009	7261	7504	7856	8006	8229	8310	8426	8578	8607	8614	8617	8621	8622	8623
00.700	6753	7040	7294	7710	7891	8145	8333	8400	8494	8614	8638	8644	8647	8650	8652	8653
00.800	6777	7065	7321	7739	8046	8175	8363	8429	8522	8643	8663	8669	8672	8675	8676	8677
00.900	6797	7087	7542	7764	8072	8286	8443	8497	8572	8667	8684	8690	8693	8695	8697	8698
01.000	6814	7331	7563	7934	8094	8308	8465	8552	8611	8688	8703	8709	8711	8713	8715	8716
01.200	7093	7365	7599	7971	8236	8414	8544	8612	8660	8720	8734	8739	8741	8743	8745	8745
01.400	7120	7392	7801	8127	8351	8445	8606	8641	8698	8747	8759	8764	8765	8768	8769	8769
01.600	7142	7614	7826	8153	8377	8524	8655	8684	8721	8768	8779	8784	8786	8788	8789	8789
01.800	7160	7636	7847	8279	8467	8546	8676	8718	8748	8786	8797	8801	8803	8805	8806	8807
02.000	7177	7654	8014	8298	8487	8607	8712	8735	8770	8802	8812	8816	8818	8820	8821	8821
04.000	7707	8089	8257	8569	8698	8776	8848	8864	8877	8896	8903	8907	8908	8909	8910	8910
06.000	7767	8152	8424	8685	8815	8865	8910	8920	8930	8944	8950	8953	8954	8955	8956	8956
08.000	7982	8319	8551	8768	8871	8911	8947	8955	8963	8975	8981	8983	8984	8986	8986	8986
10.000	8012	8455	8651	8830	8913	8945	8973	8980	8987	8998	9004	9006	9007	9008	9008	9009
20.000	8251	8631	8794	8955	9007	9030	9046	9049	9054	9063	9068	9070	9070	9071	9072	9072
30.000	8427	8748	8883	9011	9052	9069	9082	9085	9090	9098	9101	9103	9104	9104	9105	9105
40.000	8564	8835	8946	9049	9082	9096	9107	9109	9113	9121	9124	9126	9126	9127	9127	9128

Index